#19·65

STRATHCLYDE UNIVERSITY LIBRARY

30125 00076696 3

KV-038-688

Books are to be returned on or before
the last date below

28 APR 1976	10. AUG. 1979	
22. DEC. 1976	17. APR. 1980	
-9. MAR. 1977	2 4 MAR 1994	
-4. APR. 1977		
5/5/77	1 0	JAN 1996
-9. MAY 1977		
12. MAY 1978		
23 APR. 1979	LIBREX —	

#925851

SECOND JOINT CONFERENCE ON SENSING OF ENVIRONMENTAL POLLUTANTS

Washington, D. C.

December 10-12, 1973

ANDERSONIAN LIBRARY
WITHDRAWN FROM LIBRARY STOCK
UNIVERSITY OF STRATHCLYDE

Sponsored by

American Chemical Society
American Institute of Aeronautics and Astronautics
American Meterological Society
Department of Transportation
Environmental Protection Agency
Institute of Electrical & Electronic Engineers
Instrument Society of America
National Aeronautics and Space Administration
National Oceanographic and Atmospheric Administration

PUBLICATION POLICY

Technical papers may not be reproduced in any form without written permission from the Instrument Society of America. The Society reserves the exclusive right of publication in its periodicals of all papers presented at the Annual ISA Conference, at ISA Symposia, and at meetings co-sponsored by ISA when the Society acts as publisher. Papers not selected for such publication will be released to authors upon request.

In any event, following oral presentation, other publications are urged to publish up to 300-word excerpts of ISA papers, provided credits are given to the author, the meeting, and the Society using the Society's name in full, rather than simply "ISA".

For further information concerning publications policy write to:

Publications Department
Instrument Society of America
400 Stanwix Street
Pittsburgh, Pa. 15222

Library of Congress Catalogue Card Number 74-75508
ISBN 87664-226-1

Copyright 1973

INSTRUMENT SOCIETY OF AMERICA
400 Stanwix Street
Pittsburgh, Pennsylvania 15222

D
628.5028
JOI

CONTENTS

© 1973, ISA JSP 6651

TECHNOLOGY AND GROWTH OR ENVIRONMENTAL QUALITY?*

Stanley M. Greenfield, Ph.D.
Assistant Administrator
for Research and Development

U. S. Environmental Protection Agency

It is the nature of society to grow, develop and expand. There is a constant seeking for a better life and the ability to enjoy it. It is only in the last few years, the realization has emerged that the very act of seeking and obtaining that better life may have played a role in producing the degradation in the quality of our environment. Each new advance in our technology offers the glittering promise of enrichment in one form or another, and in the process, promotes further growth and development. Populations do not remain stagnant; they increase. They may do it in spurts, they may be self-restricting for a time, or suffer some natural or man-made disaster, but inevitably they expand. An expanding population requires additional capabilities to provide that which at the very least maintains the standard of living at a constant level. A society doesn't voluntarily step back from its level of achievement. A society will not stand for a burgeoning population that lowers that level of achievement, hence the pressure is extreme for the technology that will permit it to both grow and develop.

And so the cycle continues. A society demands advances to improve its lot, and having achieved it, grows, and in gorwing demands more. What then is the ultimate fate of such a society? While man has the ultimate capacity to destroy his world catastrophically or otherwise, the probability of this occurring in the immediate future is rather remote. That does not mean that we should ignore that possibility, but rather that we should build the capacity to assess the impact of man's actions and provide the institutions to take the necessary steps.

A good case in point would be the current energy crisis. This crisis is man-made and it can be man-alleviated. Energy production and consumption produces environmental impacts of varying degrees; the solution of these environmental problems does not necessarily mean only a cutback in the consumption of energy but also an improvement in technology. This country urgently needs a high quality R&D organization working on an expanded and better balanced effort for all sources of energy. In addressing the energy crisis itself, we can certainly bring about selected improvement in environmental quality. Whatever the case, the development and implementation of an energy system designed to achieve and maintain a degree of self-sufficiency, clearly must be sensitive to the effects that the system will have on health, welfare, national economy, and ecosystems. For if this sensitivety is rationally incorporated at the outset into the development and implementation process, then I would think that our domestic resources could be broadly utilized in harmony with the environment.

Stopping the growth of society as a means of insuring environmental quality in the energy case or any other is patently ridiculous in view of the fact that growth also has a role to play in maintaining or improving the quality of life. Similarly, depending on technology to ultimately save us in and of itself is equally ridiculous in the absence of other measures to insure a sane approach to our future.

Overconfidence in our technology leads to faulty judgments. As Lee Dubridge pointed out several years ago, we have become so accustomed to the almost magical capabilities of technology that we expect instantaneous solutions to all problems, no matter how complicated. This is unreasonable, even when the problems are purely technical. When complex social, political, and ethical considerations are additional important factors affecting technical decisions, then rosy expectations are unreal.

It is clear then that there is no single approach that offers a panacea to the problem of growth and environmental quality. Rather we must be concerned with understanding the problem sufficiently to allow us to state our goals for the future, and decide on the optimum strategies for their achievement.

In this regard let us very briefly examine growth as a cause of environmental problems and controlled growth as a potential solution.

It is easy to assume that the impact of man on the

---
*Keynote Address, 2nd Joint Conference on Sensing of Environmental Pollutants, December 10, 1973

environment is automatically destructive and hence more growth means more impact and therefore more environmental degradation. Certainly, increased growth must result in increased environmental impact, but we must know what kinds of impact will occur before deciding whether or not man has unbalanced, improved or degraded the environment. We must consider the environment in the broad sense of quality of life, rather than simply as a wilderness or a nature preserve in making this judgment. In the broader view it is difficult to support the absolute position that high levels of growth and development are the specific causes of environmental degradation. Rich countries are not typically dirtier or more unpleasant than poor countries; countries with higher population densities are not even typically dirtier or more unpleasant than countries with lower population densities. If this assertion sounds surprising and seems to run counter to well-known examples, it may be because a subtle but critical distinction is usually missed in discussions of this issue; namely, the distinction between the level of a social variable such as economic activity or population and its rate of growth. India's problem is not so much the number of people as it is the rate at which these numbers grow, forcing much current income and energy to be invested simply to keep pace leaving little societal energy to generate beneficial growth and development; if population growth could be stopped, India would not have as severe a population impact problem. Japan's problem is not so much that production is now high as it is that by consciously forcing rapid growth and narrowly defined economic output, other aspects of economic welfare have been allowed to decline; as Japan lets its narrowly defined economic growth rate fall a little and turns to solving neglected environmental problems it will find its high output levels more a help than a hindrance. Policies which produce high, unbalanced, growth rates can indeed be a cause of both environmental and energy problems; high output per se, cannot.

Ironically, most of the environmental doomsayers seem to be more concerned about high levels of economic activity and population, than with high growth rate -- except as high growth rate hastens the day when high levels will produce collapse.

It is clear from what I just said, that unbalanced and excessive economic growth rates can cause environmental problems, among others, and that an optimal environmental policy will require some slowing and redirection in the rate of growth. So the issue now is whether the best way to obtain the appropriate amount of slowing and redirection is to attack growth directly or to work to solve the environmental problems which are facing us and let these efforts impact the composition and rate of growth as they will. For example, the production, transportation, consumption, and waste disposal involved in energy cycles impose high environmental costs on society. Should we, therefore, try to limit the consumption of energy, on the grounds that doing so will reduce these environmental impacts in a cost-effective manner, knowing that doing so will have some impact on the rate of growth of energy consumption? Clearly we must do both.

It seems reasonable to accept the premise that, within this increasingly populated world, one of the prime initiators of environmental change is the scientific or, more specifically, the R&D community. Most new knowledge, inventions and innovations are discovered by scientists in research and development laboratories. And it is from this new knowledge and its application that ideas, products and services which alter the character of our environment emerge.

It would now seem that a growing proportion of the new developments launched by scientists have the character of boomerangs. One of the areas of increasing concern for us, as we peer into the future, is how much the efforts of future scientists will be influenced by the time they must devote to ducking the returning boomerangs or devising systems, techniques, hardware and software to insulate society and protect it from the impacts of earlier technologies.

We already know that we live in a world of increasing specialization. But the benefits of specialization must be balanced by a clearer recognition of the interdependence between society and science and technology, and scientists must take into account all the possible effects of new developments before they are introduced into society. The magnitude of the problem indicates clearly that neither the government nor the private sector can assume the entire burden. We believe that the effort must be a joint undertaking as is exemplified by the representation at this conference.

Government programs should be designed to stimulate industrial R&D by defining problems and identifying potential solutions. Direct R&D support by the government should be provided only when it is clear that the right combination of incentive and capability to develop and apply the needed technology, in a timely way, is not present in industry. Government participation is especially needed in three kinds of environmental R&D: (1) high-cost and risky demonstrations; (2) work to generalize new and demonstrated control systems by development of design data needed to apply them to a variety of situations and extrapolate use to other industries; and (3) exploratory R&D to identify technically feasible concepts for future monitoring and control systems for problems that cannot be economically solved with available techniques.

All of this costs money. However, to get to the point where there would be zero risk from pollution would cost even more. Society may choose to accept some level of risk rather than incur the much larger costs of zero risk. Society has already decided that it will incur costs -- how much and for what purpose hasn't yet been decided. But I am certain there will be trade off in many directions, i.e., higher costs of increased control, the realities of extended time schedules to meet promulgated standards, revised standards, facing the realities of meeting the energy shortage as well as re-evaluated measurement techniques. I think it is fair to say that the average American is just beginning to recognize that there are difficult choices to be made, that some of his cherished patterns of behavior may have to be modified. He understands more than ever that there are fundamental interrelationships among the environment, the economy, energy, transportation, land use, and so on. In a sense, having embarked on an ecology kick, society is only now beginning to recognize the truth of the basic ecological teaching -- that everything is related in a system and that every choice involves a trade off.

This would seem to make it attractive to focus on long-term R&D objectives and avoid the near term. In fact, the selective relaxation of some standards, or extension of time schedules would seem to me to offer us a needed reprieve in R&D to acquire the data for more precise standard setting, more effective monitoring techniques, and the development of adequate control systems.

With specific regard to monitoring, which is essentially the theme of this conference, it is interesting to note that nearly one-half of the papers to be presented in the next three days deal in some way with remote sensing instrumentation. It is becoming clear that to perform the Nation's environmental monitoring task adequately, a combination of contact along with remote instrumentation is absolutely necessary. Remote monitoring as an adjunct to contact monitoring provides a cost effective method to survey large geographical areas. Remote instrumentation mounted on flying platforms provides valuable information in support of contact monitoring, such as surveys for monitoring networks, data for model verification and "quick looks" at environmental quality violations.

EPA, NOAA, NASA and many other federal agencies are now actively engaged in research and other related activities so that we can acquire sensors, instrumentation and methodology for environmental appraisal. It is necessary to have this sound environmental knowledge in order to intelligently develop, apply and defend our structure of regulations. From reliable data which can only be derived through the use of adequate techniques, we can establish realistic and enforceable environmental quality standards. The type of data needed by the federal government covers sources of pollutants, characteristics of pollutant emissions, characteristic and physical transformation which these pollutants undergo as they travel through the various aspects of the media; to identify, understand and quantify the pathways and kinetics of the pollutants through these pathways, and to measure the exposure and dose to the most sensitive receptor populations.

It is necessary to determine the total exposure or dose of the receptor. To accomplish this, a pollutant oriented, integrated monitoring network is necessary. This type of monitoring system provides necessary information for exposure from air, water, food, dust and other factors. There are certain steps which must be considered in designing an overall monitoring system. These are as follows: determine the goals and objectives of the overall monitoring system and the role subsystems play toward meeting these goals and objectives; identify the most sensitive populations at risk; identify the threshold at which effects are first detected; identify the significant sources of a pollutant, including man-made and natural; relate a source of the pollutant to its exposure pathways; develop models which include supporting information such as meteorological, hydrological, etc.; develop and implement an overall quality assurance program to verify every aspect of the monitoring network; utilizing optimal sensors, implement the network; and, provide a system whereby data can be sensitized into information that is in a usable format.

From my perspective, the monitoring of the future will be forced to go beyond the limitations of wet chemistry techniques and in situ sensors. It must also capitalize on the remote sensing R&D and experience of other federal agencies. Our approach has been that of adapting results of the NASA ERTS, the Department of Interior EROS, and related capabilities pursued by other federal agencies, the Corps of Engineers, Department of Commerce, the National Science Foundation -- RANN and R&D Incentives, for example.

The remote identification of specific local instances of pollution has been accomplished from aircraft through visual identification and photographic documentation of sewage effluents and industrial outfalls. Recent years have brought the evolution of new instrumentation with a unique capability of monitoring pollution. These instruments which show great promise include the imaging scanner, the infrared spectrometer, the correlation radiometer, the microwave radiometer, lasers and other remote sensors. It is expected that the testing and evaluating of existing sensors and the developing of new ones, particularly in the NASA, DoC, DoD, and DoI programs, will result in the full appreciation of remote sensor technology for monitoring the environment. Let me also add a word of caution. I admit to being a skeptical proponent of what I call remote sensing. I don't for a moment expect it to do everything. But every method of monitoring the environment becomes more specific if done in conjunction with the information that aircraft and satellite-borne instrumentation discloses.

What remains to be learned about the protection of our environment is staggering. More research is needed in all areas. There is temptation to recoil at the complexity of the environmental problems facing us. Yet we cannot research these problems piecemeal. They must be considered with the broadest possible perspective, taking into account all environmental implications. Otherwise, our solutions might cause still other -- perhaps more serious -- problems.

In closing, I would like very much to let you know that we recognize that we are not alone in this pollution research and technology business. We must and will cooperate with other government agencies and the private sector to carry out this most important task of cleaning up the environment, while meeting the other needs of society. We need your support, ideas, and good will. One agency cannot do the job alone. Without public interest and backing, in fact, the results of our research and technology could never be translated into action programs for public benefit. So we are dependent upon the public and upon you. We know too well, from our experience of the past three years, how enormous the task is that lies ahead of us. Within our means to do so, we welcome your participation and your assistance. Thank you very much.

© 1973, ISA JSP 6652

NATIONAL AND INTERNATIONAL PROGRAMS FOR MONITORING THE GLOBAL ENVIRONMENT

Dr. Edward S. Epstein*
Associate Administrator
Environmental Monitoring and Prediction
National Oceanic and Atmospheric Administration

I am pleased at the opportunity to address you at this Joint Conference on the Sensing of Environmental Pollutants. In describing the organization of this Conference in the program announcement, your general chairman noted that the program is both broad in scope and narrow in focus. There is a continuing need to be both comprehensive and substantive -- to be broad and yet focus -- and this is a perplexing challenge to those who wish to deal with the problems of our environment. The problems are all-encompassing and the solutions must be focused ones -- but the solutions do not lie in the observations or the sensing themselves, but in the vigilant assessments which are the proper endproduct of monitoring.

I will try to provide a particular perspective this afternoon by sketching bits and pieces of major national and international programs in global environmental monitoring which provide, to greater or less degree, for the necessary coordination of environmental observations, the processing and management of data and information, and the preparation of warnings, prediction, and assessments of environmental conditions.

The scope of concern over our environment has expanded. We hope that this expanded concern is in concert with the extent and intensity of human activity that imposes the stresses on the environment. Certainly our awakening concerns must be global. There is a need to know the status and trends of conditions for many environmental factors, in many media, and over many space and time scales. On the basis of our concerns and the absolutely clear realization of the global nature of the problem and the global and even cosmic consequences of our actions or inactions, we must piece together the national and international commitments to attack these challenges in a coordinated way, to collect the necessary data and acquire the necessary knowledge, and make the necessary assessments and judgments.

National programs of course underpin international efforts and are the basic building blocks of an integrated global system. The design of any global system must start with these national programs that nations implement mostly in their own interest. Then the international gaps can be filled according to a proper master design as all nations become aware that in so doing they can realize greater benefit from their inputs. The inputs and interests of the developed and developing countries will be different, but they are compatible, and successful international efforts can be made.

Let me review several U.S. programs that involve large-scale environmental monitoring. As you are aware, programs to collect, process and analyze data on natural physical environmental phenomena are much further developed than those that involve pollutants. These efforts provide an important base, however, upon which an expanded global monitoring system for pollution and pollutant effects can be built -- not only a technical base, but a format and design whose better features can be emulated, and whose mistakes we can try to avoid. The U.S. National Program for Monitoring Atmospheric Conditions is by far the most sophisticated and mature and demonstrates the several elements of a functioning monitoring system: observation, processing and analysis; communication, and product preparation (i.e., assessment) and dissemination. The National Weather Service relies upon a wide range of observing facilities to provide observational coverage of large areas of the earth. We may take as a lesson that it is through a complex mix of many techniques and technologies, old and new, that global weather observations are obtained.

Weather and hydrological conditions are observed and reported at more than 1000 land stations; some few are automated sensing facilities -- most, however, rely on observers' judgments and very simple instruments. On the ocean, volunteer cooperative observers aboard more than 2,000 ships transmit marine weather observations based on, mostly, relatively unsophisticated instrumentation. At the other end of the spectrum, radar and satellites have now become vital new tools of the system. The sophisticated technologies of geostationary and polar-orbiting satellites provide a near-continuous capability to detect and track weather phenomena, and to provide vertical profiles of temperature, cloud imagery, upper wind

*Distinguished Speaker

data, and sea surface temperatures and related oceanic circulation features. With time, technology is producing even newer monitoring equipment and techniques. We have environmental data buoys with capabilities of reliable continuous observation from remote oceanic areas and powerful new remote sensing devices. This mix of facilities and various sensors are organized into many different network configurations depending upon the type of observation and the need for its timely availability at analysis and processing centers. This mix of observational approaches, I suggest, will be a characteristic of any global environmental monitoring system we happen to come up with in the future. I also contend that the optimum system is such a mix.

I will skip over questions of communications at this time. But remember, without efficient and effective communication, if the observations cannot play a timely role in analysis and assessment, they are of little value.

Federal programs to monitor the marine environment are not nearly as well defined and organized as those in the atmosphere. A number of federal agencies have responsibilities for particular facets of the oceanic environment -- NOAA, EPA, Coast Guard, Geological Survey, NASA, Corps of Engineers, NSF, and others. There exist interagency efforts to clarify these responsibilities, but clearly we have not yet achieved a coordinated ocean monitoring program. Some, but not all the parts exist.

There are a number of diverse monitoring activities now underway in the oceans. Most are limited, however, in either time or space and I will not go into them in much detail. One major program is underway within the National Marine Fisheries Service of NOAA -- the Marine Resources Monitoring, Assessment, and Prediction Program, or MARMAP.

The overall objectives of MARMAP are to develop techniques for obtaining accurate measures of the abundance and geographic distribution of living marine resources; to assess the productive capacity of these resources and develop models for predicting future yields; to monitor seasonal and annual fluctuations in fishery stocks; and to relate fluctuations to environmental factors and fishing pressure. The MARMAP initiative involves resource surveys that monitor catch effort, mortality, fecundity, growth, and migration to allow continuing assessments of the conditions of commercially important stocks.

An initial MARMAP Survey for fish larvae was conducted in 1973 in the waters from Cape Cod to the Caribbean and yielded several surprises. In addition to the expected plankton and larval fishes, significant amounts of plastic and tar contaminants were collected in nets towed at the surface. The results of these initial efforts indicated the situation may be more critical than had been expected and demonstrated the need for large-scale environmental monitoring to detect not only the sources and histories of these pollutants, but potentially harmful effects on the oceans. These findings further suggest that in addition to such gross observations, an improved capability must be developed to detect dissolved constituents such as heavy metals, hydrocarbons, and petroleum residues in situ. Incidentally, upon first occurrence, the tar and plastics were disregarded as a nuisance that contaminated nets and samples until their widespread occurrence and significance were realized.

Let me now turn to the international arena where national programs can be tied together and augmented. A number of global monitoring programs are coordinated under the auspices of the U.N. specialized agencies, such as the World Meteorological Organization (WMO), the World Health Organization (WHO), the United Nations Educational, Scientific and Cultural Organization (UNESCO), Food and Agriculture Organization (FAO), and the International Atomic Energy Agency (IAEA). These efforts are the next step in program organization toward an integrated international global monitoring system.

The meteorologists again have the jump as far as having developed an effective global monitoring system. The World Meteorological Organization has produced in its world weather watch and global atmospheric research program remarkable examples of what can be achieved. The world weather watch has the essential components of a monitoring scheme -- a global observing system, a global telecommunication system and a global data processing system (something <u>is</u> done <u>with</u> and <u>to</u> the data that is collected). These components of the world weather watch are not the national programs themselves, but rather the world weather watch builds on national activities and supplements them to provide a system that could hardly be conceived in any other way. The world weather watch is built upon technologies that are developed in the first instance for national programs, but soon find their ways into the international arena. Our satellite program has been developed to serve national needs, but that same program has been exported and its benefit felt keenly in some of the least developed parts of the world through the automatic picture transmission programs. Even the global atmospheric research program, the research arm of the world weather watch, has been able to involve in many of its facets the less technologically advanced countries. In the international theater programs of training and education and assistance are significant elements of any successful program.

One of the more exciting aspects of GARP is that it is leading to a truly international system of satellites. The plan calls for five geostationary satellites; two from the United States and one each from Russia, Japan and the European Space Research Organization (ESRO) which will provide not only exciting observational coverage of the globe, but will have communication capabilities for relaying other environmental data from Earth based sensors and among processing centers. Tying to these communications links is another mode in which developing countries can become involved with new technologies.

One important part of the world weather program is the global network of atmospheric baseline and regional air pollution monitoring stations. The WMO has established criteria for atmospheric baseline stations that call for their location in remote areas away from the impact of human activities that are expected to remain relatively pristine for the foreseeable future. At the moment the list of parameters that constitutes the basic monitoring program is limited to carbon dioxide, turbidity and precipitation chemistry. But an expanded program is being developed. At the moment, we are operating three baseline stations: at Mauna Loa, Hawaii; at Point Barrow, Alaska; and in Antarctica. We hope soon to be operating a station in American Samoa. However, only one of these stations -- Mauna Loa -- is operating at full steam and taking the whole set of observations that one would like to see. It is significant that this single observatory, which has been in existence for less than twenty years, is really the sole source of reliable information on what is happened in the last two decades to atmospheric carbon dioxide levels. It is also one of the few sources of reliable information we have on the record of atmospheric turbidity. It seems somewhat casual, as though we really didn't care, for us to rely so strongly on a single station to give us some handle on our climate and its potential for alteration. I must add that several other countries have indicated their intent to establish atmospheric baseline stations. It is possible that eventually a global network of fifteen to twenty-five such stations might evolve. In addition, approximately forty-five countries have either established, or plan to establish, regional air pollution stations for a total network of more than seventy stations.

The counterpart of the WMO on the oceanography side is the Intergovernmental Oceanographic Commission (IOC) which is organized under UNESCO. The IOC, though slow to get started, has been moving lately in an encouraging way toward establishing monitoring activities. IGOSS, an Integrated Global Observing Station System, is a joint activity with WMO, a pilot project to collect, communicate and process bathythermograph data from ships of participating countries so that the thermal structure of the upper layer of the ocean can be analyzed. It is hoped that the success of this pilot project will lead to broader projects for monitoring pollution over large areas of the ocean. Initially the focus is to be on petroleum products. Other contaminants will be included depending upon sensor and platform capabilities. But I must emphasize in this context that there is great need for new and improved sensing capabilities for the marine environment. When fully implemented, and assuming the technology keeps pace and the costs do not, IGOSS will provide systematic observations of a broad range of physical, chemical and selected biological parameters.

Let me now turn to one final international program, the UNEP, UN Environmental Program. This is a program which grew directly from the UN Conference on the Human Environment held in Stockholm just a year and a half ago. The purpose and concept of UNEP is to act to tie together the many national and international programs related to environmental concerns that now exist, and to stimulate awareness and knowledge and management that is conducive to man's living on his one and only Earth. The action plan which was proposed by the Stockholm Conference and subsequently approved by the UN General Assembly provides the basic framework for international efforts. Its three main functional areas are: Earthwatch, environmental management, and supporting measures. Within Earthwatch, details of which are now being put together, there are four kinds of environmental activities: monitoring, evaluation and review, research, and information exchange.

Earthwatch is the integrating mechanism for implementing comprehensive multidisciplinary global environmental assessment. It provides for the international collaboration among nations, for the sharing of facilities and observational platforms, and for the assistance to developing countries to allow their full participation in global efforts. Earthwatch is built upon existing national and international capabilities and serves to integrate these capabilities using the environment fund to fill gaps where necessary.

To assist in implementing the action plan, an intergovernmental working group on monitoring will meet in February 1974. Within this country, in preparation for this meeting we have developed our own picture of how the Earthwatch monitoring program should be organized. The approach has been to conceive of a global monitoring program that would be responsive to particular objectives, including the establishment of capabilities for:
The surveillance of human health, natural disasters, and food contamination.
The assessment of man's impact on climate, the oceans, and ecosystem stability and modification and the evaluation of the impact of land-use practices.

In order to implement a comprehensive program, two separate streams of action are recognized:
First stream actions for which facilities and technology are available, the approach to the problem has been fairly well defined, and adequate scientific knowledge is available, and
Second stream actions involving program development, additional research, and new technologies prior to implementation of an expanded program.

Within this framework and to meet these objectives, the U.S. is proposing a strategy that we hope will lead to the development of a multidisciplinary data base for comprehensive global environmental assessments. Priority actions include a global network of Earthwatch reference sites, cooperative impact monitoring activities, selected indicator monitoring programs, and a coordinated international system of pollutant analysis facilities and data and information management centers.

It is too early to know what really will come of the U.N. Environment Program and its efforts to pull together national and international activities into a common global monitoring thrust. The problems to be overcome will be political, scientific, technical, and cultural. New capabilities for the remote and in situ sensing of environmental constituents are required. Observational criteria and instrument intercalibration techniques must be developed and applied on a worldwide basis.

Of particular importance, scientists and technicians from the lesser developed regions of the world must be trained to allow them to participate in monitoring activities. Also required is a new utilitarian technology -- uncomplicated instrumentation; rugged sensors; and simplified analytical techniques -- that can be used by people with relatively little technical training under less than optimum conditions.

The question can not be "can we put it all together?" We must! Without such global monitoring of the environment, we are, in effect "flying blind" in the future. That, to me, is unacceptable.

© 1973, ISA JSP 6654

MEASUREMENT OF THE ABUNDANCE OF SEVERAL NATURAL STRATOSPHERIC TRACE CONSTITUENTS FROM HIGH ALTITUDE AIRCRAFT*

Crofton B. Farmer
Odell F. Raper
Peter W. Schaper
Rudolf A. Schindler
Robert A. Toth
Jet Propulsion Laboratory
Space Sciences Division
4800 Oak Grove Drive
Pasadena, California 91103

ABSTRACT

A summary report of the initial results obtained from near infrared observations of the stratosphere from the Anglo-French SST Concorde will be presented, together with the most recent results from previous flights aboard an Air Force NC-135. The measurements were made with a fast Fourier interferometer spectrometer operating in the 1.2 to 7.5 micron range of the infrared with a spectral resolution of 0.25 cm^{-1}. For the Concorde experiments, flight times and trajectories were selected which allowed the Sun to be viewed near the horizon with the relative solar elevation angle held constant throughout the measurements. Mixing ratios as low as a few parts per ten billion for the trace constituents in the absorption path were determined by spectroscopic analysis of the data. Results to be reported include the identification of features due to N_2O, NO, NO_2, CO, CO_2, CH_4, H_2O and indications of their latitudinal variations. New values for the upper limit of concentration levels for other trace gases of importance to pollution studies, such as HCl and H_2CO, have been determined and will be discussed.

INTRODUCTION

This report covers the more recent results from the continuing analysis of high altitude data obtained onboard a USAF NC-135 aircraft in February and March of this year — for which a preliminary report was given earlier[1] and in addition the results from 3 flights onboard the prototype Anglo-French Concorde 001 on June 13, 14, and 16. At the time of this writing, preparations are being made for an additional six flights onboard Concorde 002, and it is hoped that preliminary results from these flights, which are scheduled for October and November, 1973 will be available for the oral presentation in December.

Data were taken in the upper troposphere and lower stratosphere using a high speed stepping interferometer, which covers the wavelength region from 2 to 8 microns with a resolution of 0.25 cm^{-1}. The interferometer itself has been described by Schindler[2], and the Concorde experiments have been described by Farmer, at. al[3]. The bands used for the spectral analysis of the data for the various stratospheric constituents are shown in Table 1, together with the literature references from which information concerning line positions, widths, and strengths were obtained.

DISCUSSION OF RESULTS

Nitrous Oxide

Stratospheric nitrous oxide has been measured using in situ methods by Schütz el at.[4], by Harries[5] in the far infrared, and by Goldman, et al.[6],[7], who observed the ν_3 fundamental at 2224 cm^{-1} in the near infrared region. The data of Schütz, et al. and Harries are in reasonable agreement, and indicate an N_2O mixing ratio of about 0.250 ppmv at the tropopause, decreasing gradually to about 0.220 ppmv at 15 km. Harries' data cover a latitude range from 5°S to 48°N and show no significant latitude variation.

The results of Goldman et al. are derived from two separate sets of observations. The more recent (1972) set, covering the greater range of stratospheric altitudes, shows a variation of from 0.33 ppmv at the tropopause falling to about 0.15 ppmv at 18 km. This is in comparison with a value of 0.11 ppmv at 13 km from the 1968 observations. The large difference at the lower altitude between the two observation periods may perhaps be the result of temporal or seasonal variations.

Out results were obtained from both the ν_3 and $2\nu_1$ (2560 cm^{-1}) bands for data obtained at 12 km and, in addition, the $\nu_1 + \nu_2$ (1880 cm^{-1}) band for observation at 16 km. From these we derive a value of 0.20 ± 0.02 ppmv for the N_2O mixing ratio, with no discernable variation between altitudes of 12 and 20 km in the latitude range from 33°N to 68°N. This result refers only to the period of the year

*This paper presents the results of research carried out at the Jet Propulsion Laboratory, California Institute of Technology, under Contract Number DOT-AS-20094 sponsored by the Department of Transportation as part of the Climatic Impact Assessment Program, by agreement with the National Aeronautics and Space Administration (Contract NAS 7-100).

Superior numbers refer to similarly-numbered references at the end of this paper.

between March and June. It should be pointed out however that these observations have so far been limited to two fixed altitudes so that, taking into account the variation of tropopause height with latitude, the results could hide any small systematic increase in the mixing ratio with latitude at the tropopause height, if there were a compensating fall-off in the mixing ratio with height above the tropopause. In general the results are in good agreement with those of Shütz et al. and Harries. In view of its probable role as the principal source of NO in the lower stratosphere (see, for example, Nicolet and Peetermans[8]) it is of fundamental importance that the detailed seasonal and spatial variations of N_2O be determined to a greater degree of refinement than is possible from the observations made to date.

Nitrix Oxide

Among the expected trace molecular constituents of the lower stratosphere, nitric oxide is of particular importance because of the role it is thought to play in the net depletion of ozone in this region. A number of recent papers have been devoted to this question, but at the time when concern over the possible perturbation of this catalytic balance was first expressed no measurements of the abundance of NO in the lower stratosphere had been made. It therefore became a prime molecular species for which to search in the continued high resolution spectroscopic investigation of the stratosphere.

Nitric oxide was first detected from sunrise spectra taken during a flight on March 11, 1973 over Albuquerque, New Mexico; the result has been reported by Toth, et al[9]. The observed NO features, in the R and Q branches of the 1-0 fundamental at 1876 cm^{-1}, were very weak and close to the detection limit. Furthermore, the initial spectra were taken over a restricted range of observation conditions so that it was not possible to extract with any accuracy the variation of NO mixing ratio with altitude. The derived total molecular cross section gave, on the assumption of constant mixing, a mean apparent mixing ratio of 1.0 ± 0.2 ppbv*. The NO features as they appear in the stratospheric spectrum are shown in Fig. 1; the region illustrated is dominated by the ν_2 band of H_2O vapor and $3\nu_2$ band of CO_2.

On March 16, 1973, Schiff[10] obtained in situ data from a balloon-borne chemiluminescence instrument at a latitude of 33°, which yielded an NO mixing ratio of 0.1 ppbv, with an estimated accuracy of ± 60 percent. This result referred to the altitude range from 17.4 to 23 km. Subsequently, on May 14, 1973, Ackerman and Girard[11] observed the 1-0 NO band with a balloon-borne grille spectrometer over Southern France (latitude ~ 43°N). The Girard instrument was capable of a higher spectral resolution than that used by Toth et al. and this factor, together with the advantage gained from the balloon altitudes, enabled these observers to determine for the first time with precision the NO mixing ratio and its vertical profile. Ackerman and Girard report values which vary from 0.07 ppbv at 16.5 km to 4.4 ppbv at 37.5 km. For these balloon observations Girard chose to examine the region of the 1-0 band between 1900 and 1910 cm^{-1}.

We have made two additional observations of NO on the evening of June 13th with the instrument mounted in the French Concorde prototype 001, flying at an altitude 16 km between 45° - 50°N latitude. At solar zenith distances of 91.6° and 90.5°, which correspond to tangential ray altitudes of 12.8 and 15.2 km, mixing ratios of 0.2 and 0.4 ppbv, respectively, were obtained.

The results summarized here show marked differences among the individual observations, which were made under widely differing conditions of location, season and time of day. It is clear that more extensive observations of NO are required before any firm description of its spatial and temporal variations can be given.

Nitrogen Dioxide

Nitrogen dioxide was first observed in the grating spectra of Goldman et al.[12] covering the ν_3 band (at 1617 cm^{-1}). A quantitative analysis of these data, and additional lower resolution spectra covering the much weaker $\nu_1 + \nu_3$ band (2306 cm^{-1}), was carried out by Ackerman and Muller[13] who derived mixing ratios ranging from an upper limit of 2.3 ppbv at 16.1 km to an observed value of 4.0 ± 1.5 ppbv at 28.3 km. Harries[5], from emission spectra covering the rotational lines in the 20 to 30 cm^{-1} region, reported an average mixing ratio of 20 ppbv for NO_2, but pointed out that the accuracy of his measurement was severely limited by spectral assignment uncertainties. Harries has subsequently revised his value for the average mixing ratio of 5 ppbv[14].

In the rotational region the NO_2 lines observed and reported to date are severely blended with the very much stronger lines of more abundant species such as H_2O, N_2O, O_3, O_2 and the spectrum in this region is consequently difficult to analyze. The ν_3 band at 1617 cm^{-1} is strong and, at stratospheric altitudes, is relatively free from interference from the overlying 6.3 μ water vapor bands. The third band which can be used for infrared observations of NO_2 ($\nu_1 + \nu_3$ at 2906 cm^{-1}) is some 30 times weaker than the ν_3 band and is severely blended with lines of the CH_4 bands in the 3 μ region. Thus it would appear that the most favorable band for the investigation of NO_2 is ν_3 although, until recently, the analysis of spectra obtained in this region (see, for example,

*This preliminary value was subsequently corrected by more extensive analysis to a revised value of 0.8 ± 0.2 ppbv.

Murcray, et al.[15]) has not been attempted, perhaps as a result of difficulties associated with rotational quantum assignments for the band.

From spectra obtained from the Concorde observations, which covered the region of the $\nu_1 + \nu_3$ band, we were able to detect two features due to NO_2 which, after a thorough comparison with high resolution CH_4 spectra, were found to be sufficiently isolated to be amenable to unambiguous analysis. From these a tentative mixing ratio was derived for NO_2 of 1.8 ± 0.5 ppbv over the 15 to 20 km altitude range. The importance of this result lies in the fact that it was obtained simultaneously with the observations of NO (see previous section). The combination of these results yields a value for the ratio of NO_2 to NO (at 16 km), for sunset conditions, of $(NO_2/NO) \approx 4.5$.

Water Vapor

Stratospheric water vapor has been measured by Goldman et al.[16] from a region of the pure rotational band between 24 and 29 microns, and by Harries[5] in the far infrared between 300 and 1000 microns (10 to 30 cm^{-1}). The results of Goldman, et al. indicate a change in the mixing ratio with altitude, with a broad minimum at 15 km and a maximum around 25 km; at the maximum, the values range from 6.9 to 9.7 ppmv. Harries, however, found a constant mixing ratio for water vapor of 3 ppmv in the stratosphere up to altitudes of 60 km. These results can be compared with the earlier values of Houghton (see, for example, Pick and Houghton[17]) and Mastenbrook[18] who give mixing ratios of 5 ppmv and 3-4 ppmv, respectively.

Our results (Fig. 2) using principally the ν_2 band of H_2O centered at 1594.73 cm^{-1}, indicate a constant mixing ratio of 2.5 ± 0.30 ppmv over the altitude range from 14 to 20 km, with slightly higher values between 14 km and the tropopause; no significant variation with latitude was observed. The relatively small uncertainty quoted with this value is a result of the wide range of line intensities observed in the spectra, which permit close fitting of the observed and calculated absorption. The values derived from the ν_2 band were verified by additional analyses carried out for many other regions where the spectrum is characterized by water vapor features, viz. the $2\nu_2$, ν_1, and ν_3 bands.

Carbon Monoxide

Carbon monoxide has been measured over the altitude range from 8 to 12.5 km by Seiler and Warneck[19] using in situ methods, and by remote sensing of the near infrared (1-0) band from 12 to 15 km by Murcray et al.[20] Seiler and Warneck's data show a sharp decline in the CO mixing ratio at the tropopause, leveling off in the stratosphere to a constant value of 0.04 ppmv. From their data Murcray, et al. obtained a similar value for the mixing ratio, and extended the measurements to 15 km.

Our observations, again for the 1-0 (2143 cm^{-1}) band, confirm the constant vertical mixing profile above the tropopause, but show a marked decrease in mixing ratio with latitude, from 0.068 ppmv at 33°N to 0.021 ppmv at 76°N. These results, with the associated measurement uncertainties, are summarized in Fig. 3, from which it would appear that the Seiler and Warneck values, obtained at 50°N latitude, are consistent with this latitude variation, whereas the Murcray et al. value (30°N) is considerably lower.

Carbon Dioxide

In recent years carbon dioxide has been measured by numerous investigators, using a variety of experimental techniques and with remarkably concordant results. These measurements have shown in general that CO_2 is uniformly mixed vertically from the surface to as high as 62 km, with reported values for the mixing ratio ranging from 310 to 340 ppmv. Our value of 325 ± 20 ppmv, obtained from analysis of the $3\nu_2$ band at 1932.5 cm^{-1}, is in good agreement with these previous measurements, and no variation with latitude could be discerned.

The important problem with respect to CO_2 is not so much the further refinement of the absolute value of its concentration as the determination of any long term variation or increase. This has been estimated as being of the order of 1 ppm per year, the significance of the magnitude of such a change being the concommitant change to the radiation balance which it might ultimately cause. At present none of the remote observational techniques discussed here is capable of achieving the absolute accuracy required for such a measurement, although it may perhaps be possible to extract long term variations by statistical methods from data taken over a period of some months.

Methane

Stratospheric methane has been measured by Ackerman and Muller[21] and Low and McKinnon[22], by analysis of the near infrared (ν_3) band at 3018 cm^{-1}, and by Kyle, et al.[23] using the ν_4 band in the region of 1300 cm^{-1}. Kyle et al. presented their data in terms of the "quantity of methane per air mass" above different levels in the stratosphere, and reported a decrease by a factor of 2 to 5 in this quantity at 30 km over that at sea level. Ackerman and Muller's values indicated a nearly constant mixing ratio (of about 2 ppmv) to 25 km, decreasing to 0.75 ± 0.4 ppmv at 33.6 km. Lowe and McKinnon also reported constant mixing to 25 km, but their mean value of 0.94 ± 0.16 ppmv for the mixing ratio was lower by a factor of 2 than that of Ackerman and Muller in the same altitude range. Although Lowe and McKinnon's data were collected over a considerable range of latitudes, they saw no evidence for any latitude variation in the methane mixing ratio.

Our data also indicate constant vertical mixing for methane between 11 and 20 km, but with some latitude dependence. The results are summarized in Fig. 4. From the ν_3 band we obtained mixing ratios varying from 1.2 ± 0.05 ppmv at 17°N to 0.60 ± 0.04 ppmv at 76°N. This result is seemingly in disagreement with the latitude independence reported by Lowe and McKinnon; the difference may not be of any significance, however, in view of the limited data available to date, although it should be noted that a latitude and seasonal variation of stratospheric CH_4 is not unexpected considering its biogenic origin.

Other Molecules

A number of other species, such as H_2O_2, H_2CO, HCl, SO_2, N_2O_5, and NO_3 have been postulated as important trace constituents in the stratosphere, but have not yet been detected in infrared spectra. For several of these species, little or no laboratory data on the spectral assignments and line strengths exist — due either to the transitory nature of the species or the complexity of their spectral features — and as a consequence information concerning them cannot be extracted from the observational data. For others however, the required spectral information does exist and the reason that no detection has been made in these cases is that their abundances (if present at all) are below the current limits of detectability of the instrumentation. In this latter case, it is important for atmospheric modeling purposes that upper limits be reported which specify the minimum detectable amounts of the applicable species which could have been detected under the existing experimental conditions, and hence that the amounts actually present are below these limits.

We have recently undertaken the laboratory spectroscopic investigation of the ν_1 and ν_5 H_2CO bands in the 2800 cm^{-1} region and have made several spectral quantum assignments for lines in this region. Analysis of stratospheric spectra is complicated in this region as a result of severe blending with several weak bands of methane, and this has entailed further work on these bands in order to determine which H_2CO lines are best suited to analysis. Although this work is not complete, by applying the results so far obtained to the analysis of our stratospheric spectra we can report that to date no positive detection has been made, and can place an upper limit on H_2CO of 2 ppbv. Similarly, the spectra have been searched for evidence of the presence of HCl (in the region of the 1-0 band at 2886 cm^{-1}). In this case the line intensities are well known and an upper limit to the HCl concentration in the lower stratosphere is placed at 0.1 ppbv.

References

(1) Farmer, C. B., Raper O. F., Schaper P. W., Schindler, R. A., and Toth R. A., AIAA/AMS Conference on Environmental Impact, June 1973 (no reprints available).

(2) Schindler, R. A., Appl. Opt. 9, 301 (1970)

(3) Farmer, C. B., Toth R. A., Schindler, R. A., and Raper, O. F., Proceedings of 2nd Conference on CIAP, DOT-TSC-OST-73-4, 65 (1973)

(4) Schütz, K., Junge, C., Beck, R., and Albrecht, B., J., Geophys. Res., 75, 2230 (1970).

(5) Harries, J. E., Nature, 241, 515 (1973).

(6) Goldman, A., Murcray, D. G., Murcray, F. H., Williams, W. J., Kyle, T. G., and Brooks, J. N., J. Opt. Soc., 60, 1466 (1970).

(7) Goldman, A., Murcray, D. G., Murcray, F. G., and Williams, W. J., J. Opt. Soc., 63, 843 (1973)

(8) Nicolet, M., and Peetermans, W., Ann. Géophys. 28, 4, 751 (1972).

(9) Toth, R. A., Farmer, C. B., Schindler, R. A., Raper, O. F., and Schaper, P. W., Nature, 244, 7 (1973).

(10) Schiff, H. I., (in press).

(11) Ackerman, M., and Girard, M., Nature, in press.

(12) Goldman, A., Murcray, D. G., Murcray, F. H., and Williams, W. J., Nature, 225, 443 (1970).

(13) Ackerman, M., and Muller, C., Nature, 240, 300 (1972).

(14) Harries, J. E., private communication.

(15) Murcray, D. G., Murcray, F. H., Williams, W. J., Kyle T. G., and Goldman, A., Appl. Opt. 8, 2519 (1969).

(16) Goldman, A., Murcray, D. G., Murcray, F. H., Williams, W. J., and Brooks, J. N., Applied Optics, 12, 1045 (1973).

(17) Pick, D. R., and Houghton, J. T., Quart. J. Roy. Met. Soc., 95, 535 (1970).

(18) Mastenbrook, H. J., J. Atmos. Sci., 28, 1495 (1971).

(19) Seiler, W., and Warneck, P., J. Geophys. Res., 77, 3204 (1972).

(20) Murcray, D. G., Goldman, A., Murcray, F. H., Williams, W. J., Brooks, J. N., and Barker, D. B., "Proceedings of the Second Conference on the CIAP," U.S. Department of Transportation, November 14-17, 1972, p. 86.

(21) Ackerman, M., and Muller, C., "Proceedings of the Joint Mtg. of the Amer. Geophys. Union and the American Meterological Soc.," Aug. 15-17, 1972, p. 12-1.

(22) Lowe, R. P., and McKinnon, D., Can. J. Phys., 50, 668 (1972).

(23) Kyle, T. G., Murcray, D. G., Murcray, F. H., and Williams, W. J., J. Geophys. Res., 74, 3421 (1969).

Table 1. List of References Used to Determine Spectral Parameters for Data Analysis

Molecule	Band	Center Frequency	References		
			Vibration-Rotation Parameter	Band Strength	Line Width (air)
CO_2	$3\nu_2$	1932.477	a	b	c
N_2O	ν_3	2223.756	d	e	f
	$\nu_2 + \nu_3 - \nu_2$	2209.521	d	e	f
	$2\nu_1$	2563.358	g	h	f
	$\nu_1 + \nu_2$	1880.271	i	-	f
H_2O	ν_2	1594.73	j	j	j, k
	$2\nu_2$	3151.631	l	k	j, k
	ν_1	3657.054	l	k, m	j, k
	ν_3	3755.924	l	n	j, k
CO	1-0	2143.274	p	q	r
	Hot Bands	-	s	t	-
CH_4	ν_3	3018	u	v	w
O_3	$\nu_2 + \nu_3$	1726.44	-	x	-
	$\nu_1 + \nu_3$	2110.79	y	x	-
	$\nu_1 + \nu_2 + \nu_3$	2785.236	z	x	-
NO	1-0	1876	aa	ab	ac*
NO_2	ν_3	1616.846	ad, ae	af	ag
	$\nu_1 + \nu_3$	2906.074	ah	af	ag

*Self broadened

References for Table 1

a) A.G. Maki, E.K. Plyler, and R.J. Thebault, J. of Res NBS. 67A, 219 (1963)

b) R.E. Ellis, and B. Schurin, Appl. Opt. 8, 2265 (1969)

c) G. Yamamoto, M. Tanoka, and T. Aoki, JQSRT 19, 350 (1969)

d) J. Pliva, J. Mol. Spec. 12, 360 (1961)

e) J.E. Lowder, JQSRT 12, 873 (1972)

f) R.A. Toth, J. Mol. Spec. 40, 605 (1971)

g) E.D. Tidwell, E.K. Plyler, and W.S. Benedict, J. Opt. Soc. Amer. 50, 1243 (1960)

h) J. Vincent – Geisse, Am. de. Phys. 10, 69B (1955)

i) E.K. Plyer, E.D. Tidwell, and A.G. Maki, J. Res NBS. 68A, 79 (1963)

j) W.S. Benedict, R.F. Calfee, Essa Professional paper 2, U.S. Department of Commerce, Washington D.C., June 1967

k) R.A. Toth, JQSRT, to be published (1973)

l) L.A. Pugh, Ohio State Disserlation (1972)

m) H. J. Babrov, and F. Casden, J. Opt. Soc. Am. 58, 179 (1968)

n) D.M. Gates, R.F. Calfee, D.W. Hansen, and W.S. Benedict, Natl. Bur. Stand. Monograph 71 (1964)

o) Y. Ben-Aryeh, J. Opt. Soc. Am. 60, 570 (1970)

p) E.K. Plyler, L.R. Blaine, and E.D. Tidwell, J. Res. NBS 55, 183 (1955)

q) D.E. Burch, and D. Williams, Appl. Opt. 1, 587 (1962)

r) G.D.T. Tejwani, JQSRT 12, 123 (1972)

s) V.G. Kunde, NASA Publication X-622-67-248, June 1967

t) L.A. Young, and W.J. Eachus, J. Chem. Phys. 44, 4195 (1966)

u) W.L. Barnes, J. Susskind, R.H. Hunt, and E.K. Plyler, J. Chem. Phys. 56, 5160 (1972)

v) T.G. Kyle, AF 19(628)-5706 Scientific Report No. 1, October 1968

w) G.D.T. Tejwani, and P. Varanasi, J. Chem. Phys. 55, 1057 (1971)

x) D.J. McCoa, and J.H. Shaw, J. Mol. Spec. 25 (1968)

y) S. Trajmar, and D.J. McCoa, J. Mol. Spec. 14, 244 (1964)

z) D.E. Snider and J.A. Shaw, J. Mol. Spec. 44, 400 (1972)

aa) J.H. Shaw, J. Chem. Phys. 24, 399 (1956)

ab) T.C. James, J. Chem. Phys. 40, 762 (1964)

ac) L.L. Abels, and J.H. Shaw, J. Mol. Spec. 20, 11 (1966)

ad) Hurlock, Dissertation Ohio State 1970

ae) R.A. Toth — Analysis of Hurlocks spectra
W.S. Lafferty — Analysis of Hurlocks spectra

af) A. Guttman, JQSRT 2, 1, (1962)

ag) G.D.T. Tejwani, J. Chem. Phys. 57, 4676 (1972)

ah) M.D. Olman, and C.D. Hause, J. Mol. Spec. 26, 241 (1968)

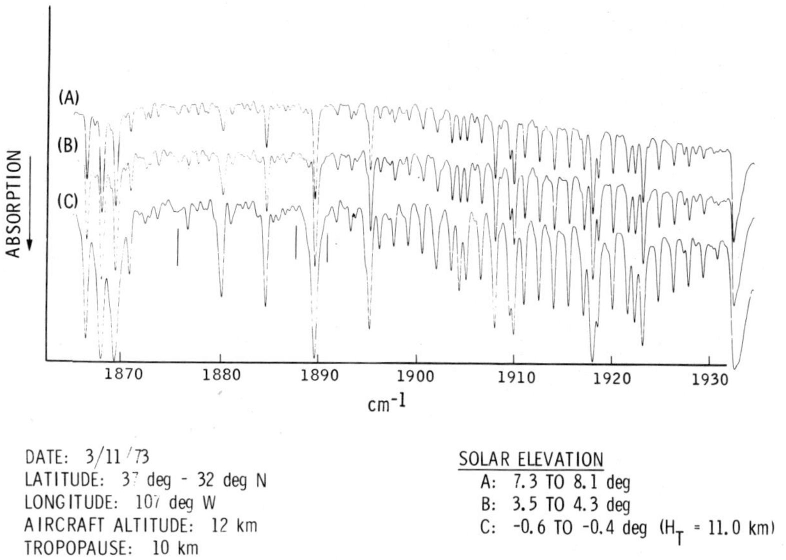

Figure 1. Stratospheric Solar Spectra in the Region of the 1-0 Band at 1876 cm^{-1}. Three of the weak NO Features are Indicated by Vertical Bars.

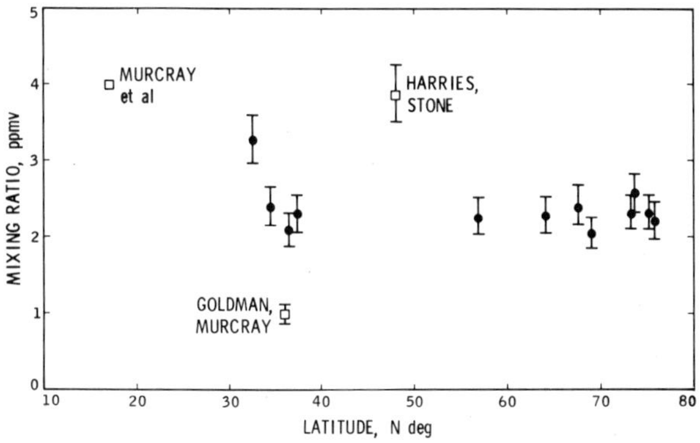

Figure 2. Stratospheric Latitude Variation at H_2O at 16 km Altitude

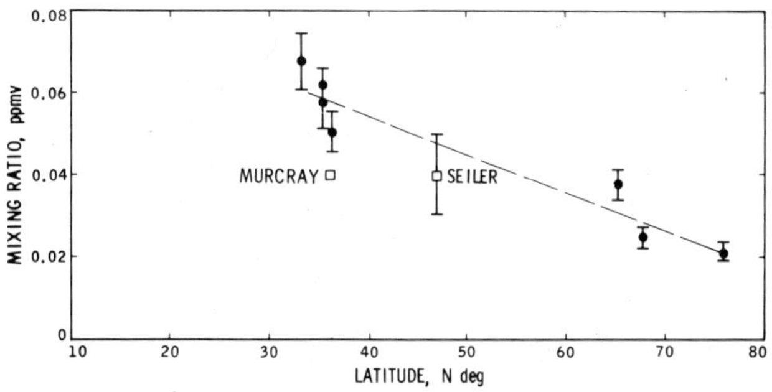

Figure 3. Latitude Variation of CO at 16 km Altitude

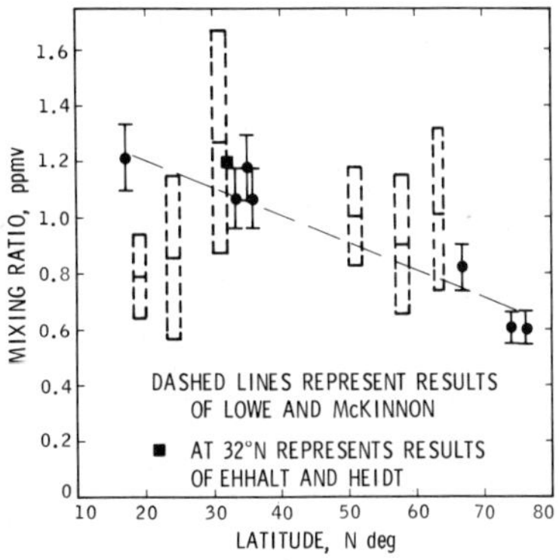

Figure 4. Latitude Variation of CH_4 at 16 km Altitude

©1973, ISA JSP 6655

THE REMOTE MEASUREMENT OF TRACE ATMOSPHERIC SPECIES BY CORRELATION INTERFEROMETRY I. CARBON MONOXIDE AND METHANE*

Harold W. Goldstein, M. H. Bortner, Robert N. Grenda and Arieh M. Karger
General Electric Company, P.O. Box 8555, Philadelphia, Pa. 19101
Robert Dick and Frank David
Barringer Research Ltd., Toronto, Ontario, Canada
Peter J. LeBel
NASA Langley Research Center, Hampton, Va. 23365

ABSTRACT

A correlation interferometer has been developed for the measurement of carbon monoxide and methane at 2.35 micrometers in the troposphere and in the stratosphere. This instrument has been tested in laboratory tests, solar-looking outdoor tests, and downward-looking airplane-based tests. The aircraft tests were flown on a Falcon fanjet provided by The Canada Centre for Remote Sensing over both polluted and unpolluted regions of North America. The results of these various tests are discussed. Based on the results obtained for carbon monoxide and methane, a study was undertaken to investigate the feasibility of measuring other atmospheric trace species by correlation interferometry. Results of the feasibility study for carbon dioxide, water vapor, ammonia, nitrous oxide, nitric oxide, nitrogen dioxide, sulfur dioxide, and several hydrocarbons are presented.

INTRODUCTION

The development of methods for the remote measurement of species present in trace amounts in the atmosphere has been a major effort over the past several years, especially for those species generally classified as pollutants. Effects of pollutants occur from ground level well up into the stratosphere with the effects being most significant near the ground. Thus any remote measurements should include effects of pollutants at and near ground level. Further, since concentrations of pollutant species vary, it is necessary that measurements should cover a wide range of optical thicknesses. Requirements for both tropospheric and stratospheric measurements have been given by the RMOP report[1]. The concentrations of the various pollutants vary drastically from the unpolluted levels (some of which are only roughly known) to levels of the order of a hundred times that of the background. Thus a wide range must be covered in any measurement. The present work includes the testing of a correlation interferometer for such measurements of carbon monoxide and methane and an analysis of the feasibility for the use of the correlation interferometer for such measurements of several other pollutants.

*Work supported in part by NASA Contract NAS1-10139.

TECHNIQUES

The correlation interferometry technique has been described in detail elsewhere[2,3,4,5]. Briefly, it compares interferograms with several interferograms for which the amount of the gas being measured is known. The differences among the interferograms is attributed to the amounts of the gas of interest and of the other gases in the comparison interferograms. Only that small portion of the interferogram most sensitive to the gases being measured is used and the various parts of the interferogram are weighted to minimize the effects of interferents and maximize that of the gases being measured. No attempt is made to generate or otherwise work with a spectrum.

The portion of the interferogram used is that having a difference in the path lengths of the two arms of the interferometer between 2.7 to 3.95 mm. The optical filter has a peak transmission at 4280 cm^{-1} and a half-width of 10.71 cm^{-1} at 50% transmission. A PbS detector operating at 195K and having an N.E.P. of 1.6×10^{-11} watts $hz^{-1/2}$ was used. The instrument had a 6.6 cm aperture and a 7° field-of-view. This gave approximately a 1200 foot field-if-view on the ground when the aircraft was flying at 10,000 feet. The measurement time for one scan was 1 second. Most of the data were treated by co-adding a number of scans.

ENGINEERING MODEL TESTING

The engineering model correlation interferometer was subjected to a series of tests in the laboratory and outdoors where it could be tested using real atmospheric amounts of test gases and interferents. These tests were designed to demonstrate the ability of the instrument to measure CO and CH_4 in the presence of interferent gases and to determine instrument calibration correlation functions which could be applied to future measurements to indicate the CO and CH_4 in the instrument optical path.

A multiple gas cell test facility was constructed to permit known amounts of CO and CH_4 to be introduced into the optical path of the instrument. Three test cells were aligned to form a cell with total length of 1.5 m but these separate chambers could individually be filled with any required

"Superior numbers refer to similarly-numbered references at the end of this paper."

amount of gas. The test gases were obtained premixed with nitrogen and analyzed for composition.

The laboratory tests were conducted using CO and CH_4, and were intended to indicate whether the instrument was functioning properly and to determine the noise level in the measurements. The tests indicated a noise level of about 0.02 atm-cm of CO or about 10% of a nominal atmospheric amount.

The outdoor tests were conducted with the instrument looking through the same gas cells at a ground glass diffuser which was illuminated with sunlight. This was adjusted to give a source illumination corresponding to that which the instrument would see if it were in an aircraft looking down at sunlit earth with an albedo of 30% to 40%. Measurements were made during the course of a day during which the atmospheric air mass varied from 1 to about 3. Also during the measurements known amounts of CO and CH_4 were introduced into the cells to provide a matrix of test cases which covered a range of CO from about 0.08 to 1.6 atm-cm and CH_4 from about 1.0 to 5.5 atm-cm. In conjunction with these measurements solar absorption spectra were obtained with an Idealab IF-6 Fourier spectrometer and these were used to determine the amount of CO and CH_4 in the atmospheric path when the correlation interferometer measurements were being made.

As a result of these measurements correlation functions for CO and CH_4 were determined for use in analysis of the later flight data. Data not used to develop the correlation functions were used as test data and showed agreement to better than 10% except for those cases where the gas burdens required extrapolation beyond the conditions for which the correlation functions were obtained. Since, in practice, correlation functions can be obtained for any reasonable ranges of gas burdens, this is not a major problem.

AIRCRAFT-BASED MEASUREMENTS

Measurements were made from an aircraft provided by the Canada Centre for Remote Sensing. The aircraft was a Falcon fanjet which had a downward-facing quartz window in the floor. The instrument looked through this window straight down with a 7° field-of-view. The view was followed on a TV monitor and recorded on tape. The tests were made to test the operation of the instrument in flight using a downward-looking mode under a variety of conditions. The gas measured was that in the path from the aircraft to the ground (unless clouds interfered) and from there to the sun.

Two flights were made. The paths of the two flights are shown in Figure 1. On the first flight, data for a number of scans were co-added and recorded on paper tape. On the second flight, the data were recorded on magnetic tape and subsequently processed by co-adding various numbers of scans. Data from the outdoor and laboratory measurements were used for calibration. The instrument appeared to operate properly during both flights.

The data obtained on the first flight are shown in Table I for the measurements of carbon monoxide and methane. The values obtained are reasonable. The area around Petawawa was largely cloud covered. It would thus be expected that the CO would be much lower than elsewhere since only that in the atmosphere above the clouds was measured. The values in the more urban areas were highest with the variations in CO being appreciably greater than those in CH_4. The first seven of the measurements listed were made from 28,000 feet. The last two were made from 10,000 feet, while the eighth was made while descending between these two altitudes. Consequently, considering the total path and assuming a constant mixing ratio over the atmosphere in order to give an approximate idea of the concentrations of CO and CH_4 in ppm, the CO varied from .048 to .315 ppm and the CH_4 varied from 1.09 to 2.19 ppm.

ANALYSIS OF FEASIBILITY FOR OTHER GASES

An analysis was made to determine the feasibility of using the correlation interferometry technique for the measurement of atmospheric amounts of various other gases - NH_3, NO, NO_2, N_2O, and SO_2. This was accomplished by (1) experimentally obtaining spectra of these gases as well as of any gases which spectrally interfere (those of water were obtained theoretically for various amounts of water); (2) combining spectra in designated combinations and generating interferograms; and (3) using some of these interferograms to generate weighting functions testing the accuracy of tests made on the other interferograms. By comparing interferograms from step 2 with and without a given gas, a determination was made of the best spectral and interferogram path difference regions, to give the greatest sensitivity and accuracy for that gas. Figures 2 and 3 show interferograms with and without each of several gases. The results given above for CO were obtained by use of that part of the interferograms between 0.27 and 0.39 cm. The effect of CO on this region is shown in Figure 2. From this it can be seen that only very small effects are needed for the measurement. Thus the effects of the other gases on the interferograms, most of which for normal atmospheric burdens of the gases are larger, should be suitable for the needed measurements. Figure 3 shows similar interferograms for nitric oxide. It can be seen that the effects are largest in the 6 to 7 mm range of path differences although the regions around 3.0 and around 9.3 mm are usable. It can be estimated from these curves that the method has a sensitivity of .003 atm-cm for NO.

From the tests of the interferograms by the three-step analysis described above for a variety of conditions, the following indications were obtained.

NO measurements for .0160 atm-cm could be made with an accuracy of 10% if all interferents are considered and if the interferograms used to generate the weighting functions cover the entire range of gas burdens encountered in the measurements.

The optimum delay region to use for NO is that of 5.8 to 7.2 mm path difference.

The sensitivity of the technique for NO is

approximately .002 atm-cm which corresponds to a mixing ratio of .0005 ppm for a path through the atmosphere with a 20 km grazing altitude.

SO_2 measurements for .0036 atm-cm could be made with an accuracy of the order of 30% if all interferents are considered and if the interferograms used to generate the weighting functions cover the entire range of gas burdens encountered in the measurement.

The optimum delay regions to use for SO_2 is that of 8.0 to 9.5 mm path difference.

The sensitivity of the technique for SO_2 is approximately .001 atm-cm which corresponds to a mixing ratio of .0003 ppm for a path through the atmosphere with a 20 km grazing altitude.

Similar results have been obtained for NH_3, NO_2, N_2O, CO_2, and H_2O, in addition to the previous results for CO and CH_4.

The inclusion of the effect of species not included in these used to generate the weighting functions caused errors of up to 35%. However, since the technique should use the most advantageous delay region, this can be considerably reduced, so that the accuracy can be within 10% for NO and 30% for SO_2.

REFERENCES

(1) NASA Langley Research Center, "Remote Measurement of Pollution," NASA SP-285, August, 1971.

(2) Bortner, M.H., F.N. Alyea, R.N. Grenda, G.M. Levy and G.R. Liebling, "Analysis of the Feasibility of an Experiment to Measure Carbon Monoxide in the Atmosphere," GENERAL ELECTRIC CO. REPORT NASA CR-2303, March, 1973.

(3) Bortner, M.H., R. Dick, H.W. Goldstein, R.N. Grenda and G.M. Levy, "Development of a Breadboard Model Correlation Interferometer for the Carbon Monoxide Pollution Experiment," GENERAL ELECTRIC CO. REPORT NASA CR-112212, March, 1973.

(4) Grenda, R.N., M.H. Bortner, P.J. LeBel, J.H. Davies and R. Dick, "Carbon Monoxide Pollution Experiment - (I). A Solution to the Carbon Monoxide Sink Anamoly," AIAA Paper #71-1120, presented at the Joint Conference on Sensing of Environmental Pollutants, Palo Alto, November, 1971.

(5) Goldstein, H.W., M.H. Bortner, R.N. Grenda, A.M. Karger and P.J. LeBel, "Correlation Interferometric Measurement of Trace Species in the Atmosphere," AIAA Paper #73-515, Presented at the International Conference on the Environmental Impact of Aerospace Operations in the High Atmosphere, Denver, June, 1973.

TABLE I

MEASURED COLUMN DENSITIES (atm-cm)

Approximate Location	CO	CH_4
Petawawa	.074	1.41
North Bay	.179	1.31
Sudbury	.216	1.31
Owen Sound	.275	1.43
Perth	.148	1.57
Cornwall	.486	2.19
Montreal Area	.479	2.01
Ottawa (North)	.149	1.35
Ottawa	.389	1.39
Ottawa	.431	1.38

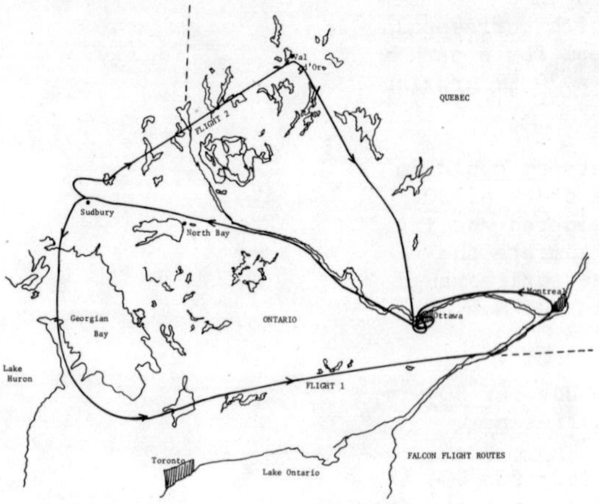

Figure 1. Falcon Flight Routes.

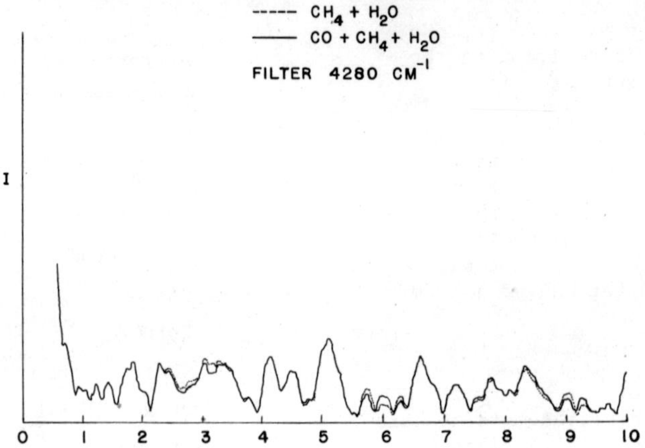

Figure 2. Interferograms Showing the Effect of CO.

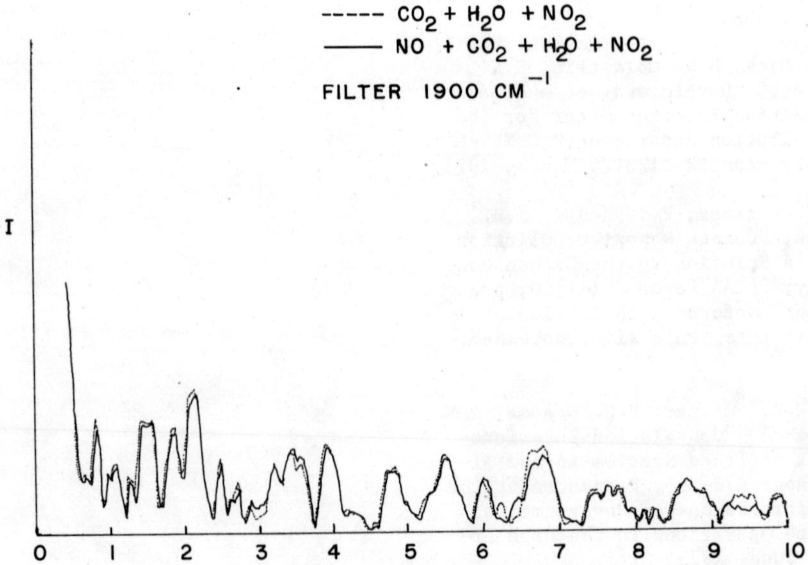

Figure 3. Interferograms Showing the Effect of NO.

© 1973, ISA JSP 6656

ULTRA NARROW BAND INFRARED FILTER RADIOMETRY

Aidan E. Roche
Alan M. Title
Lockheed Palo Alto Research Laboratory
Palo Alto, California

I. INTRODUCTION

Interference filters for infrared wavelengths have generally been limited to bandwidths of the order tens of cm^{-1} (typically 500Å at 5 μm) and they have been almost entirely used in a single, fixed frequency mode. Therefore, even though the filter photometer has a large inherent throughput advantage over grating and prism spectrometers (Hunten), it could not compete either with the frequency scanning facility of these instruments nor the fact that up to now the high resolution (< 2 cm^{-1}) required for infrared atmospheric emission and absorption measurement has only been achievable by non-filter devices. Now, recently developed solid spaced Fabry-Perot filters are available which have extremely narrow bandwidths (< 0.2 cm^{-1}), and high transmission (75%), which work over a range approximately ±10% of center wavelength, and which in addition can be fine scanned over hundreds of Angstroms, at any chosen operating wavelength in the working range, simply by tilting the filter several degrees with respect to the incident radiation axis (see Eather and Reasoner, 1969).

These filters can thus be used to great advantage for airborne (including spacecraft) atmospheric infrared measurements where size, weight, and reliability are prime consideration and must be weighed against resolution and sensitivity requirements. For the detection and measurement of atmospheric minor constituents and pollutants such as NO, NO_2, N_2O, O_3, CH_4 and CO, all of which have distinct emission features in the 3 to 7μm range, a spectral bandwidth of the order of a few Angstroms is desirable for unambiguous resolution of specific emission features against ambient atmospheric backgrounds. Resolution this high (~ 0.2 cm^{-1}) is currently the limiting value achieved by non-filter instruments flown to date, and can generally be achieved only by the methods of fast fourier transform spectroscopy. A single solid Fabry-Perot filter, on the other hand, centered for example at 5.8μm and having a 2Å bandwidth, would have a useful operating range of 5.2 to 6.4μm and by the selection of blocking filters and tilt-tuning it could be used to measure line emission in bands of NO, NO_2, CH_4, H_2O and NH_4. The single, broad-range tilt tuned filter thus replaces the prism, gratings and mirrors and associated scanning drive systems of spectrometer and interferometers, resulting in improved reliability and throughput and resolution comparable to the most sophisticated instruments now being flown.

The design and laboratory measured performance of several recently fabricated ultra narrow band filters of this type will be described (centered at 1.1, 3.35, and 5.0μm) and an estimation will be made of the sensitivity of a scanning photometer using the 5μm filter for the measurement of stratospheric NO.

II. FILTER PERFORMANCE

1. General Characteristics

Solid-spaced Fabry-Perot filters can be made either in an all-deposited configuration in which a dielectric layer is vacuum deposited as the spacer between the reflecting layers (mirrors), or a spacer can be ground and polished from a solid piece of material and the reflecting layers can be coated over the parallel surfaces of the wafer. Bandwidths < 300Å at infrared wavelengths, however, cannot in practice be achieved with all deposited filters due to the inherent non-uniformity in the evaporated spacer layers, as the thickness is increased for these wavelengths. The Perkin Elmer Corp. devised a way for making fused-silica - spaced filters for wavelengths up to 1μm so that wafers as thin as 50μm with surface uniformity λ/100 or better can be achieved (Austin, 1973). The measured transmission band profile of a fused-silica -spaced integral filter, designed for 10830Å operation, is shown in Figure 1. Fabry-Perot filters, of course, produce a channel spectrum in which the separation between transmission bands, or the free spectral range is

$$\lambda_s = \lambda_m^2/2\,\mu d, \qquad (1)$$

μ being the refractive index of the spacer, λ_m the wavelength of peak transmission, d its thickness. Two transmission bands are shown in Figure I for the 10830Å filter to illustrate both the bandwidth (Δλ = 0.3Å FWHM) and free spectral range (λ_s = 10Å) indicating a finesse [(λ_s)/(Δλ)] of 33. The evaporated layers forming the reflectors for the filter generally show reflection curves which are flat over a wavelength range ±10% of the center wavelength and outside this range drop off rapidly. A single narrow transmission band can thus be isolated at any point in the operating range by choosing a blocking filter which cuts off appropriately to either side of the transmission band. The availability of blocking filters is thus a non-trivial consideration in the design of narrow band filters for infrared wavelengths and λ_s must be chosen or arranged large enough to be compatible with the available blocking filters.

When the filter is tilted an angle θ to the incident light, the wavelength λ_o of peak transmission of each band is shifted to a lower wavelength by an amount $d\lambda$ which for small angles is given by

$$d\lambda = \lambda_o \, \theta^2/2 \, \mu^2. \qquad (2)$$

Thus, having isolated a given transmission band using an appropriate blocker, the filter can be fine-tuned within the range of the blocking filter bandpass by tilting an appropriate number of degrees. In this way, a scan can be made across an emission line profile and background subtraction obtained.

Equation (2) shows that the angular sensitivity of a Fabry-Perot filter is in proportion to the inverse square of the refractive index, indicating that the solid spaced filter is much less angular sensitive than air-spaced filters, and has a throughput μ^2 times that of the air gap device. This angular sensitivity also means that the filter bandwidth is dependent on the acceptance cone angle of the systems. (See Title, 1971.)

Figure 2 shows the results of a laboratory experiment in which the 10830Å filter, shown in Figure 1, was used in a tilting mode in an f/60 beam to scan the He(2^3P-2^3S) feature produced in a Helium discharge. Complete resolution of the two lower wavelength lines of the triplet (separated by approximately 1Å) is seen to have been achieved. The same filter used for atmospheric measurement could be used to advantage in resolving the upper atmospheric Helium 10830Å feature against the strong OH (5-2) emission which lie within a few Å of the Helium line. The OH line emissions could, of course be simultaneously measured.

2. The 5μm Filter

To extend the solid-spaced design to 5μm, essentially the same techniques were used in polishing and coating the spacers, except that of course the spacer material had to be infrared transmitting at 5μm, whereas fused silica cuts off in the region of 4.5μm. Several materials were examined including Germanium, Sapphire, and Magnesium Fluoride, and finally YTTRALOX (a combination of YTTRIUM and THORIUM OXIDE) was chosen for the prototype filter. Transmission band profiles of this filter are shown in Figure 3. The design specifications were for a 0.5" diameter filter having a 300Å for spectral range, a 5Å bandwidth and better than 50% transmission, centered at 5μm. The only installation available to us upon completion of the filter with sufficiently high resolution to measure the filters transmission band profile was the Kitt Peak grating solar spectrograph and this was operating at the time at a maximum wavelength of 4.6μm, which applies to the measurement shown in Fig. 3. The bandwidth is seen to be 8Å, the free spectral range 300Å (a finesse of 37.5) and the peak transmission ~ 58%. This bandwidth is orders of magnitude better than anything currently available in such a filter, and is expected to be even better at the 5μm design wavelength. All indications are thus that the accurate translation of the design into a working filter is relatively straightforward once the spacer material properties are well understood. A 5Å bandwidth was chosen for the prototype so that a 300Å free-spectral range could be achieved which can be relatively easily blocked by all-dielectric filters. A 1Å bandwidth could just as easily have been realized if the resulting 60Å free spectral range (assuming the same finesse) could be tolerated. One way of retaining very narrow bandwidths but increasing the effective λ_s to a figure large enough to be accommodated by available blockers is to operate two or more filters in tandem, as described by Austin (1973). The ratio of λ_s is chosen so that the transmission peaks coincide at some multiple of the larger λ_s, thus increasing the effective λ_s but retaining the $\Delta\lambda$ of the narrowest filter. This technique is being used by us in the design of an effective 1Å filter for 3.35μm operation.

3. The 3.35μm Filter

The third infrared filter, still under fabrication, is actually a combination of two filters. One has a 5Å bandwidth and 150Å λ_s, and this filter has been completed and successfully tested at Kitt Peak, the tests including a tilt measurement in which the filter was tilt-scanned over a free spectral range with no significant deterioration in its performance. The second filter will have approximately 1Å bandwidth and a λ_s designed so that the two-filter combination has an effective λ_s > 300Å. The resulting combination, a 1Å bandwidth device, will be used for measurement of lines in the 3.35μm band of methane in Comet Kohoutek from a NASA airplane, and later for atmospheric methane. Fig. 4 refers to the 5Å filter.

III. PASSIVE SENSING OF STRATOSPHERIC NO_x USING A 5.5μm NARROW BAND FILTER

To estimate the detection sensitivity of a filter photometer, we chose a problem of considerable current interest to CIAP -- the altitude profile measurement of NO in the stratosphere. We consider a filter which is very similar in performance to our 5μm device, but which has a 2" or greater aperture, and is centered at 5.2μm so that it can encompass line emission measurements in the 5.3 NO band.

We have calculated the emission intensity for the most populated lines in the 5.3μm band of NO to be of the order of 10^{-25} w/molecule at 220°K. This translates into an expected atmospheric NO column emission intensity of the order of 5×10^{-12} w/cm^2/sr for horizontal viewing at 30 km altitude and an NO mixing ratio of 2×10^{-10}. We assume the lines to be collision dominated with half-width less than 1Å so that essentially all the line radiation is encompassed by a 1Å bandwidth filter. For a 5 cm aperture and a 3° field of view, the required detector NEP would be approximately 3×10^{-13} watts (Hz)$^{-1/2}$, a figure well within the sensitivity of state-of-the-art cryogenically cooled instruments.

To estimate the background levels seen by the instrument within the 1Å filter bandwidth, we assume the main contribution to arise from particulate layer thermal emission. Overlap from emission in the line wings of abundant ambient molecules such as H_2O is considered to be negligible, since we can choose our NO line to be well isolated from such emissions, and we can also scan across the equivalent of many line-widths to estimate the contribution from adjacent emitters. Based on observations of Murcray et al. (private communication), we estimate that 5.3μm particulate background to be of the order of 3×10^{-12} w/cm^2/sr/Å at a 30 km stratospheric altitude. This figure is more then an order of magnitude lower than our estimated NO intensity, and thus should present no problem.

An NO mixing ratio of 10^{-10} should thus be detectable using such a narrow band tilt scanned filter instrument. Mounted in a balloon, altitude profile information could be obtained during ascent and descent modes for horizontal NO column densities. Theoretical estimates put the NO mixing ratio in the range 10^{-8} - 10^{-10} (Crutzen, 1972) while the most recent spectroscopic and chemiluminescent measurements indicate levels between 10^{-10} and 10^{-9}, below 25 km. Our estimated sensitivity should thus be sufficient to detect NO at levels below the smallest theoretical and reported values. The same filter could be used to detect NO_2 at 6.2μm with comparable sensitivity.

Acknowledgements:

We wish to acknowledge Mr. Edward Strouse of the Perkin Elmer Corporation who worked with patience and perseverence to meet our requirements on the infrared filters, and Dr. Donald Hall of the Kitt Peak National Observatory who played a major part in the filter transmission band profile measurements.

References:

Austin, R.R.; Electro-Optical Systems Design 6, 32, 1973.

Crutzen, P.J.; Physics and Chemistry of Upper Atmospheres, Ed. B.M. McCormac (D. Reidel Pub. Co.) 110, 1973.

Eather, R.H. and D.L. Reasoner; Applied Optics 8, 227, 1969.

Hunten, D.M.; Space Sci. Rev. 6, 493, 1967.

Murcray, D.M.; private communication.

Title, A.M.; New Techniques in Space Astronomy, pp. 325-332, 1971.

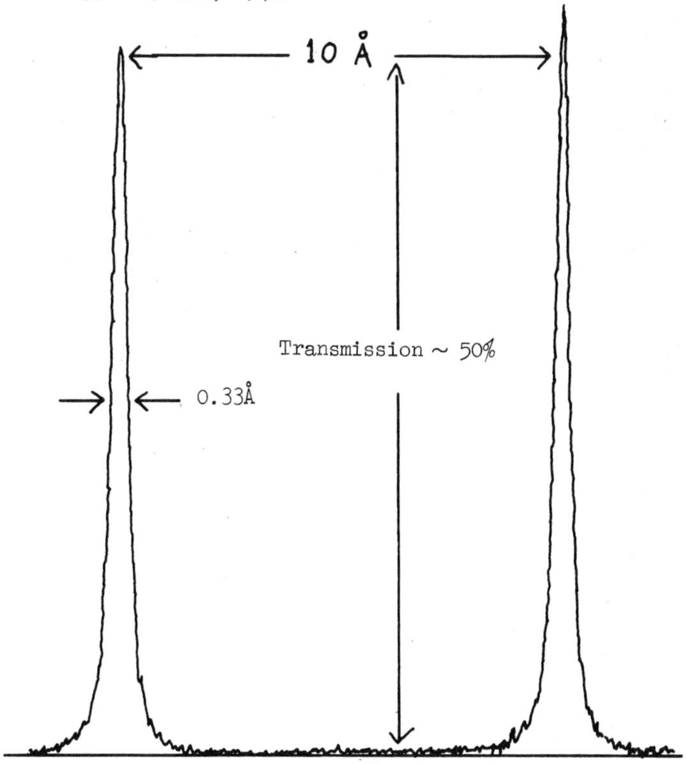

Fig. 1 Measured transmission band profiles of a 10830Å fused silica filter.

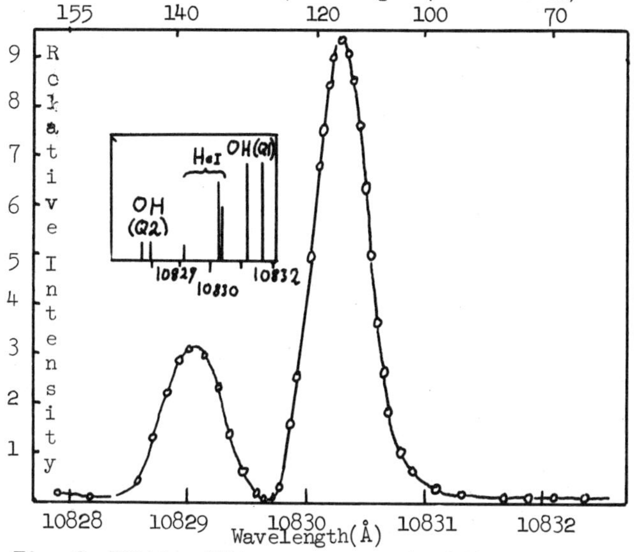

Fig. 2 Tilting filter measurement of the HeI [2^3P-2^3S] triplet feature 1Å. Laboratory Source using the 10830Å filter.

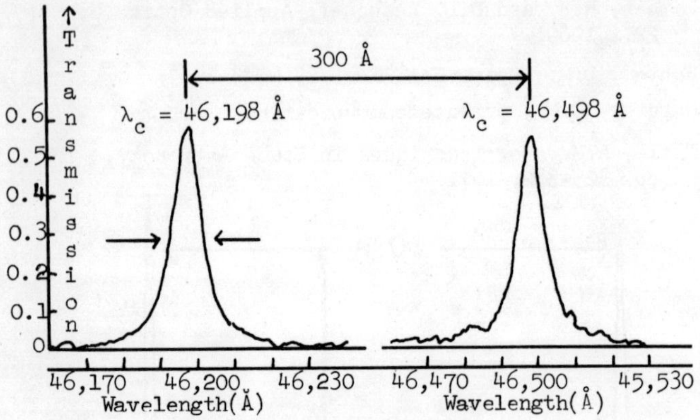

Fig. 3 Measured transmission band profiles of a 5.0μm central wavelength filter at 4.6μm YTTRALOX substrate.

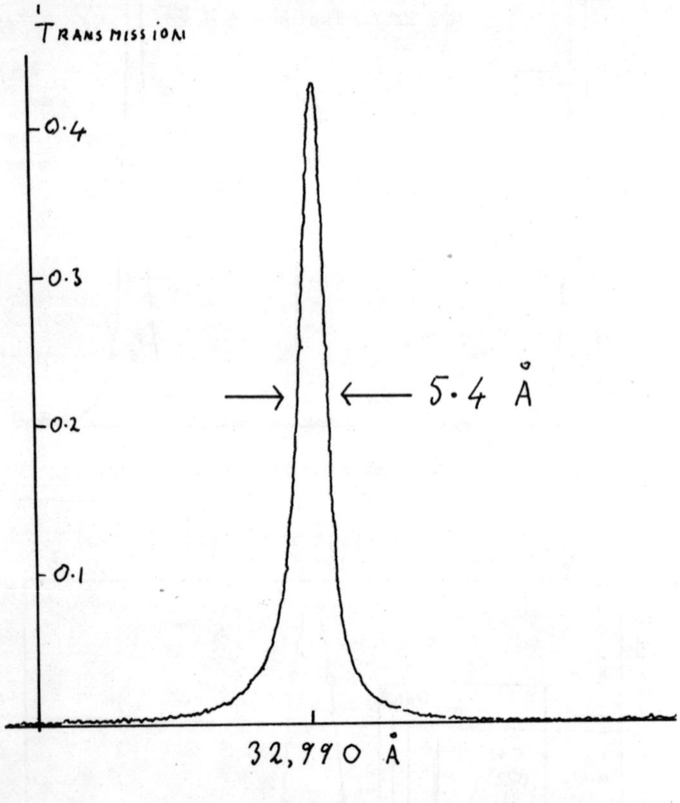

Fig. 4 Measured transmission band profiles of a 3.35μm central wavelength filter at 3.29μm.

©1973, ISA JSP 6658

FURTHER DEVELOPMENTS IN CORRELATION SPECTROSCOPY FOR REMOTE SENSING AIR POLLUTION

A. R. Barringer, J. H. Davies and A. J. Moffat
Barringer Research Limited
304 Carlingview Drive
Rexdale, Ontario, Canada

ABSTRACT

The majority of gaseous pollutants exhibit optical absorption bands in some portion of the spectrum. If one correlates the absorption spectrum of the gas being measured against a stored replica or mask of that spectrum, quantitative detection is possible. The term correlation spectroscopy has been coined to describe this technique. When natural background radiation is used as for example ultraviolet and visible radiation from the sky, remote monitoring of certain species such as SO_2 and NO_2 is possible.

One of the principle types of measurements that can be made using the remote sensing spectrometric techniques for pollution monitoring is that of vertical burden measurements. This measurement refers to the amount of gas that lies in a column above a unit area of ground and can be expressed in milligrams per square meter indicating the mass in milligrams of the gas which lies in the vertical column of air above a square meter of ground. Measurements of vertical burdens can be made with correlation remote sensors placed in moving vehicles looking vertically upwards.

INTRODUCTION

The instrument techniques and application methods of correlation spectroscopy have been under development for several years and status papers have been presented from time to time [1, 2, 3, 4, 5].

The purpose of this paper is to discuss a specific application, namely the measurement of total vertical burden of SO_2 and NO_2 in the atmosphere and the determination of total gaseous pollutant mass flow using a remote sensing correlation spectrometer. The sensor is mounted in a vehicle and the zenith sky is used as the source of illumination.

The sensor described herein is designed to detect and measure pollutant clouds of SO_2 and NO_2 in the atmosphere and utilizes the characteristic molecular absorption of solar energy in the ultraviolet for SO_2 and in the blue visible for NO_2. Figures 1 and 2 show the SO_2 and NO_2 absorption spectra utilized by the COSPEC family of SO_2/NO_2 remote sensors. Since the technique depends on the removal of radiant energy at wavelengths specific to the molecular species of interest and quantifies the measurement by comparing the energy content at these wavelengths with the energy content at wavelengths where the molecules do not absorb, or absorb less intensely, it follows that a detailed knowledge of the spectral content of the background skylight is an essential input to the instrument design. The present paucity of reliable high resolution solar spectra which applies to a suitably wide range of sky and cloud conditions is a problem of considerable magnitude in the design of high sensitivity versatile remote sensors. An ideal remote sensing gas analyzer of the COSPEC type should be capable of measuring the total vertical burden of a chosen gaseous species in the polluted lower layers of the atmosphere with acceptable absolute accuracy which is unaffected by changing sky conditions, e.g., changes in sky brightness and spectral distribution caused by changing sun position, cloud characteristics, smoke and haze etc. This paper describes recent test results of an experimental high sensitivity single gas correlation spectrometer referred to herein as the COSPEC IV. A brief description of the instrument would perhaps be worthwhile.

THE INSTRUMENT DESIGN

As shown in Figures 3, 4, and 5, the sensor contains two telescopes to collect light from a distant source, a grating spectrometer for dispersion of the incoming light, a disc-shaped exit mask or correlator and an electronics system. The correlator functions as a high contrast reference spectrum for matching against the incoming spectra and is comprised of arrays of circular slits photo-etched in aluminum on quartz. The slit arrays are designed to correlate sequentially in positive and negative sense with absorption bands of the target gas by rotation of the disc in the exit plane. The light modulations are detected by photo-multiplier tubes (PMTs) and processed in the electronics to produce a voltage output which

"Superior numbers refer to similarly-numbered references at the end of this paper."

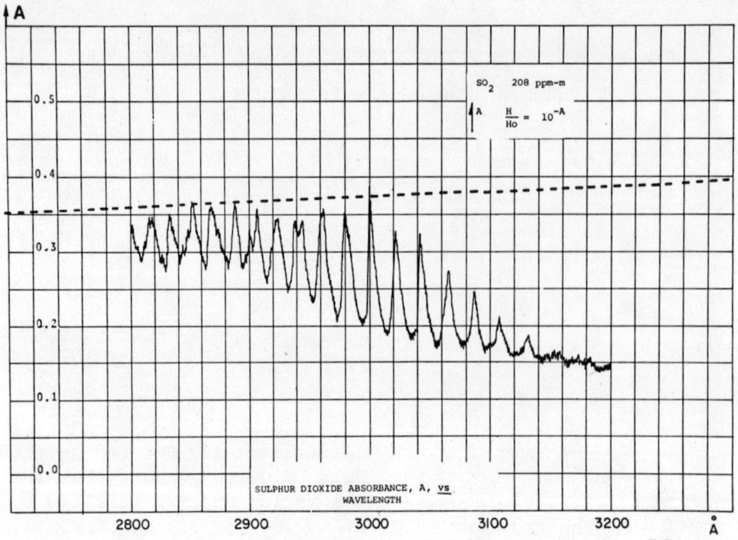

FIG. 1

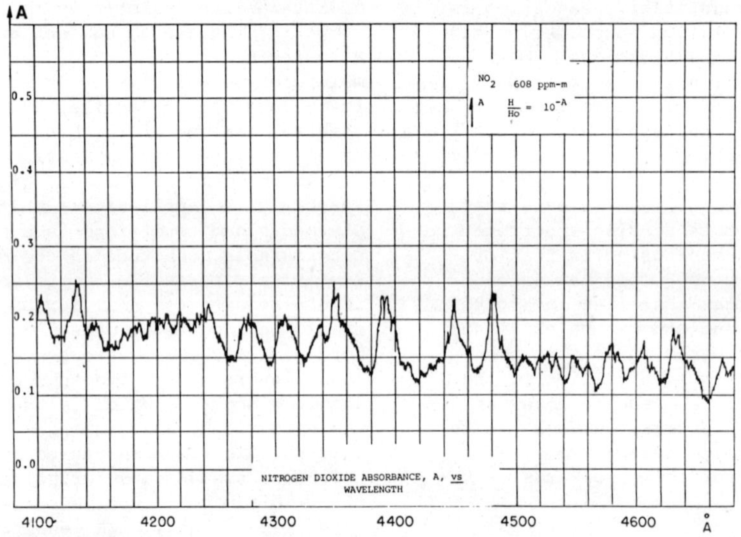

FIG. 2

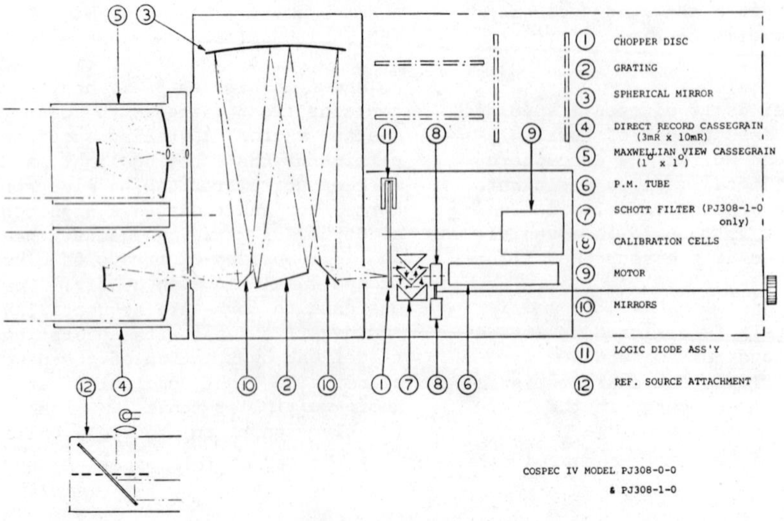

FIG. 3

is proportional to the optical depth or burden (ppm-meters) of the gas cloud under observation. The system automatically compensates for changes in average source light intensity in each channel.

The direct recording telescope (see 4 of Figure 3) provides an angular field of view of approximately 3 m radians in azimuthal and 10 m radians in elevation dependent upon the focus setting of the secondary mirror of the telescope. An image of the source is produced at the entrance slit of the spectrometer box.

The Maxwellian telescope (5) provides an angular field of view of one degree and the Cassegrain primary aperture is imaged on the entrance slit by the field lens. A relay lens then images the field lens onto the grating. This arrangement gives an evenly illuminated aperture at the entrance slit thereby eliminating any structure contained within the field of view.

The direct recording arrangement is used when a very small field of view is required such as when observing distant stack plumes. The Maxwellian system is used where high angular resolution is not necessary such as when performing vertical-look traverses in the passive mode.

Shown at (12) Figure 3 and (4) Figure 5 is the Reference Source Attachment. This is a screw-on accessory that may be used to provide a stable instrument zero. The source is a 12 volt, 55 watt quartz iodine auto-lamp.

Mid-Optics

The spectrometer per se is a 1/4 meter f/4 Ebert-Fastie system with the grating shown at (2) of Figure 3. Adjustment of the position of the spectra, i.e., wavelength interval at the exit aperture is achieved by adjustment of the micrometer (1) of Figure which controls the grating angle. Use of an Ebert-Fastie configuration provides a very compact, rugged design with very low aberrations and excellent focal plane characteristics.

Detector Optics

The rotating disc (1) Figure 3 and Figure 4 contains in its optical pattern the information for processing the SO_2 (UV) or NO_2 (visible) spectra. The inner band of large slits is an optical code providing a position reference for the electronics. The spectrum is chopped by the disc and detected by the photomultiplier tube (PMT).

Shown at (7) of Figure 3 is a Schott type UV reflection filter which rejects all stray light which falls outside the SO_2 absorption region. (A filter is not required for NO_2 monitoring).

Calibration cells are provided for insertion at position (8) Figure 3 by a rotatable shaft. Two cells can be inserted sequentially for a two point calibration.

Sensor Response and Signal Processing

Sensor response follows the Beer-Lambert law of absorption

$$H(\lambda) = H_o(\lambda)\ \exp\ (-\sigma(\lambda)n.x)$$

where

$H_o(\lambda)$ spectral irradiance of the light source ($W\ cm^{-2}\ .\ nm^{-1}$)

$\sigma(\lambda)$ absorption coefficient of the gas of interest ($cm^2\ molecule^{-1}$)

n number of absorbants per cubic centimeter (cm^{-3} molecule)

x pathlength (cm)

Each mask or slit array is designed to sample either high or low absorption bands in the spectrum of the target gas and as the slit arrays pass in sequence through the focal plane the radiant energy reaching the photo-cathode from each pair of high/low absorption masks is compared in a continuous repeated fashion. Fluctuations in the remote light source intensity are taken care of by sampling the PMT output for one mask and using the integrated signal in an automatic gain control (AGC) loop to normalize the signal amplitude in the presence of rapidly changing light conditions.

Considering only one pair of high absorption/low absorption masks with identical slit arrays the response R_s can be shown to be [4]

$$R_s = \underbrace{\alpha I_1 \left\{1 - \Psi(\xi) + \alpha I_1 \Psi(\xi)\right\}}_{\text{no gas offset}} + \underbrace{\alpha I_1 \Psi(\xi) \left\{1 - e^{-(a_2 - a_1)(\xi)cL}\right\}}_{\text{gas signal}} \quad (1)$$

Where I_1 is the PMT plate current

α is a gain factor associated with circuit impedance and duty cycle on the disc

c is gas concentration in ppm-m by volume

L is the pathlength through the gas cloud in meters.

a_1 and a_2 are the average absorption cross section per ppm-m of the chosen gas for mask (1) and (2) respectively. The value of a_1 and a_2 depend of course not only on the molecule but on the slit width, number of slits and slit position relative to the spectrum. $\Psi(\xi)$ is the ratio of the power passing through mask (2) to the power passing through mask (1) if the target gas were not present. ξ is the wavelength co-ordinate of a representative feature of the mask (completely arbitrary), i.e., could be the beginning wavelength of the first slit in mask (1). In this manner the various factors affecting the instrument response are divided into two main categories; one relating to the chosen gas and the other $\Psi(\xi)$ dependant on the background spectral radiance, filter function and interferring gases. Note that $\Psi(\xi)$ appears in the gas signal term as a scaling factor.

In the COSPEC II, III, IV line of sensors the no-gas offset term is cancelled or nearly so, with the use of a second pair of high/low absorption masks of appropriate design. The second pair of masks developes a no-gas offset signal component opposite in phase to that of the first pair and a gas component which is in-phase with that of the first pair. Subtraction of the non-gas signal and addition of the true gas signal components is performed in a simple integrator. In this manner the effect of $\Psi(\xi)$ variations is reduced to a second order effect only and is a scaling factor in the gas term rather than an additive non-gas signal. As will be shown later the residual non-gas component of signal is still a significant source of error for the most demanding remote sensor applications and it is this aspect of sensor design which is receiving current research and development effort. The response of the instrument to varying quantities (ppm-meters) of the target gas is linear to all intents and purposes from zero to about 600 ppm-meters of gas burden (see Figure 6). This is adequate for the majority of mobile pollution monitoring tasks.

The multiplex capability of the rotating disc also permits the monitoring of more than one molecular species. Although this paper is concerned only with the single gas sensor, and with SO_2 in particular, two-channel instruments (COSPEC MK II and MK III) have been developed for simultaneous monitoring of SO_2 and NO_2.

SENSOR CALIBRATION

Wavelength Calibration

A pen ray mercury vapour lamp was used to calibrate the grating angle in terms of micrometer readings as shown in Figure 7. Mercury lines at 2537Å, 3132Å and 3341Å were scanned across the correlator masks with the first positive correlation peak at the shortwave end taken as the reference position. The normal working range of COSPEC is shown bracketted in Figure 8. Excellent linearity is exhibited throughout the calibration range.

The design position of the grating is shown at 0.671 which corresponds to the wavelength at the start or short wavelength edge of the first slit of Mask No. 1. The slit arrays in each mask are designed to operate at the wavelengths noted in Table I.

SO_2 Correlations

Figure 9 shows a typical grating scan using a smooth tungsten-quartz-iodine light source and a reference cell containing 68 ppm-meters of SO_2. Note that as the grating is rotated the absorption bands in the SO_2 spectrum correlate in a sequentially positive and negative sense with the masks, i.e., slit arrays, etched in the correlator disc. Note that although any one of the positive correlation peaks can be used to provide a positive sensor response to SO_2, the preferred position is the design position shown at 1. The importance of using the correct grating position becomes evident when considering Figure 10.

Sky Correlations

This scan was taken with completely overcast skies and shows a considerably more rapid fall-off at the shortwave end than would be the case for clear skies and higher sun altitude. The sinusoidal correlation pattern is in this case due primarily to the presence of strong Fraunhofer features ever-present in the solar spectrum plus a relatively minor influence from background SO_2 in the atmosphere. Note that the design position of the grating now coincides with a negative Fraunhofer correlation peak. Because of the spectral intensity of the Fraunhofer features, large non-gas signal offsets can be produced by relatively small shifts in grating position. In the design of the correlation masks it is a primary objective to maximize the sensor response to the target gas spectra and minimize the response to Fraunhoffer and other interferring spectral features. As is to be expected, the best solution is a compromise. Also because of the unchanging nature of Fraunhofer features in the solar spectrum (Fraunhofer in-filling effects have so far been found to be small) it is possible to nullify the unwanted sensor output signals produced by Fraunhofer effects by means of an internal offset voltage. The reliability of this method is enhanced by choosing a grating position which yields an acceptable gas sensitivity and which corresponds to a positive or negative Fraunhofer correlation maxima. In this way the sensor is made less susceptible to errors induced by spectral shifts, temperature changes, shock and vibration etc.

Effect of Cloud and Solar Altitude

COSPEC was set up in the laboratory bench to view the eastern sky through the laboratory window. The window was quartz and the look-up angle of the sensor was $21^{\circ}40'$. A series of data runs was made over a three day period with fixed grating setting, sensor integration time and recorder reflection sensitivity. Samples of the chart data are shown in Figures 11 to 16.

Figure 11 shows the effect of the sun passing directly through the line of the sight of the sensor in the presence of heavy but broken cumulous clouds. Under these conditions of course the irradiance increases very rapidly, particularly at the longer wavelengths, as the sun moves into the field of view. In practice the sensor's line of sight is directed upwards at the

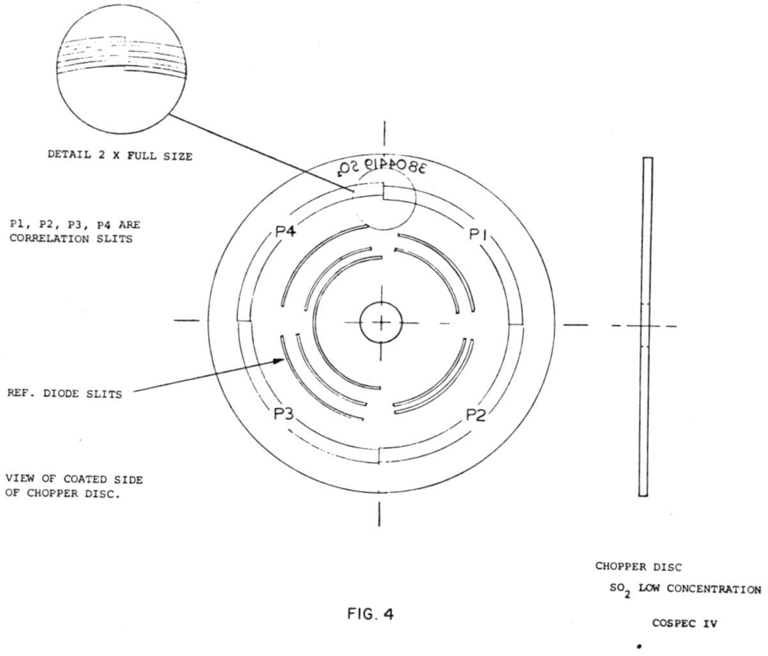

FIG. 4

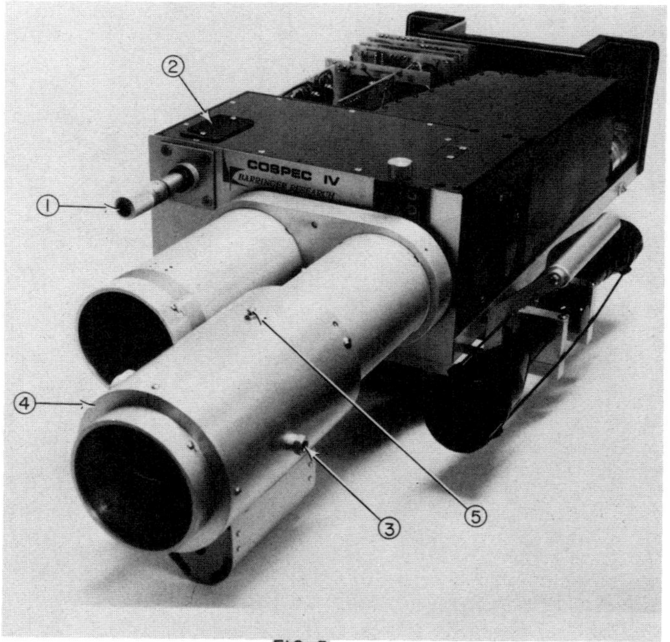

FIG. 5

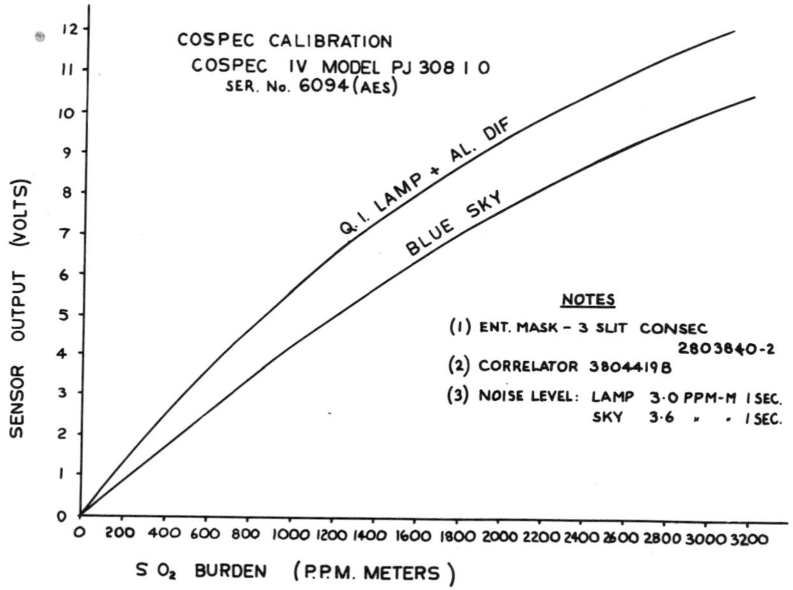

FIG. 6

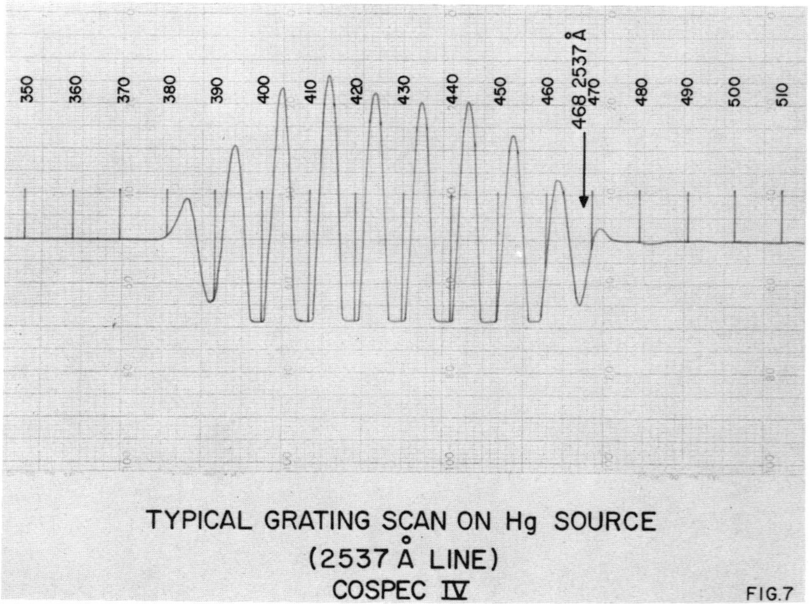

FIG.7

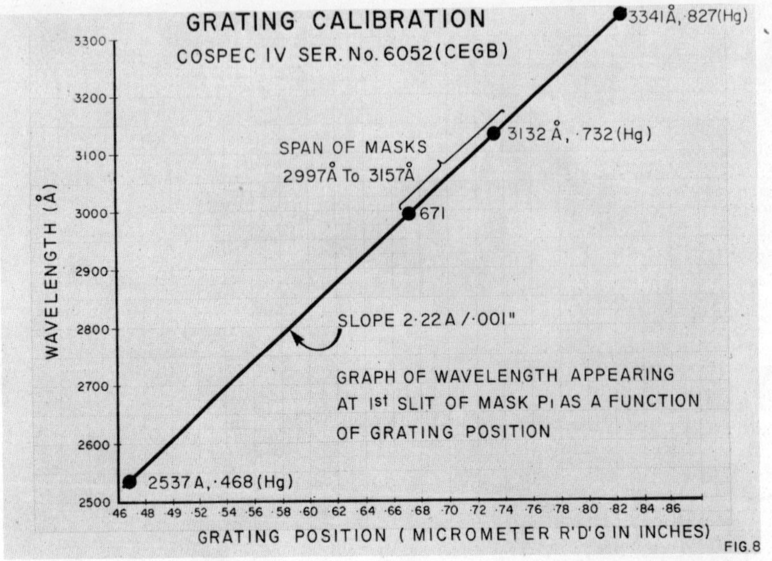

FIG.8

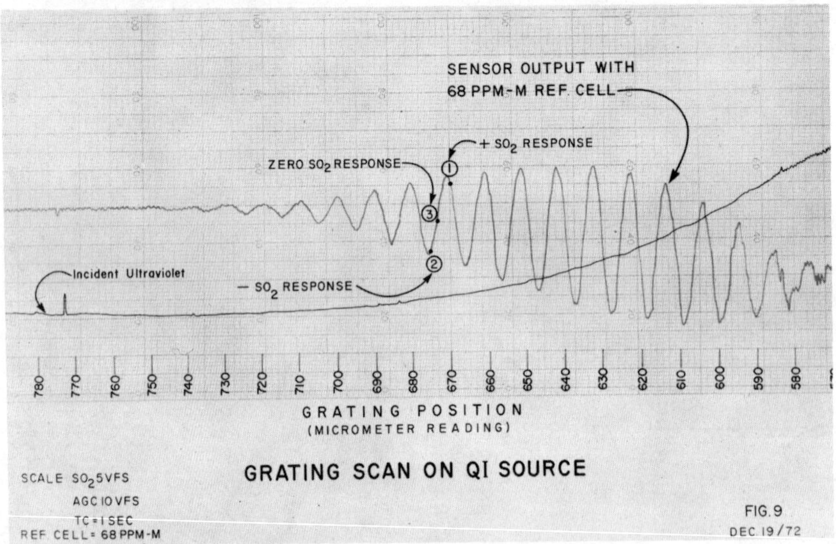

GRATING SCAN ON QI SOURCE

FIG.9

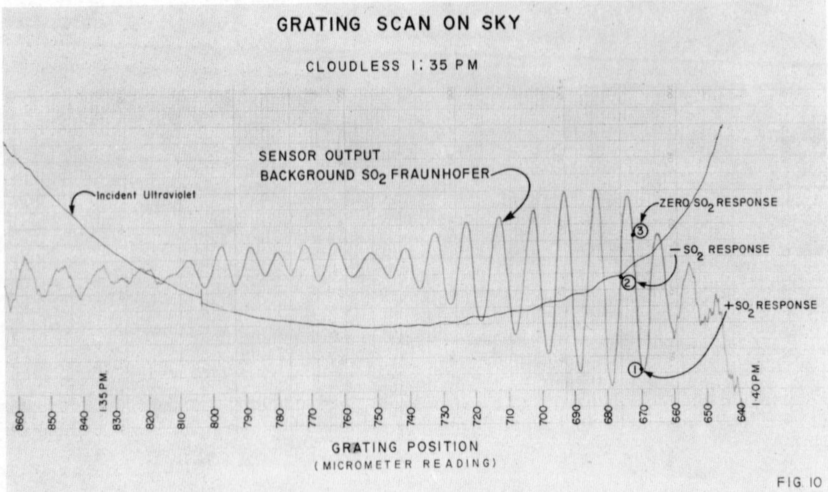

FIG.10

Mask #1

Wavelengths Å

Beginning	End
2997	3005
3017	3025
3038	3046
3059	3067
3084	3092
3103	3111
3125	3133

Mask #2

Wavelengths Å

Beginning	End
3008	3016
3029	3037
3050	3058
3071	3079
3092	3100
3114	3122
3137	3145

Mask #3

Wavelengths Å

Beginning	End
3008	3016
3029	3037
3050	3058
3071	3079
3092	3100
3114	3122
3137	3144

Mask #4

Wavelengths Å

Beginning	End
3017	3025
3038	3046
3059	3067
3084	3092
3103	3111
3125	3133
3145	3157

TABLE 1

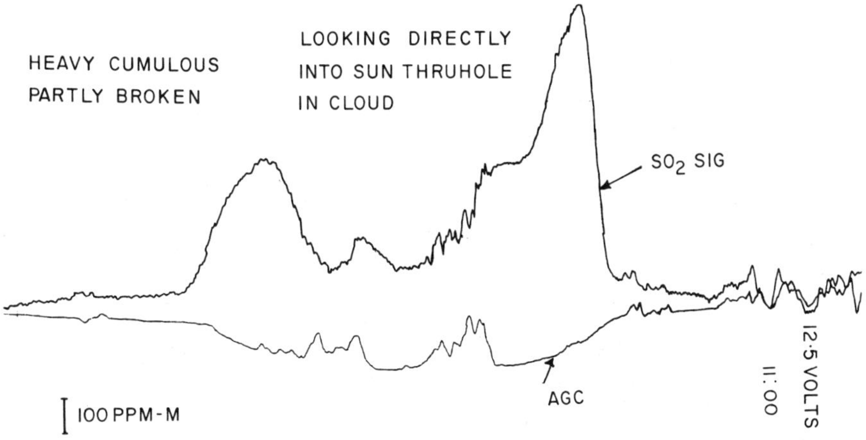

COSPEC IV RESPONSE TO DIRECT VIEWING OF SUN

(WITHOUT CORRECTION)

FIG. 11

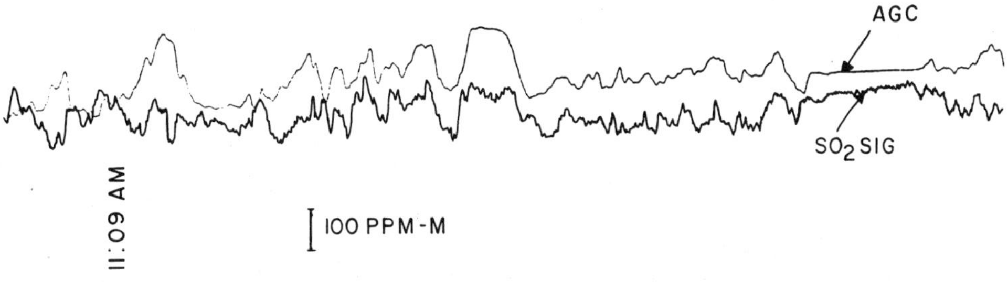

COSPEC IV RESPONSE TO HIGH CONTRAST

IN CLOUD BRIGHTNESS

(SILVER LINING EFFECT)

(WITHOUT CORRECTION)

FIG. 12

zenith sky during a typical ground mobile survey and direct viewing of the sun seldom occurs. As one might expect the very severe changes in source intensity and spectral distribution produced an erroneous SO_2 signal equivalent to many hundreds of ppm-meters. As the sun moved out of the sensor's direct line of sight, broken cumulous cloud with very bright peripheries interspersed with patches of bright blue sky moved through the sensor's field of view. The resulting noisey signal as shown in Figure 12 lasted for approximately fifteen minutes. As the sun moved further away the sharp contrast in cloud brightness faded leaving only the contrast between the while cloud and bright blue patches of sky with the resulting record as shown in Figure 13. Figure 14 shows a typical recorder trace for a clean cloudless sky. The remaining noise is almost entirely photon noise and for a given instrument depends on the ambient light level and the electronic integration time. All Figures 11 through 16 were obtained with an integration time of one second. From clear blue sky to variable light cloud cover there is no noticeable increase in noise as seen by comparing Figures 14 and 15. With heavy overcast skies (Figure 16) signal changes due to variations in brightness are not large but the photon noise level is some 3 - 4 times higher than in the blue sky case. Since there is little change in sensor sensitivity to SO_2 from one type of sky to the other, the effect of a heavy overcast sky is to reduce the system S/N ratio by a factor of 3 - 4.

Table 2 and Figure 17 show data from one half day of uninterrupted measurement. From 12:00 noon to 3:15 P.M. the sky was completely free of cloud. The wind was blowing from the WNW steadily at 20 to 28 mph which should be expected to result in low SO_2 background levels.

With negligible changes in SO_2 background satisfactory performance requires that the SO_2 signal channel maintain a flat baseline with changing sun angle until the ultraviolet light level falls at the end of the day to the point where the system S/N ratio is no longer adequate. In Figure 17 the SO_2 signal channel showed a positive change of 18 ppm-meters in the first 30 minutes, falling back to the initial value after 1 1/4 hours followed by a second excursion to 18 ppm-meters. By 3:00 P.M. the solar altitude had fallen to 8 degrees and the UV level was dropping rapidly particularly at the shortwave length end of the passband. The resulting rapid change in spectral distribution is the reason for the sudden positive change in the SO_2 baseline. In Table 2 the data in the column headed "Chart Scale" is a measure of the sensor response to the 68 ppm-m reference cell. Note that the sensor SO_2 sensitivity remained constant within 10% over a solar altitude change of from 25° to 8°. After 3:30 P.M. the sensor response to the reference cell rapidly diminished to zero. Also note that the sensor noise level increased rapidly after the sun fell below 8°. At mid-day the sensor noise level was 12.5 ppm-meters at 1 second (i.e. about 4.5 ppm-meters at 8 seconds of integration time.)

Figure 17 also shows a solar altitude curve for latitude 50°N and Figures 18 and 19 show solar altitudes for various latitudes throughout the year.

<u>Vertical Burden Measurements</u>

A series of sky measurements were made using the 90° mirror attachment as shown in Figures 20, 21 and 22. Sky conditions were far from ideal with 10/10 cloud cover and low light levels. The results however are quite interesting. The COSPEC line of sight was adjusted to the following zenith angles.

θ	sec θ (air mass no.)
0	1
48°15'	1.5
60°	2.0
66°25'	2.5
70°30'	3.0

where sec θ or air mass number is taken as the ratio of the distance through a homogeneous layer of polluted atmosphere along a slant path to the distance through the same atmosphere along a vertical path.

Note that Figure 20 was obtained using a grating setting of 0.675 which corresponds to a negative SO_2 response as shown in Figure 9 at 2 and approximately at the mid point between +ve and -ve Fraunhofer correlation peaks as shown at 2 in Figure 10. In Figure 20 the response indicates increasing SO_2 burdens with increasing slat path distance, which is linear to approximately two air masses. Extrapolation to a zero air mass, i.e., zero SO_2 indicates a total vertical burden of 53 ppm-meters. At the time of observation the altitude of the cloud base was at 1060 meters.

If we now make the simplifying assumptions that most of the light reaching the sensor has traversed the polluted layer only once, that most of the atmospheric SO_2 is below the cloud layer and ignoring temperature and pressure variations within the layer, the average SO_2 concentration in the layer is, to a first approximately, 50 ppb.

Figure 15 is a similar chart record with the grating set for a positive SO_2 response at 0.670 as shown at 1 on Figure 9 and a negative Fraunhoffer response as shown at 1 in Figure 10.

Figure 16 was obtained with the grating set to 0.6725 for zero SO_2 response (3 Figure 6) and positive Fraunhofer response (3 Figure 7).

<u>Correction for Cloud Effects</u>

As is evident in Figures 11, 12 and 13 the mask design and signal processing as described do not fully compensate for the extreme changes in sky conditions associated with direct viewing of the sun or the bright sun-lit edges of broken cumulous clouds. As mentioned previously direct viewing of the sun is not really an operational requirement

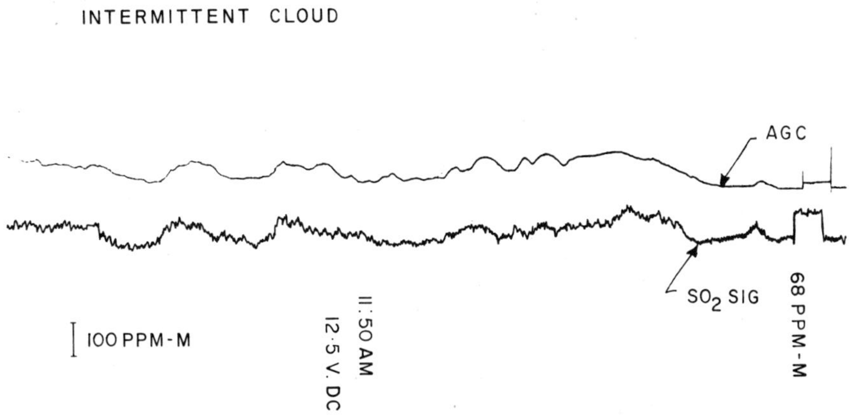

COSPEC IV RESPONSE TO CONTRAST ON BLUE SKY / WHITE CLOUDS
(WITHOUT CORRECTION)

FIG. 13

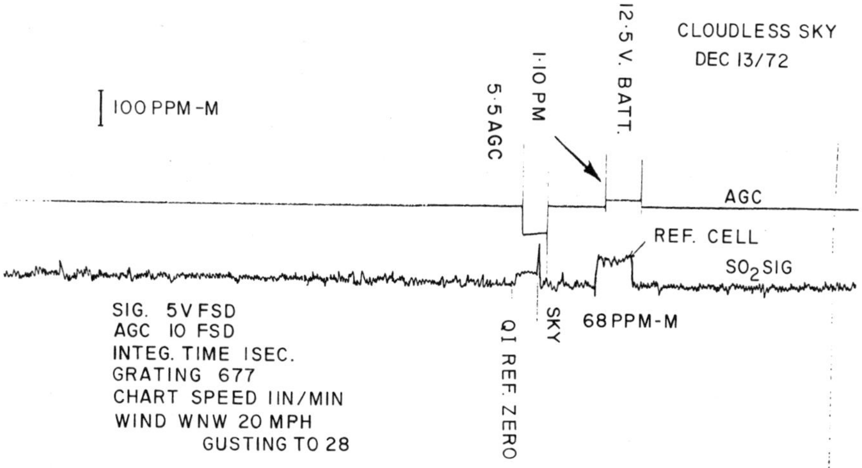

COSPEC IV RESPONSE TO CLOUDLESS SKY

DEC. 13·1972 1:10 PM TORONTO

FIG. 14

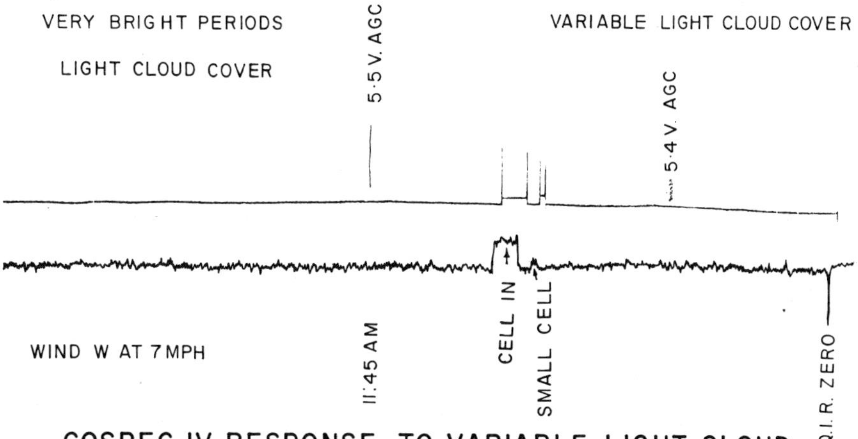

COSPEC IV RESPONSE TO VARIABLE LIGHT CLOUD

DEC. 14/72 11·45 AM TORONTO

FIG. 15

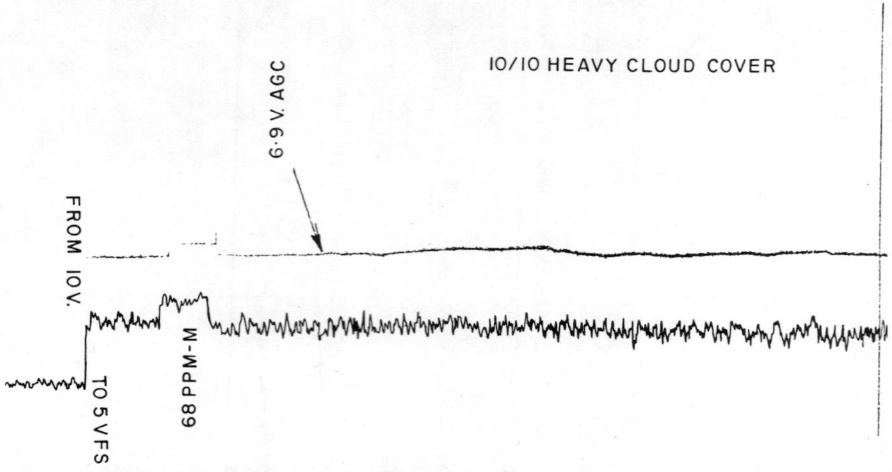

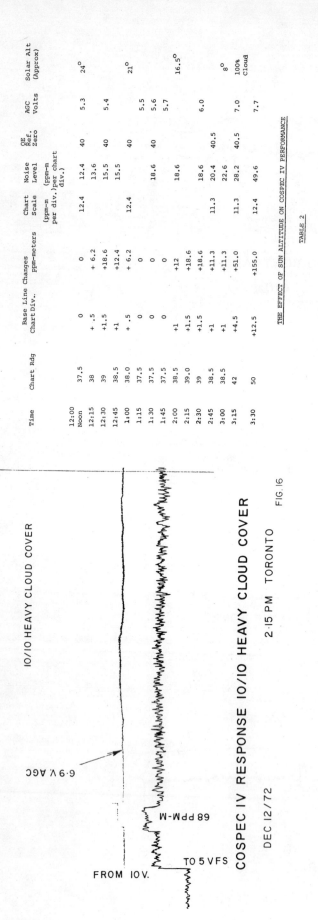

COSPEC IV RESPONSE 10/10 HEAVY CLOUD COVER

DEC 12/72 2·15 PM TORONTO

FIG. 16

Time	Chart Rdg	Base Line Changes Chart Div.	Base Line Changes ppm-meters	Chart Scale (ppm-m per div.)	Noise Level (ppm-m per chart div.)	QE Ref. Zero	AGC Volts	Solar Alt (Approx)
12:00 Noon	37.5	0	0	12.4	12.4	40	5.3	24°
12:15	38	+ .5	+ 6.2		13.6			
12:30	39	+1.5	+18.6		15.5	40	5.4	
12:45	38.5	+1	+12.4		15.5			
1:00	38.0	+ .5	+ 6.2	12.4		40		21°
1:15	37.5	0	0				5.5	
1:30	37.5	0	0		18.6	40	5.6	
1:45	37.5	0	0				5.7	
2:00	38.5	+1	+12		18.6			16.5°
2:15	39.0	+1.5	+18.6					
2:30	39	+1.5	+18.6		18.6		6.0	
2:45	38.5	+1	+11.3	11.3	20.4	40.5		
3:00	38.5	+1	+11.3		22.6			8°
3:15	42	+4.5	+51.0	11.3	28.2	40.5	7.0	100% Cloud
3:30	50	+12.5	+155.0	12.4	49.6		7.7	

THE EFFECT OF SUN ALTITUDE ON COSPEC IV PERFORMANCE

TABLE 2

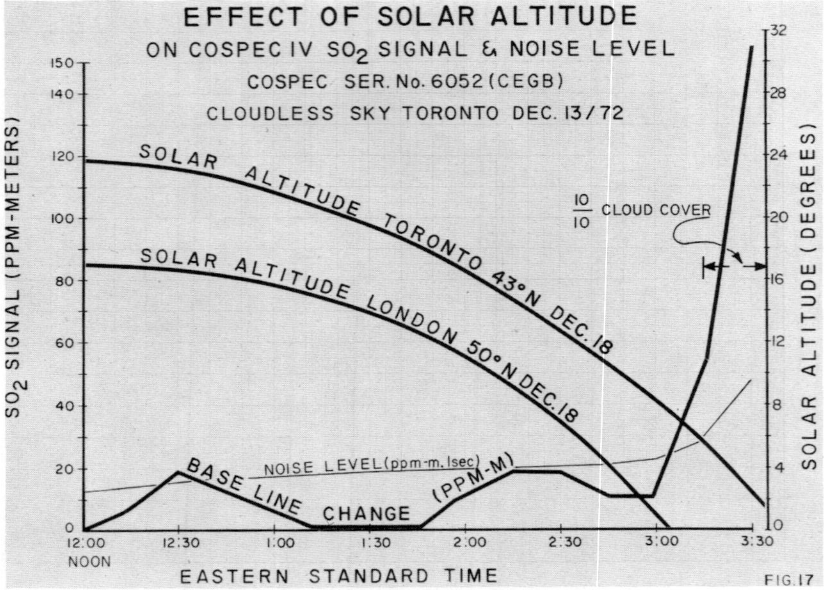

FIG. 17

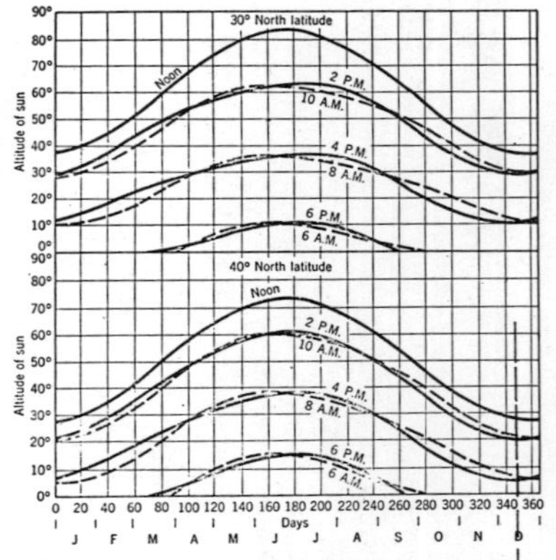

Fig. 3(a). Solar altitude at various hours of the day throughout the year at north latitude 30° and 40°. (F. Benford, *Illum. Eng.*, 42, 527 [1947].)

FIG. 18

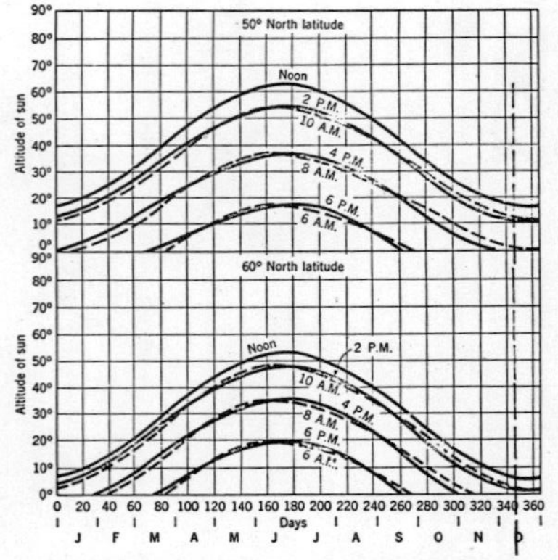

Fig. 3(b). Solar altitude at various hours of the day throughout the year at north latitude 50° and 60°. (F. Benford, *Illum. Eng.*, 42, 527 [1947].)

FIG. 19

since the sensor is used almost always in a vertical upward-looking mode. It should be emphasized here perhaps that the automatic gain control system compensates for changes in total solar radiance received over the entire dynamic range of sky brightness changes in a fully satisfactory manner. The AGC does not compensate however for redistribution of the spectral energy without the sensors spectral passband.

Such spectral changes which can result from highly contrasting cloud formations within the sensor's field of view, places a serious operational limit on applications where high sensitivity is mandatory under a wide range of sky conditions.

The reason for incomplete cancellation of the non-gas ($\Psi(\xi)$,) component of signal is not yet fully understood but recent work is encouraging. Generally speaking changes in spectral distribution within the field of view of an upward-looking sensor can be caused by differences in the scattering properties of the relatively clean upper atmosphere and the intervening cloud layers of varying opacity. Turbid air masses produced by photochemical smogs also tend to attenuate the shortwave radiation most severely. These scattering effects are generally broadband in a spectral sense and can be expected to result in at least quasi linear changes across the spectral bandwidth. Non-linear changes can also occur due to changes in ozone absorption and Fraunhofer line in-filling associated with photochemical activity. These latter effects however appear to be quite small and are probably insignificant.

A number of different approaches are now being investigated as possible means of reducing the sky-to-cloud signal offset. These include development of new correlation masks designed to reduce the sky/cloud offset signal and improved signal processing techniques for more effective cancellation of linear gradient offsets.

Another possibility is a spectral gradient measuring circuits which are capable of providing a separate slope-calibrated signal for subsequent correction of the gas channel data. Figure 23 shows results from tests of an experimental gradient compensation circuit in this case for the NO_2 region. This work is very current at the moment and will be reported on more fully at a later date. The upper curve is the slope channel signal obtained when the COSPEC field of view was scanned across broken cumulous cloud interspersed with bright blue sky. The lower curve is the gas signal with slope correction added. The slope channel signal was uncalibrated but the gas + slope channel is scaled at 1.57 ppm-m NO_2 per minor division. Note that scanning from blue sky across white/grey cloud and back to blue sky resulted in a baseline deviation of 10 ppm-meters. This represents a reduction of about 75% from the 35 - 40 ppm-meters of offset signal typically produced by clouds without slope correction.

Mobile Monitoring

At the present time the primary application of the remote sensing correlation spectrometer is in mobile SO_2 and NO_2 pollution surveys of industrial complexes and urban or metropolitan areas. Mobile survey methods and data reduction techniques are discussed in considerable detail in Reference 6. A typical presentation of data is shown in Figure 24. Briefly the instantaneous signal output from COSPEC can be considered a measure of the total vertical burden of SO_2 overhead in ppm-meters or in units of mass per unit area. When plotted as a function of vehicle position, a profile of pollutant distribution can be obtained. When combined with wind speed and direction the total pollutant mass rate of flow from individual point sources or area (distributed) sources can be measured as shown in the lower profile of Figure 24. This type of survey is usually performed with the sensor reading obtained on the upward side of the surveyed area used as the reference or datum level.

CONCLUDING COMMENTS

The sensitivity and accuracy of COSPEC IV in the mobile upward-looking mode depends on the spectral characteristics of the overhead sky. As might be expected the sensitivity and accuracy are best under conditions of bright blue sky or blue sky and variable light cloud with sensitivity limited only by photon noise. In a cloudless sky and outside the sensor's field of view the position of the sun has little effect on sensor performance until the sun altitude falls to less than about 8 degrees. Changing sun position only has the effect as expected of altering the photon noise level. As the cloud thickness and contrast in sky/cloud spectral distribution increases significant error signals can result. Largest errors occur when the sensor line of sight is aligned within a few degrees of the sun and severe forward scattering occurs from the bright edges of broken cumulous cloud. Typical sensor noise levels for SO_2 are shown in Table 3. Note that when the overhead sky contains little or only moderate variations in spectral distribution the system sensitivity is determined by the photon noise level and thus can be optimized with the use of a 4 second integration time. Experience has shown this to be satisfactory for surveys conducted at normal highway speeds.

Engineering effort is now being directed to reducing the signal offset errors caused by bright broken cloud. Investigations to date indicate that it should be possible to reduce the cloud offset error very substantially and perhaps to the level of the photon noise.

Sky test results were particularly encouraging in view of the adverse conditions with December 18th representing very nearly the lowest solar altitude throughout the year.

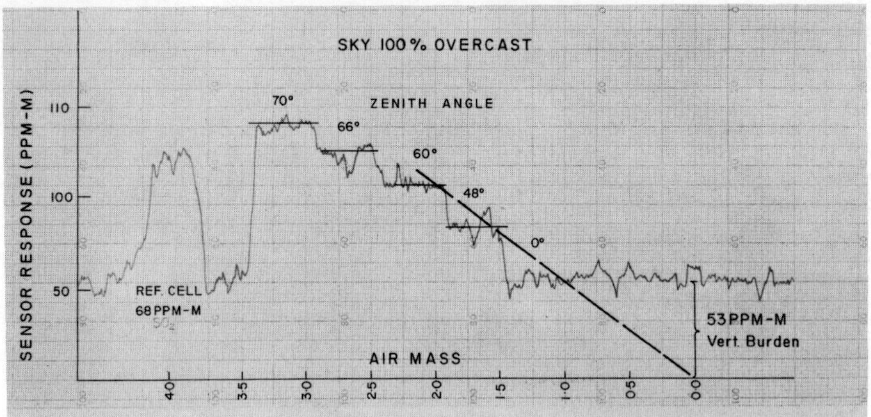

$-SO_2$ RESPONSE vs ZENITH ANGLE

TOTAL VERTICAL BURDEN DETERMINATION

FIG. 20

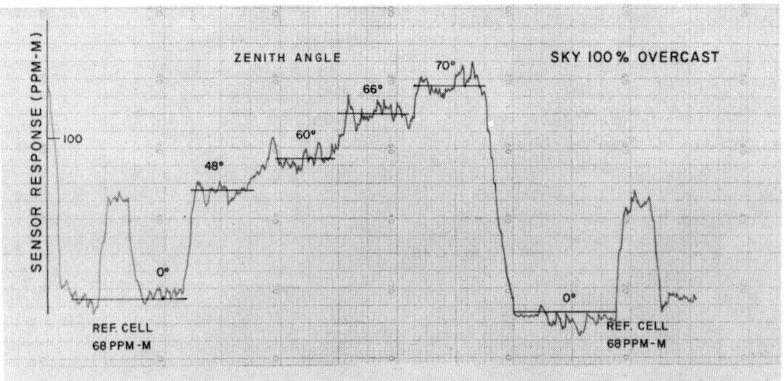

$+SO_2$ RESPONSE vs ZENITH ANGLE

COSPEC IV

FIG. 21

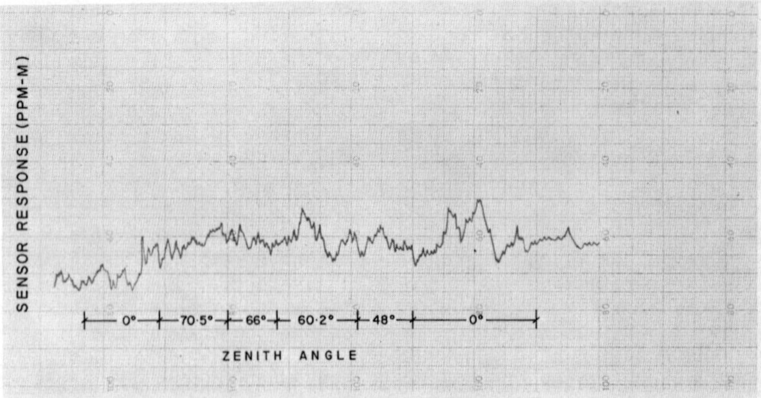

SENSOR RESPONSE vs ZENITH ANGLE

WITH

GRATING ADJUSTED FOR ZERO SO_2 RESPONSE

COSPEC IV

FIG. 22

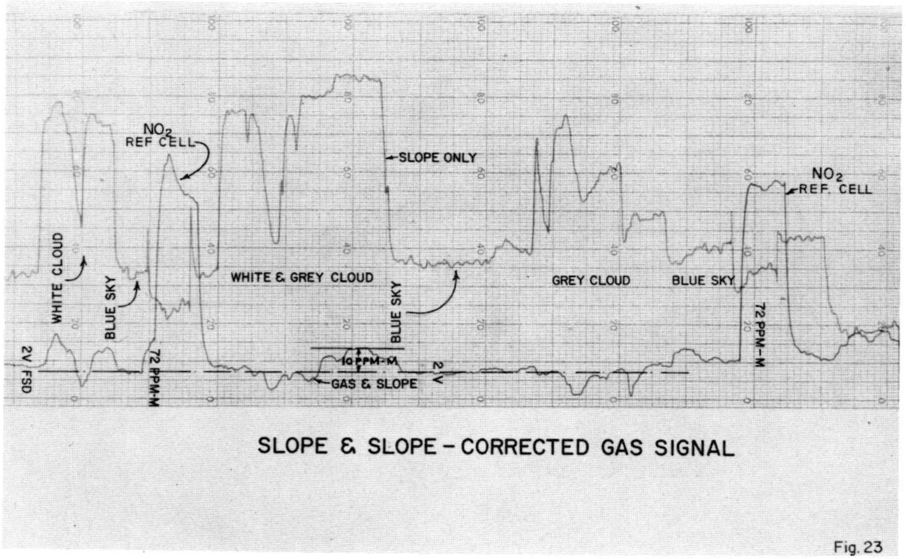

Fig. 23

Sky Condition	System Noise Level (ppm-m)	
	Integration Time 1 second	Integration Time 4 second
Blue sky	10	5
Blue sky & variable light cloud	10	5
10/10 heavy cloud	35	17
Blue sky & bright white cloud	70	70
Broken cumulous with intense bright edges (silver lining)	100	100

TABLE 3

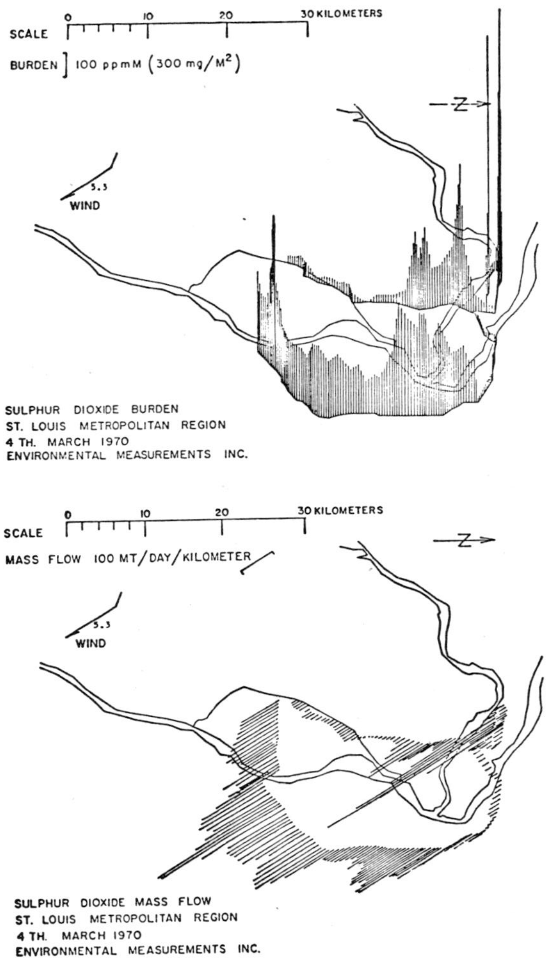

Fig. 24

The 18 ppm-meter excursions in baseline of Figure 17 is believed to be due to real changes in background SO_2.

Figure 17 shows that COSPEC IV can provide useful data for solar altitudes greater than about 8 - 10 degrees, i.e., 9:00 A.M. to 3:00 P.M. in Toronto and 10:00 A.M. to 2:00 P.M. in London, England in mid December. At 60° north latitude (Stockholm, Oslo) the solar altitude remains below 10° for much of November, December and January. COSPEC would therefore not be useful for measuring SO_2 in the passive mode at these latitudes during the mid-winter months.

The vertical burden measurements look particularly interesting. Although the tests were not comprehensive and weather conditions were far from ideal the results were most encouraging. The approximate 50 ppb determination appears reasonable in view of the fact the wind was blowing steadily from the west south-west which would bring SO_2 from the heavily industrialized area of Detroit, Michigan, a distance of some 200 miles. Further experiments are required under a range of sky conditions, to fully evaluate this technique.

REFERENCES

1. Barringer, A.R. and Schock, J.P., Progress in the Remote Sensing of Vapours for Air Pollution, Geologic and Oceanographic Applications, Proceedings 1966 Symposium on Remote Sensing of Environment, University of Michigan, pp. 779 - 792.

2. Williams, D.T. and Kolitz, B.L., Molecular Correlation Spectroscopy, Applied Optics, Vol. 7, pp 607 - 616, April, 1968.

3. Barringer, A.R., Chemical Analysis by Remote Sensing, 23rd Annual ISA Instrument Automation Conference, New York, New York, October 1968.

4. Millan, M.M., Townsend, S.J. and Davies, J.H., Study of the Barringer Remote Sensor Correlation Spectrometer, University of Toronto, Institute of Aerospace Studies, Toronto, Report 146, 1969.

5. Barringer, A.R., and Davies, J.H., Investigations in Correlation Spectroscopy and Interferometry for the Remote Monitoring of Air Pollution, Joint Conference on Sensing of Environmental Pollutants, Palo Alto, November, 1971.

6. Langan, L., Remote Sensing Measurements of Regional Gaseous Pollution, Joint Conference on Sensing of Environmental Pollutants, Palo Alto, California, November 8th - 10th, 1971.

© 1973, ISA JSP 6659

AIRBORNE AND BALLOONBORNE SPECTROSCOPY FOR THE STUDY OF ATMOSPHERIC GAS POLLUTANTS

Marcel Ackerman
Dirck Frimout
Institut d'Aéronomie Spatiale de Belgique (IASB)
3, Avenue Circulaire, Bruxelles, Belgium

Jean-Claude Fontanella
André Girard
Raymond Gobin
Louis Gramont
Nicole Louisnard
Office National d'Etudes et de Recherches Aérospatiales (ONERA)
92320, Châtillon, France

ABSTRACT

Infrared absorption spectrometry was recently used from aircraft and balloon for analyzing concentrations of minor constituents of the stratosphere, using the sun as a source, seen tangentially to the earth. Grille spectrometers with a resolution limit close to that of their gratings were used in connection with sun seekers. Experiments were performed on Concorde 001 prototype and on a 300 000 m^3 balloon at 40 km altitude. The vertical profile of NO between 20 and 40 km altitude was determined for the first time. Number density of NO_2 at 16 km was found to be $10 \pm 3\ 10^8$ cm^{-3}. Other results obtained in the 1300 cm^{-1} absorption band of HNO_3 are presented.

INTRODUCTION

Eventuality of stratospheric pollution by SST aircraft is a subject of many discussions and controversies. It has been pointed out that active chemical processes take place throughout the stratosphere. They involve species of which the mixing ratio by volume are lower than 10^{-9}. So the accurate knowledge of the stratosphere is of major interest to define the possible role of these elements.

Two experiments were performed during 1973 to detect and measure the concentration profile of some species (NO, NO_2, HNO_3, HCHO, SO_2) within the stratosphere.

They consist in infrared absorption grille spectrometry from a high flying platform using the setting or rising sun as a source. Data are obtained by studying the variations with altitude of the solar spectrum in narrow wavelength intervals. When the spectra are observed at large solar zenith angles (> 90°) the air mass crossed by the solar energy is significantly increased and the method reaches its ultimate sensitivity.

The first spectrometer was flown at 40 km altitude in a 320 kg gondola carried by a 300,000 m^3 balloon from Aire sur l'Adour (S.W. France), May 14, 1973. The second spectrometer was flown on Concorde 001 at 16 km altitude and several experiments were achieved during June and July 1973. These two experiments are quite complementary. In the first one the altitude is high enough to minimize absorption by other constituents such as CO_2 or H_2O ; the different measurements versus solar angles make it possible to get a concentration profile along altitudes lower than the balloon altitude. The second one allows an accurate measurement of the concentration around the flight altitude. They were not simultaneous but rather close in place and time.

Minimum Detectable Densities

The simplest method of analysis is based on the measurement of the equivalent width of a spectral line, W, which is the integrated absorption over the frequencies

$$W = \int_0^\infty A_\nu \, d\nu$$

were A_ν is the monochromatic absorption by the element.

Under the restriction that this line is suitably isolated, the equivalent width is independent from the slit function and if this slit function is much larger than the linewidth it is possible to write

$$W \simeq \int_{\nu - \Delta\nu}^{\nu + \Delta\nu} A_\nu \, d\nu$$

where $\Delta\nu$ is the slit function halfwidth.

Figure 1 indicates absorption line-growth for different total number of molecules and for two altitudes :

- 45 km (Doppler profile)
- 15 km (Lorentz profile).

In fact, the atmospheric infrared line shapes are given by the Voigt profile, which is a convolution of the Lorentz and Doppler shapes. It is possible to see that in the case of the stratosphere the linewidth is approximately 0.005 cm^{-1}, and one must remember that the best balloonborne spectrometer has a slit function width of only about 0.1 cm^{-1}.

So it is not easy to find a suitably isolated line. Many absorption lines from CO_2, N_2O, H_2O, CH_4, etc. may be inside the selected spectral interval and partially or totally mask the absorption to be measured. From this point of view the AFCRL atmospheric lines compilation(1) is of great help in the spectral interval of highest interest.

The equivalent width w for an isolated line is dependent of :

- N, the total number of absorbing molecules through the optical path,
- S(T), the integrated absorption cross-section at temperature T,
- α (p, T), the profile halfwidth at temperature T and pressure p.

In some cases the spectroscopic constants of the molecule have already been measured, and it is possible to compute S(T) and α (p, T).

The inverted profile of concentration versus altitude may then be deduced from the measured equivalent width versus solar zenith angle by computing, with an atmospheric model, the equivalent width versus grazing altitude of solar rays for different pollutant concentration profiles, and comparing it with the measurements.

A linear approximation may be accepted for w if $\frac{N\bar{S}}{\bar{\alpha}} < 0{,}1$, $\bar{S}$, $\bar{\alpha}$ being mean values over the optical path.

Then W = $N\bar{S}$, and inside the stratosphere $\bar{S} \sim S_{220^\circ K}$. So the minimum detectable total amount of molecules N_{min} is linearly dependent of the smallest detectable area A. This area is a simple function of two instrumental parameters :

- $\Delta\sigma$, the slit function width,
- R, the signal to noise ratio.

Finally :

$$N_{min} = \frac{1}{S}\,\frac{\Delta\sigma}{R}.$$

Table I lists the N_{min} values for NO and NO_2 molecules and the minimum detectable concentration in following conditions :

- grazing ray is 20 km altitude,
- the pollutant is uniformly mixed in the atmosphere,
- signal/noise ratio is 500 for NO and 100 for NO_2,
- $\Delta\sigma = 0.1\ cm^{-1}$ for both cases.

A more elaborate method of analysis consists in computing a synthetic spectrum inside the interval explored by the instrument. This method must be used if the resolution is not sufficient to suitably isolate a characteristic line.

Lastly, if the spectroscopic constants of the molecules are not known the spectra must be analysed only by comparison with laboratory spectra. This is an empirical, temporary method, and we used it only for HNO_3 and SO_2 molecules.

We have remarked on the importance of the instrumental parameters. In classical spectrometry it is well known that resolution and luminosity are absolutely related, while they are dependant upon separate parameters with Fourier spectroscopy and grille spectrometry.

Fourier transform spectrometry is generally the most attractive method. However when operated from flying platforms (aircraft and balloon) with severe requirements for resolution and reliability, many delicate problems have to be solved. On the other hand, a good stability is required for the phenomena under study during the whole interferogram recording time. This condition is difficult to fulfil since solar zenith angle is changing with time. Lastly, what is strictly required to detect each component is the record of only one or a few absorption lines, so the multiplex advantage of Fourier spectroscopy is not in this case of great interest. With a grille spectrometer it is possible to reach the theoretical resolution of the grating without serious instrumental problems.

Instrument Description

The two spectrometers used for airborne and balloonborne experiments are essentially a grille spectrometer associated with a sun tracker. In both cases the optical configuration of the monochromator is the same :

- Littrow mounting with an off-axis parabolic mirror, whose focal length is 600 mm ;
- Jobin-Yvon grating with 59,63 grooves/mm and 65 × 65 mm ruled area ; the incidence is around 65 to 70° ;
- the grille is deposited on a 15 × 15 mm calcium fluoride window and the minimum step is 0.2 mm ; so the resolution is very close to the theoretical limit and the half-width instrumental profile is about 0.1 cm^{-1} ;
- chopping is ensured by vibration of the collimating mirror ; the vibration frequency is 180 Hz, which is the resonance frequency of the mechanical mounting ; a servo system including this mechanical mounting keeps the vibration amplitude stabilized ; in this way the useful signal is 360 Hz-chopped ;
- the selected spectral range is scanned continuously ; the scanning speed is about one spectral element per second ;
- spectral calibration is performed with absorption lines of CO_2 and H_2O ; interpolation is made possible by simultaneous recording of the spectrum and position grating marks ;
- an optical auxiliary mounting is built to control any possible deformation of the mechanical system.

The other parts of the apparatus (sun tracker, amplifiers, optical photometric platforms) are different in the two experiments.

Concorde Instrumentation

The optical configuration is presented fig. 2. One of the aircraft scuttles is a 140 mm diameter calcium fluoride window. The image of the sun is obtained on the grille by means of lenses L_1 and L_2. The sun tracker P allows the mirror M to maintain the solar radiation on the spectrometer axis. At the output the image of the sun is focused on the detector area.

The detectors are :

- liquid nitrogen cooled SAT InSb photo-element for spectral range < 5,3 μm,
- for larger wavelengths, liquid nitrogen cooled Mullard HgTe-CdTe photoelement.

The total instrument is kept within a nitrogen atmosphere in order to avoid any parasitic absorption.

The flight trajectory is computed so as to maintain the sun direction approximately in the direction of the instrument.

The heliostat accuracy is 3' and the field of view is 3°. Measurements are made while the sun zenith angle is varying from 80° to 92°. Nine experiments have been performed on board the SST prototype Concorde 001, 6th June and during the 9-26 July period. The flight altitude was around 16 km. Seven spectra were obtained at sunset between the northwest of Spain and the west of Cornwall (fig. 3) ; two spectra were obtained at sunrise between the west of Cornwall and the Bay of Biscay.

Balloonborne Instrumentation (fig. 4)

The telescope is of Cassegrain type. Its diameter is 300 mm, the focal length is 4 meters. It conjugates the sun and the grille. The sun rays are reflected by the grille inside the monochromator.

To maintain the instrumental performance during the whole flight it is necessary to balance the mechanical deformations ; so the whole optical path inside the monochromator is servo controlled. Two annex grilles are used, alongside the main grille, to obtain this result. The first one is illuminated by an incandescence filament. The rays path is close to that of the sun rays. The flux is received by the second grille and focused on a germanium photodiode detector. The mirror vibrating at frequency f induces a photodiode signal with frequencies f and 2f. The f component is a characteristic of some irregularity : this component, called "error signal",constitutes the information to be sent to the servo system, which maintains the instrument properly tuned. Synoptic diagram of the total chain is shown figure 5.

The servo system consists in :

- a preamplifier,
- a synchronous demodulator (the reference signal comes from the excitation circuit oscillator of the parabolic mirror),
- a low frequency filter,
- a logical comparator which drives a motor bound to the mechanical mounting of the parabolic mirror.

The logical comparator turns the error signal into a logical information in order to act only in discontinuous mode :

- as long as the error is less than a fixed limit, the role of the servo system is null ;
- as soon as the error reaches the limit the servo system acts and the motor rotates.

Two different detectors can be used :

- photovoltaïc indium antimonide with a transformer,
- gold doped germanium with a bias of 48 volts.

The signal from the detector is amplified by a preamplifier Barnes DP7 with a gain of 5000. This preamplifier has a very good stability over a very wide temperature range ; it is mounted as close as possible to the detector.

Synoptic of the electronic chain is shown figure 6.

In the electronic box there is first a switch, disconnecting the signal of the preamplifier from the electronics during about 15 sec at one of the inversions of the grating drive motor. At that moment calibrating signals are fed to the electronic chain. There are signals at different levels coming from the excitation circuit oscillator of the parabolic mirror, and shifted in the correct phase by a phase shifter. There are consequently a zero and three signals calibrating the three channels of the electronics.

In the normal position of the switch the signal from the preamplifier is fed to a bandpass filter with a centerfrequency of 360 Hz (twice the oscillating frequency of the parabolic mirror). The bandwith is 30 Hz. It is an active filter with separate band and frequency controls.

First, there are three non inverting amplifiers with gains of 3, 9. and 27 respectively. The following circuit is the synchronous detection. The signal is rectified with a phase depending on the reference signal coming through a phase shift circuit. Before the flight, the phase relation between the reference signal and the incoming signal is matched in order to have a maximum positive d.c. output. This reference signal is provided from the vibration of the mirror, with a frequency doubler.

After the demodulator, a low frequency filter with a cutting frequency of 3 Hz and a voltage limiter to 2 volts delivers a low impedance signal to the telemetry. Other parameters are also measured during the flight. The outside pressure is measured with a CEC pressure gauge. This information is multiplexed with two temperature measurements (one on the spectrometer and one in the electronics box) in the low speed commutator and sent to the telemetry.

The sun pointing system used for this experiment is the "Astrolab" from former Compagnie des Compteurs, already used for other experiments and described before.

The principle of pointing can be summed up as follow. The gondola is uncoupled on the three axes. In a first mode, the gondola is oriented along a fixed axis referred to the earth. The sensors are a gyroscope and a magnetometer. The couples are provided by inertia wheels. At the end of the acquisition mode a rendez-vous is realized with the sun. At the moment of switching from acquisition to pointing mode by a timer, the sun has to be in the field of the sun sensors. From that moment the pointing information is given by the sun sensors.

RESULTS

Only nitric oxide was studied with both balloon (from 1900 to 1915 cm^{-1}) and Concorde spectrometers (from 1885 to 1915 cm^{-1}) ; absorption lines of CO_2 and H_2O were also identified ; so a profile can be deduced for these molecules from 15 km to 40 km by division of the atmosphere in successive layers of three kilometers thickness. Other molecules (nitrogen dioxide, sulfur dioxide, nitric acid, etc.) were only studied from aircraft ; for these species, only the number density at the flight altitude may be defined.

Nitric oxide

The analysis is made chiefly with the 1909.14 cm^{-1} and 1914.99 cm^{-1} lines (fig. 7). Although the first is partially mixed with CO_2 1909.21 cm^{-1} line, corrections were made and simultaneous results with the two lines are correct, within the limit of accuracy. The integrated line absorption cross sections for 220°K are listed in Table II[2].

The number density of NO versus altitude was inferred from the equivalent width of the lines and are presented figure 8 with other experimental data. The limits of error correspond to the maximum scattering of the original data points.

These results are only obtained with the linear approximation.

Computed by the more elaborate method, these values will be probably slightly increased.

An example of spectra obtained on Concorde at 1890 cm^{-1} is presented figure 9. We can see that the spectral element width is 0.1 cm^{-1}. The signal/noise ratio is better than 100 with the Concorde spectrometer and 500 with the balloon spectrometer.

We see on figure 10 that the integrated amount of NO measured in balloon along the optical path for a grazing ray altitude of 16 km. is $8 \pm 2\ 10^{16}$ moles cm^{-2}. The measured value from Concorde at solar zenith angle of 90° is 6 ± 2.10^{16}. This value is larger than expected from balloon measurements. This fact may be explained by the photochemistry of NO, the path from the sun to Concorde being more illuminated with UV than the path between the balloon and Concorde. Analysis of the 1909,512 and 1912,519 cm^{-1} lines of CO_2 yields to a constant mixing ratio of $3 \pm 1\ 10^{-4}$ in agreement with the generally accepted value.

Nitrogen dioxide

The spectra were taken from 1603 cm^{-1} to 1617 cm^{-1}. The integrated line absorption cross sections for 220°K are listed in Table III[3]. Figure 11 is an example of result. The number density deduced at 16 km is $10 \pm 3\ 10^{8}\ cm^{-3}$.

Nitric acid

Spectra were taken in the 1300 cm^{-1} region. The empirical method was used, comparing the amplitude of lines with laboratory spectra. The analysis is made with the Q Branch. An exemple of result is presented figure 12. Number density is not still available at the state of the analysis.

CONCLUSION

Identification and measurements on gases need high resolution spectra on a few specific lines for each component. The results presented here clearly show the ability of the spectral method used to solve the specific problem of remote detection of minor gaseous components in the stratosphere :

- high reliability for airborne experiments ;
- high luminosity due to independence of parameters governing resolution and available energy ;
- each scanned spectral line gives an information by itself so that the phenomena under study have only to remain steady during a very short interval of time.

Many other spectra obtained with Concorde Instrumentation have yet to be analysed, and there remains a lot of work to do.

The results obtained with NO and NO_2 show that simultaneous measurements on these species are of great interest for the knowledge of photochemical processes.

Lastly any gain on spectral resolution will allow improvement on the minimum detectable concentration.

Further experiments are planned for next year with improved instrumentation.

ACKNOWLEDGEMENT

This work has been sponsored in part by the COVOS (Groupe d'Etudes sur les Conséquences des Vols Stratosphériques) contract n° 304/73, by the CNES (Centre National d'Etudes Spatiales) and by the CIAP, Office of the U.S. Department of Transportation.

REFERENCES

[1] Mac Clatchey et al. - Infrared Compilation (1973).

[2] S. Cieslik and C. Muller - Aeronomica Acta A nr 114 (1973).

[3] C.A. Barth - J. Geophys. Res. 69 3301 (1964).

[4] L.G. Meira Jr. - J. Geophys. Res. 76, 202 (1971).

[5] L.C. Hale in Mc Elroy - Proceedings of the survey. Conf. DOT- T SC OST 72-13. Cambridge, Ma, 1972.

[6] C. Muller (private communication).

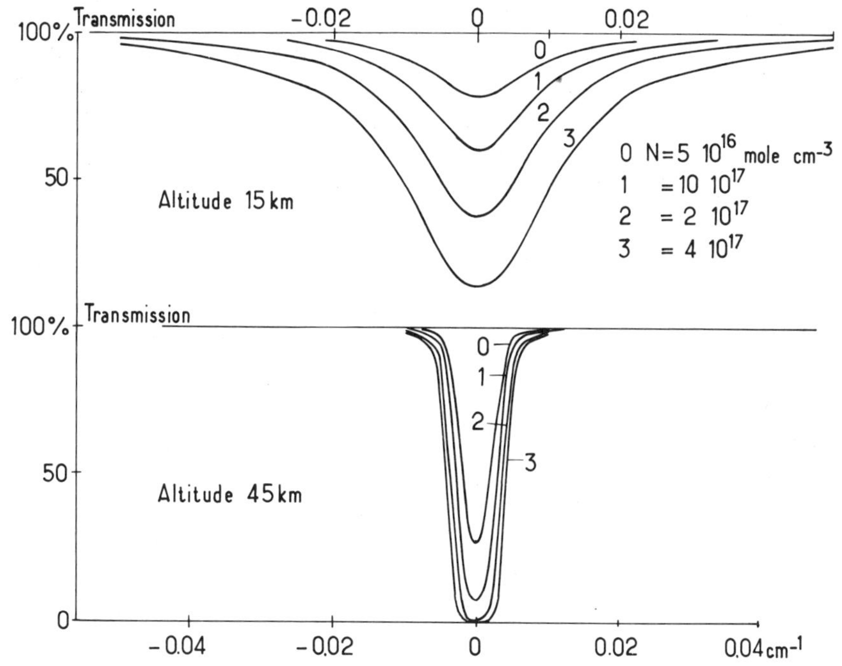

Fig. 1 - Absorption line growth with Doppler and Lorentz profiles for various optical thicknesses.

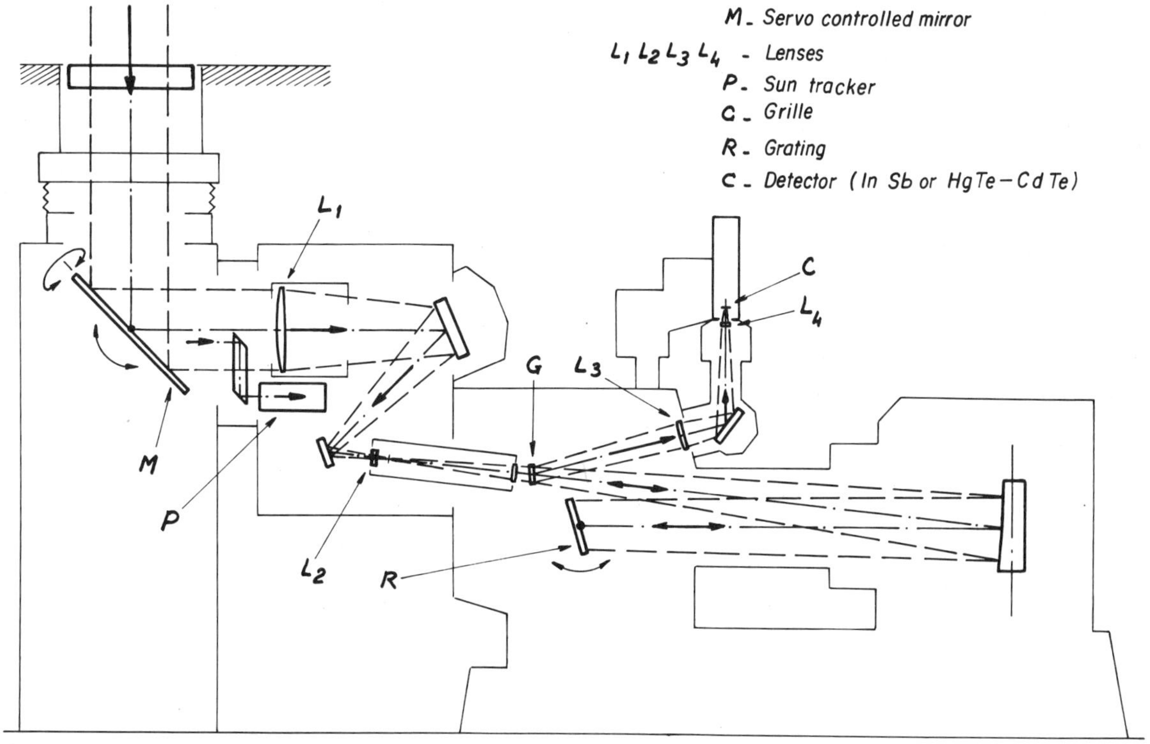

Fig. 2 - Optical diagram for the Concorde spectrometer.

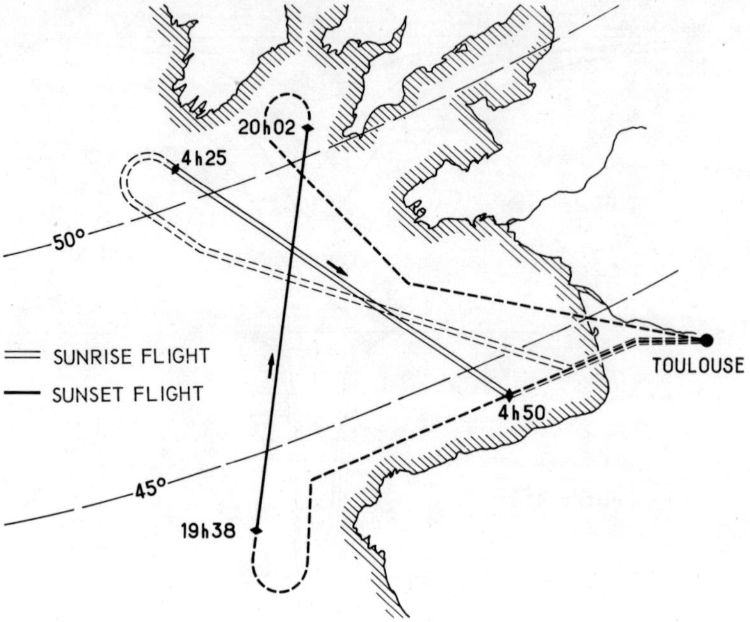

Fig. 3 - Concorde 001 environmental flights. June-July 1973.

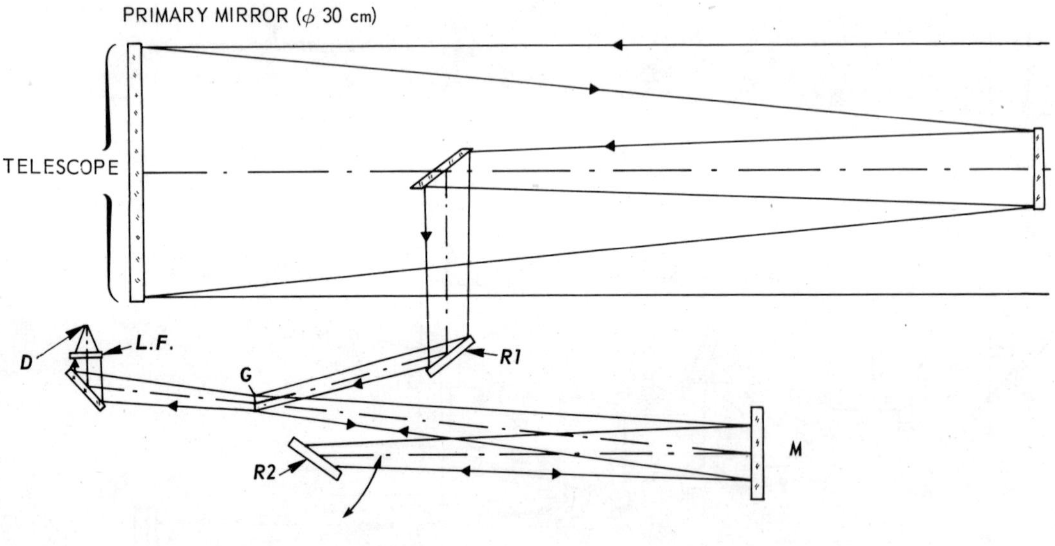

Fig. 4 - Optical diagram for the balloonborne spectrometer.

D - InSb detector

L.F. Lens + filter

G - Grille

R1 - pre-dispersing grating

R2 - Chief grating

M - Off axis mirror (F = 600 mm)

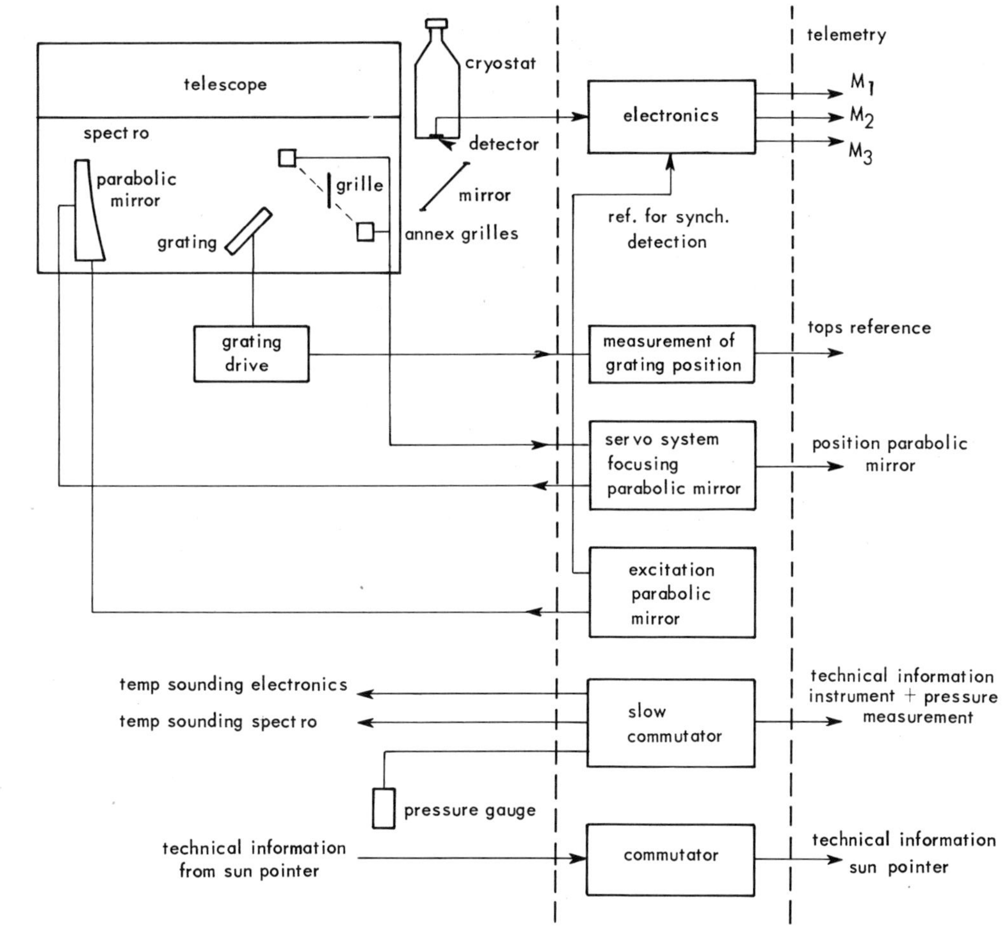

Fig. 5 - Synoptic diagram of the total chain.

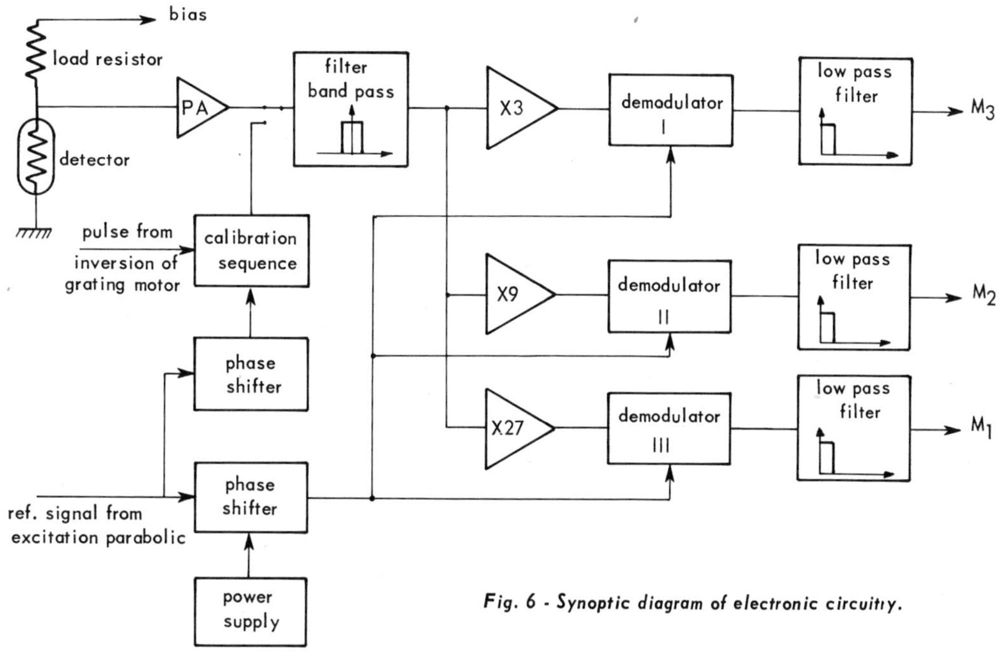

Fig. 6 - Synoptic diagram of electronic circuitry.

Fig. 7 - NO stratospheric absorption spectrum. Balloonborne experiment (May 14, 1973).

Solar zenith angle 88° / 93°

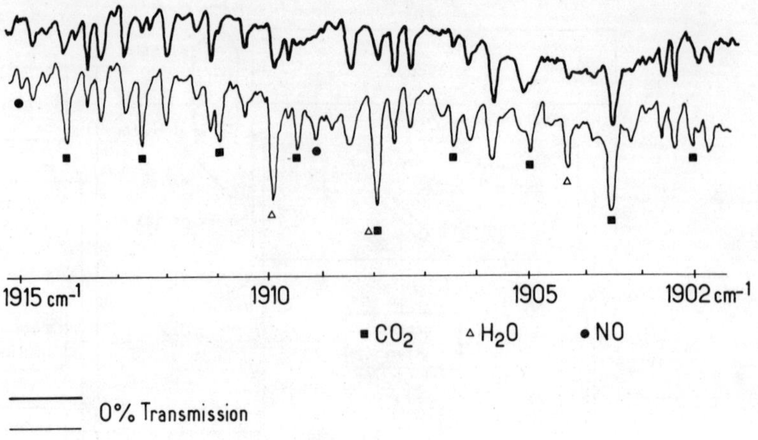

Fig. 8 - NO number density versus altitude. Experimental values of the number density of nitric oxide versus altitude in the chemosphere. The results published by Barth[(3)], Meira[(4)], Hale and Tisone[(5)] are shown, together with those presented here for the stratosphere.

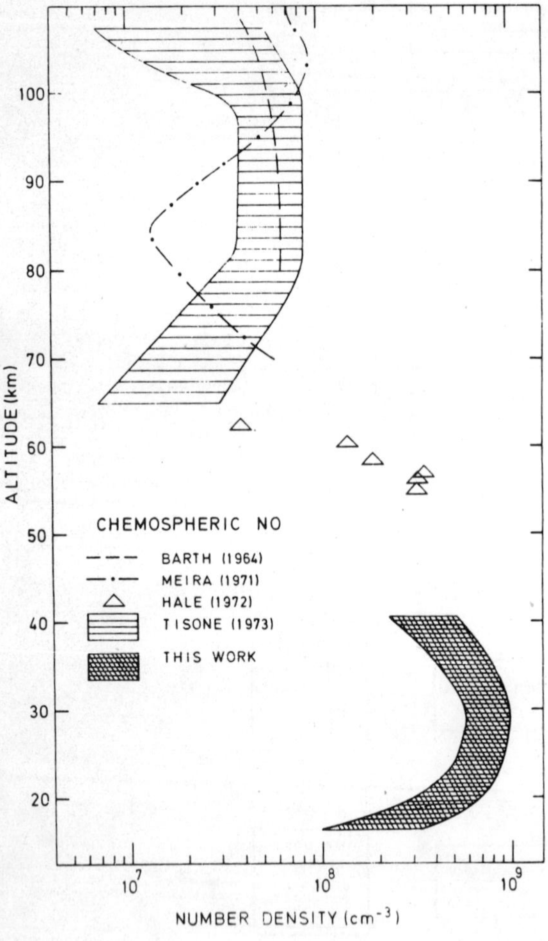

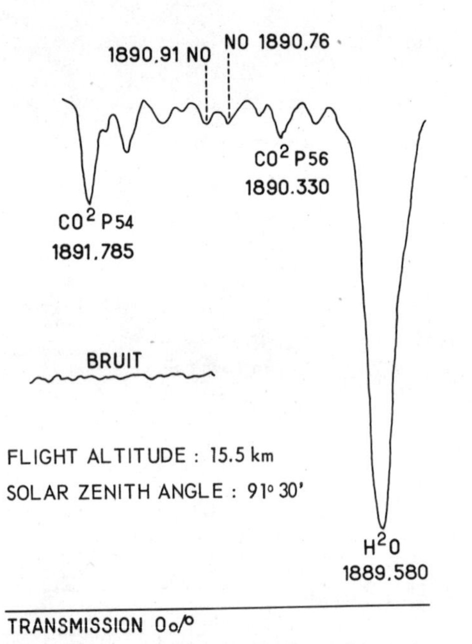

Fig. 9 - NO stratospheric absorption spectrum. Concorde 001 experiments, July 13, 1973, sunrise.

Fig. 10 - Integrated amount of NO measured along the optical path from aircraft A and from balloon B.

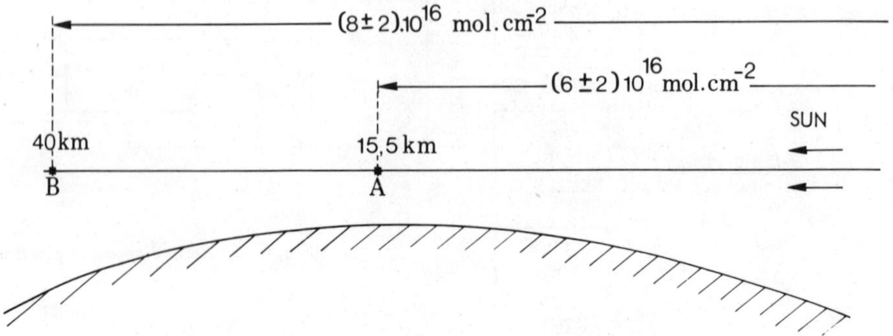

Fig. 12 - HNO_3 stratospheric absorption spectrum.
Concorde 001 experiments.
Flight altitude : 15.5 km
Solar zenith angle 90°20' (A) and 86°50' (B)

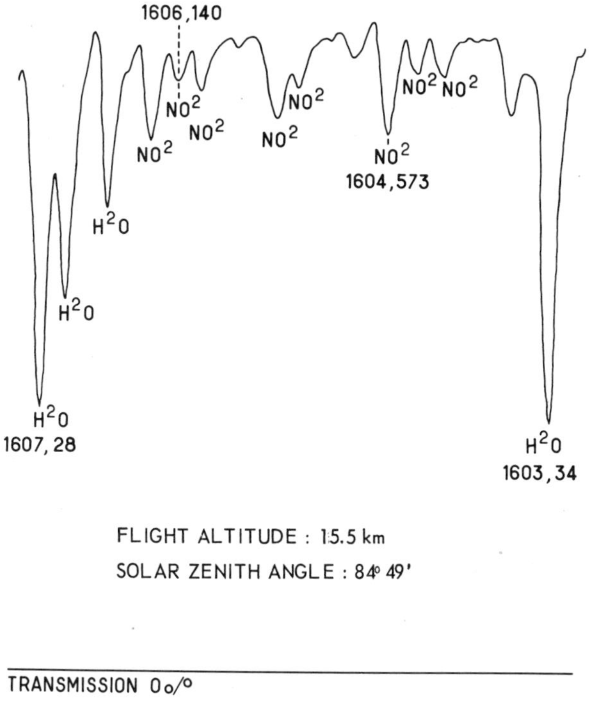

Fig. 11 - NO_2 stratospheric absorption spectrum.
Concorde 001 experiments, July 17, 1973, sunset.

Table I

Molecule	Minimum detectable number of molecules	Minimum detectable concentration (constant mixing ratio)
NO	2 10^{15} mole cm^{-2}	2.5 10^{-11}
NO_2	1 10^{15} mole cm^{-2}	1.3 10^{-11}

Table II

σ cm^{-1}	S cm $mole^{-1}$ 220° K
1890,71	1.31 10^{-19}
1909,14	1.23 10^{-19}
1914.99	9.09 10^{-20}

Table III

σ cm^{-1}	S cm $mole^{-1}$ 220° K
1604.154	0.544 10^{-18}
1604.350	0.344 10^{-18}
1606.140	0.316 10^{-18}

© 1973, ISA JSP 6661

ENERGY DISPERSIVE X-RAY (EDX) SPECTROSCOPY

D. Thomas Carlton
Technical Manager
Magnavox Environmental Systems
The Magnavox Company
Fort Wayne, Indiana

ABSTRACT

The applications for EDX in the general fields of Pollutant Monitoring and Environmental Sciences are endless, but in order to utilize any system to its fullest potential, a thorough understanding of its operation is necessary. Thus, this paper presents first a brief discussion of the basis of X-Ray Spectroscopy and the traditional WDX technique, then a discussion of the EDX technique and its advantages, concentrating on its physical chemistry and operational aspects.

BACKGROUND

X-Ray Spectroscopy

X-Ray Spectroscopy, or the measurement of secondary X-Rays, has been a valuable tool for the analytical chemist for a number of years, almost since the time that the atomic structure was thoroughly understood. It was found that the electron clouds surrounding the nucleus of an atom are in a series of shells, with decreasing binding energies as the shells are formed farther from the nucleus. As shown in Figure 1, the shells are designated K, L, M, N, etc. with the K-shell being closest to the nucleus.[1] It was further noted that if an atom were exposed to a sufficiently high energy source of photons (either X-rays or Gamma rays) or high energy particles (alpha or beta particles or electrons), the various shell electrons of the sample would be ejected, creating secondary X-rays or photons. The ejected electron would be replaced by an electron from a shell farther away from the nucleus, with the creation of a secondary X-ray that had a characteristic energy equal to the change in the atomic energy level for that electron participating in the event.

Providing that the primary source of energy is of a sufficiently high level, the probability is highest that a K X-ray will be produced. The probability increases as the energy approaches the absorption edge of each element from the high energy side (see Figure 2).[1] The probability, however, drops to zero when the energy is slightly below the absorption edge for that particular element. At that point, the probability for exciting a secondary L X-ray then becomes highest as the energy approaches each of the three L absorption edges.

There are two characteristic K X-rays (K-alpha and K-beta), and three characteristic L X-rays (L-alpha, L-beta, and L-Gamma). M and N X-rays are also created, but for most purposes, K and L X-rays are used for analysis. A spectrum of the elements present can be obtained by observing all of the K and L secondary X-rays emitted from an unknown sample.

WDX Spectroscopy

Until recently, Wave Length Dispersive X-ray Spectroscopy (WDX) has been the classical means of measuring X-rays. This technique, utilizing the phenomenon of diffraction, is based on the fact that if an X-ray strikes a crystal with a particular lattice structure at a precise angle, the X-ray emerges as a diffracted beam. By knowing the crystal lattice spacing and the diffraction angle, the effective wave length of the diffracted beam becomes a factor of its energy. By sweeping a series of angular positions with a detector and one or more crystals to cover the range of elements anticipated, an X-ray spectrum can be obtained based on the wavelength at which the diffraction occurred. The detectors are usually gas-filled proportional counters or scintillator/photo tube counters. This method is quite time-consuming for multi-element analysis, since each element analyzed is being counted only while the detector and crystals are at the precise angle of each element's specific wavelength.

EDX SPECTROSCOPY

General Description

A relatively new analytical tool is now available to the analytical scientist in the form of Energy Dispersive X-ray Spectroscopy (EDX). Essentially, EDX differs from WDX in that it measures the energy of the secondary X-ray directly. This has become possible with the advent of the lithium drifted silicon diode detector and small, fast, dedicated mini-computers. The mini-computer, which is an integral part

Superior numbers refer to similary-numbered references at the end of this paper.

of the EDX, functions both as a multi-channel analyzer and a data manipulation device through various software programs that are part of the total system.

The operational results of EDX is a direct energy spectrum of the elements analyzed. The principle advantage is that all unknown elements in the sample are measured simultaneously, thus giving larger counts for a multi-element analysis in a much shortened time frame. Moreover, EDX eliminates the complex mechanical linkage required by WDX to rotate the crystals and detectors through the angular positions.

Principles of EDX Operation

A source of primary excitation X-rays is directed towards the sample containing the unknown material. The primary X-rays can either be bremsstrahlung radiation (X-ray tube) or mono-energetic X-rays (radio isotope). The secondary X-rays created by the sample are columnated and directed through a beryllium window to strike the Si(Li) detector diode. The detector and a FET preamplifier are operated in a liquid nitrogen atmosphere to minimize thermal and electrical noise. The X-rays that strike the detector ionize the silicon atoms in the detector, creating electron hole pairs and free electrons proportional to the energy of the incoming X-rays.(2) Since 3.8 electron volts (eV) are required to ionize a single silicon ion, the number of electrons that are freed is equal to the energy (eV) of the incoming X-rays divided by 3.8.

Since the diode is operated under a high bias, the freed electrons migrate to the contact of the opposite charge and a small signal is created. The signal is preamplified by the FET and is further amplified and shaped into a usable signal in the electronics portion of the system. The signal is integrated on a capacitor that is under a constant current, and the resultant electronic charge is stored in a preselected memory bin. The signal is stored on a time basis that is directly proportional to the original energy of the X-ray. The amplifier portion can handle up to 20,000 signals per second of various energy levels. The X-ray spectrum is counted over a preselected time base such as 100-500 seconds and stored in a memory for later display. The electronics are such that if two signals enter the detector at one time, the pulse will be rejected and compensation for "live time" counting will be performed.

Available Computer Programs

As mentioned earlier, the computer portion of the EDX performs both multi-channel analysis and data manipulation. The data manipulations are performed on two built-in memories. Some of the programs available with the equipment are:

a) OVERLAPPING - Overlapping one memory group with another for the purpose of visual comparison.
b) ADDITIVE TRANSFER - Transferring any memory group to any other memory group for visual display or print-out.
c) TOTAL INTENSITY - Intensifying any selected portion of the spectrum for print-out or display.
d) STRIPPING - Stripping or subtracting of any group from any other group with negative number suppression.
e) EXPANSION - Expanding the visual spectrum display in either the horizontal or vertical direction.
f) TELETYPE PRINT-OUT - Either printing out the information on the display or punching out a tape in ASCII CODE from any data region of the memory.
g) MARKERS - Provides a visual K, L, or M marker on the display for visual interpretation of the spectrum.
h) LINEAR CONCENTRATION - The concentration of up to five unknown materials as a ratio of one built-in standard.
i) LEARN - Programming of up to 51-step operational routine for automatic processing. This capability exists on both memories.
j) REPEAT - Automatically repeating the previous operational routine for either of the two memories.
k) RATIO - Ratioing any region of the spectrum to any other region of the same spectrum.
l) SMOOTH - Produces Five-Point Least Squares Convolution smoothing of the spectrum in any memory region.
m) ANGSTROMS - Displays the region of the peak identification after conversion from KeV to Angstroms.

Sensitivity

Such factors as sample matrix, detector resolution, and excitation energy play key roles in sensitivity determinations.

Detector resolution is usually specified as the width of the peaks in a spectrum. More specifically, it is usually applied to the full width at half maximum (FWHM) of the K-alpha manganese line at 5.9 KeV using an Fe^{55} radio isotope. With the present state-of-the-art, most companies have standard detectors with 150-165 eV resolution.

The excitation energy can be adjusted to any level compatible with the analysis as an operator controlled function of the power supply. The primary X-rays thus can be bremsstrahlung with single or multiple target tubes, filtered or unfiltered, or can be monoenergetic using any number of radio isotopes with sufficient half-lives to be suitable to that particular application.

The sample matrix is extremely important and can be analyzed as either a liquid or a solid. The system has demonstrated sensitivities averaging 0.06 micrograms per square centimeter. (see Table 1). Thus, if we used an initial sample of 100 milliliters, chemically precipitated the heavy metals on a one square centimeter filter paper, the detection limits would be 0.06 parts per million or 60 parts per billion.

Table 1. SENSITIVITIES

Material	Detection Limits (Micrograms per CM^2 filter paper)
Iron	0.04
Cadmium	0.13
Copper	0.05
Arsenic	0.04
Zinc	0.03
Lead	0.07
Chromium	0.06
Nickel	0.05

APPLICATION

Since EDX performs elemental analysis on all materials above sodium (atomic number 11), it is almost boundless in its application for detecting and analyzing specific environmental pollutants. (It is applicable to the analysis of airborne particulates as well as to dissolved or particulate water pollutants.) The only considerations must be sound engineering judgements as to the condition of the sample presented to the detector and the most effective tailoring of the instrument variables to yield data that is both valid and useful to the environmentalist.

It would be impossible to list all of the applications of the EDX system to environmental control, but some of the more obvious are: monitoring of dissolved and suspended heavy metals in both fresh water and sea water and its sediments; analysis of animal feed (such as calcium and phosphorous), in broiler feeds, and arsenic in medical feeds; analysis of heavy metals in mussel tissue; analysis of respirable mine dusts; monitoring of particulates in air; and analysis of sulphur and nickel-to vanadium ratio in oils.[3]

Essentially, the environmentalist today is faced with an extremely dynamic and complex situation that can be rectified only with a complete understanding of the problem. EDX is merely a tool, but when utilized properly, it can provide some of the critical answers.

REFERENCES

(1) Russ, John C., "EXAM Method", EDAX Laboratories, 1972.

(2) Hime, P., IAS-263, Dow Chemical Corporation, 5 January 1973

(3) "Application Tips", Q/M 24, Finnigan Corporation, 30 November 1972

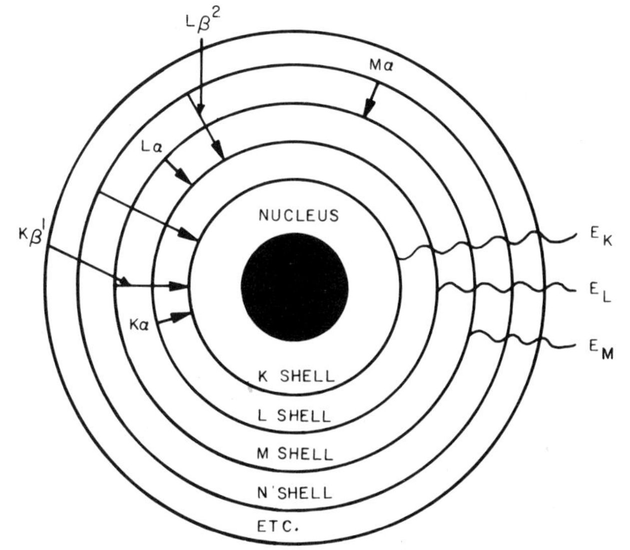

Figure 1. Electron Transitions (Major Emission Lines)[1]

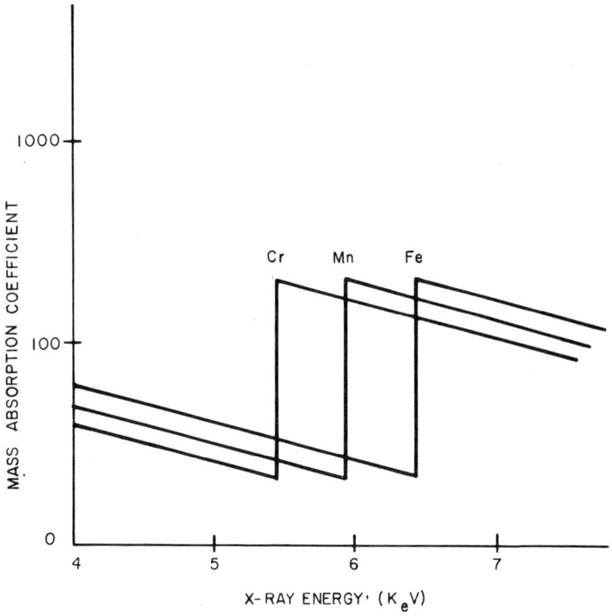

Figure 2. Absorption Edges

©1973, ISA JSP 6664

MULTIWAVELENGTH LASER INDUCED FLUORESCENCE OF ALGAE *in-vivo*: A NEW REMOTE SENSING TECHNIQUE

P. B. Mumola*
and
Olin Jarrett, Jr., and C. A. Brown, Jr.
NASA Langley Research Center
Hampton, Virginia 23665

ABSTRACT

In order to accurately determine the quantity of chlorophyll *a* in living algae by fluorescence spectrometry, either remotely or in the laboratory, the fluorescence excitation cross section must be known. Laboratory fluorescence studies of a number of different algae species representative of the various color groups were performed using phytoplankton supplied by the Virginia Institute of Marine Science. These measurements indicate distinct maximum spectral excitation regions which differ from one color group to another. Within each color group, however, the fluorescent properties were nearly identical, regardless of species. These two key features, namely: (1) the similarity of fluorescent properties within a color group, and (2) the distinct spectral differences between color groups, make possible the simultaneous determination of chlorophyll *a* content of an unknown mixture of phytoplankton and the distribution of chlorophyll *a* among the various color groups. The multiple wavelength LIDAR equations are developed and chlorophyll *a* concentrations are calculated using matrix inversion techniques. The design and application of a flight-qualified multiwavelength laser fluorosensor for performing remote measurements from an airborne platform are described. The application of this technique to wide area surveys of water resources is discussed.

INTRODUCTION

Perhaps the most important photochemical reaction known to man is photosynthesis — a process carried out in all living terrestrial and aquatic plants. Photosynthesis requires, among other things, the presence of chlorophyll *a*. Sorenzen(1) and El-Sayed(2) have found highly significant correlations between chlorophyll *a* concentration and primary productivity in marine surface waters. The concentration of chlorophyll *a* has therefore become a key indication of basic biological processes and can serve as a yardstick for water quality assessment.

The concentration of chlorophyll *a* may be determined by a variety of chemical and/or spectroscopic techniques. Yentsch and Menzel(3) have developed a method using fluorescence analysis of acetone extracts of natural phytoplankton chlorophyll. Sorenzen(4) has made continuous measurements of *in-vivo* chlorophyll concentrations using a modified commercial fluorometer. Both methods require the acquisition of water samples for off-line or flow-through analysis. This requirement becomes a severe burden when one is concerned with measurements over any sizable body of water. Airborne remote sensing techniques offer a possible solution to this problem. A variety of remote sensors have already been developed(5-9) to exploit the optical properties of chlorophyll in living algae and phytoplankton. The absorptive and fluorescent properties of chlorophyll extracts have been studied to some degree in the laboratory(3). Optical investigations of chlorophyll in living algae and phytoplankton have been generally qualitative in nature and recourse is often made to correlative wet chemical analysis to verify results and calibrate sensors.

This paper presents a theoretical and experimental analysis of laser induced fluorescence for remote detection of living phytoplankton in natural waters. The fluorescent properties of various species of phytoplankton representative of the different color groups are described. Laboratory measurements of fluorescence excitation cross sections will be discussed and quantitative data presented. The overlapping excitation spectra demonstrate the requirement of multiwavelength laser excitation in order to make quantitative measurements of chlorophyll *a* concentration when more than one color group is present in the water (the general case). A prototype airborne laser fluorosensor is described. Preliminary field data taken from both a fixed-height platform and a helicopter are presented.

FLUORESCENCE OF LIVING PHYTOPLANKTON

Speculations as to the reasons for the common occurrence of red algae in deep water, brown algae at middepths, and green algae near the surface have been discussed in the literature for years. Originally, two alternative explanations were offered: (1) that this vertical distribution has been determined by an adjustment by the plant to the intensity

*Present address: Perkin-Elmer Corporation, Electro-Optics Division, Wilton, Connecticut 06897

Superior numbers refer to similarly numbered references at the end of this paper.

L-9170

of the prevailing light, and (2) that this vertical zonation is due to the adaptation of the algae to the spectral distribution of the prevailing light[10]. Currently the importance of both types of adaptation is recognized.

Adaptations permitting algae to survive well below the sea surface have apparently been achieved in two ways: by the development of photosynthetically active accessory pigments which are complementary in color to the prevailing light (such adaptations are exemplified by deep-growing red algae), or by an increase in the chlorophyll concentration which enhances absorption in the middle of the spectrum (such adaptations have been made by deep-growing green algae). Engelmann[10] provided evidence to show that light absorbed by the phycobilins (the red and blue-green pigments in the red and blue-green algae) was utilized in photosynthesis. Haxo and Blinks[11] have observed that the direct sensitization of chlorophyll and carotenoids plays only a subordinate role in oxygen production in the red algae, and that photosynthesis is most effectively sensitized by the phycobilins whose maximum absorption occurs in the 495-630 nanometer (nm) region. The available evidence indicates that the phycobilins are not direct sensitizers, but that energy absorbed by the phycobilins is transferred to chlorophyll[12].

The radiant energy absorbed by and transferred to chlorophyll may be used in several ways, among which are photosynthesis and fluorescence. To some degree, these are competing processes. If photosynthesis is slowed down by external conditions such as low CO_2 or nutrient concentrations, the fluorescent output is enhanced. Increasing photosynthesis is accompanied by decreasing fluorescence[13]. Therefore, one might expect to see corresponding differences between daytime and nighttime fluorescence of living algae.

Photosynthetic organs of plants always contain an assortment of pigments. These can be divided into three major classes: (1) chlorophylls and (2) carotenoids, both water insoluble, and (3) phycobilins, water soluble (because of their attachment to water soluble proteins)[14]. Chlorophyll *a* is found in all photosynthesizing cells, and it is the fluorescence of this molecule which we will be principally concerned with in the remainder of this article. In living algae, the fluorescence of chlorophyll *a* is peaked at 685 nm. Since different algae possess different pigment combinations, one can intuitively expect to see variations in their fluorescence excitation and emission spectra. The characteristic chlorophyll *a* fluorescence peak at 685 nm is universally present though additional fluorescence of other pigments may also appear as shown in Figure 1. The wavelength for inducing chlorophyll *a* fluorescence varies from one phytoplankton color group to another, as shown in Figure 2. The excitation and emission spectra of different species within one color group,* however, are remarkably similar, as indicated in Figure 3. It is this similarity within a given color group coupled with the distinct differences between color groups that is exploited in this remote sensing technique.**

FLUORESCENCE CROSS SECTION MEASUREMENTS

One method for measuring fluorescence cross sections of dissolved molecules in solution is to refer all measurements to those of a material of known fluorescent cross section. Rhodamine B is one common standard often used. The absorption and fluorescence of Rhodamine B was examined using a Cary Model 14 spectrophotometer and a Hitachi-Perkin Elmer MPF-2A fluorescence spectrophotometer. An absorption measurement was made of a 10^{-5} molar solution of Rhodamine B in pure ethanol. The peak absorption occurs at 545 nm using a 1-cm quartz cuvette. The absorption cross section, σ_A, was calculated from the transmittance using Beer's Law to be

$$\sigma_A\,(545) = 3.588 \times 10^{-16}\ \text{cm}^2 \text{ (Rhodamine B)} \qquad (1)$$

for a bandwidth of 5 nm.

The excitation and emission fluorescence spectra were measured for various concentrations of dye to determine a region over which the fluorescent power was linearly dependent on concentration. A solution of 10^{-7} molar concentration was found to be well within the linear range. The emission intensity $E(\lambda)$ of the dye was recorded and normalized so that

$$\int_0^\infty E(\lambda)\, d\lambda = \phi \qquad (2)$$

where ϕ is the quantum yield for fluorescence and λ is the emission wavelength. The value for ϕ commonly used is 0.73[15]. The area under the emission curve shown in Figure 4 was found to be 2185 nm × intensity units or

$$\int_0^\infty E(\lambda)\, d\lambda = 2185 \text{ nm} \times \text{intensity units} = \phi = 0.73 \qquad (3)$$

The effective dye fluorescence cross section for a bandwidth of 5 nm was thus calculated to be:

$$\sigma_f\,(545) = 2.584 \times 10^{-17}\ \text{cm}^2 \qquad (4)$$

This cross section can now be used in computing values for the phytoplankton for a 5-nm bandwidth.

---

* The term "color group" is used here to denote the apparent color of a pure mass culture of a given algae or phytoplankton. For a very few phytoplankton species, this may be taxonomically inaccurate, but the spectroscopic characteristics are more easily generalized by this definition.

** The technique to be described is by no means limited to remote sensing applications. Laboratory spectrofluorometers can be used to analyze water samples using the same approach; that is, multiwavelength excitation.

If the fluorescent solution is optically thin implying low absorption, N_f, the fluorescent emission detected by the photomultiplier can be expressed as follows:

$$N_f = C_1 \, N_{EX}(\lambda_{EX}) T_M(\lambda_{EM}) q(\lambda_{EM}) \sigma_A(\lambda_{EX}) E(\lambda_{EM}) \Delta\lambda_{EM} \; n \quad (5)$$

where:

N_f = fluorescent emission (photons/sec)

C_1 = geometrical constant of the optical system

N_{EX} = excitation photon count rate (s^{-1})

λ_{EX} = excitation wavelength (nm)

T_M = transmission of the monochromator

q = photomultiplier photocathode quantum efficiency

λ_{EM} = emission wavelength (nm)

σ_A = absorption cross section (cm^2)

E = emission intensity

$\Delta\lambda_{EM}$ = bandwidth of the emission monochromator (nm)

n = molecular density of the fluorescent medium (cm^{-3})

Expressed in terms of p_1 power (watts), it is noted that:

$$N_f = \frac{p_{EM} \lambda_{EM}}{hc} \quad (6)$$

$$N_{EX} = \frac{p_{EX} \lambda_{EX}}{hc} \quad (7)$$

and

$$q = \frac{C_2 \, S}{\lambda_{EM}} \quad (8)$$

where h is Planck's constant and S is the photocathode radiant sensitivity (amperes/watt). For the photomultiplier used (RCA C31025C), S is nearly constant over the range of 400-800 nm. Thus, Equation (5) can be rewritten in terms of power as follows:

$$p_f = C_3 \, p_{EX}(\lambda_{EX}) T_M(\lambda_{EM}) n \, \sigma_A(\lambda_{EX}) E(\lambda_{EM}) \Delta\lambda_{EM} \lambda_{EX} \lambda_{EM}^{-2} \quad (9)$$

where $C_3 = C_1 \, C_2 \, S$. The effective fluorescence cross section can be defined as:

$$\sigma_f(\lambda_{EX}) = \sigma_A(\lambda_{EX}) E(\lambda_{EM}) \Delta\lambda_{EM} \quad (10)$$

A similar equation to (9) can be developed for the phytoplankton (designated by primed quantities) and ratio of the two equations used to calculate the unknown cross section of chlorophyll a in the phytoplankton as referenced to that of Rhodamine B as follows:

$$\sigma_f'(\lambda_{EX}') = \frac{p_f' \, n}{p_f \, n'} \frac{p_{EX}(\lambda_{EX})}{p_{EX}'(\lambda_{EX}')} \frac{T_M(\lambda_{EM})}{T_M'(\lambda_{EM}')} \frac{\lambda_{EX}}{\lambda_{EX}'} \left(\frac{\lambda_{EM}'}{\lambda_{EM}}\right)^2 \sigma_f(\lambda_{EX}) \quad (11)$$

Figure 3 shows the results of such computations for five different species of golden-brown phytoplankton. Figure 5 shows results for phytoplankton species representative of the four different color groups. From these two figures, several key features are evident. Above 440 nm, Figure 3 indicates a remarkable similarity of fluorescence cross sections, both in shape and magnitude, for different species of golden-brown phytoplankton studied.

Figure 5 shows typical fluorescence excitation spectra for phytoplankton from the four color groups and indicates that each group has a spectral region wherein excitation is most efficient. From these data, the fluorescence cross section matrix required for the multiwavelength LIDAR equation can be generated.

These data are summarized in Table I at the four wavelengths used in the James River flight tests. The cross sections, the standard deviation, and number of samples are shown for the four phytoplankton color groups.

From Figure 5, several characteristics of the color groups may be observed. Red phytoplankton is most sensitive to green light, while blue-green phytoplankton is most sensitive to red light. Green and golden-brown phytoplankton have nearly equal sensitivity (cross sections) in the region near 520 nm, while at 470 nm, the green phytoplankton are more sensitive than the golden-brown.

Several conclusions can be drawn from the above observations. Chlorophyll a fluorescence at 685 nm depends on both the excitation wavelength and the color group of the phytoplankton containing the chlorophyll a. Therefore, no single excitation wavelength can be chosen to excite chlorophyll a fluorescence uniformly in all phytoplankton. Since the excitation spectra overlap one another, one cannot excite uniquely any single phytoplankton color group. Therefore, fluorometric determination of chlorophyll a concentration in living phytoplankton cannot be made with a single excitation wavelength without a priori knowledge of the color groups of phytoplankton present and their relative densities. A technique for fluorometric determination of chlorophyll a concentrations in an arbitrary mixture of phytoplankton color groups using a multiple wavelength excitation scheme is presented in the following section.

MULTIWAVELENGTH LASER RADAR ANALYSIS

In this section the relationship between the concentration of chlorophyll a and the fluorescent power detected at 685 nm at an arbitrary laser excitation wavelength is derived. The laser and detector are assumed to be on a platform at an altitude of at

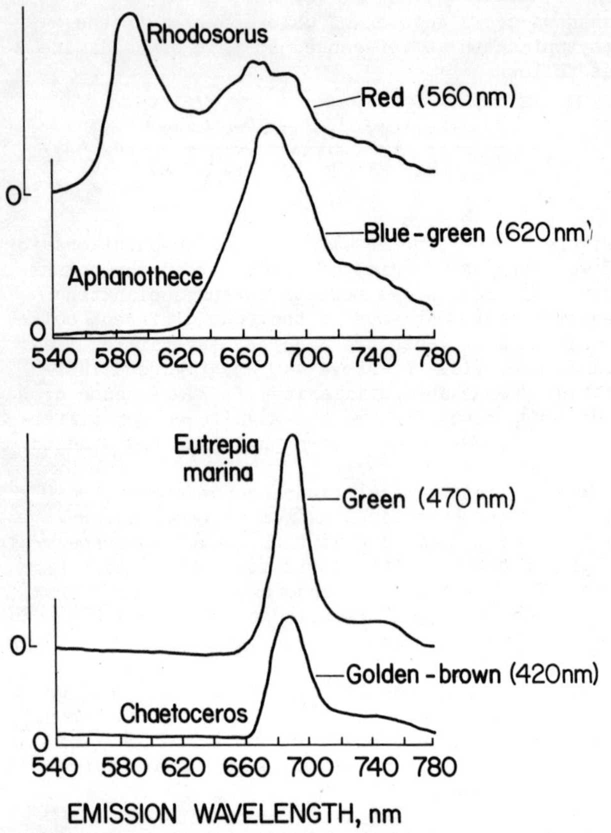

Figure 1. Typical emission spectra of golden-brown, green, red, and blue-green phytoplankton. Each sample excited at the wavelength shown.

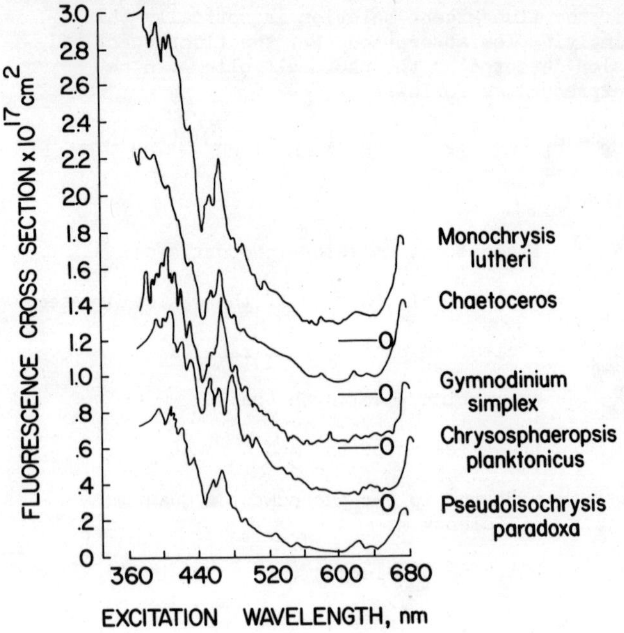

Figure 3. Fluorescence cross sections of chlorophyll a for five different golden-brown phytoplankton species for emission at 685 nm with 5 nm bandwidth.

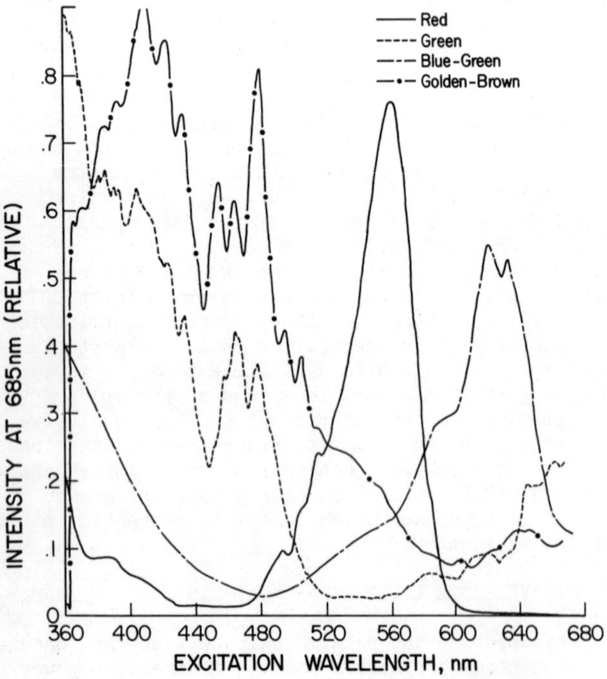

Figure 2. Typical excitation spectra of golden-brown, green, red, and blue-green phytoplankton observed at an emission wavelength of 685 nm.

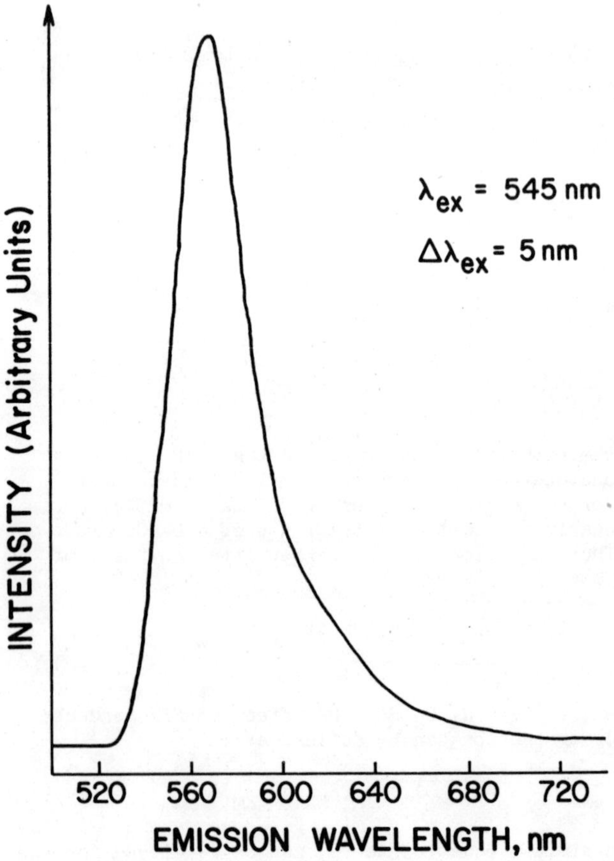

Figure 4. Emission spectra of 10^{-7} molar solution of Rhodamine B in ethanol.

least 100 meters above the sea surface. The analysis is first carried out assuming only one phytoplankton color group is present (or dominant) in the water. A multiwavelength matrix solution is then developed in order to generalize the analysis to include the presence of an arbitrary mixture of color groups.

Consider the measurement geometry shown in Figure 6. The laser and receiver are coalined and nadir pointing at an altitude R above the sea surface. The receiver field of view, θ_r, is somewhat larger than the laser beam divergence, θ_l. The radiant output power of the laser incident on the water surface is essentially the same as that measured just beyond the laser output mirror since atmospheric attenuation is negligible over such short ranges ($R \lesssim 1$ km). This assumption is not valid for high moisture and fog conditions. The laser beam divergence (which can be controlled through collimation) is assumed to be several milliradians. The irradiance produced by the laser, $H_l (W/m^2\ nm)$ in the wavelength interval $\lambda_l \pm 1/2\ \Delta\lambda_l$ can be expressed as:

$$H_l = \frac{p_o}{\Delta\lambda_l\ \pi(1/2\ \theta_l\ R)^2} \qquad (12)$$

where $\Delta\lambda_l$ is the laser emission bandwidth, p_o is the laser output power (w) at the laser wavelength λ_l. Approximately 96-98% of this irradiance penetrates the surface and propagates downward, the remainder being reflected by the water surface. As the light propagates downward, it is both absorbed and scattered. The sum effect is expressed as the optical attenuation, $\alpha(m^{-1})$, which is wavelength and turbidity dependent. Over the visible spectrum, α can vary from less than 0.1 m^{-1} in clear waters to greater than 5.0 m^{-1} in coastal waters and estuaries.

Chlorophyll a fluorescence is maximum at the wavelength $\lambda_f = 685$ nm and its bandwidth is approximately 10 nm. The subscripts "l" and "f" refer to laser and fluorescence parameters, respectively. The amount of laser induced fluorescence in the wavelength interval $\lambda_f \pm 1/2\ \Delta\lambda_f$ that emanates from a unit area of a thin slab of thickness dz at a depth z below the surface and reaches the surface is

$$dW_f(\lambda_f)\Delta\lambda_f = 1/2\ \sigma_f(\lambda_l)\, n\, H_l(\lambda_l)\Delta\lambda_l \cdot \exp[-(\alpha_f + \alpha_l)z]dz \qquad (13)$$

where $\sigma_f(\lambda_l)$ is the cross section for fluorescence (at λ_f) due to excitation at λ_l, n is the density of chlorophyll a (molecules/m^3), and α_f and α_l are the attenuation coefficients of the water at λ_f and λ_l, respectively. In this analysis, n, α_f, α_l are assumed to be constant with depth. The total laser induced radiant emittance from the air-water interface in the wavelength interval $\Delta\lambda_f$ about λ_f is therefore:

$$W_f(\lambda_f) = 1/2\ \frac{\Delta\lambda_f\ \sigma_f(\lambda_l)n\ H_l(\lambda_l)}{\Delta\lambda_l[\alpha_f + \alpha_l]} \qquad (14)$$

where it has been assumed that the laser radiation has been totally attenuated in the water (i.e., Eq. (13) has been integrated for $0 \leq z \leq \infty$). The emittance from the surface is assumed to be Lambertian. The total fluorescent power received by the detector in the spectral interval $\Delta\lambda_D$ (the receiver bandwidth) about λ_f is given by:

$$p_r = \frac{\xi}{4}\Delta\lambda_D\ \pi\ \theta_r^2\ A\ T_f\ W_f(\lambda_f) \qquad (15)$$

where ξ is the receiver optical efficiency, A the effective area of the receiving telescope primary mirror, and T_f is the transmittance of the air-water surface at λ_f (assumed to be = 1.0). Substituting Equations (12) and (14) into (15) yields the following expression for the fluorescent power received:

$$p_r = \frac{A}{2R^2}\frac{\Delta\lambda_D}{\Delta\lambda_f}\left(\frac{\theta_r}{\theta_l}\right)^2 \frac{\sigma_f(\lambda_l)n\ p_o(\lambda_l)}{(\alpha_f + \alpha_l)} \qquad (16)$$

This result is valid for a single phytoplankton color group. Practical limits for $(\Delta\lambda_D/\Delta\lambda_f)$ and $\theta_r/\theta_l)$ are unity.

Since the various phytoplankton color groups exhibit different spectral properties (though all possess chlorophyll a), one can expect the fluorescence cross section for each group to be different. The optimum excitation wavelength is different for each color group and their excitation spectra overlap. This overlap indicates that it is not possible to excite only one group of phytoplankton in the presence of other groups. The power which is received from a mixture of j different groups due to excitation at λ_i would be:

$$p_r = \frac{\xi A}{2R^2}\frac{\Delta\lambda_D}{\Delta\lambda_f}\left(\frac{\theta_r}{\theta_l}\right)^2 \sum_j \frac{\sigma_{fj}(\lambda_i)n_j\ p_o(\lambda_i)}{(\alpha_f + \alpha_i)} \qquad (17)$$

where the j subscript indicates the color group number and n_j is the density of chlorophyll a in phytoplankton color group j. This represents one equation with j unknowns (unless only one color group is present or dominant). In order to solve for n_j, one needs j - 1 additional equations. These equations can be obtained by varying the laser wavelength and measuring the power received $p_r(\lambda_j)$ due to that excitation. Laboratory fluorescence studies indicate four distinctly different fluorescence excitation spectra corresponding to the blue-green, green, golden-brown, and red phytoplankton. Therefore, setting j = 4, one obtains in matrix form:

$$\begin{bmatrix} p_r(\lambda_1) \\ \vdots \\ p_r(\lambda_4) \end{bmatrix} = \frac{\xi A}{2R^2}\frac{\Delta\lambda_D}{\Delta\lambda_f}\left(\frac{\theta_r}{\theta_l}\right)^2 \times$$

$$\begin{bmatrix} \frac{p_o(\lambda_1)}{\alpha_f + \alpha_1} & 0 & 0 & 0 \\ 0 & \frac{p_o(\lambda_2)}{\alpha_f + \alpha_2} & 0 & 0 \\ 0 & 0 & \frac{p_o(\lambda_3)}{\alpha_f + \alpha_3} & 0 \\ 0 & 0 & 0 & \frac{p_o(\lambda_4)}{\alpha_f + \alpha_4} \end{bmatrix} \begin{bmatrix} \sigma_{f1}(\lambda_1) & \cdots & \sigma_{f4}(\lambda_1) \\ \vdots & \ddots & \vdots \\ \sigma_{f1}(\lambda_4) & \cdots & \sigma_{f4}(\lambda_4) \end{bmatrix} \begin{bmatrix} n_1 \\ n_2 \\ n_3 \\ n_4 \end{bmatrix} \qquad (18)$$

Defining: $$\kappa = 1/2\ \xi A \frac{\Delta\lambda_D}{\Delta\lambda_f} \left(\frac{\theta_r}{\theta_1}\right)^2 \qquad (19)$$

and writing P and X to represent the 4 by 4 matrices for output powers divided by attenuation coefficients, and fluorescence cross sections, respectively, one gets for the densities:

$$\begin{bmatrix} n_1 \\ \vdots \\ n_4 \end{bmatrix} = \frac{R^2}{\kappa} X^{-1} P^{-1} \begin{bmatrix} p_r(\lambda_1) \\ \vdots \\ p_r(\lambda_4) \end{bmatrix} \qquad (20)$$

The attenuation coefficients required in the solution must be measured by some technique (either in-situ or remotely). Such attenuation data are essential for accurate measurements of chlorophyll a density. One should note, however, that $\alpha_f > \alpha_1$ for all laser excitation wavelengths. The value of α_f (at 685 nm) is apparently less sensitive[16] to variations in turbidity than at shorter wavelengths. Finally, the laser output power at the four chosen wavelengths can be measured on a shot-to-shot basis or, if the laser output is sufficiently repeatable, less often. Thus, with these data, the multiwavelength LIDAR matrix equation can be solved to yield the chlorophyll a density and the relative concentration of the chlorophyll a among the various phytoplankton color groups.

RADIOMETRIC SIGNAL/NOISE ESTIMATES AND PROTOTYPE DESIGN

The fluorescent power received from a mixture of several phytoplankton color groups due to excitation at several different wavelengths can be used to calculate the total chlorophyll a density and the relative concentrations of the various color groups. Since the power received due to fluorescence is wavelength dependent, so too will be the signal-to-noise ratio. Before the signal estimates are made, the noise sources will be discussed.

During daylight operation the major noise source is the surface reflected solar irradiance, $H_{\lambda s}$ (w/m^2 nm). The power received due to this source is:

$$p_B = \frac{\xi}{4} H_{\lambda s} \Delta\lambda_D \theta_r^2 A\rho \qquad (21)$$

where ρ is the average surface reflectance (typically 0.02 - 0.05). The solar irradiance is naturally dependent on weather and angle from zenith. At zenith, under clear sky conditions, $H_{\lambda s}$ = 1.427 W/m^2 nm. The intensity at sunrise or sunset is roughly 300 times less than this zenith value. In order to minimize this noise source, the receiver bandwidth $\Delta\lambda_D$ and/or the receiver field of view θ_r can be reduced. Both terms also appear in the expression for the power received due to fluorescence (Eq. (16) or (17)).

During nighttime conditions the major source of noise is that due to surface reflected laser light which passes through the receiver interference filter. This small but finite transmission can be considerable in comparison with the fluorescence, especially when the chlorophyll a concentration is low ($\lesssim$ 1 mg/m^3). The power received due to laser light reflection of the surface can be written as:

$$p_R = \frac{\xi\, p_o\, \rho(\cos\psi)\, A\, T_x}{\pi R^2} \qquad (22)$$

where T_x is the receiver filter transmittance at the laser wavelength and ψ is the angle of incidence with the water surface measured from the normal (in our case $\psi = 0$ for calm water). Obviously, one can reduce p_R by reducing T_x. Values of 10^{-4} to 10^{-8} for T_x can be achieved with single or multiple interference filters. The prototype system was designed around the operational parameters shown in Table II.

For purposes of estimation, assume that only a single color group of phytoplankton, say, golden-brown, is present in the water; thus the simplified LIDAR Equation (16) can be used.

Let $$\lambda = 470 \text{ nm}$$

$$\alpha_f = \alpha_1 = 2.0 \text{ m}^{-1}$$

$$n = 1 \text{ mg/m}^3 \ (= 6.75 \times 10^{17} \text{ molecules/m}^3)$$

and $$\sigma_f\ (470 \text{ nm}) \approx 3 \times 10^{-22} \text{ m}^2$$

then $p_r = 4 \times 10^{-8}$ w

Referring to Equations (21) and (22):

$$p_B \approx \begin{cases} 10^{-10} \text{ watts (dawn or dusk)} \\ 10^{-8} \text{ watts (solar zenith)} \end{cases} \quad (23)$$

$$p_R \approx 10^{-10} \text{ watts for } T_x = 10^{-6} \quad (24)$$

Then for radiometric considerations

$$S/N \equiv \frac{p_r}{p_B + p_R} \quad (24)$$

Thus, for the prototype system described, S/N at the most severe background lighting conditions is approximately 4. This radiometric analysis does not include considerations for noise from signal detection and processing electronics.

MULTIWAVELENGTH LIDAR INSTRUMENT

Figure 7 shows a schematic of the airborne LIDAR system which has been designed and fabricated at Langley Research Center to demonstrate the multiwavelength concept of chlorophyll a detection. The laser used in the system is a unique four-color dye laser pumped by a single linear xenon lamp. Figure 8 shows a cross-sectional view of the laser head. It consists of four elliptical cylinders spaced 90° apart with a common focal axis. The linear flashlamp is placed along this axis and its radiant emission is equally divided and focused into the dye cuvettes located on the surrounding focal axes. Fluorescent dyes form the active medium for the four separate dye lasers. A rotating intracavity shutter permits only one color at a time to be transmitted downward to the water.

The resulting fluorescence from the chlorophyll a is collected by a 25.4-cm-diameter Dall-Kirkham type telescope. The signal is then passed through a narrow bandpass filter centered at 685 nm and on to the photomultiplier (PMT) tube (RCA, 8852). The PMT signal is digitized by a waveform digitizer and stored on magnetic tape for later analysis. The dyes and the water for the flashlamp are kept at a uniformly cool temperature by the refrigerator. The high voltage supply, charging network, coaxial capacitor, trigger generator, and a spark gap, along with a central control system, complete the package.

FIELD TESTS

Field tests have been performed to evaluate the capabilities of this technique. Experiments have been conducted from a fixed-height platform (George P. Coleman Bridge, Yorktown, Virginia) 30 meters over the York River. This site was selected because it was convenient to both Langley Research Center and the Virginia Institute of Marine Science (VIMS). Ground truth data (chlorophyll a concentration, salinity, and phytoplankton species identification) were supplied by VIMS using standard water sampling techniques. The attenuation coefficient (at 632.8 nm) and temperature of the water were measured on site by Langley personnel.

Measurements were made every half hour on the evening of July 9, 1973, and data are shown in Figure 9 along with ground truth data supplied by VIMS. The ground truth values represent total chlorophyll a concentration. The fluorescence measured over this period was analyzed by reducing the matrix technique to a 2 by 2 analysis. This was necessary due to the fact that three of the wavelengths used were too closely spaced, thus causing several columns of the cross section matrix to be nearly identical which led to large inversion errors. This technique seemed acceptable since ground truth data indicated a predominance of golden-brown and green phytoplankton during the entire measurement period. The data plotted in Figure 9 were therefore obtained by restricting the matrix analysis to the golden-brown and green phytoplankton by using only two laser excitation wavelengths.

Different dye solutions were then substituted to more closely match the excitation regions of the different color groups. These new wavelengths, shown in Table I, were then employed in further field tests.

On July 25, 1973, the LIDAR system was flight tested over the James River between Hampton Roads and the Chickahominy River. Flight altitude was 100 meters and flight speed was 120 km/hr. The flight path is shown in Figure 10 along with the chlorophyll a concentration fluoremetrically measured over the 138-km round-trip flight. During each flight leg, the laser was fired at a rate of 0.5 pps. The data plotted in Figure 10 represent average total chlorophyll a concentration over each leg determined by a 3 by 3 matrix analysis. The reason for restricting the analysis to a 3 by 3 rather than a 4 by 4 matrix was due to the apparent failure of the interference filters to sufficiently block the backscattered radiation from the 598.7 nm laser. This laser, using an ethanol solution of Rhodamine 6G, was chosen to optimally excite the red phytoplankton. Ground truth data showed an absence of red phytoplankton in the water. The erroneous data were eliminated by restricting the matrix to a 3 by 3. Improved interference filters can be used to block the backscattered radiation produced at 598.7 nm, thereby permitting the data analysis technique to include the complete 4 by 4 matrix. However, the data obtained using the 3 by 3 matrix show that the chlorophyll a measured on the NW leg agrees reasonably well with the data obtained on the SE leg.

Figure 11 shows the concentration of chlorophyll a in golden-brown, green, and blue-green phytoplankton as determined by the 3 by 3 matrix analysis and averaged over each flight leg. The data taken during the NW and SE flights show some variance. This may be caused by the imprecise overlap of the flight paths and the small laser beam footprint (1 m) on the water surface. Since the phytoplankton often cluster in patches or "windrows," the localized measurements at different locations are sensitive to the local variations in phytoplankton concentration.

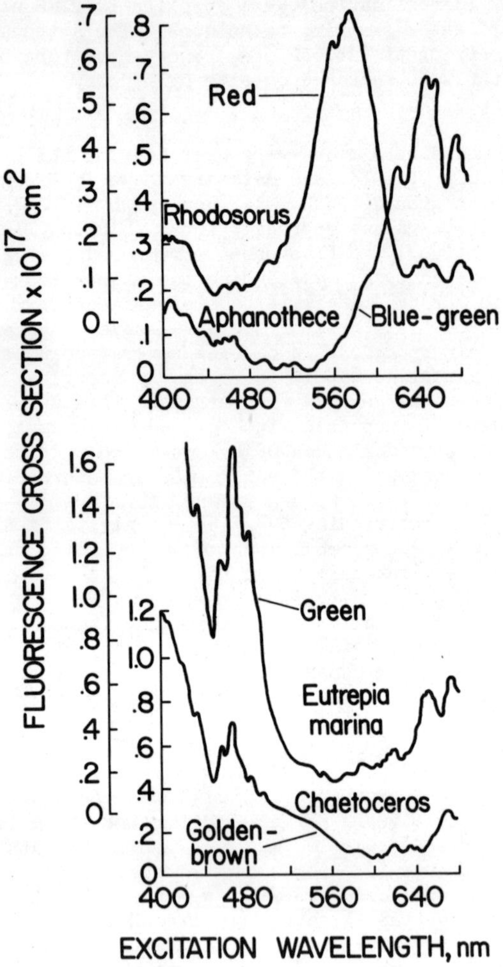

Figure 5. Fluorescence cross sections at 685 nm (5 nm bandwidth) of isolated phytoplankton species representative of the green, golden-brown, red, and blue-green color groups.

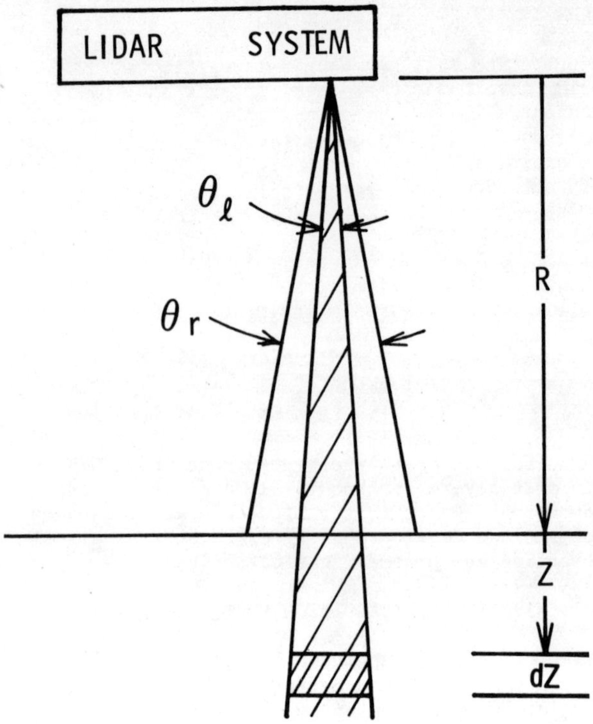

Figure 6. LIDAR system measurement geometry.

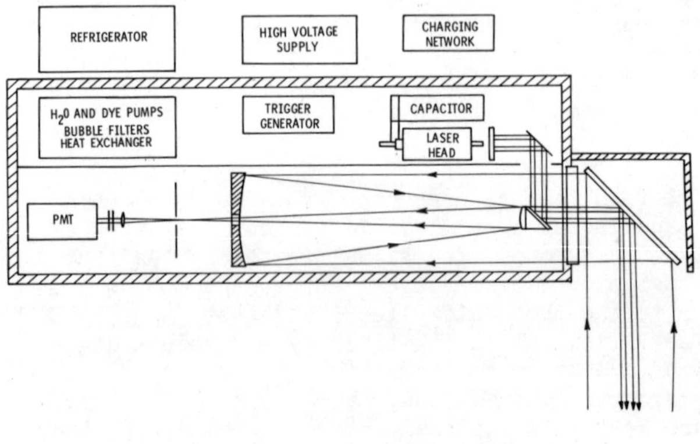

Figure 7. LIDAR system schematic.

SUMMARY AND CONCLUSIONS

A prototype multiwavelength laser fluorosensor is being developed to remotely measure chlorophyll a concentration of living phytoplankton in natural waters. The flurosensor uses the LIDAR technique to excite and detect chlorophyll a fluorescence at 685 nm. Spectral dependence of the fluorescence cross sections on excitation wavelength for isolated phytoplankton species, representative of the various color groups, was used to advantage in the solution of the multiwavelength LIDAR equation to calculate the chlorophyll a concentration when more than one color group is present. Although the fluorescence cross sections of only a fraction of the known phytoplankton species were used (696 species have been listed for lower Chesapeake Bay waters[17]), the similarity in fluorescence cross sections of the species within a color group and the variations in excitation spectra among color groups appear consistent enough to merit development of a prototype chlorophyll a fluorosensor. The prototype system described herein, operating with a minimum energy of 0.6 mJ/pulse, can detect $< 1\ mg/m^3$ of chlorophyll a from 100 meters above the water during daytime operation.

The prototype LIDAR fluorosensor has been operated from both fixed and airborne platforms. A specially developed four-color dye laser which employs a single flashlamp was used. The dyes are in four separate dye cuvettes, and a rotating intracavity shutter permits a single laser pulse to be transmitted at the rate of 0.5 pulse per second. The energy/pulse ranges from 0.6 mJ to 7.15 mJ for four discrete wavelengths between 454.4 nm and 617.8 nm. These values are within the eye safe limits for operational altitudes of 100 meters.

Until light attenuation coefficients for the water at the excitation and fluorescence wavelengths can be remotely measured, the LIDAR fluorosensor will be dependent upon in-situ measurements of these parameters which are linear terms in the LIDAR equation. It may be possible to remotely obtain estimates of the attenuation coefficients by measuring the backscattered signal from the laser pulses. The LIDAR fluorosensor system described herein, when completely developed, offers a technique for rapidly assessing the chlorophyll a content for large areas of water.

ACKNOWLEDGMENTS

The authors are grateful for the cooperation of Drs. Franklin Ott and Robert Jordan of the Virginia Institute of Marine Science for obtaining phytoplankton samples and ground truth data, and to L. G. Burney, C. S. Gilliland, and B. T. McAlexander of NASA Langley Research Center for their assistance in the design, fabrication, and testing of the LIDAR flight instrument.

REFERENCES

(1) Sorenzen, C. J., "The Biological Significance of Surface Chlorophyll Measurements," LIMNOLOGY AND OCEANOGRAPHY, 15, 479,480 (1970).

(2) El-Sayed, S. Z., "Phytoplankton Production of the South Pacific and the Pacific Sector of the Antartic," SCIENTIFIC EXPLORATION OF THE SOUTH PACIFIC, National Academy of Sciences, Washington, D.C. (1970).

(3) Yentsch, C. S., and D. W. Menzel, "A Method for the Determination of Phytoplankton Chlorophyll and Phaeophytin by Fluorescence," DEEP SEA RESEARCH, 10, 221-231 (1963).

(4) Sorenzen, C. J., "A Method for the Continuous Measurement of in vivo Chlorophyll Concentration," DEEP SEA RESEARCH, 13, 223-227 (1966).

(5) Arvesen, John C., Ellen C. Weaver, and John P. Millard, "Rapid Assessment of Water Pollution by Airborne Measurement of Chlorophyll," PROC. OF JOINT CONF. ON SENSING OF ENVIRONMENTAL POLLUTANTS, Palo Alto, Calif., November 8-10, 1971, AIAA Paper No. 71-1097.

(6) Hickman, G. D., and R. B. Moore, "Laser Induced Fluorescence in Rhodamine B and Algae," PROC. OF 13TH CONF. ON GREAT LAKES RESEARCH, Buffalo, New York, March-April 1970.

(7) Hemphill, W. R., George E. Stoertz, and David A. Markle, "Remote Sensing of Luminnescent Materials," PROC. OF THE SIXTH INTERNATIONAL CONF. ON REMOTE SENSING OF THE ENVIRONMENT, University of Michigan, 1969.

(8) Bressette, W. E., "The Use of Near Infrared Photography for Aerial Observation of Phytoplankton Blooms," PROC. OF THE SECOND ANNUAL REMOTE SENSING OF EARTH RESOURCES CONF., Tullahoma, Tenn., March 1973.

(9) Clarke, George L., Gifford C. Erving, and C. J. Sorenzen, "Spectra of Backscattered Light From the Sea Obtained From Aircraft as a Measure of Chlorophyll Concentration," SCIENCE, 167, 1119-1121 (1970).

(10) Englemann, T. W., "Farbe und Assimilation," BOT. ZEITUNG, 41, 1-13, 17-29 (1883).

(11) Haxo, Francis T., and L. R. Blinks, "Photosynthetic Action Spectra of Marine Algae," J. GEN. PHYSIOL., 33, 389-422 (1950).

(12) Daysens, L. N. M., "Transfer of Excitation Energy in Photosynthesis," KEMINK EN ZOON N.V., 1-96 Utrecht (1952).

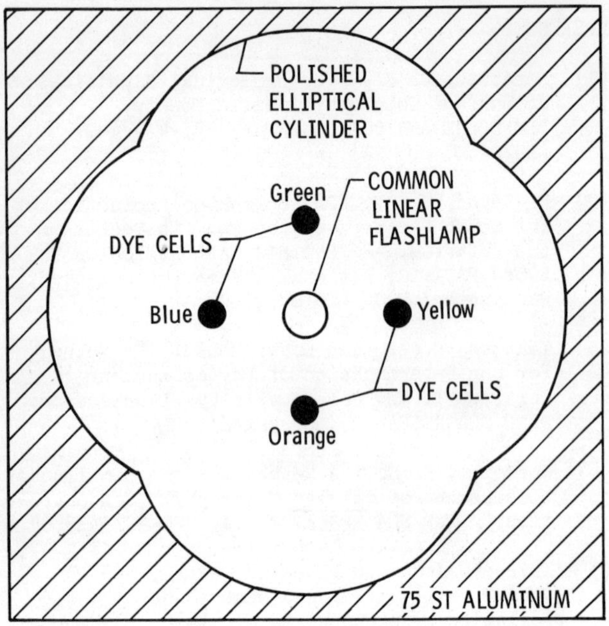

Figure 8. Cross-sectional view of the four-wavelength dye laser head.

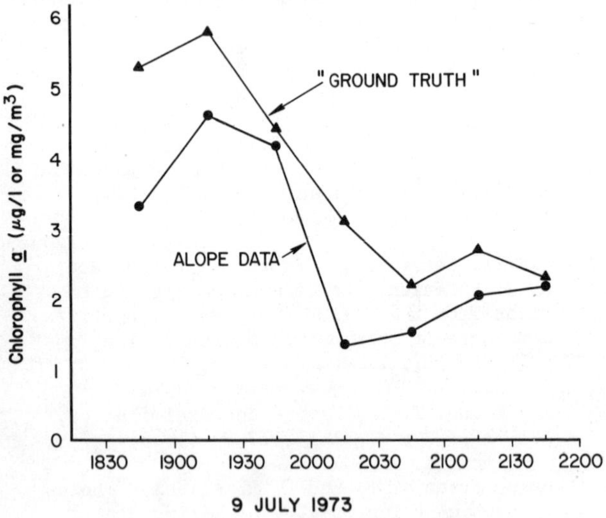

Figure 9. Bridge data (preliminary).

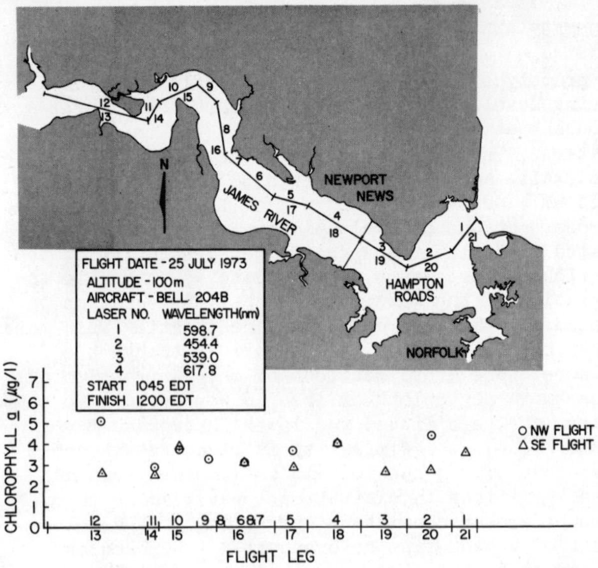

Figure 10. Flight data (preliminary).

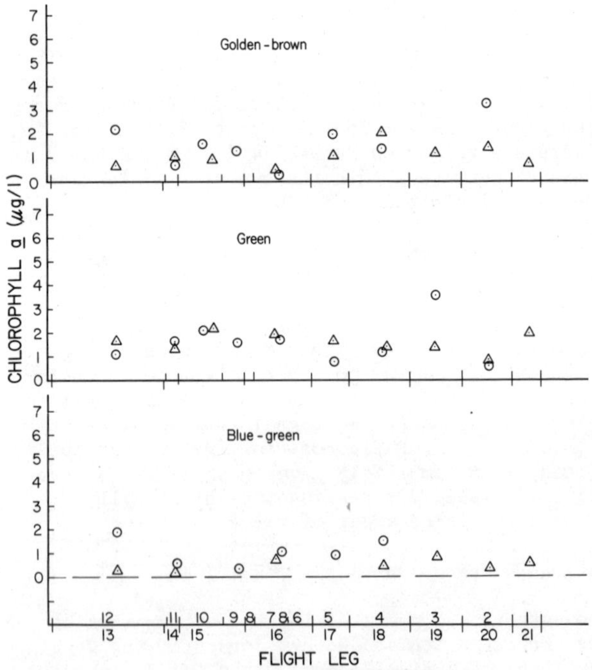

Figure 11. Chlorophyll a concentration in golden-brown, green, and blue-green phytoplankton over the flight path (see Fig. 10).

(13) Franck, J., C. S. French, and T. T. Puck, "The Fluorescence of Chlorophyll and Photosynthesis," J. PHYS. CHEM., 45, 1268 (1941).

(14) Rabinowitsh, E., and Govindjee, PHOTOSYNTHESIS, Ch. 9, John Wiley and Sons, Inc., New York (1969).

(15) Parker, C. A., PHOTOLUMINESCENCE OF SOLUTIONS, Elsevier, New York (1968).

(16) Tyler, J. E., and R. W. Preisendorfer, "Transmission of Energy Within the Sea," THE SEA, M. N. Hill (ed.), 433, Interscience, New York (1962).

(17) Wass, Marvin L., et al., "A Checklist of the Biota of Lower Chesapeake Bay," SPECIAL SCIENTIFIC REPORT NO. 65, The Virginia Institute of Marine Science (1972).

TABLE I. FLUORESCENCE CROSS SECTIONS

	Wavelength	454.4 nm	539.0 nm	598.7 nm	617.8 nm
Golden-brown	$\sigma_f \times 10^{-17}$ cm^2	0.6533	0.1511	0.0810	0.1103
Five samples	SD, %	47.6	51.5	92.1	71.2
Green	$\sigma_f \times 10^{-17}$ cm^2	1.1515	0.1392	0.1635	0.2324
Four samples	SD, %	44.7	79.7	43.7	41.2
Red	$\sigma_f \times 10^{-17}$ cm^2	0.133	0.4498	0.4472	0.2547
Three samples	SD, %	74.2	72.1	55.6	49.9
Blue-green	$\sigma_f \times 10^{-17}$ cm^2	0.0707	0.0497	0.1517	0.2388
Five samples	SD, %	114.2	113.4	53.8	67.2

TABLE II. LIDAR SYSTEM PARAMETERS

$\xi = 0.25$	$R = 100$ m
$A = 5 \times 10^{-2}$ m^2	$p_o(\lambda_i) \approx 5 \times 10^3$ W at each wavelength
$\theta_r = 10$ m rad	$\Delta\lambda_D = 5$ nm
$\theta_1 = 10$ m rad	$\Delta\lambda_f \approx 20$ nm

© 1973, ISA JSP 6666

DETECTION OF WATER POLLUTION SOURCES WITH AERIAL IMAGING SENSORS*

Charles L. Rudder and Charles J. Reinheimer
McDonnell Aircraft Company
McDonnell Douglas Corporation
St. Louis, Missouri 63166

Aerial remote sensing methods can provide effective and economic means for monitoring real and potential spills of oil and hazardous materials. The advantages of airborne techniques have long been recognized by the military in reconnaissance applications. In particular, large areas can be overflown in a short period of time, and access can be gained into industrial sites where ground truth teams or in situ sensors may be disallowed. With properly selected airborne sensors, data can be collected for detailed analysis of industrial operations, the location of outfalls, the determination of run-off patterns, identification of spill materials, and the surveillance of numerous other environmental problems that would be difficult, if not impossible, to accomplish on the ground in a timely manner. When an episode occurs, an aircraft can be dispatched to the scene for acquisition of data for enforcement purposes. Often times, due to time and location constraints, aerial surveillance is the only means to assess the extent of damage. Although these advantages are made possible by the aircraft, the viability of remote sensing is attributed to the data collection systems.

The need for sensors capable of directly detecting the many varied pollutants has caused a great deal of emphasis in developing sensor technology for data collection. However, field use of sensors requires exploitation of sensing technology which includes data collection, data reduction and analysis, and finally information extraction. Once sensor systems are developed, tested, and deployed, the most important part of remote sensing is in the extraction of information from that data collected. This step places man as part of the system to make decisions. The sensor design engineer often ignores the associated system requirements. Typically, new sensors are tested under controlled or at least known conditions so that that information extraction can be treated as a test result. This paper addresses the very important requirement for imagery interpretation keys used by interpreters to extract information from data collected with imaging sensors. Examples of aerial imagery are discussed to emphasize the value of such keys.

To realize the full potential of remote sensing, the complete inventory of existing sensors and newly developed sensors may have to be exploited. Because the types of industries and associated pollutants are so varied, the problem of data collection and interpretation can be quite complex. This can be explained best by first looking at a simple explanation of the physics of remote sensing.

Since the sensor is airborne, its only mechanism for detecting the pollutant is the collection and recording of electromagnetic waves scattered or emitted from that material. These EM waves carry information about the scattering material which discriminates it from the surrounding material. This process is easily recognized when the sensor is a camera and the data are displayed as a photograph. The unskilled interpreter can look at an aerial photograph and recognize rivers and lakes and even oil refinery storage tanks (if he knows he is looking at an oil refinery). How does this novitiate arrive at his conclusion? He simply looks at the geometrical shapes defined by the recorded contrasts, compares these shapes with his mental storehouse of knowledge, and makes his decision. He really doesn't have to understand any physics or chemistry.

To elucidate further, Figure 1 is presented as an example. The factory displayed in the aerial photograph is a titanium plant adjacent to a major river. The whole manufacturing facility is shown, including the several outfalls. The unskilled interpreter probably would say little about the factory except that it looks complicated and appears to be dumping something into the river. The skilled interpreter, using direct analysis could offer an abundance of information on conditions of the facilities and other apparent information, but he too could not identify the content of the outfalls as hazardous or non-hazardous. Both interpreters have evaluated the geometries and contrasts to arrive at conclusions. This is the essence of "direct analysis."

More information could be gleaned from Figure 1 if the interpreter had an imagery interpretation key for titanium plants designed for environmental use. Such a tool would describe the entire plant operation with many photographic views of each type of facility. A chemical description of the operation also would be given so that products and wastes would have their origins and destinations identified. With a key the interpreter may be able to determine that at location (A) the outfall contains ore-gangue; cooling basin overflow river water is being dumped at (B); at (C), (D), (E) and (G), process water with various pollutants is discharged; and at (F) the discharge contains titanium dioxide, ferrous sulphate, and sulfuric acid. (The information was gained by ground truth collected during an EPA sponsored study.) To arrive at such conclusions the interpreter would employ "deductive analysis" since no direct data are available. Deductive analysis requires tracing the outfall's content back to its source.

**This work was sponsored in part by the U.S. Environmental Protection Agency under Contract Numbers 68-10-0140 and 68-01-0178.*

FIGURE 1 TITANIUM PLANT SHOWING OUTFALLS

FIGURE 2 THERMAL INFRARED IMAGE OF STEEL MILL AND INDUSTRIAL WASTES

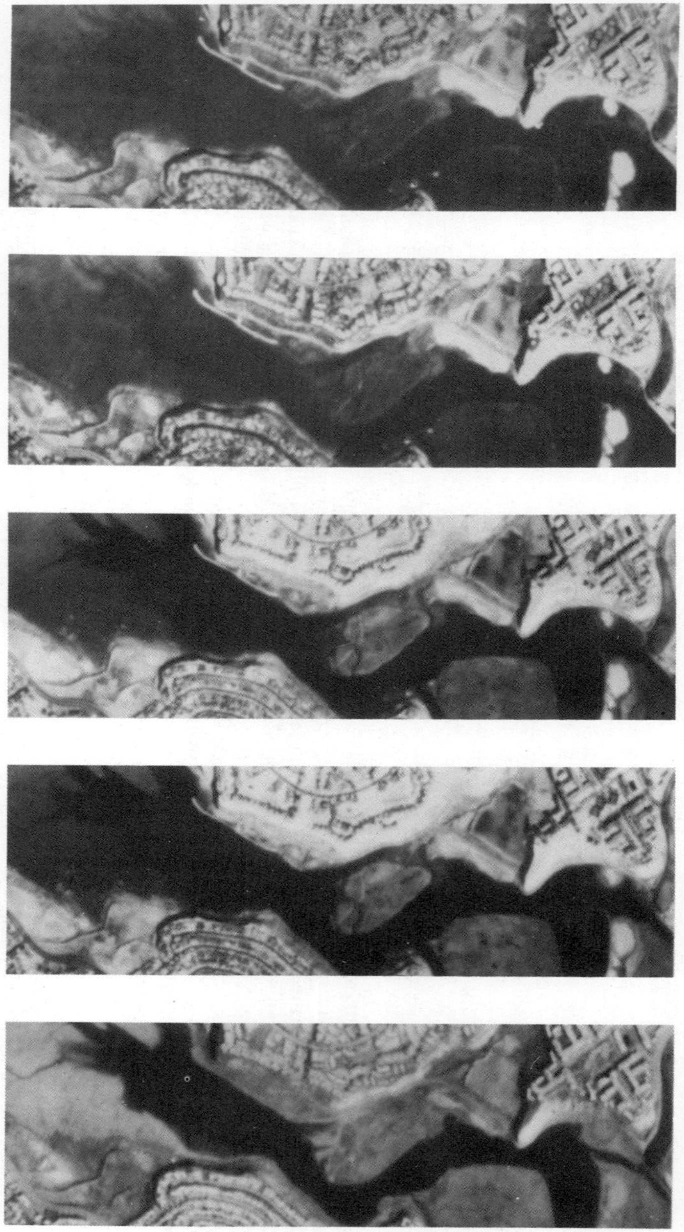

FIGURE 3 MULTISPECTRAL SCANNER IMAGERY OF A RIVER BASIN

By using other sensors, additional data can be collected to detect pollution. For example, thermal imagery is the result of detection of radiation in the infrared (IR) spectrum. The IR sensor provides data outside the spectral range of photographic cameras. Therefore, the sensor complements cameras by supplying added spectral information capability. An example of thermal infrared detection of pollution being discharged from a steel mill is shown in Figure 2. Hot drainage is observed in the ditches leading from the plant to the adjacent river. Not only is the thermal pollution evident, but its presence and apparent source suggests that it may be a hazardous material. If such data were collected concurrently with photographic data and a properly constructed imagery interpretation key were available, then the material emitting heat could possibly be identified.

By detecting discrete spectral bands in the visible and infrared spectrum, a very crude spectrographic analysis can be accomplished. Figure 3 shows five channels of imagery from a multispectral scanner. The scene is a portion of a river basin downstream from industrial activity. Comparison of contrasts with the knowledge of spectral signatures of materials reveals evidence of residual industrial pollution in the river basin. However, without a special interpretation key providing guidance in making a comparative analysis, such information could not be obtained. Such a key requires corroborative ground truth in its formulation to establish confidence in information extraction. The illustration demonstrates that the spectral characteristics of an area can be detected for evaluation and interpretation by direct analysis when the proper key is available.

The foregoing discussion points out the strong potential value of aerial remote sensing for the detection of oil and hazardous materials. Although sensors can collect the necessary data, the actual detection is not accomplished until that data are analyzed and interpreted. This task is not a simple one. Types of pollution are quite varied and can originate in many different kinds of industries ranging from uncomplicated municipal sewage treatment plants to complex chemical plants and oil refineries. Furthermore, data collected with different types of sensors require different rules for analysis (as illustrated by Figures 1 through 3). Therefore, an EPA interpreter is confronted with a nearly impossible task unless he has "guidebooks for analysis" to aid in his job. These guidebooks, or keys, are actually essential to the field use of the environmental remote sensing system.

The design of an interpretation key for environmental applications must emphasize the need to detect conditions that create pollution. An approach to such design would be to describe the manufacturing processes for the selected industry so that all products, by-products, and waste materials can be traced from their origins to their final disposition. Such information would have to be presented through processing diagrams (i.e., flow charts), photographs of facilities from several aspect angles, and verbal descripions of facilities and events in a language needed for recognition by an interpreter. If special sensors are used, then interpretation information regarding methods of analyzing the data collected by such sensors is required. A key containing such information would permit identification of spill sources within the confines of the industry premises. Furthermore, deductive analysis of outfalls would be more reliable when direct analysis isn't possible.

McDonnell Aircraft Company, under contract to the EPA, used the approach described above to develop an Aerial Spill Detection Key for Petroleum Refineries. The selection of this industry allowed a reasonable demonstration of the approach since the refinery processes are quite complex in their photographic rendition. Figure 4 shows a flow diagram of a typical petroleum refinery similar to that which may be found in a military key. Features that are not shown are the origin, flow, and disposition of waste materials. Process descriptions, simplified flow diagrams, and aerial photographs for each phase of the operation depicted in Figure 4 with the additional environmental data were included to provide the interpreter with the required information.

The aerial surveillance system used to collect the data for the development of the key consisted of a cartographic camera, a multiband camera array, and low performance commercial aircraft. These are shown in Figure 5. The cartographic camera was chosen to provide baseline imagery for standard photogrammetric (cartographic when needed) and interpretation analysis. The multiband camera was used as a special sensor to collect spectral information. This was accomplished by judiciously selecting photographic film and filter combinations to record only that light which would emphasize the spectral reflectance of petroleum waste material or spilled oil. Comparison of the simultaneously obtained photographs from the four cameras of the array provided a crude spectral analysis. Figure 6 shows the spectral sensitivity curves of the black and white films used. Color film was also used to obtain color discrimination for cueing of suspect pollution areas. The direct analysis techniques for the multiband imagery provided in the resulting key are correlated with the analysis of the baseline imagery.

To illustrate features of the key that are essential to effective imagery analysis, several sample figures have been selected for discussion. Figure 7 shows two views of a lubricating oil refinery area. The oblique view is used to orient the analyst (trained or untrained). The specific processing areas are identified as deasphalting at A, solvent extraction at B, dewaxing at C, and clay treating at D. Each area has unique features that can be readily identified. The vertical view of the same scene, shown in Figure 7b, permits detailed analysis of the area. Features such as storage tank conditions and revetment inadequacies can be detected and could signal potential spill sources. Multiband imagery of the area (not shown) could reveal spilled material.

A particular processing area of the lubricating oil refining is described in detail by using a simplified flow diagram (Figure 8). This figure describes the solvent extraction process and provides a functional background for recognizing facilities. It also identifies those materials that could be spilled at this location. Such a diagram accompanied by textual information precedes representative imagery in the key. A view that allows detailed direct analysis is the stereo pair. Figure 9 is an example showning the solvent extraction process where settling tanks and furnaces are located at positions A and B respectively. This type of display presents a three dimensional view of objects when used with any stereo optical device. Sizing and volumetric data can be obtained in addition to terrain features that may establish the run-off patterns.

The products of the petroleum refinery are stored in tanks of many shapes and sizes. The storage area may occupy up to 75 percent of the refinery area. It is in this area that the potential for spills is most prevalent. Normally, the analyst would be concerned with recognizing the types of storage tanks and, by inference, the stored material; the condition of storage tanks; and the condition and adequacy of the revetments. It is important to note that various industries have different practices in storing material so that the interpretive inferences for petroleum refinery storage would be inapplicable to other industries.

The potential pollution source in a storage tank area shown in Figure 10 is typical of data needed in an environmental key. The leaking tank in the ground view is readily observed at position (A) in the stereo pair. Of course, the analyst would use a stereo viewer to provide an enlarged three dimensional image to perform his task.

The type of storage tank when correlated with location and related features often provides a reliable clue to the identity of the tank contents. However, shortage of space or increased demand for specific products may dictate that tanks be used for products other than those for which the tank was designed. Ideally, low volatile crude oil is stored in fixed roof cylindrical tanks. To reduce evaporation, the more volatile products are best stored in breather tanks, floating roof, spherical, or spheroidal tanks. The tanks designed primarily for storing gases under pressure are the butane, spherical, and spheroidal tanks.

The oblique photograph shown in Figure 11 shows a typical mix of tanks commonly seen in a refinery. Examples include a group of fixed roof cylindrical tanks at (A), spherical tanks at (B), vertical butane at (C), horizontal butane at (D), and floating roof cylindrical tanks at (E).

The multiband imagery shown in Figure 12 is the final example of imagery used in the petroleum refinery key. These simultaneous photographs show a waste area that is typical of refineries. Often times such areas are located near the water treatment ponds that lead to outfalls. These parts of the refinery compound can be suspect as pollution sources. However, the goal of the industry is to properly treat, confine, and eventually safely remove the waste material. The numbers appearing beneath each picture indicate the film and filter combination (the four digit number specifying film type and the two digit number specifying the filter). The information provided in the key tells the interpreter how to make contrast comparisons to perform a crude spectral analysis for materials identification.

The Aerial Spill Detection Key that was developed was used to evaluate imagery of several refineries taken with the same aerial camera systems. With some time spent in familiarization, the interpreters found the key to the invaluable. Clearly, the benefits of the aerial technique were demonstrated during the aerial surveillance program. Ground truth teams collected correlative data which confirmed the effectiveness of the imagery interpretation. Also this coordinated correlative study emphasized the tremendous time advantage and data collection capability inherent in the airborne remote sensing method. At the very least, such techniques complement ground surveillance by locating suspicious areas so that field investigation teams operating on the ground need not spend time observing "clean" areas.

The importance of environmental keys has been demonstrated. The successful implementation of aerial remote sensing for detecting and identifying pollution sources through field investigations necessitates effective information extraction methods and tools.

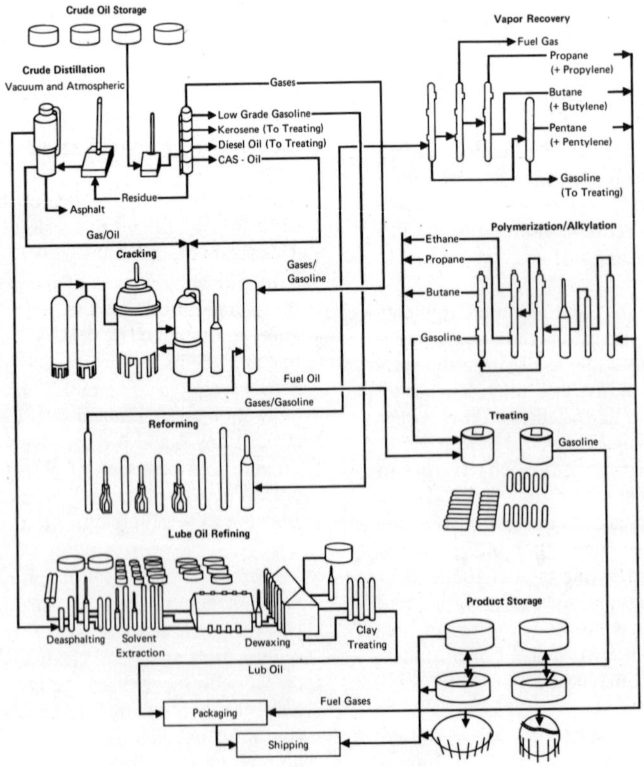

FIGURE 4 FLOW OF TYPICAL REFINERY

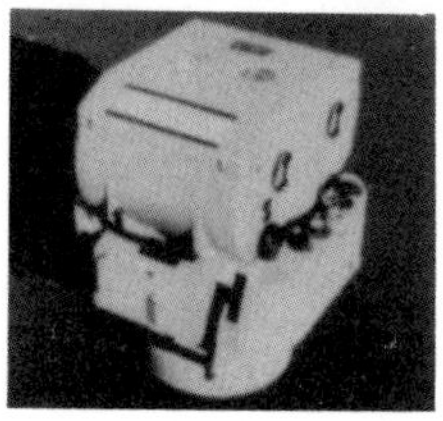

Zeiss RMK 1523 Camera

Hasselblad
Camera Array

Aero Commander

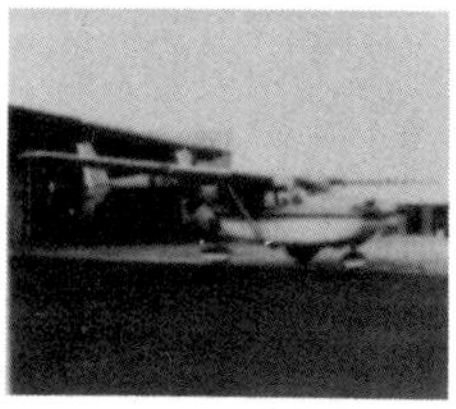

Cessna 336

FIGURE 5 AERIAL SURVEILLANCE SYSTEM

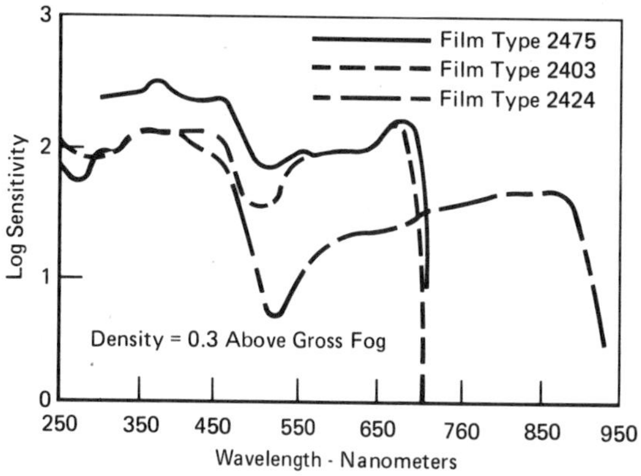

FIGURE 6 SPECTRAL SENSITIVITY CURVES FOR PHOTOGRAPHIC FILMS

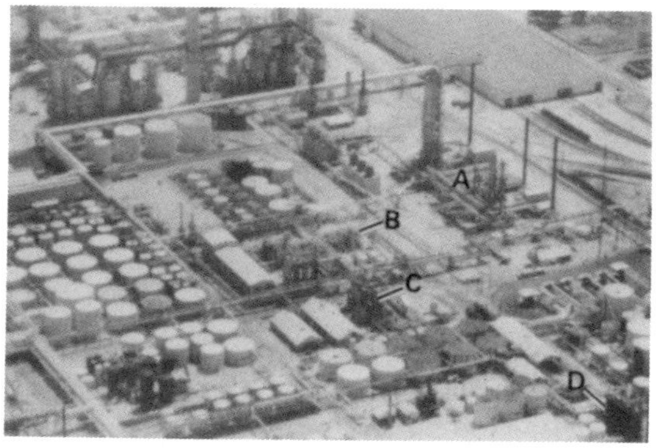

FIGURE 7a LUBRICATING OIL REFINING - OBLIQUE VIEW

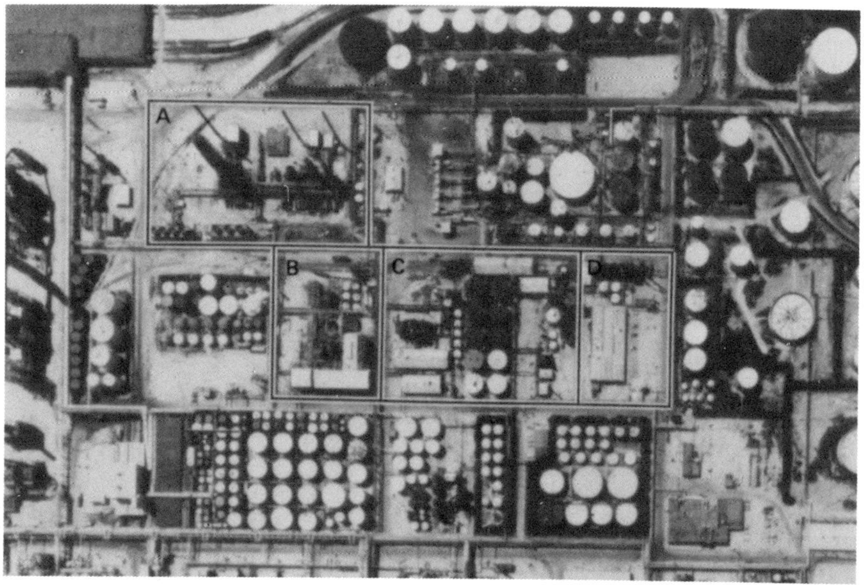

FIGURE 7b LUBRICATING OIL REFINING - VERTICAL VIEW

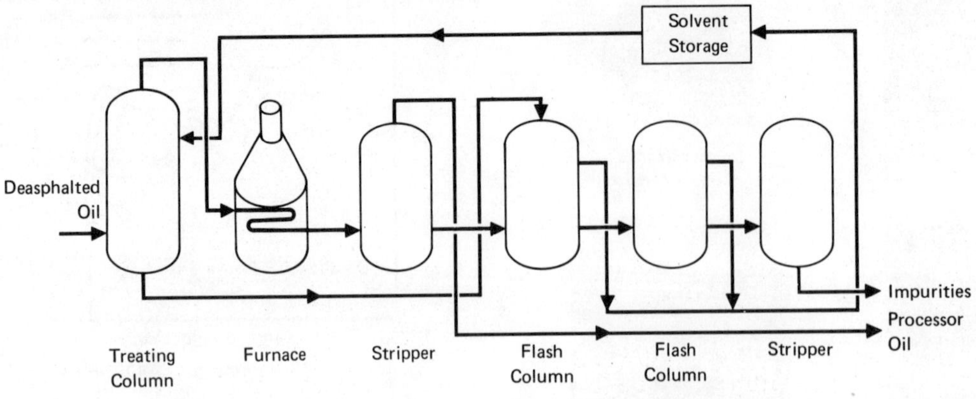

FIGURE 8 SIMPLIFIED FLOW OF SOLVENT EXTRACTION

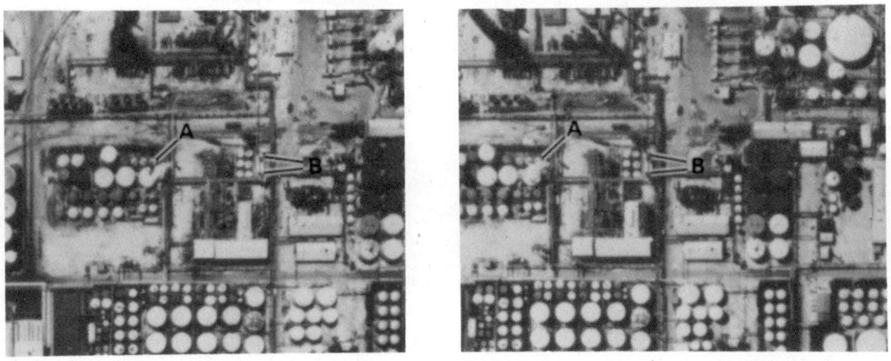

FIGURE 9 LUBRICATING OIL REFINING, SOLVENT EXTRACTION

FIGURE 10 STORAGE, LEAKING TANKS

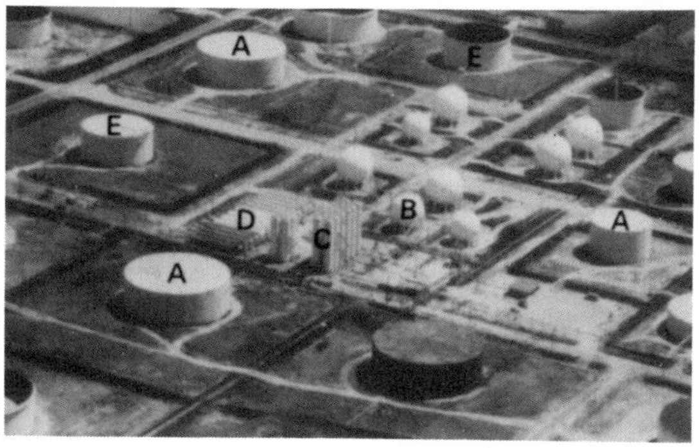

FIGURE 11 STORAGE TANKS

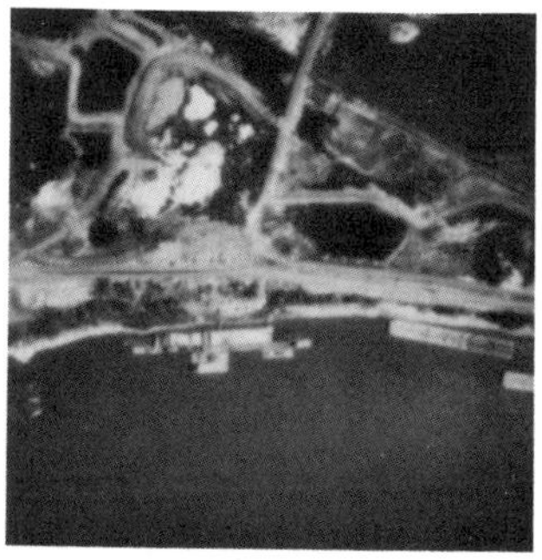

a) 2403/99

b) 2403/35

c) 2424/99

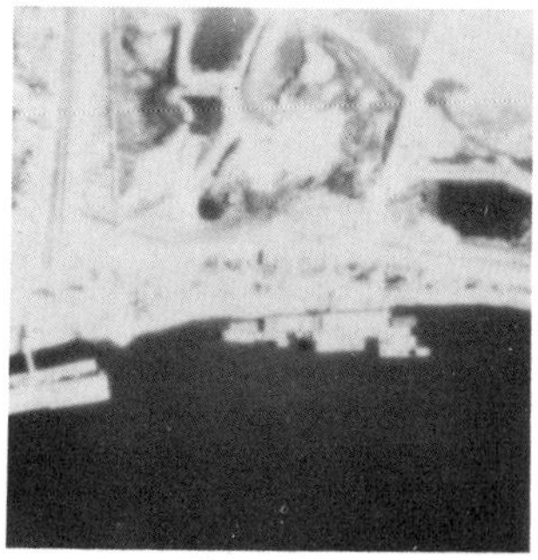

d) 2424/35

FIGURE 12 PETROLEUM WASTE AREA

© 1973, ISA JSP 6667

STANDARD METHODS FOR ANALYSIS AND INTERPRETATION OF LIDAR DATA FOR ENVIRONMENTAL MONITORING

S. H. Melfi*
NASA - Langley Research Center
Hampton, Virginia 23365

I. INTRODUCTION

Remote monitoring of the atmosphere, in many cases, provides a capability to obtain unique measurements horizontal and vertical over a wide area. Most remote monitoring techniques depend on the sensing of absorption and/or scattering of electromagnetic radiation, and are usually divided, for discussion purposes, into passive and active techniques. Passive techniques rely on radiation already present in the atmosphere; whereas active techniques utilize energy that is transmitted into the atmosphere as an integral part of the monitoring methodology.

This paper will discuss one of the active remote monitoring schemes, LIDAR. LIDAR is an acronym for LIght Detection And Ranging. It is similar in principle to microwave RADAR but uses a pulsed laser as the source and an optical telescope as the receiver. The laser propagates a pulse of energy into the atmosphere which interacts with the aerosols and molecules and which in turn scatter light in all directions. A small portion of radiation is backscattered toward the receiver. Monitoring the backscattered energy in time and wavelength provides information on both the range and the identity of the scattering species.

LIDAR observations of elastic scattering from aerosols and more recently Raman scattering from molecules have been performed in the atmosphere with favorable results. The research effort has progressed to the point that standard methods of data analysis and interpretation are not only desirable but necessary. Standard techniques are needed both for the intercomparison of data obtained by the various LIDAR users, and for a consistent interpretation by the many actual and future user groups. It is also indicated that the standardization of techniques should: result in a unified effort in the future of LIDAR, provide indications of various problem areas, and promote a broader acceptance of LIDAR results both by the user and scientific community.

*Presently with the U.S. Environmental Protection Agency, National Environmental Research Center, Las Vegas, Las Vegas, Nevada 89114

The following sections will emphasize the need for standardization and will include a description of LIDAR in broader terms in Section II followed by selected applications of the technique in Sections III, IV, and V. Section III will be devoted to the application of LIDAR to remote aerosol monitoring, covering both its application to the dynamics of the atmosphere and the interpretation of LIDAR data in terms of aerosol characteristics and concentration. Section IV will discuss the application of LIDAR to the remote measurement of visibility, stressing the measurement of Raman scattering by the nitrogen molecule. Finally, Section V will discuss methods of remotely monitoring molecular concentrations by observing Raman scattering.

II. DESCRIPTION OF THE LIDAR TECHNIQUE

A LIDAR system consists of a laser and telescope whose optical axes are either coincident or aligned parallel. Figure 1 is an artists conception of a LIDAR system with the laser pulse shown at one instant of time. As the pulse propagates through the atmosphere, the telescope collects the backscattered energy scattered by the atmospheric constituents. The energy scattered back is primarily at the laser wavelength, λ_o, but a small portion is shifted in wavelength by the molecules (denoted as λ_R). This shifted radiation is called Raman scattering. An approximate spectrum from the atmosphere is shown in Figure 2. The signal at λ_o is due to both Rayleigh scattering by the molecules and Mie scattering by the aerosols, whereas the λ_R's are specific to a particular Raman scattering molecule. The exciting wavelength, λ_o, is normally not coincident with a molecular absorption. For any λ_o chosen, the λ_R's satisfy the condition $\lambda_o^{-1} - \lambda_R^{-1}$ = a unique constant for each molecular scattering species. Selection of wavelengths is accomplished either by a monochromator or narrow band interference filters placed at the output of the telescope. Sensitive photomultipliers detect the filtered radiation.

The general LIDAR equation for a signal detected by the photomultiplier is given as:

$$V(\lambda',R) = \frac{c\gamma(\lambda')\ S(\lambda')\ E(\lambda_o)\ Ar}{2R^2} \times f(\lambda',R)\ q(\lambda_o,R)\ q(\lambda',R)\ , \quad (1)$$

where $V(\lambda',R)$ is the voltage across the photomultiplier load resistor, r; γ is the receiver optical efficiency; S is the spectral sensitivity of the photomultiplier; E is the laser pulse energy; A is the receiver area; R is the range to the scattering volume; f is the total backscattering function of the atmosphere; q is the transmissivity from the LIDAR to the scattering volume; and λ_o and λ' are the laser and sensing wavelengths, respectively. Equation (1) indicates that the signal received, $V(\lambda',R)$, is proportional to laser pulse energy, E, and backscattering function, f, which includes both backscatter cross-section and concentration of scatterers, and inversely proportional to the square of the range.

Considering elastic scattering ($\lambda' = \lambda_o$), which includes molecules and aerosols, Equation (1) can be rewritten as:

$$V(\lambda_o,R) = \frac{c\gamma(\lambda_o)\ S(\lambda_o)\ E(\lambda_o)\ Ar}{2R^2} \times \left[f_M(\lambda_o,R) + f_A(\lambda_o,R)\right] q^2(\lambda_o,R) \tag{2}$$

where f_M and f_A are the backscatter function for molecules and aerosols, respectively.

Considering Raman scattering from molecules, Equation (1) becomes:

$$V(\lambda',R) = \frac{c\gamma(\lambda')\ S(\lambda')\ E(\lambda_o)\ Ar}{2R^2} \times f_R(\lambda',R)\ q(\lambda',R)\ q(\lambda_o,R) \tag{3}$$

where f is the backscatter function for Raman scattering. In general, the scattering function is defined as: $f = \sigma N$ where σ is the backscatter cross-section and N is the number density of scatterers.

III. LIDAR MEASUREMENT OF AEROSOLS

The presently accepted methods for the measurement of aerosol loading in the atmosphere rely on in-situ samplers and occasionally balloon-borne photoelectric counters. While these techniques are adequate for point measurements and vertical profiles, they lack the capability of sensing a wide area that LIDAR can provide.

The LIDAR equation for aerosol monitoring, as given in Equation (2), indicates that the scattering at λ_o is due to both aerosols and molecules. The interpretation of the data at λ_o relies heavily on a standard method of separating the aerosol from the molecular component. In the lower atmosphere (i.e. the mixed layer), the aerosol scattering is typically of the order or larger than the molecular scattering. In this case the LIDAR measurement may be nearly proportional to the aerosol component. However, this is the region in the atmosphere of the strongest and most variable attenuation. The attenuation effects can be approximately eliminated by monitoring Raman scattering from nitrogen and ratioing the elastic scattering signal to the Raman nitrogen signal. The ratio of Equation (2) to Equation (3) demonstrates this:

$$\frac{V(\lambda_o,R)}{V\left(\lambda_{N_2},R\right)} = \frac{\gamma(\lambda_o)\ S(\lambda_o)}{\gamma\left(\lambda_{N_2}\right) S\left(\lambda_{N_2}\right)} \times \frac{q(\lambda_o,R)}{q\left(\lambda_{N_2},R\right)} \frac{\left[f_M(\lambda_o,R) + f_A(\lambda_o,R)\right]}{f_R\left(\lambda_{N_2},R\right)} \ . \tag{4}$$

Since the difference in λ_o and λ_{N2} is small, the atmospheric transmissivity at the two wavelengths are approximately equal, and Equation (4) can be rewritten as:

$$\frac{V(\lambda_o,R)}{V\left(\lambda_{N_2},R\right)} = k \left[1 + \frac{f_A(\lambda_o,R)}{f_M(\lambda_o,R)}\right] \ . \tag{5}$$

The quantity in the brackets in Equation (5) is referred to as the aerosol scattering ratio. The constant, k, can be evaluated by normalizing the signal ratio in Equation (5) to unity in the clean region of the atmosphere above the mixing layer. An example of a measurement of this type is shown in Figure 3.[1] The LIDAR data in Figure 3 is normalized to unity at an altitude of 2.3 km. Excursions above unity indicate regions of aerosol scattering. The LIDAR data is compared in Figure 3 to an independent measurement of water vapor mixing ratio (defined as the ratio of the weight of water to the weight of dry air in unit volume of the atmosphere).

LIDAR measurements in the lower atmosphere can be used to continuously monitor the height of mixing, low level aerosol pollution, and atmospheric phenomena such as plume trajectories.

Stratospheric aerosol monitoring by LIDAR is complicated by two factors: (1) Aerosol scattering is typically less than molecular scattering, and (2) Raman nitrogen scattering is typically too weak to be measured; therefore, it is unavailable for comparison purposes. A standard method for LIDAR analysis of stratospheric aerosol data is similar to that for the lower atmosphere, except that a model for attenuation or transmissivity must be used along with a molecular density profile from an independent measurement. The model of transmissivity along with the moledular number density provides a measure of the expected molecular LIDAR return if no aerosols are present in the atmosphere. A ratio of the data to the expected molecular signal provides the scattering ratio from Equation (2) as:

$$R_S = k' \frac{R^2 V(\lambda_o,R)}{\left[f_M(\lambda_o,R)\ q^2(\lambda_o,R)\right]_e} \doteq \left[1 + \frac{f_A(\lambda_o,R)}{f_M(\lambda_o,R)}\right] \tag{6}$$

Superior numbers refer to similarly-numbered references at the end of this paper.

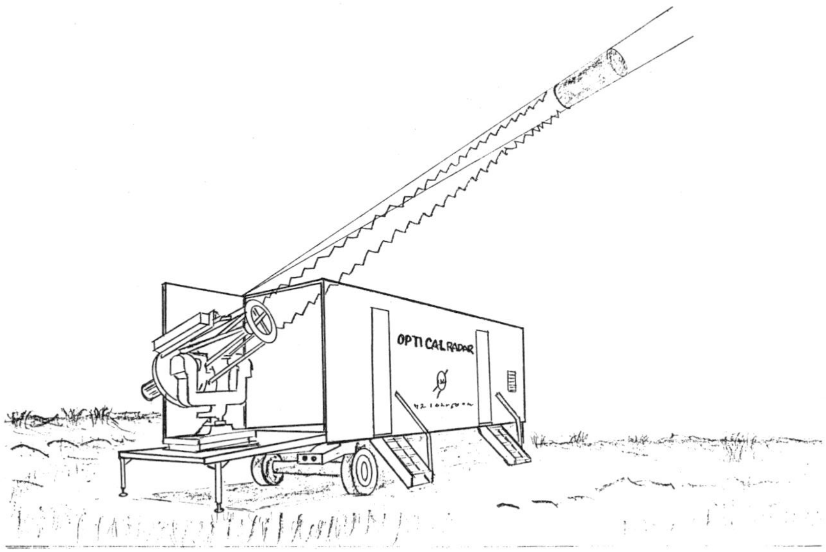

FIGURE 1. AN ARTIST'S CONCEPTION OF A LIDAR SYSTEM

RAMAN RETURN WAVELENGTHS

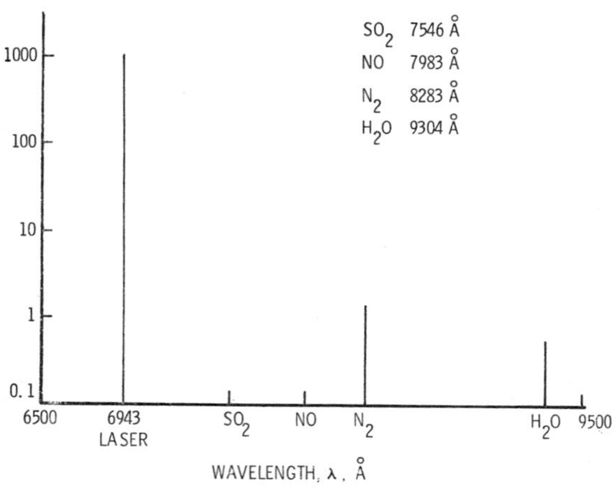

FIGURE 2. AN APPROXIMATE SPECTRUM FROM THE ATMOSPHERE EXCITED BY RADIATION AT 6943Å.

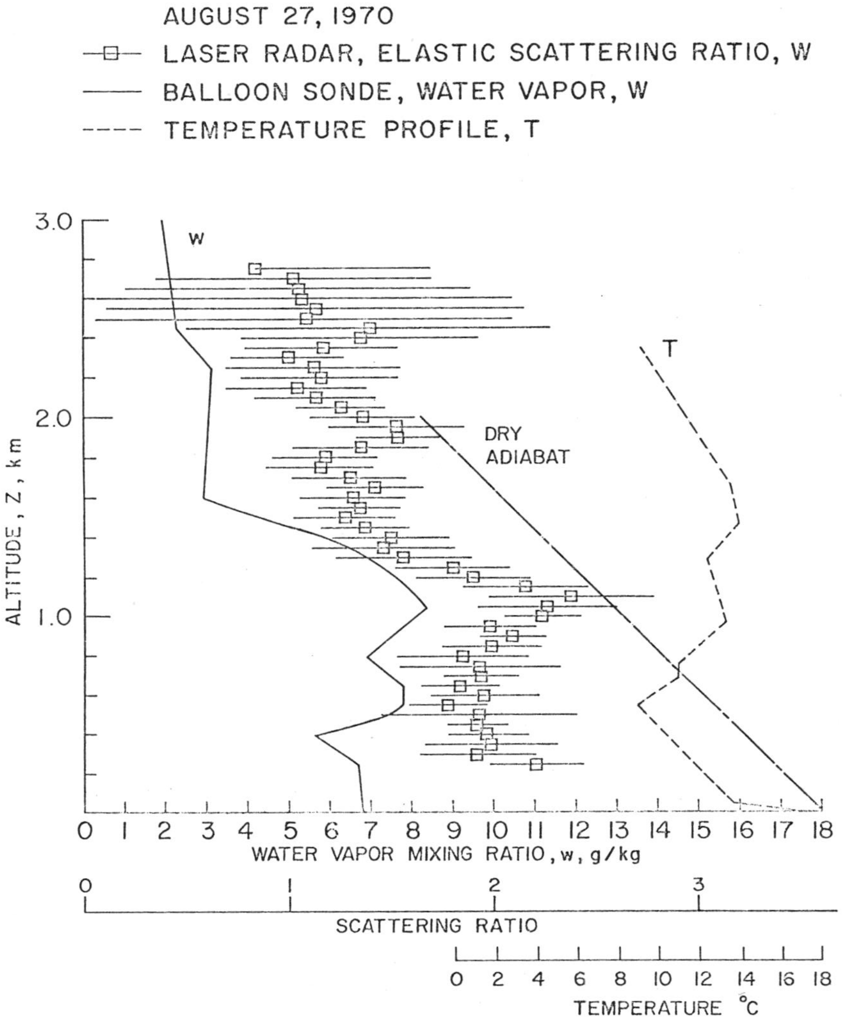

FIGURE 3. ELASTIC SCATTERING RATIO AS A FUNCTION OF ALTITUDE COMPARED WITH STANDARD BALLOON-SONDE DATA (MELFI, APPL. OPTICS II, 1605, (1972).

where $\frac{\left[f_M(\lambda_o,R)\ q^2(\lambda_o,R)\right]_e}{R^2}$ is proportional to the expected molecular LIDAR signal. The constant k', is evaluated by fitting the expected molecular signal to the minimums of the LIDAR data. The minimums are assumed to be atmosphere regions free of aerosols.

Equation (5) can be rearranged to give the aerosol scattering function as:

$$f_A(R) = \left[R_S - 1\right] \sigma_M N_M(R). \quad (7)$$

Aerosol scattering function is proportional to aerosol concentration if the aerosol scattering characteristics such as size, distribution, and composition remain essentially constant with altitude. An example of a comparison of aerosol scattering function with an independent balloon-borne in-situ particle counter is shown in Figure 4.[2] The good agreement of the results of the two techniques shown in Figure 4 indicate the usefulness of LIDAR in measuring aerosol concentration in the stratosphere. Comparisons of this type can also be used to determine models of aerosol size, distribution, and index of refraction of stratospheric aerosols.

Raman nitrogen scattering can also be used to obtain a remote measurement of visibility as described in the next section.

IV. LIDAR MEASUREMENTS OF REMOTE VISIBILITY

The measurement of visibility is important because reduced visibility is one of the most visible indications of pollution and is an inherent impediment to navigation. At present, the standard techniques include visual observation and ground-level, double-ended transmissometers. Visual observations are limited to the availability of objects on the horizon and transmissometers to their location at ground level. LIDAR provides a single-ended measurement of visibility in all directions both at ground level and aloft. LIDAR measurements for visibility at λ_o provide good range capability but are difficult to interpret due to fluctuations in aerosol scattering. Even though observations of Raman nitrogen scattering are typically range limited, the interpretation of the results are straightforward.

Equation (3) indicates that the Raman nitrogen return is proportional to nitrogen concentration and two-way transmissivity, $q_{\lambda_o}\ q_{\lambda_R}$. Since the nitrogen concentration in the lower atmosphere is well known, the Raman nitrogen return can be used to measure transmissivity. Again, a standard method of interpretation of data is needed. Normalization of the signal at R to the signal at R_o provides for a measure of the two-way transmissivity for ranges between R_o and R as [from Equation (3)]:

$$\frac{V\left(\lambda_{N_2},R\right)}{V\left(\lambda_{N_2},R_o\right)} = \frac{N_{N_2}\left(R\right)}{N_{N_2}\left(R_o\right)}$$

$$\times \quad q\left(\lambda_o,R_o \text{ to } R\right)\ q\left(\lambda_{N_2},R_o \text{ to } R\right), \quad (8)$$

or rearranging:

$$\overline{q}\ (R_o \text{ to } R) = \frac{N_{N_2}\left(R_o\right)}{N_{N_2}\left(R\right)}\left[V\left(\lambda_{N_2},R\right) \Big/ V\left(\lambda_{N_2},R_o\right)\right]^{1/2}, \quad (9)$$

where $\overline{q}\ (R_o$ to R) is defined as the effective transmissivity.

Since $q = \exp -\left[\int_{R_1}^{R_2} \beta(R')\ dR'\right]$,

an approximate extinction coefficient, β, over the range increment, ΔR, can be obtained from Equation (9) as:

$$\overline{\beta}\left(\overline{\lambda},\ R + \frac{\Delta R}{2}\right) \quad (10)$$

$$= \frac{\log \overline{q}\left(R_o \text{ to } R + \Delta R\right) - \log \overline{q}\left(R_o \text{ to } R\right)}{\Delta R}.$$

In regions of low visibility, where attenuation is independent of wavelength, the following equation relates visibility or visual range with extinction coefficient.[3]

$$\text{Visual range} = \frac{3.912}{\text{extinction coefficient}} \quad (11)$$

An example of Raman nitrogen data taken at the zenith, normalized to the molecular profile, and the resultant visibility is shown in Figures 6 and 7, respectively. This technique has applications to regions of variable visibility where the measurement is critical, such as airport glide slopes.

In addition to remote visibility, Raman scattering can be used to monitor molecular concentrations. This application is described in the next section.

V. LIDAR MEASUREMENTS OF MOLECULAR CONCENTRATIONS

Active optical sensing with LIDAR of molecular concentrations provide remote measurements over a wide area not possible with in-situ samples. Figure 2 indicates that most atmospheric molecules have Raman lines specific to that particular molecule. Interference filters centered on particular Raman lines provide a remote range resolved measurement of the molecular signal from the appropriate molecule.

An example of the Raman spectrum from an oil-fired smoke plume located at a range of 20 m from a LIDAR system is shown in Figure 7.[4] This figure shows that the Raman lines can be separated experimentally. Field observations have been performed remotely, monitoring SO_2 in the plume of a generating plant. Signal strength (photon counts) is shown in Figure 6 as a function of range.[5] The plume was located $\simeq$ 210 m from the LIDAR, which corresponds to the high signal of channel 5. Figure 7 is a comparison of Raman SO_2 signal and power plant output as a function of time. The above

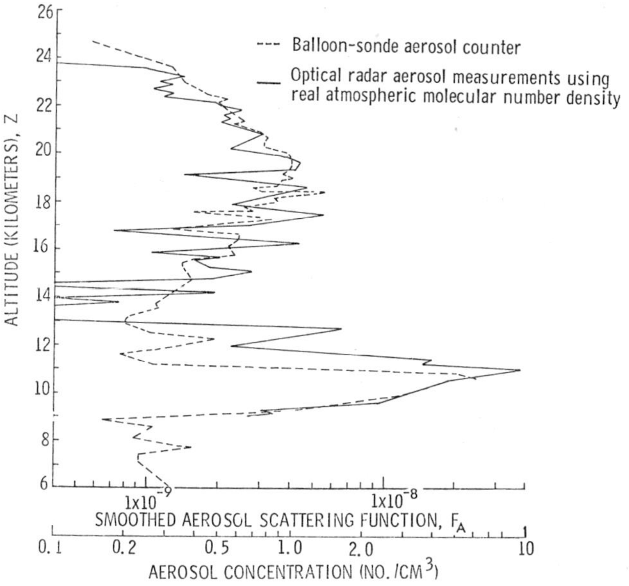

FIGURE 4. COMPARISON OF NORMALIZED LIDAR DATA WITH BALLOON-BORNE AEROSOL COUNTER (LARAMIE COMPARATIVE EXPERIMENT, DOT CONTRACTOR REPORT, MARCH 15, 1973).

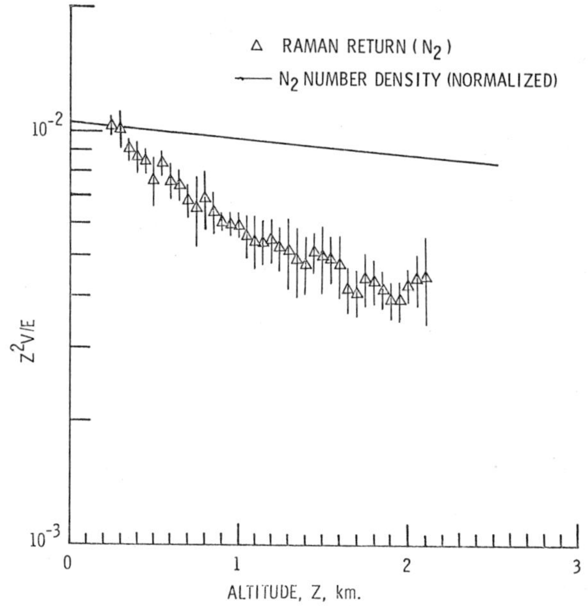

FIGURE 5. COMPARISON OF RAMAN LIDAR NITROGEN SIGNAL WITH NITROGEN MOLECULAR PROFILE.

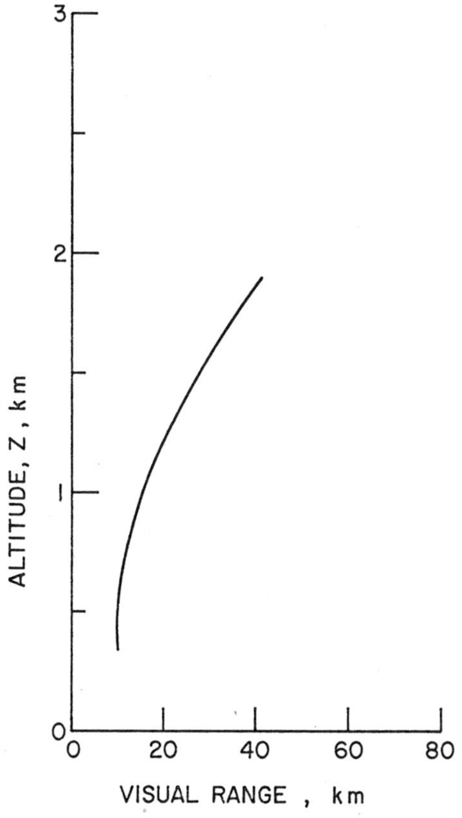

FIGURE 6. VISUAL RANGE AS A FUNCTION OF ALTITUDE, DERIVED FROM RAMAN NITROGEN DATA SHOWN IN FIGURE 5.

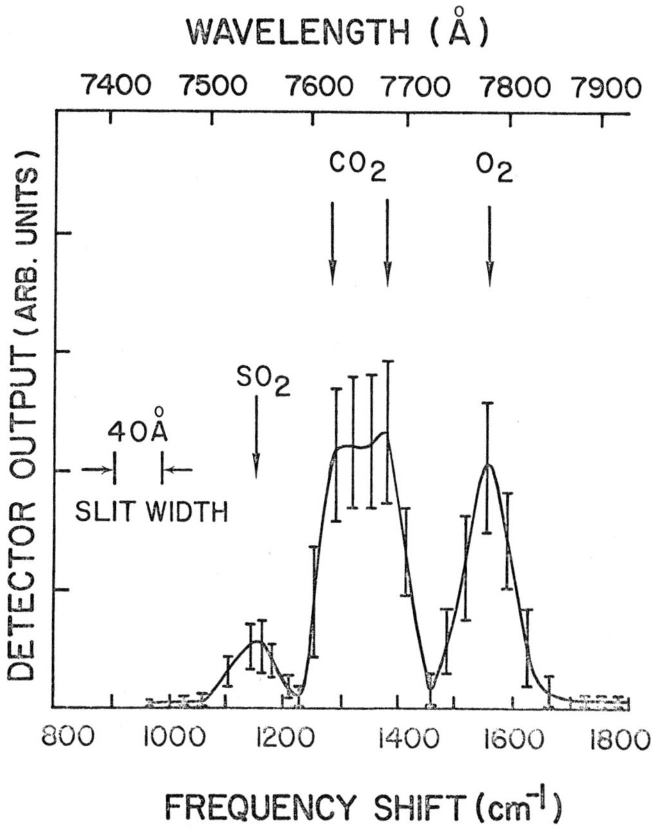

FIGURE 7. RELATIVE DETECTOR OUTPUT OF RAMAN RETURNS AS A FUNCTION OF FREQUENCY WITH OIL PLUME POLLUTING THE ATMOSPHERE (KOBAYASI AND INABA, APPL. PHYS. LETT. 17, p. 139, 1970).

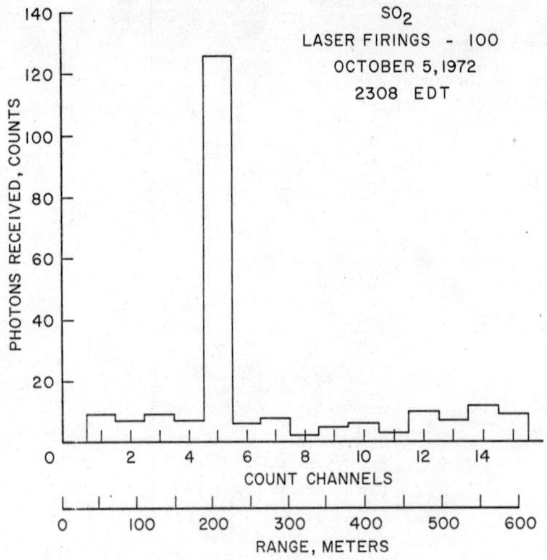

FIGURE 8. THE NUMBER OF PHOTON COUNTS RECEIVED AT 7546Å (V_1 LINES OF SO_2) FOR 100 LASER FIRINGS (MELFI, ET AL., APPL. PHYS. LETT., APRIL 15, 1973).

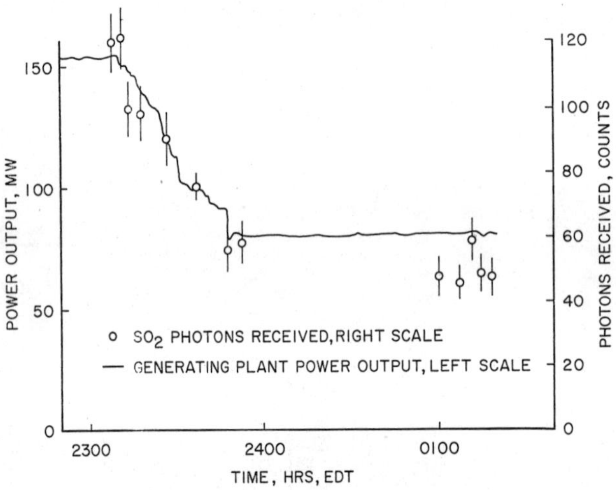

FIGURE 9. THE COMPARISON OF GENERATING PLANT POWER OUTPUT WITH THE OBSERVED SO RAMAN SCATTERING FROM THE PLUME (MELFI, ET AL., APPL. PHYS. LETT., APRIL 15, 1973).

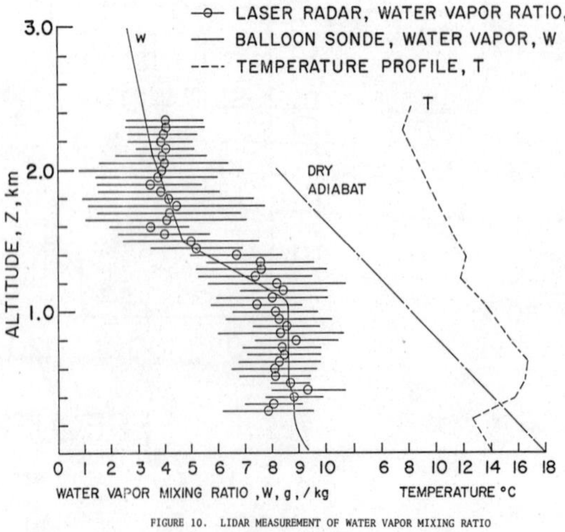

FIGURE 10. LIDAR MEASUREMENT OF WATER VAPOR MIXING RATIO COMPARED WITH STANDARD BALLOON-SONDE DATA (MELFI, APPL.OPTICS II, 1605, 1972).

measurements are qualitative. A standard method of quantifying Raman measurements of molecular concentration is needed.

Equation (3) indicates that the Raman signal is proportional to various system parameters, molecular concentration, and the transmissivity terms. Again a standard data analysis is indicated. As was done in previous section, relating the Raman molecular signal to the Raman nitrogen signal provides a measure of, in this case, molecular mixing ratio as:

$$\frac{V(\lambda',R)}{V\left(\lambda_{N_2},R\right)} = \frac{\gamma(\lambda')\ S(\lambda')\ q(\lambda',R)}{\gamma\left(\lambda_{N_2}\right) S\left(\lambda_{N_2}\right) q\left(\lambda_{N_2},R\right)} \frac{f(\lambda',R)}{f\left(\lambda_{N_2},R\right)} . \tag{12}$$

Again, most system parameters cancel in the ratio, and those that do not need only to be known relatively, the transmissivity at the two wavelengths are approximately equal; therefore, Equation (12) indicates that:

$$\frac{V(\lambda',R)}{V\left(\lambda_{N_2},R\right)} = k'' \frac{N_{molecule}}{N_{N_2}} . \tag{13}$$

An example of LIDAR data treated in this manner for the remote measurement of water vapor mixing ratio is shown in Figure 10.[1] The LIDAR measurement is compared to an independent measurement of water vapor mixing ratio from a standard balloon-sonde. The good agreement suggests that Raman LIDAR will be very useful in remote monitoring of molecular concentrations in the atmosphere. The most serious limitation of the technique is the small cross-section for Raman scattering. Therefore, it can only presently be used at short ranges, typically on the order of kilometers.

VI. SUMMARY

In this paper an attempt has been made to indicate the versatility and usefulness of LIDAR as a technique in remotely monitoring the atmosphere. Standard methods of data analysis have been proposed for aerosol, visibility, and molecular concentration measurements. These standard techniques rely heavily on Raman nitrogen scattering at short ranges and modeling of the expected molecular LIDAR return at greater ranges. It should be repeated that standard methods of data treatment are necessary both for intercomparison of data between LIDAR research groups and to ease the effort of various user groups utilizing the results of LIDAR observations.

LIDAR is a very active research area, and further advances in its application can be expected because of advances in laser and photo-detector technology. The advanced techniques expected in the future include differential absorption and fluorescence measurements of molecular concentrations. As these applications of LIDAR progress, monitoring of ambient pollutant concentrations may become practical.

REFERENCES

1. Melfi, S. H.: Applied Optics II, 1605, (1972).

2. "Laramie Comparative Experiment." DOT Contractor Report, edited by Dynatrend, Inc., March 15, 1973.

3. Fleagle, R. C. and Businger, J.A.: An Introduction to Atmospheric Physics, Academic Press, New York, N.Y., 1963.

4. Kobayasi, T. and Iuaba, H.: Applied Physics Letters, 17, p. 139, (1970).

5. Melfi, S. H.; Brumfield, M. L.; and Storey, R. W., Jr.: Applied Physics Letters, April 15, 1973.

©1973, ISA JSP 6668

VISIBILITY SENSORS IN YOUR AIR QUALITY PROGRAM

David H. George
National Weather Service, NOAA, Sterling, Va.

Karl F. Zeller
Air Pollution Training Institute, EPA, Raleigh, N.C.

ABSTRACT

Sensor technology plays a critical role in observations of secondary meteorological air quality parameters. Observations of secondary parameters - solar radiation, atmospheric moisture and pressure, precipitation, and visibility — serve very important technical needs.

The general public, however, relies upon but one parameter for its judgment about the nation's atmospheric health. That single, most aesthetic air quality parameter, is visibility. The opacity of the atmosphere is taken as an indicator of the quantity of suspended particulate matter and/or photochemical oxidants in our air. As such, the ability to see or not see distant objects is everyman's measure of air quality.

This paper presents a status report of visibility sensor technology and application. Sensor development in-progress is outlined and expected results are presented. The paper begins with a discussion of three dominant technical definitions of atmospheric visibility and their relationship to environmental aesthetics. Atmospheric sensing program applications are referenced for their guidance upon visibility sensor technology. Visibility sensing techniques are studied and basic theory, assumptions, strong points and weak points of each are examined. Off-the-shelf visibility sensors and their characteristics are also presented along with some useful program application for each sensor.

The future of visibility sensing is discussed. Most promising for atmospheric applications are infrared and laser backscatter signal analysis. The development of more versatile instruments indicates that one sensor may eventually supply simultaneous information for several program applications. Instrument-sensed visibility may also develop as an index of public air quality awareness.

INTRODUCTION

Our nation's concern over the quality of the air we breathe has caused a flurry in the design and implementation of air pollution measurement programs. As pointed out by McCormick[10], the many factors involved in program implementation cause a marked diversity in measurement practices from agency to agency. It is sometimes hard, if not impossible, to keep up with developments in sampling and analysis instrumentation and techniques used in these programs. We have noticed the deficiency of readily available information on visibility and visibility instrumentation in general. This is particularly true in the air pollution field. A quick inspection of major meteorological and air quality instrument catalogs will bear this out.

Meteorological parameters can be divided into two categories for air pollution applications — primary and secondary. This division is related to the dispersive character of the atmosphere. The primary parameters, wind velocity and atmospheric stability, are directly related to dispersion while secondary parameters are not.

Observations of secondary meteorological parameters — solar radiation, atmospheric moisture and pressure, precipitation, and visibility — serve three important air quality needs:

a) Describing regional or local environmental conditions. Precise description requires that new and unique information be derived from individual parameters or from related combinations of parameters.

   For instance, in order to describe the air over an entire metropolitan region, a large number of monitoring stations are required to produce an adequate temporal and spacial distribution; the more refined the requirements, the larger the number of monitoring stations needed and the greater the cost. Instead, the sensing of secondary meteorological parameters might be used to replace several point monitoring stations once the secondary data is correlated with air quality data.

b) Anticipating conditions favorable to the formation of certain pollutants. A knowledge of initial and evolving parameter conditions is essential to the application of forecasting techniques.

   As an example, visibility data may be desirable in order to correlate measured air quality

Superior numbers refer to similarly-numbered references at the end of this paper with visibility reduction when photochemical reactions are apparent.

c) Conducting experimental studies to evaluate down-wind dispersion of pollutants in areas of complicated terrain. Many popular dispersion models are based on a gaussian dispersion model described by Gifford[6] and by Turner[16]. Stability curves derived from experiments under ideal flat terrain conditions are used for many topographic conditions. These models must then be "calibrated" in order to fit measured data. Perhaps here visibility measurements could be used for study of dispersion parameters both in the vertical and horizontal.

From the layman's standpoint, visibility is probably the most important parameter for evaluating air quality. Whether he realizes it or not, the layman is sensing the opacity of the atmosphere when he looks out his window to see the Washington Monument, Mount Olympia, Pike's Peak, the Statue of Liberty or whatever the dominant local landmark might be. Visibility as a measure of atmospheric opacity is perceived to be an indicator of the quantity of pollution in the air.

As an aesthetic parameter, visibility is the public's tool for measuring the success of any atmospheric get-well program. And as it has been demonstrated by Medalia[11] [12], for example, citizen awareness is not related to community production of pollution. Increased awareness also seems to produce a lowered tolerance to pollution.

Besides the layman's unhappiness over visibility reduction, two other important aspects must be kept in mind: the handicap to commercial and private aircraft and in some cases, automobiles, and the reduction in transmission of solar energy to the ground. The latter effect which tends to maintain a stable atmosphere longer than normal is characteristic of most large scale smog episodes. There is also controversy over the impact of visibility reducing particles on the global energy balance. The widespread transport of visibility reducing pollution is well documented, most recently and clearly by Hall, et. al.[7]. With these considerations in mind, we will present some basics of visibility measurement and introduce several state-of-the-art instruments which might be included in a comprehensive air quality program.

VISIBILITY DEFINITIONS

There exist three predominant definitions of visibility. Each has a specific, intended purpose.

Prevailing Visibility

Defined as the greatest visibility equalled or exceeded throughout half of the horizon circle [14], prevailing visibility is the measurement made by the human meteorological observer. It is a subjective measurement which depends upon individual ability to both detect and recognize distant objects seen against the horizon sky by day or to detect distant lights at night. It involves all of the subjective characteristics and considerations which make the human observer an excellent sensor of aesthetics, but a terrible sensor of quantitative scientific phenomenon.

Prevailing visibility is taken routinely at U.S. National Weather Service Observation stations and military installations that have observers on duty. Prevailing visibility observations made under smoke and haze conditions are sometimes correlated with wind direction measurements to produce a "pollution rose." This pollution rose is then used to locate possible pollution sources.

Visual Range

The visual range is that distance at which the target versus background contrast becomes just great enough for a detector to sense the presence of the target. Visual range is the antithesis of prevailing visibility — it is completely objective. It is a purely physical relationship which assumes a fixed level of detector performance in defining visual range V, as

$$V = \frac{1}{\sigma} \ln \frac{|c|}{\varepsilon} \qquad (1)$$

where σ is the measured atmospheric light extinction coefficient, $|c|$ is the measured relative contrast between object and background and ε is the measured contrast threshold of the detector. Equation (1) is the source of integrity from which most sensors seek to establish their theoretical bases. Many currently available visibility instruments use a visibility versus instrument output (in terms of σ) based upon equation (1). Visual range is not at all concerned with perceived aesthetics. Frequently equation (1) is taken as

$$V = \frac{2.9}{\sigma} \qquad (2)$$

if the target object is to be both detected and identified or as

$$V = \frac{3.9}{\sigma} \qquad (3)$$

if the target need only be detected.

Sensor Equivalent Visibility

Proposed by George and Lefkowitz[5], this definition seeks to combine the subjective measure of prevailing visibility with the objective measure of visual range. Sensor equivalent visibility is defined simply as any equivalent of human visibility which is derived from instrument sensed measurements. This involves both temporal and spatial integration of instrument measurements. In a sense, sensor equivalent visibility is equation (1) aestheticised by data processing techniques. A unique sensor equivalent visibility exists to meet the requirements of each program application. For example, data from three horizontally separated visibility instruments might be collected over a time interval, weighted to favor the newest data, and processed to arrive at an instrumental visibility which would approximate a human prevailing visibility evaluation. The time interval, instrument separation and data weighting would be chosen to match human performance over the visibility values of interest.

PROGRAM APPLICATIONS

Four distinct visibility sensor applications can be identified within air quality programs. Not every program will contain all four applications and many contain only two or three. In programs where research is conducted along with control, it is customary to have separate staffs perform these functions.

The goal of research is to develop understandings and to determine relationships among meteorological variables. Very short sampling intervals are required. Instrument error minimization is of prime importance. Redundancy is an essential element in research applications.

Climatology establishes baseline data and datum points from which environmental change may be evaluated. Occasionally, the importance of climatology is underestimated. We suspect that such underestimates demonstrate a lack of appreciation for project goals and occur frequently in programs where goals have been defined before atmospheric conditions have been adequately cataloged. Long term trends in visibility are useful for evaluating our efforts to improve air quality only if we know the state of pre-existent conditions. In climatological applications, small short term fluctuations are de-emphasized in favor of representative, longer term variations.

Operating programs seek to apply the knowledge gained from research results and from climatology to air quality management procedures and to anticipate or forecast air pollution conditions. Short term fluctuations play an important role in operation applications and for this reason, instrument response is an important consideration as is instrument siting for representative data measurements.

The purpose of control is to record and document environmental conditions to insure adequate compliance and enforcement of regulations. Scientific traceability of instrument theory is very important as is the time processing of output data. Repeatability and reliability are most important instrument considerations here.

VISIBILITY SENSING TECHNIQUES

Light Extinction

Light extinction is the most theoretically direct technique. Light extinction is the reduction of apparent light intensity after a known initial intensity is transmitted through an atmospheric volume (figure 1). It is described by

$$x = 1 - \varepsilon^{b/V} \tag{4}$$

where x = extinction, V = visibility, b = distance between light source and light receiver, and ε = an assumed detector threshold. Assumptions are made that single scattering by particles causes loss of transmittance and that energy absorption and multiple scattering are negligible. Extinction techniques require double-ended instrumentation with relatively large distances between the projector and detector. Hence, they are rather expensive. Extinction devices do, however, provide the most stable, accurate and precise of instrumental visibility measurements. Extinction is best used in research applications where theoretical traceability and accuracy are of prime concern. These techniques are also very useful in climatological or bench mark applications where long-term stability is required.

Backscattered Light

Backscatter is related empirically to light extinction. The relationship has been developed and tested by several researchers(3) (4) (8) (17). The amount of backscattered light S is related to visibility by

$$V = \frac{\varepsilon}{S^y} \tag{5}$$

where y is near about 1.5.

Backscattered light measurements may be related to extinction measurements by

$$R = \left[\frac{1}{V} t^{V/\sigma}\right]^{2/3} \tag{6}$$

Basically, the variable sensed is the apparent intensity of light scattered backward after an initial intensity is transmitted into a volume of atmosphere (figure 2). As with extinction theory, negligible absorption is assumed. Backscattered light is highly dependent upon wavelength and particle size. Broad spectrum light sources are therefore necessary for accurate visibility measurement, while narrow spectrum sources yield particle volume information. Since both source and receiver are collocated, only one installation point is required. The effective sampling volume of backscatter instruments is small relative to extinction devices since most of the backscattered light emanates from very near the light source. Since backscatter instruments may be made relatively small and hence reasonably portable, they are useful when operations and control programs require some instrument mobility.

Forward Scattered Light

Forward scatter may also be related empirically to light extinction. A detector senses the apparent intensity of light scattered in the forward direction when a known intensity is transmitted through

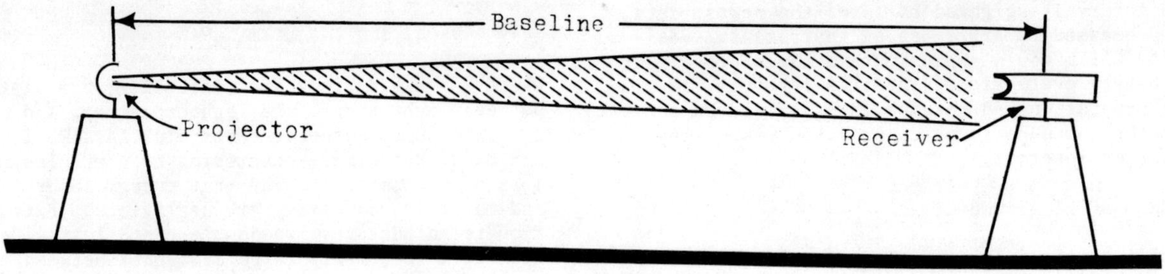

FIGURE 1. The light extinction technique requires considerable real estate and very solid projector/receiver mounting pads.

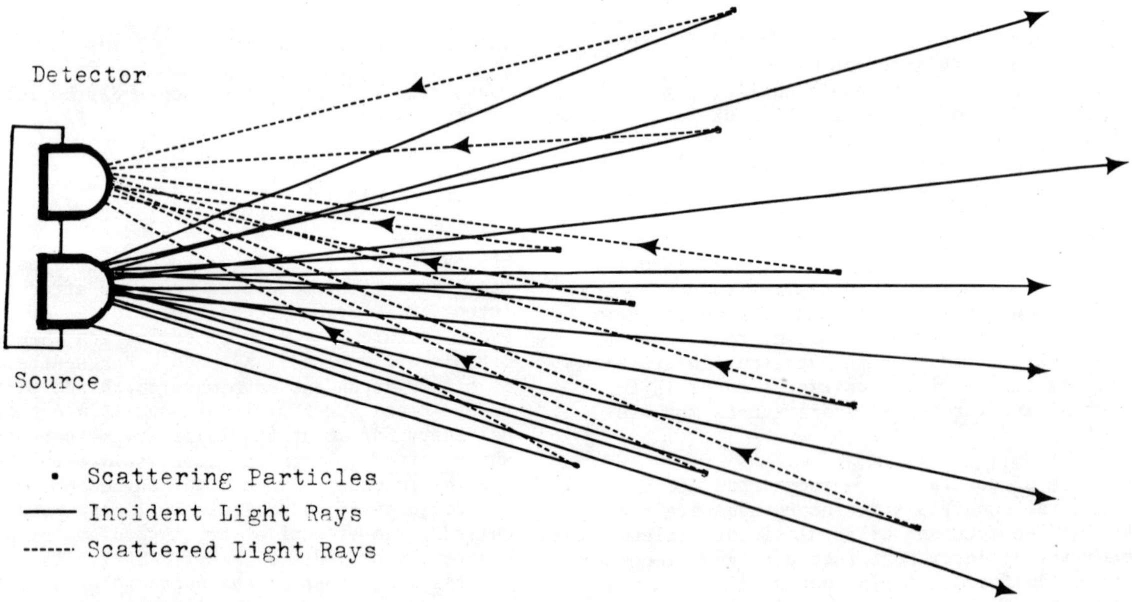

FIGURE 2. Schematic of backscatter sampling volume. About 90% of the backscattering occurs in the first 30% of the volume.

an atmospheric volume (figure 3). The intensity of forward scattered light is many times that of backscatter until visibility becomes very low. The general form of visibility derivation from forward scattered energy is given by

$$V = \frac{\varepsilon}{\sigma} c(f) \qquad (7)$$

where c(f) is an instrument related constant.

Total Scattering of Light

Total scattered light is defined as the apparent intensity of light scattered in all directions when a known intensity is transmitted through an atmospheric volume. Total light scattering is empirically related to light extinction by equation (1). The assumption is made that errors of back and forward scattering cancel. The generalized form of visibility determined from total scattered energy produces the relationship identical to that for forward scattered light,

$$V = \frac{\varepsilon}{\sigma} c(t) \qquad (8)$$

where c(t) is an instrument related constant.

Considerable work has been done by Charlson, et. al.[1] and by Noll, et. al.[15] to relate aerosol mass concentration and prevailing visibility to total scattering coefficient. Total scattering coefficient is best related to prevailing visibility when relative humidity is less than about 70% [13] and when the scattering particles are not large[9].

The relationship between backscatter and total scatter is seen in figure 4. Note that each scattering curve has its own scale, shown in relative units.

OFF-THE-SHELF INSTRUMENTS

The following compilation of currently available off-the-shelf visibility instruments is designed to present a picture of state-of-art technique applications discussed earlier.

Extinction Instruments

These instruments, frequently referred to as "transmissometers", are essentially photometers having a fairly long separation between light source and detector. A transmissometer is relatively expensive to purchase. Installation is also expensive because the double-ended nature of the instrument requires considerable signal and power cabling as well as two very substantial mounting pads. Transmissometers are, however, extremely stable and durable instruments whose measurements are based upon sound, proven theory which has been verified by numerous empirical studies. Transmissometers are best suited to research applications where a traceable, accurate reference standard is needed and to climatology where long term stability is desired. Three of the more modern transmissometers are the Skopograph (figure 5), manufactured under license from the Impulsphysics Association, and the German-manufactured SM-4 and Eltro units. Each of these sensors employs solid state electronics and long-life components.

Backscatter Instruments

By far the most widely used backscatter sensor is the Videograph, which like the Skopograph, is a product of Impulsphysics (figure 6). The sensor has been used and improved upon for several years by government agencies in Europe and North America.

The Videograph is particularly useful for climatological applications where long-term, unattended and low-maintenance operation is required. It is also useful for control applications where a traceable, sound theory is desired, or when the instrument may be moved to a new site every few weeks and portability is an important consideration.

Forward Scatter Instruments

The two most common forward scatter instruments available are the Fumosens, by Impulsphysics, and the EG&G Model 207 Forward Scatter Meter. The Fumosens is quite small, inexpensive and is highly portable. It is best adapted to control applications where the goal is to alert personnel that atmospheric contamination levels have been exceeded. That is, it is best used as an alarm device.

The EG&G 207 is rather large and bulky (figure 7). It has been used in mesoscale observing networks and is relatively proven in the field. The EG&G 207 is best suited to operational programs where visibility is one of several inputs used to evaluate current conditions and to estimate those a few hours hence.

Total Scatter Instruments

Meteorology Research, Inc., manufactures the two most commonly used total scatter instruments, the Integrating Nephelometer and the Fog Visiometer. Both are relatively low cost, highly portable and durable instruments. The Fog Visiometer (figure 8), being the newer of the two, has a shorter field use history, but like the Nephelometer, appears to have proven itself. The Integrating Nephelometer (figure 9) is best suited to research applications where its fine scale measurements can be of most benefit while the Fog Visiometer is best used in operational applications.

Table 1 lists several available off-the-shelf visibility instruments which are currently used in air quality programs. We have not attempted to catalog all available instruments, however we do feel that the ones listed are certainly representative of what is available in the market place.

FUTURE INSTRUMENT CAPABILITIES

Great amounts of activity are being devoted toward extension of visibility sensing capabilities. Several manufacturers and researchers are working

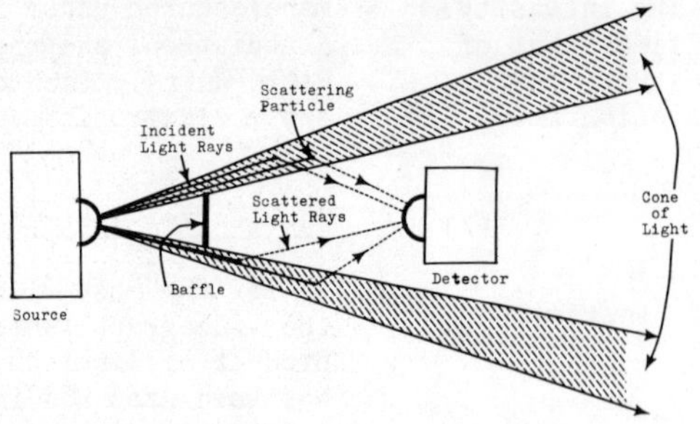

FIGURE 3. Forward scatter schematic showing a typical light path from source to scattering particle to detector.

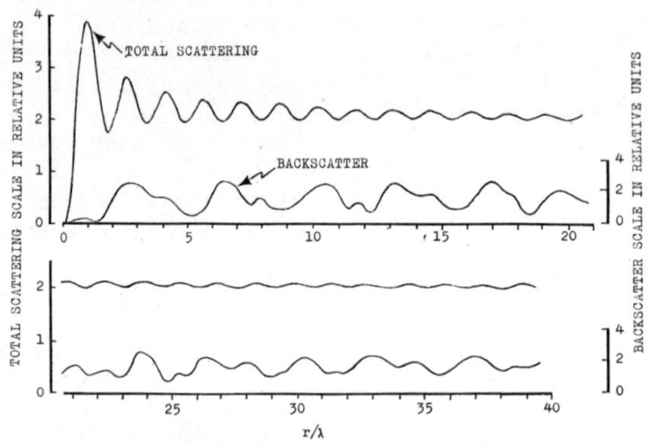

FIGURE 4. The relationship of total scattering and backscatter to particle radius r versus incident light wavelength λ. Single spherical water particles are assumed. From (2).

Figure 5. The Impulsphysics Skopograph transmissometer at the National Weather Service Test & Evaluation Laboratory. The projector is on the left, detector on the right.

Figure 6. The Impulsphysics Videograph backscatter instrument at Fairbanks, Alaska.

Figure 7. The EG&G 207 Forward Scatter Meter.

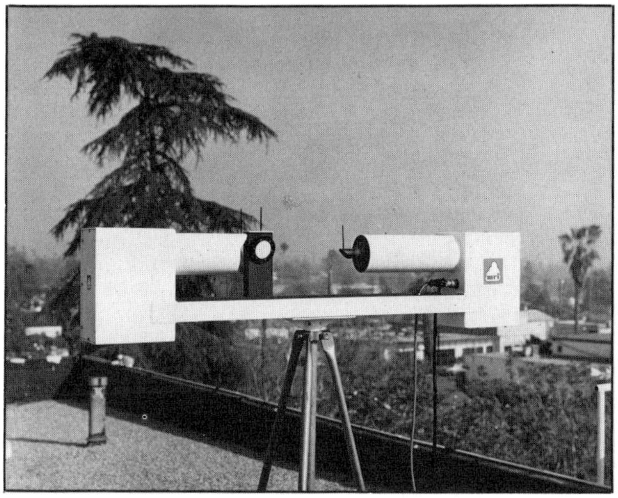

Figure 8. The MRI Fog Visiometer 1580.

Figure 9. The MRI Integrating Nephelometer.

to relate contemporary instrumental outputs to pollutant levels. Results of these efforts will continue to be presented in operationally oriented trade and technical journals.

Other work is under way to investigate backscattered energy using laser sources. Both visible and non-visible light is being used to research the relationship of scattering to particle size, shape, mass, velocity and composition. Multiple wavelengths are used to differentiate between several types of particles and gasses. These techniques involve the selection of two or more wavelengths each having different absorbing and scattering coefficients so that their interaction with atmospheric particles produces different backscatter characteristics. The differentially backscattered energy is then related to that particular contaminant. Similar techniques are being tried using infrared energy sensing.

In addition to identifying contaminants, laser ranging will be able to locate areas of smoke, dust and water vapor which are prime sources of public awareness. In the future, not only will visibility as sensed by the public be better observed instrumentally over wide areas, but the effects of individual atmospheric constituents which reduce visibility and hence environmental aesthetics will be isolated and quantified, thus providing for improved environmental control. Visibility measuring instruments will become more common to air quality programs as we become more sensitive to public perception of visibility as an indication of air quality.

REFERENCES

(1) Charlson, R. J., N. C. Ahlquist and Helmut Horvath, "On The Generality of Correlation of Atmospheric Aerosol Mass Concentration and Light Scatter," ATM. ENV., Vol 2 (1968), 455-464.

(2) Collis, R. T. H., "Lidar," ADVANCES IN GEOPHYSICS, Academic Press, 1969.

(3) Curcio, J. A., and G. L. Knestrick, "Correlation of Atmospheric Transmission With Backscattering," J. OPT. SOC. AMER., Vol 48 (1958), 686-689.

(4) Fenn, Robert, "Correlation Between Atmospheric Backscattering and Meteorological Visual Range," APPL. OPTICS, Vol 5 (1966), 293-295.

(5) George, David H., and Matthew Lefkowitz, "A New Concept: Sensor Equivalent Visibility," INTERNATIONAL CONFERENCE ON AEROSPACE AND AERONAUTICAL METEOROLOGY, American Meteorological Society, Preprint Vol (1972), 243-250.

(6) Gifford, F. A., Jr., "Uses of Routine Meteorological Observations for Estimating Atmospheric Dispersion," NUCLEAR SAFETY, Vol 2 (1961), 45-51.

(7) Hall, F. P., C. E. Duchon, L. G. Lee and R. R. Hagen, "Long-Range Transport of Air Pollution: A Case Study, August 1970," MON. WEA. REV., Vol 101 (1973), 404-411.

(8) Hochreiter, F. H., "Analysis of a Backscatter Visibility Measuring Technique," NOAA-NATIONAL WEATHER SERVICE, TEST & EVALUATION LABORATORY REPORT NO. 3-71, August 1971, 32 pp.

(9) Horvath, Helmut, and R. J. Charlson, "The Direct Optical Measurement of Atmospheric Air Pollution," AM. INDUS. HYG. ASSOC. J., Vol 30 (1969), 500-509.

(10) McCormick, R. A., "Air Pollution Measurements," SECOND SYMPOSIUM ON METEOROLOGICAL OBSERVATIONS AND INSTRUMENTATION, Preprint Vol (1972), American Meteorological Society, 58-65.

(11) Medalia, Nahum Z., "Air Pollution as a Socio-Environmental Health Problem: A Survey Report," J. OF HEALTH AND HUMAN BEHAVIOR, Vol 5 (1964), 154-165.

(12) ______ , "Community Perception of Air Quality: An Opinion Survey in Clarkston, Washington," HEW REPORT, June 1965, 104 pp.

(13) Nelson, W. G., R. W. Boubel and W. P. Lowry, "A Comparison of Several Techniques for Estimation of Aerosol Loading and Visibility," SECOND INTERNATIONAL CLEAN AIR CONGRESS OF THE INTERNATIONAL UNION OF THE AIR POLLUTION PREVENTION ASSOCIATION, 406-410.

(14) NOAA-National Weather Service, "Federal Meteorological Handbook No. 1," Government Printing Office, Washington, D. C.

(15) Noll, K. E., P. K. Mueller and Miles Imada, "Visibility and Aerosol Concentration in Urban Air," ATM. ENV., Vol 2 (1968), 465-475.

(16) Turner, D. B., "Workbook of Atmospheric Dispersion Estimates," U.S. Public Health Service Publication No. 999-AP-26, 1967.

(17) Vogt, H. , "Visibility Measurement Using Backscattered Light," J. ATMOS. SCI., Vol 25 (1968) 912-918.

TABLE 1. Some off-the-shelf visibility instruments

INSTRUMENT	TECHNIQUE	MANUFACTURER OR REPRESENTATIVE	MEASURING PATH	STRONG POINTS	WEAK POINTS
Skopograph	Extinction	Impulsphysics, Inc. 642 Coral Drive Cape Coral, Fla. 33904	Up to 1000 Ft.	Accuracy, Durability	Critical alignment of optical units
SM-4	Extinction	Lear-Siegler, Inc. Aeorspace Elect. Grp. 4141 Eastern Ave., S.E. Grand Rapids, Mich. 49508	Up to 125 Ft.	Low cost for extinction	Critical alignment of optical units
Eltro	Extinction	ELTRO GmbH 6900 Heidelberg 1 Postfach 520 F. R. Germany	Up to 500 Ft.	Objective internal calibration	Critical alignment of optical units
Videograph	Backscatter	Impulsphysics, Inc. 642 Coral Drive Cape Coral, Fla. 33904	About 35 Ft.	Stability, Durability	Subjective calibration
EG&G 207	Forward Scatter	EG&G 151 Bear Hill Road Waltham, Mass. 02154	4 Ft.	Easy to install	Expensive
Fumosens	Forward Scatter	Impulsphysics, Inc. 642 Coral Drive Cape Coral, Fla. 33904	2 Ft.	Easy to install, Inexpensive	Long time constant
Integrating Nephelometer	Total Scatter	Meteorology Research 464 W. Woodbury Road Altadena, Calif. 91001	Few inches	Output related to aerosol mass	Moving parts, Small sample
Fog Visiometer	Total Scatter	Meteorology Research 464 W. Woodbury Road Altadena, Calif. 91001	Few inches	Easy to install, Inexpensive	Small sample

© 1973, ISA JSP 6669

A STANDARD METHOD FOR EXPRESSING INSTRUMENTAL PERFORMANCE

Robert L. Chapman
Principal Application Engineer
Beckman Instruments, Inc.
Fullerton, California

ABSTRACT

Since pollution is a widespread problem, analytical methods must be standardized in order to obtain comparable data over broad geographic areas. Toward this end, the International Electrotechnical Commission (IEC) is developing a series of recommendations for the expression of functional performance of various types of environmental quality instruments. This paper presents the basic plan and outline of the prototype document, which is currently under discussion in IEC.

INTRODUCTION

An urgent need for quantitative data regarding existing air quality and source emission levels developed with the recognition of environmental pollution as a widespread social problem, in the late 1950's. At the time, instrumental techniques for such determinations were not sufficiently developed to serve as standard reference methods; therefore, classical manual analytical methods were adopted. As instrumental development progressed, it was decided that the automatic analyzers should produce data identical to that obtained by the reference methods, so that all data would be comparable, regardless of where obtained, by whom, or by what method. Thus, the concept of "equivalency" was born.

The rules and regulations, promulgated by EPA and published in the Federal Register, state the limiting values for air quality and source emissions, based on the reference analytical techniques, "or equivalent." Attempts have been made by EPA to establish protocols whereby candidate analytical methods can be tested and certified as being equivalent to the reference methods.[1-3] To the time of this writing, however, none have been officially adopted. The lack of such documents turns out to be a road block for expeditious application of existing instrumental methods, as well as a deterent to the development of new methods.

In theory, if standard methods for testing instrument performance and expressing it in universally accepted terms were available, the governing agencies in any country could set up their requirements on such a basis and thus obtain comparable results. Developing such standards has long been the successful business of well established standards organizations such as ASTM, SAMA, ANSI, IEC, and ISO. Thus, it was a natural action for ISO and IEC to start moving early in 1971 toward the establishment of standards for analytical methods for environmental quality.

In May of 1971, the coordinating committee of the International Organization for Standardization (ISO) resolved to concern itself with air and water pollution, and to look to the International Electrotechnical Commission (IEC) for recommendations covering electrical and electronic instruments used for environmental monitoring.

In June of 1971, IEC established a Working Group for air and water quality instrumentation, whose scope is: "To prepare drafts of Recommendations for electronic instruments used in the measurement of air and water quality." (IEC/TC66/WG6--U.S. Secretariat)

In April of 1972, ISO established a Technical Committee for air quality, whose scope is: "Standardization in the field of air quality, including definition of terms, sampling of air, measurement and reporting of air characteristics." (ISO/TC146--German Secretariat) Coincidentally, ISO established a second Technical Committee for Water Quality (ISO/TC147--U.S. Secretariat).

All three groups are now actively engaged in carrying out their various tasks.

IEC RECOMMENDATIONS ON INSTRUMENTATION

The IEC Working Group was organized through central office communication channels and, through the same channels, it was resolved that:

1) IEC should concern itself with methods of determining and expressing the functional performance of the instruments used for measuring air and water quality.

Superior numbers refer to similarly-numbered references at the end of this paper.

2) IEC should not concern itself with establishing minimum performance specifications for such instruments.

3) IEC should not concern itself with the selection of preferred methods for such measurements.

4) IEC should not concern itself with establishing standards for air quality or source emissions.

It was furthermore resolved that, as a first task, the Working Group should consider the nondispersive infrared analyzer (NDIR).

An existing IEC Publication (359), titled "Expression of the Functional Performance of Electronic Measuring Equipment," served as a guide for the format of the first environmental instrumentation document. Considerable modification was required to make it specifically applicable to the nondispersive infrared analyzer. Less modification will be required to adapt this first document to other types of environmental instruments and it is expected that subsequent recommendations will follow the final version of this document fairly closely.

EXPRESSION OF THE FUNCTIONAL PERFORMANCE OF NONDISPERSIVE INFRARED ANALYZERS USED FOR THE CONTINUOUS DETERMINATION OF AIR QUALITY

This first document was circulated to the full IEC parent committee in May of 1973, and it was placed on the agenda for discussion at the IEC/TC66 Plenary Meeting scheduled for the end of October, 1973.

The scope of the Recommendation includes the application of the NDIR carbon monoxide analyzer for ambient measurements as well as emission sources. It applies to analyzers specified for installation either indoors or outdoors. It applies to the complete analyzer, including regulated power supply, whether integral or separate. It does not apply to sample handling systems or other accessories such as recorders, A to D converters, or data acquisition systems.

The object of the Recommendation is to specify the terminology and definitions related to functional performance; to unify methods used in making and verifying statements on functional performance; and to specify what tests should be performed to determine functional performance.

The document includes definitions for the 20 most common terms related to functional performance and to conditions of operation, transport, and storage. It also stipulates the statements of functional performance characteristics, required to be made by the manufacturer or testing laboratory, which include:

1) Rated analytical range(s)

2) Output signal

3) Rated ranges for sample conditions

4) Operating error (as defined)

5) Linearity

6) Interference errors

7) Repeatability

8) Zero and span drift

9) Output noise level

10) Minimum detectable concentration

11) Delay time

12) Rise and fall times

13) Warm-up time

14) Operating period

15) Effect of barometric pressure

16) Rated ranges of use and limit ranges of operation storage and transport for all influence quantities.

Included in the document are recommended standard values and ranges for influence quantities, divided into three usage groups:

I. For indoor use with careful handling
II. For less protected areas and less careful handling
III. For outdoor use and rough handling

The manufacturer selects one usage group for the basis of his statements on the performance of his instrument, and tests are performed on this basis. These influence quantities include:

1) Ambient temperature

2) Relative humidity

3) Barometric pressure

4) Solar radiation

5) Wind velocity

6) Sand or dust contents of the air

7) Salt contents of the air

8) Contaminant gas or vapor contents of the air

9) Liquid water contents of the air

10) Operating position

11) Ventilation

12) Vibration

13) Mechanical shock

14) Sound pressure

15) Supply voltage, frequency, and distortion

The final portion of the recommendation describes the tests that must be performed, by the manufacturer or test laboratory, to determine or verify each statement of performance.

In the process of developing this document, several other appropriate documents were studied and, wherever possible, such existing material was incorporated into the new Recommendation.[1-11] Thus, the definitions of terms represent a melding, or consensus, of several existing documents. Likewise, the performance characteristics, tests, etc., are, in part, adopted from existing standards.

DISCUSSION

When completed, this series of recommendations should provide the basis for official standards on required performance for environmental monitors. By judicial selection of limits for the various performance characteristics, the governing agencies can achieve the desired equivalence between various candidate instruments as well as between instrumental methods and reference methods.

The above approach is directly applicable to ambient air monitors. There are additional considerations required with regard to emission monitors.

The most common type of emission monitoring system consists of two major parts; the analytical instrument itself and a sample interface system. Both are equally important in obtaining equivalency with the reference methods. It is proposed by this author that the best procedure for assuring equivalency for any complete system, of the extractive type (i.e. the type that extracts a sample from a flue, duct, stack, or process stream and then presents it to an automatic instrument for analysis) consists of two steps:

A. Instrument
Follow the procedure recommended above for ambient air monitors.

B. Sample interface system
Compare analyses, performed by the reference method, on the stack gas directly from the stack with analyses, by the same method, on the sample gas after passing through the interface system, as it is presented to the analyzer. If there is no intolerable change in the analysis, caused by the sample interface system, it is obviously acceptable.

By following this procedure it will be possible to specify proper analytical instruments and to verify their adequacy by test laboratory operations. Only the sample interface system need be checked in the field.

In-situ stack monitors must be checked for equivalency after complete installation in the field. This is done, of course, by comparing the results it produces with the results of analyses made by the standard reference method. It would be very difficult to apply to in-situ monitors, the simpler method recommended for extractive type analyzers.

CONCLUSIONS

In order to assure comparable data on air quality and source emissions over wide geographical areas, taken with different methods, it is necessary that all of the analytical techniques be equivalent. For instrumental methods, this requires that there be some standard procedure whereby all of the critical performance characteristics can be determined, stated, and confirmed. The International Electrotechnical Commission has assumed the task of preparing a series of recommendations for the expression of functional performance for these types of instruments in order to satisfy this need. The prototype document, on nondispersive infrared analyzers has been elaborated and is in the final stages of polishing. Similar documents for other instruments will follow.

REFERENCES

(1) "Proposed Performance Specifications for Stationary Source Monitoring Systems for Sulfur Dioxide and Nitrogen Oxides and Visible Opacity," First Draft, EPA, February, 1972.

(2) "Proposed Equivalency Guidelines," First Draft, EPA, May 24, 1972.

(3) "Performance Evaluation Procedures," EPA Contract Number 68-02-0214, Original Completion Schedule Mid-1972.

(4) "National Primary and Secondary Ambient Air Quality Standards," Federal Register, Volume 36, Number 84, Friday, April 30, 1971.

(5) "Standards of Performance for New Stationary Sources," Federal Register, Volume 36, Number 247, Thursday, December 23, 1971.

(6) "Control of Air Pollution," Federal Register, Volume 36, Number 128, July 2, 1971.

(7) "Instrumentation for Environmental Monitoring-Air," Lawrence Berkeley Laboratory, University of California, May 1, 1972.

(8) "Accuracy and Sensitivity Terminology as Applied to Industrial Instruments," SAMA Tentative Standard RC3-12-1955.

(9) "Process Measurement and Control

Terminology," SAMA Standard PMC 20-2-1970.

(10) "Expression of the Properties of Cathode-ray Oscilloscopes," IEC Publication 351.

(11) "Documentation to be Supplied With Electronic Measuring Apparatus," IEC Publication 278.

© 1973, ISA JSP 6670

COMPARATIVE EVALUATION OF IN-SITU WATER QUALITY SENSORS

Barbara S. Pijanowski

NATIONAL OCEANIC and ATMOSPHERIC ADMINISTRATION
NATIONAL OCEAN SURVEY
NATIONAL OCEANOGRAPHIC INSTRUMENTATION CENTER

ABSTRACT

Five commercially available *in-situ* type water quality monitoring systems measuring conductivity, temperature, dissolved oxygen and pH were evaluated for performance under both laboratory and field conditions. The instruments included in the evaluation are: Hydrolab Corp. Surveyor; Leeds & Northrup Corp. Water Quality System; Martek Corp. Mark 3; Ocean Data Equipment Corp. Model WQMS-101A; and Whitney/Montedoro Corp. Mark II. Each is described briefly and the various sensor types are discussed. Testing methodology is described and a number of user considerations are discussed including initial checkout and calibration. Detailed test results are given for all sensors on each system concerning effects of primary power variation, time response, relative accuracy, repeatability, stability, pressure effects, and intersystem interference. Field test results in a salt water environment are presented. Overall system specifications and results are summarized comparatively by charts. Also discussed is the author's idea of the optimum characteristics for an ideal system based upon testing experience.

INTRODUCTION

In recent years, a new type of instrument package has appeared on the commercial market in response to the current interest in water pollution measurements. These packages, generally labeled as water quality monitoring systems, are comprised of a number of sensors interfaced with an electronics readout unit to provide continuous measurement of several parameters. Water quality monitoring systems can be divided into two broad categories; those which have shore- or ship-based stations and utilize a pumping system to circulate the water of interest past the sensors and those which have *in-situ* type sensors placed directly in the water of interest. For the purposes of this investigation, the *in-situ* type units were chosen.

Although advertising literature leads one to believe that a wide variety of parameters can be measured automatically and *in-situ*, in actuality there are only a few parameters for which reliable *in-situ* sensors exist. Conductivity, temperature, depth, dissolved oxygen, and pH are the sensors usually offered. On some instruments, turbidity and a variety of specific ion sensors are also available. Unfortunately, a great deal of uncertainty and complication resulting from the lack of an acceptable definition for turbidity and the "nonspecificity" for specific ion electrodes makes these sensors useful for only limited and specialized applications. All of the systems included in this evaluation project measure conductivity, temperature, dissolved oxygen, and pH. One system also measures depth and produces a salinity output derived from conductivity and temperature measurements. In most cases, the basic techniques for sensing the various parameters are similar but sensor design varies. Most of the systems also permit the addition of optional sensors for other parameters and automatic data collection systems.

SYSTEM DESCRIPTION

All systems consist of two separate units: an underwater sensor package and an electronic readout unit for placement on a dry platform. The two units are connected by means of either a single multiconductor cable or several individual and separate conductor cables. (Specifications are summarized in Table I.)

Hydrolab Surveyor

The Hydrolab Surveyor Model 6D is manufactured by Hydrolab Corporation, Austin, Texas. The underwater unit is a compact cylindrical body which houses the sensors, preamplifiers, and a magnetically coupled stirring unit and motor for the dissolved oxygen system. The sensors can be easily removed for servicing or replacement. The system is powered by mercury cells with a separate lantern battery for stirrer operation. Output is displayed at the readout unit on an analog panel meter and is also available through recorder outputs for continuous monitoring. Temperature is measured by a multiple thermistor arrangement and dissolved oxygen is sensed with a Yellow Springs Instrument Co. gold/silver polarographic cell and is compensated for temperature.[1,2] Conductivity is measured by a four electrode, a-c nickel cell and is temperature compensated to 25°C. pH is automatically temperature compensated to 25°C and is measured with conventional pH electrodes, the reference electrode being Beckman's solid-state Lazaran model.

Martek Mark 3

The Martek Mark 3 Water Quality Monitoring System is

manufactured by Martek Instrument Corporation, Newport Beach, California. The underwater unit consists of a group of individual sensor housings banded together by large hose clamps. It is connected to the terminal unit by a single multiconductor cable. The unit can be powered by rechargeable cells or from an a-c source. Sensor outputs are displayed on separate analog panel meters; analog recorder outputs are available for continuous monitoring. Temperature is measured by thermistor and dissolved oxygen by a gold/silver Clark type cell which is automatically temperature compensated. The dissolved oxygen sensor is provided with a battery-operated agitator with a water activated switch. Conductivity is measured by a four-electrode, a-c cell and pH by a potassium chloride reference cell, glass hydrogen electrode combination. pH is automatically temperature compensated. An additional feature of this system is a salinity readout for the 25- to 35-ppt range.

ODEC WQMS-101A

The ODEC Model WQMS-101A Water Quality Monitoring System is manufactured by the Ocean Data Equipment Corporation, Warwick, Rhode Island. The underwater unit is composed of a PVC pipe frame which houses the sensors; these are connected to the terminal unit by a single cable of one inch diameter. The system can be powered by an a-c source or by rechargeable (NiCd) internal batteries. The terminal unit displays parameter outputs digitally on separate solid state readouts; recorder connections are available for continuous analog outputs. Conductivity is sensed inductively by a Nusonics sensor; temperature is measured by a thermistor. Dissolved oxygen is measured by a silver/gold polarographic cell manufactured by Delta Scientific Corporation. It is automatically temperature compensated and contains a vibrator arrangement which provides the agitation necessary for dissolved oxygen measurements. pH is automatically temperature compensated and is measured by a Ag/AgCl reference cell and standard hydrogen electrode.

Leeds and Northrup System

The Leeds and Northrup System differs significantly from previously described systems in that it is not designed as a complete, compact unit but is composed of entirely separate modules for each parameter. Because of previous favorable experience with industrial instrumentation as opposed to strictly oceanographic equipment, it was decided to include this type of system in the evaluation. It was felt that Leeds and Northrup had considerable experience with the measurement of all the parameters of interest and that the company could supply a package comparable to other systems under evaluation.

The temperature system consists of a copper constantan thermocouple which provides digital output by means of the L&N Numatron, a millivolt-to-digital electronic display. Conductivity is measured by a two-electrode cell with two interchangeable cells for the range covered. The system can be automatically temperature compensated to 25°C. Dissolved oxygen is measured with a platinum-lead galvanic cell supplied by the Weston and Stack Company. The sensor receives continuous cleaning and agitation by means of a stirrer. pH is sensed with a conventional hydrogen electrode and silver/silver chloride reference cell. The output is automatically temperature compensated. The entire system is operated by a 117-vac, 60-Hz source; all monitors have recorder outputs for continous analog monitoring. Separate multiconductor cables connect each sensor to its monitor and no single frame or housing is provided as a sensor package.

Whitney Mark II

One additional system was procured to participate in this evaluation program and will be described here for completeness. The Whitney Mark II Water Quality Monitor is manufactured by the Whitney Underwater Instruments Division of Montedoro Corporation, San Luis Obispo, California. The underwater package is housed in a compact unit which contains all of the sensors, none of which is exposed directly to the environment; rather, water is pumped through the housing across the sensors. Temperature is sensed by a thermistor and conductivity by a 3-electrode cell which is temperature compensated to 25°C. The dissolved oxygen sensor is a modified Yellow Springs probe with automatic temperature- and pressure-compensated combination electrode. The system can be operated on a-c line power, from an external 12-volt battery, or by internal rechargeable batteries (Global Gel cells). A single cable connects the terminal unit to the underwater package through a winch system; analog recording can be accomplished by a cassette tape recorder which is an integral part of the system. Additional features include a bottom proximity sensor on the underwater package and chloride ion sensor. Unfortunately, various sensor and hardware failures were encountered during initial stages of evaluation so that the system was returned to the manufacturer for repair and refurbishment and was not available for complete testing. The new, improved model will undergo complete evaluation as soon as possible.

TESTING PROGRAM

An extensive testing program was employed for these instruments in order to acquire as much information as possible concerning the operating characteristics of interest to the average instrument user. Whenever practical, the systems were tested simultaneously to obtain relative data.

Initial Check

The first test was a thorough check of each system including charging and soaking the dissolved-oxygen and pH sensors and sensor calibration for each parameter according to the manufacturers' directions. The Hydrolab Surveyor was the only system which was completely operational during initial checkout. The Martek system was functional except for its dissolved-oxygen stirrer which had come out of alignment as a result of shipping. Once the latter problem was diagnosed, it was easily corrected; however, it recurred several times during subsequent testing as a result of normal handling.

Although the ODEC system was operational, inspection revealed that the analog output terminals had not been wired, making signals from the recorder output

Table I

TEST INSTRUMENT SPECIFICATIONS

Type	Hydrolab Surveyor	Martek Mark 3	ODEC WQMS-101A	Leeds & Northrup	Whitney Mark II
Temperature Range Accuracy Sensor	 -5° to 45°C ±0.2°C Thermistor	 -10°C to 60°C ±0.1°C Thermistor	 -5° to 40°C ±0.1°C Thermistor	 -5° to 50°C ±0.1°C Thermocouple	 0° to 40°C ±0.1°C Thermistor
Conductivity* Range Accuracy Sensor	ATC 0 to 100 mmho/cm ±2.5% of reading or ±5% of FS 4-electrode (a-c)	 5 to 65 mmho/cm ±0.1 mmho/cm 4-electrode (a-c)	 0 to 65 mmho/cm ±0.1 mmho/cm Inductive cell	 0 to 100 mmho/cm ±3% of FS 2-electrode (a-c)	ATC 10 to 100 mmho/cm ±0.5% of FS 3-electrode (a-c)
Dissolved Oxygen* Range Accuracy Sensor	ATC 0 to 20 ppm ±0.1 ppm Au/Ag electrode	ATC 0 to 20 ppm ±0.2 ppm Au/Ag electrode	ATC 0 to 20 ppm ±0.2 ppm Au/Ag electrode	ATC 0 to 15 ppm ±0.15 ppm Pt/Pb electrode	ATC 0 to 20 ppm ±0.2 ppm Au/Ag electrode
pH* Range Accuracy Sensor	ATC 2 to 12 ±0.1 Ag/AgCl Lazaran type reference	ATC 0 to 12 ±0.1 Calomel reference	ATC 0 to 14 ±0.1 Ag/AgCl reference	ATC 2 to 14 ±0.1 Ag/AgCl reference	ATC 2 to 12 ±0.05 Combination electrode Ag/AgCl reference
Power Requirement	Internal mercury cells	Internal rechargeable cells or 117 vac	Internal rechargeable cells or 117 vac	117 vac, 60 Hz	Internal rechargeable cells or external 12 vdc or 117 vac
Recorder Output	0 to 10 volts dc (all systems)	0 to 0.5 volt dc (all systems)	C: 0 to 3.25 vdc T: 0 to 2 vdc DO: 0 to 10 vdc pH: 0 to 7 vdc	0 to 20 ma (all systems)	C: 0 to 1 vdc T: 0 to 0.4 vdc DO: 0 to 0.2 vdc pH: 0 to 0.1 vdc
Depth Capability	100 meters	100 meters	100 meters	Not specified	100 meters
Special Features		Auxiliary salinity readout (25 to 35 ppt). Expanded scales for pH.	Digital readout (all systems)	Digital temperature readout	Digital readout (all systems). Winch system for cable.

*ATC: Automatic Temperature Compensation

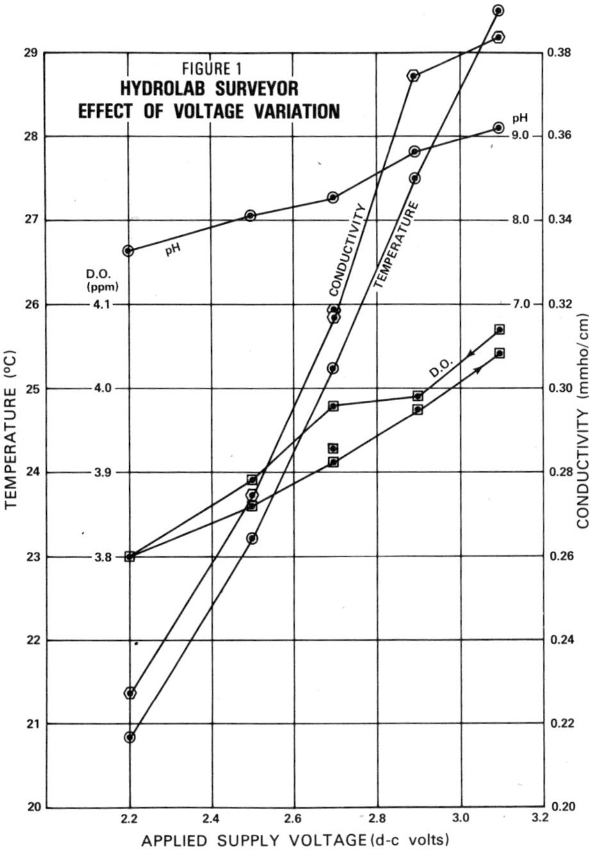

FIGURE 1
HYDROLAB SURVEYOR
EFFECT OF VOLTAGE VARIATION

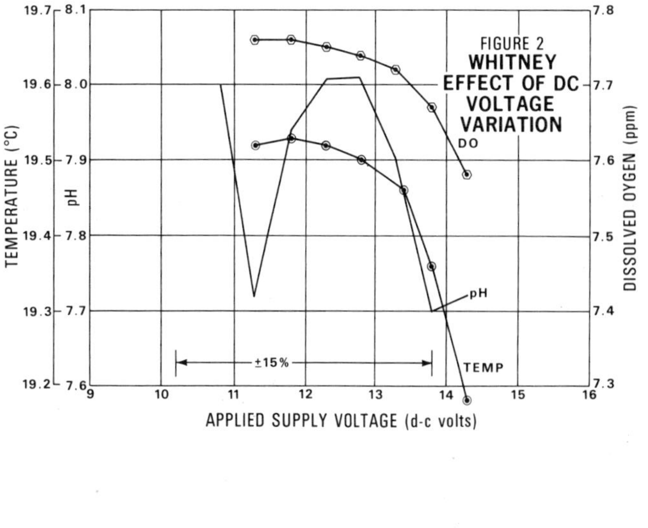

FIGURE 2
WHITNEY
EFFECT OF DC VOLTAGE VARIATION

nonavailable. This problem was corrected in-house under the manufacturer's direction. All calibration adjustments for this system were located on printed circuit boards inside the terminal unit. This required that the entire case be disassembled for any calibration adjustments. With the manufacturer's agreement, additional circuitry was incorporated to bring the adjustments to the outside of the terminal unit for greater accessibility.

After unpacking and assembling the Leeds and Northrup system in the housing designed by L&N, all systems were found to be functional if not completely practical. The "portable" terminal unit housing was a 300-pound steel box without carrying handles or wheels. Although it had a removable access panel in the rear complete with locking handle, it had no openings for cable connections into the box. Each subsystem was separate and the total system required considerable time and skill to assemble and wire because of the numerous options, manuals, and cables comprising the unit.

The terminal unit on the Whitney system was found to be poorly assembled. Among other things, internal circuit boards were loose or out of place, mounting screws and brackets were missing altogether, and a control switch was broken. The unit was returned to the manufacturer for repair and received some time later in better condition. At that time, the manufacturer also implemented a design change which involved bringing calibration adjustments to the outside of the terminal unit for accessibility.

Power Effects

On all systems, the effect of power variation was demonstrated. The a-c line voltage and frequency and/or d-c power were varied ±15% while all parameter outputs were observed. These conditions produced no effect on the Martek and Leeds and Northrup systems. Hydrolab behavior is illustrated in Figure 1. It must be noted, however, that the system is powered by mercury cells which are highly stable and drop off quickly when exhausted so that this degradation in performance would not normally be observed unless a power supply were substituted for the batteries. The Whitney system also showed considerable effects for changes in d-c power supply as illustrated in Figure 2. In this case, the implications are more serious since the system can be operated on a 12-volt battery which decays slowly or on internal batteries which have a useful life of only 3 hours.

D-c power is provided for the ODEC system by three separate supplies. Various combinations of power fluctuations on these produce no effect on temperature and conductivity readings and a maximum effect of ±0.05 ppm on dissolved oxygen and ±0.05 pH units on pH.

Response Times

Depending on the intended use of the instruments, a knowledge of response time may be essential. Tests were performed to determine the response times for the various sensors by monitoring recorder output and instrument visual display as the sensor was transferred between baths of different conditions. Temperature and conductivity response times were measured at 20°C; dissolved oxygen and pH response times were measured as functions of temperature. The results are summarized in Table II and Figures 3 and 4. The Hydrolab temperature probe displayed an unusually long total response time. Although 63% response was obtained in only 3 seconds, almost 15 minutes was required to obtain 99% response. Although not yet evaluated, the manufacturer has since developed an improved sensor with faster total response time.

Agitator Effectiveness

Valid measurements of dissolved oxygen require agitation of the water sample in contact with the oxygen sensor. During previous testing, it was found that a minimum flow rate of about 1.7 ft/sec is required before sensor output becomes independent of flow rate. All systems under evaluation provided stirring by various techniques and these were tested to determine their adequacy. All stirring rates are fixed by system design and were found to be sufficient for complete dissolved oxygen response.

Internal Interference

Since each system consisted of at least four different sensors, the possibility of intersystem interference was considered. None of the units demonstrated this type of behavior when functioning properly. However, the Martek system displayed instability in dissolved-oxygen and temperature readings when its pH system was malfunctioning.

Temperature

All systems were evaluated for accuracy. The results are summarized in Table II and Figures 5 through 8.

Conductivity

The conductivity systems were evaluated by temperature cycling in baths of various salinities and comparing conductivities with those obtained from a standard laboratory salinometer. Conductivity ratios as obtained from the salinometer were converted to conductivity and/or salinity using the NOIC accepted relationships described in Appendix A. Results for these tests are summarized in Table II and Figures 9 through 12. The Hydrolab conductivity system differs from the others in that it is automatically temperature corrected to 25°C. Since compensation is a function of both conductivity and temperature, complete compensation over the entire conductivity range is extremely difficult. It is recommended that conductivity instruments be designed to measure temperature and absolute conductivity and that the user make any corrections to his data externally. The Leeds and Northrup conductivity system was also temperature compensated when originally received. Other problems with lead wire resistance caused the compensation to be removed by NOIC, with the approval of the manufacturer, to obtain an uncorrected conductivity output.

The Martek system provides salinity data based on conductivity measurements. Salinity error as determined by the testing is presented in figure 13.

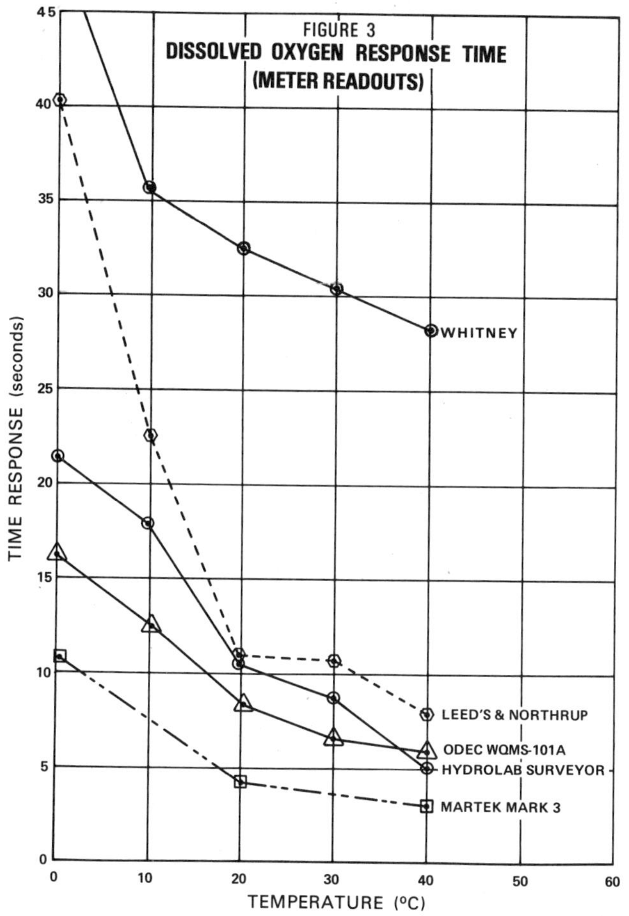
FIGURE 3
DISSOLVED OXYGEN RESPONSE TIME
(METER READOUTS)
WHITNEY
LEED'S & NORTHRUP
ODEC WQMS-101A
HYDROLAB SURVEYOR
MARTEK MARK 3
TIME RESPONSE (seconds)
TEMPERATURE (°C)

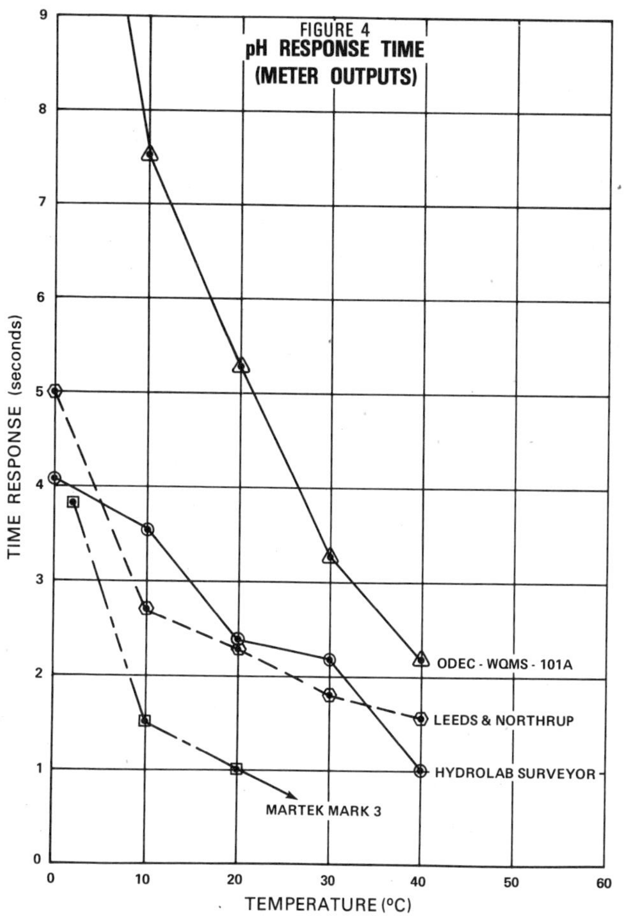
FIGURE 4
pH RESPONSE TIME
(METER OUTPUTS)
ODEC - WQMS - 101A
LEEDS & NORTHRUP
HYDROLAB SURVEYOR
MARTEK MARK 3
TIME RESPONSE (seconds)
TEMPERATURE (°C)

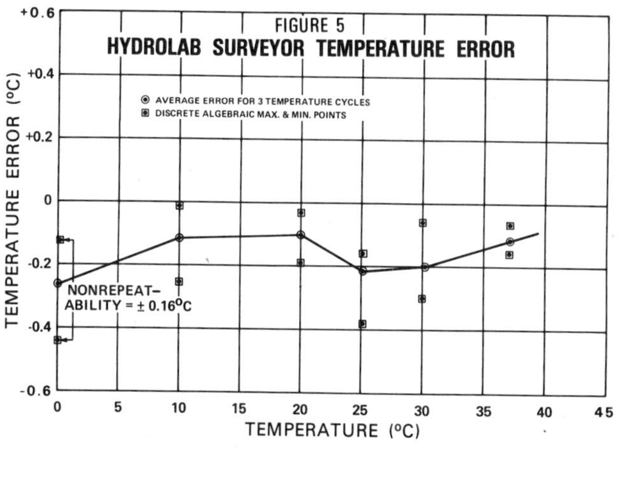
FIGURE 5
HYDROLAB SURVEYOR TEMPERATURE ERROR
AVERAGE ERROR FOR 3 TEMPERATURE CYCLES
DISCRETE ALGEBRAIC MAX. & MIN. POINTS
NONREPEAT-
ABILITY = ± 0.16°C
TEMPERATURE ERROR (°C)
TEMPERATURE (°C)

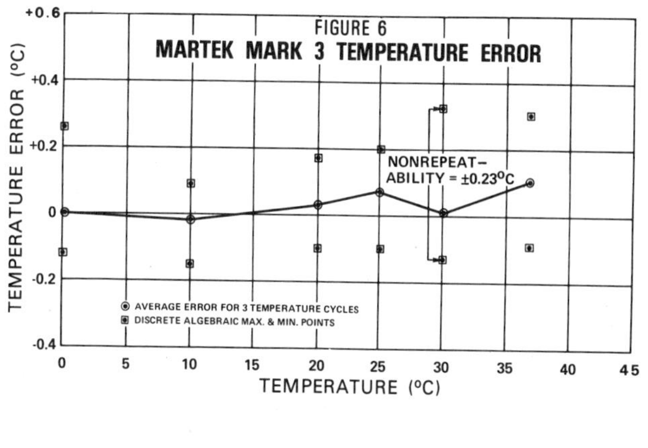
FIGURE 6
MARTEK MARK 3 TEMPERATURE ERROR
NONREPEAT-
ABILITY = ±0.23°C
AVERAGE ERROR FOR 3 TEMPERATURE CYCLES
DISCRETE ALGEBRAIC MAX. & MIN. POINTS
TEMPERATURE ERROR (°C)
TEMPERATURE (°C)

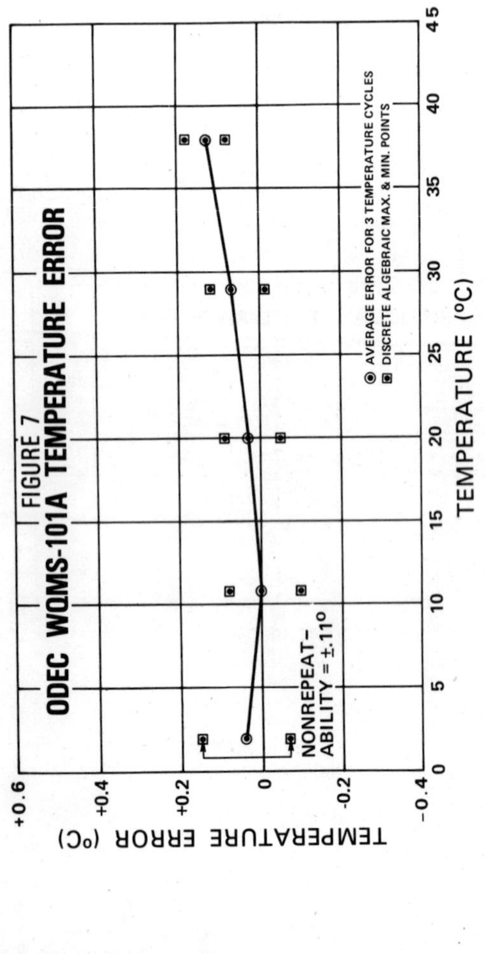
FIGURE 7
ODEC WQMS-101A TEMPERATURE ERROR
TEMPERATURE ERROR (°C)
TEMPERATURE (°C)
NONREPEAT-ABILITY = ±.11°
AVERAGE ERROR FOR 3 TEMPERATURE CYCLES
DISCRETE ALGEBRAIC MAX. & MIN. POINTS

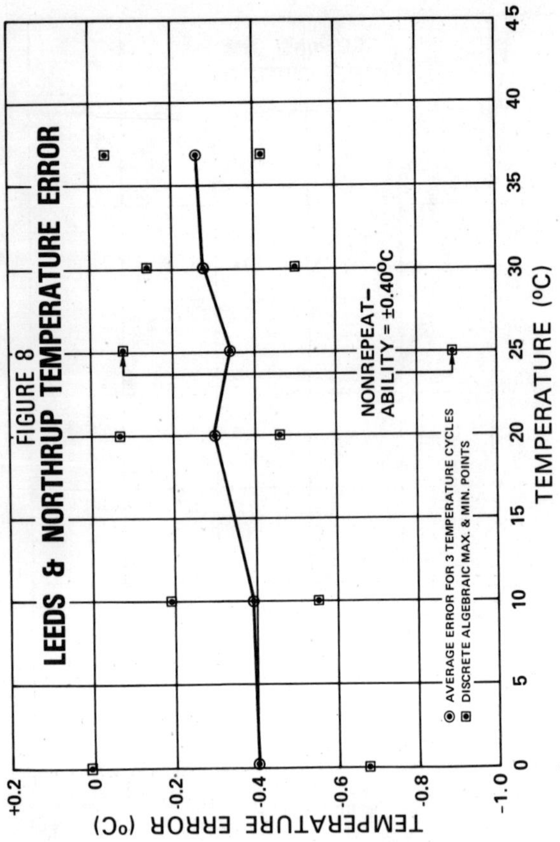
FIGURE 8
LEEDS & NORTHRUP TEMPERATURE ERROR
TEMPERATURE ERROR (°C)
TEMPERATURE (°C)
NONREPEAT-ABILITY = ±0.40°C
AVERAGE ERROR FOR 3 TEMPERATURE CYCLES
DISCRETE ALGEBRAIC MAX. & MIN. POINTS

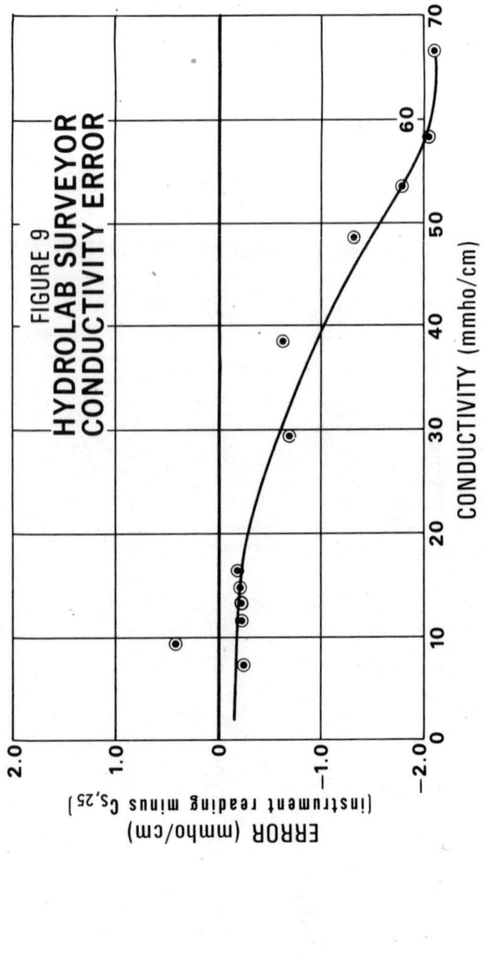
FIGURE 9
HYDROLAB SURVEYOR CONDUCTIVITY ERROR
ERROR (mmho/cm)
(instrument reading minus Cs,25)
CONDUCTIVITY (mmho/cm)

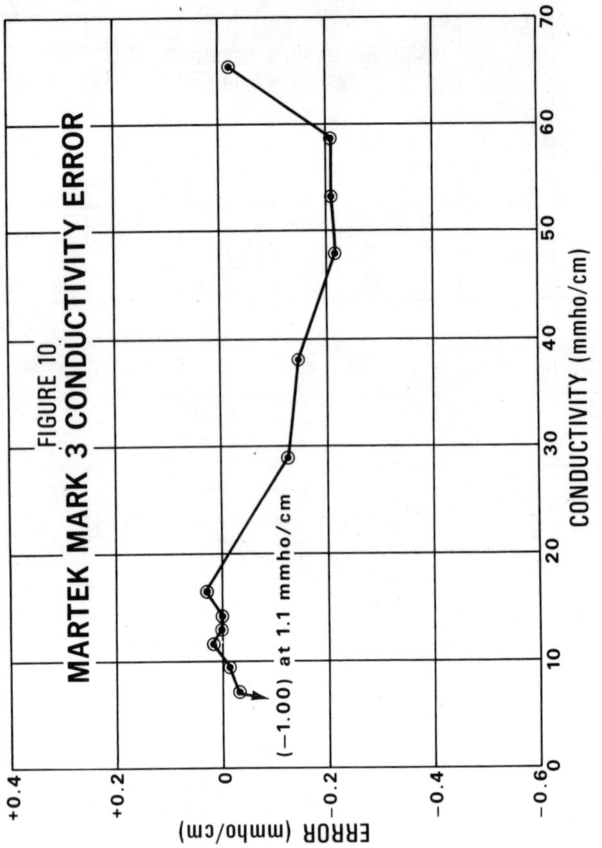
FIGURE 10
MARTEK MARK 3 CONDUCTIVITY ERROR
ERROR (mmho/cm)
CONDUCTIVITY (mmho/cm)
(−1.00) at 1.1 mmho/cm

Dissolved Oxygen

Previous testing of dissolved oxygen instrumentation has led to the establishment of a somewhat unique testing technique. "Standard" oxygen solutions are created by saturating a water bath with O_2-N_2 mixtures under controlled temperature conditions. By varying temperature and the relative composition of the gas mixtures, different oxygen levels can be generated in the bath and used as standards. It has been determined from previous work that the error for dissolved oxygen instruments is a function of both temperature and total dissolved oxygen level. Also, calibration curves are essentially linear and can therefore be established by the use of two gas mixtures rather than the five mixtures used in earlier work.[1]

Before each test cycle, a calibration adjustment was made on each sensor at 20°C, 20.95% O_2 mixture. Calibration curves were established for each system in both fresh water and salt water (0.5 ppt and 35 ppt salinities respectively) and are presented in Figures 14 through 21. It is interesting to note that both the Hydrolab and Leeds and Northrup calibrations exhibit opposite slopes for fresh and salt water.

pH

A pH sensor measures hydrogen ion concentration and consists of two electrodes: a hydrogen electrode which has a glass membrane permeable by hydrogen ions and a reference electrode which may be of several varieties. Proper operation of the reference electrode requires that physical contact be made between the solution inside the electrode and the medium in which it and the hydrogen electrode are immersed. This physical contact is maintained by some type of opening in the bottom of the electrode through which the internal solution can leak or seep out very slowly. If this junction, for any reason, becomes blocked, the pH system will not function.

Two common types of reference electrode junctions are cellulose and ceramic. It was found that the ODEC pH sensor which contained a cellulose junction displayed sensitivity to flow. This sensitivity was completely eliminated by the substitution of a similar electrode with a ceramic junction.

Proper calibration and adjustment of pH systems is generally a simple procedure but requires some knowledge of instrument design. The electrodes measure a small signal which is a function of both pH and temperature. Because of the nonlinear potential of the glass electrode, most pH systems are built with a slope adjustment as well as a calibration adjustment. The slope adjustment allows compensation for the nonlinear behavior of the electrode and must be checked or reset for different pH ranges or whenever the electrodes are changed or cleaned. The calibration adjustment is generally made at one point and compensates for the daily drift of the sensor. Standard buffer solutions available over wide pH ranges are used to create the calibration solutions; because of temperature dependence, the buffer solutions should be at the same temperature as the sample being measured. A faulty electrode system is not always obvious unless a wide pH range is tested. A cracked glass electrode can be standardized at almost any pH value. However, if the faulty electrode is placed in a solution with a pH which is different by several units, it will continue to indicate the standardization pH. Therefore, it is wise to calibrate by using two different buffer solutions. If the reference electrode is a fault, the system response may drop to zero or may produce a highly unstable output.

pH testing was carried out to the temperature and pH limits specified for each system. The instruments were temperature cycled from 0 to 40°C in baths of pH 4, 8, and 10. Results for the Martek and ODEC systems are presented in figures 22 and 23. Single readings at 25°C were made at the extreme high and low pH specified. As expected, error is greatest at the extremes. In all cases, the systems were calibrated at pH 7 and slope adjustments were set for the pH 4-10 range. The errors summarized in Table II are those obtained at the extremes.

All readings for the Martek system were obtained on the full-scale meter because the expanded scales did not appear to be functioning properly. However, after the data had been taken, an overlooked adjustment was discovered which corrected the expanded-scale problem when properly set. Testing will be repeated to obtain additional data using these expanded scales. (It should be pointed out that the preliminary instruction manual was not explicit regarding the method for adjusting the instrument in the case of the expanded scales. A company representative provided a step-by-step calibration procedure which is intended for inclusion in the final manual.)

Complete data are not available at the time of this writing for pH on the Hydrolab and Leeds and Northrup systems. A defective glass electrode was discovered in the Hydrolab system after some testing had been carried out. The glass electrode in the Leeds & Northrup system was also found to be faulty but this failure was induced by improper handling during testing. The connections on the nonsensing end of the L&N electrodes are designed to be splash proof rather than waterproof. Consequently, submersion in the test baths eventually caused failure. Tests will be completed when replacement parts are delivered.

ENVIRONMENTAL EFFECTS

Environmental tests were performed on the terminal units of each system. Temperature and humidity were varied to the limits specified by the manufacturers or between 0 and 40°C temperature and 48 to 97% relative humidity, if not specified. Worst-case results are summarized in Table II.

FIELD TESTING

Since these systems are all intended for use in field operations, it was felt important to observe their performance under such conditions. All of the systems were included in the testing except the Whitney which was still under repair by the manufacturer. The site chosen was one mile off the coast of Freeport, Grand Bahama Island at a depth of 50 feet. The sensor packages were fastened to a support frame anchored to the bottom and terminal units were placed inside an underwater habitat close by. The instru-

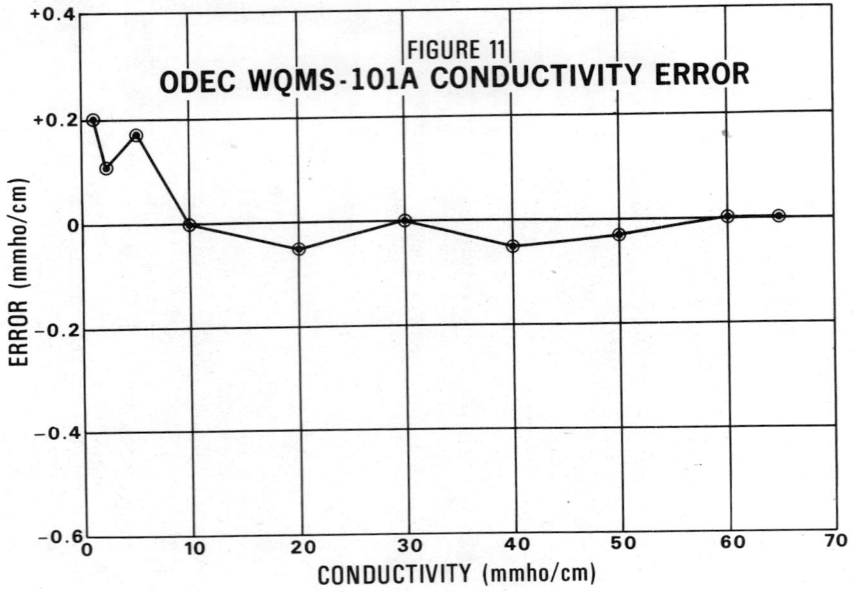

FIGURE 11
ODEC WQMS-101A CONDUCTIVITY ERROR

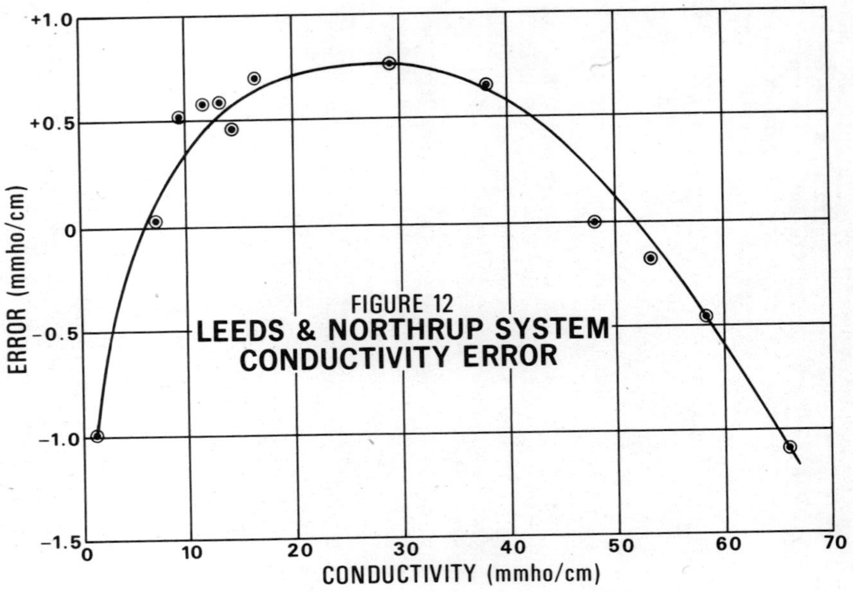

FIGURE 12
LEEDS & NORTHRUP SYSTEM CONDUCTIVITY ERROR

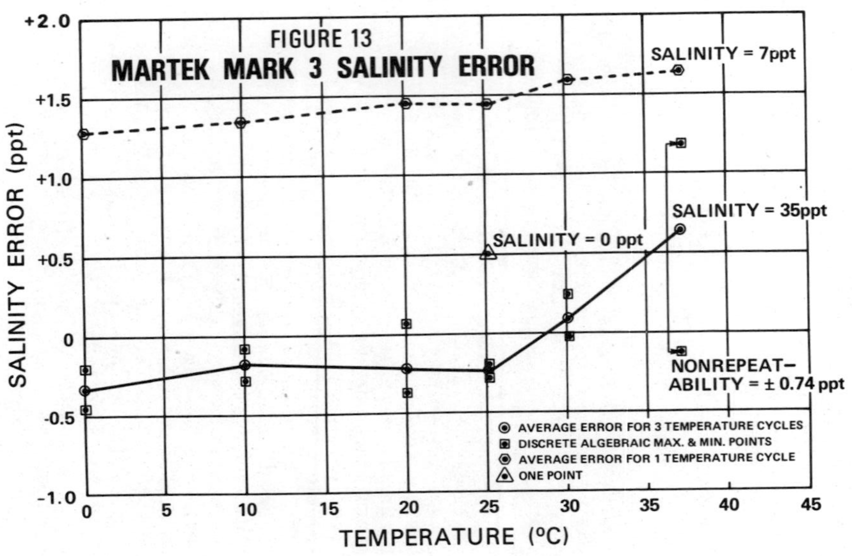

FIGURE 13
MARTEK MARK 3 SALINITY ERROR

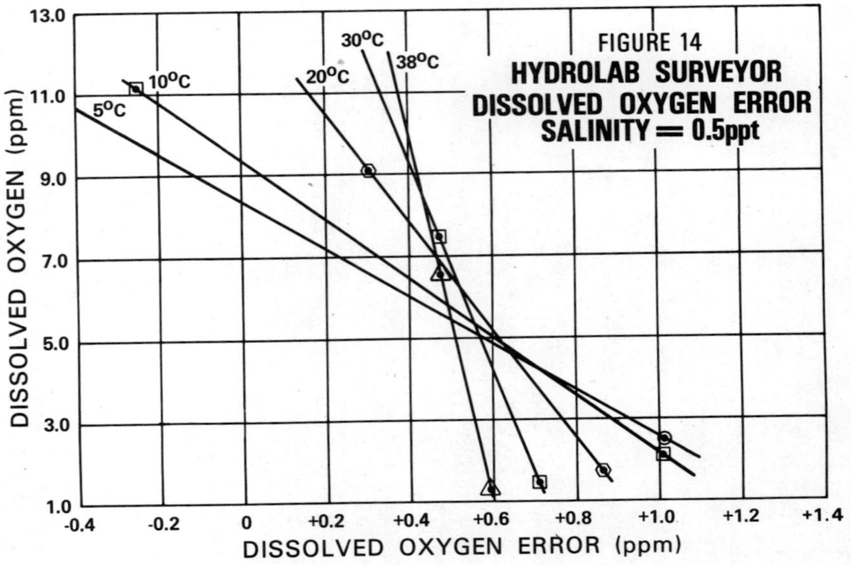

FIGURE 14
HYDROLAB SURVEYOR DISSOLVED OXYGEN ERROR
SALINITY = 0.5ppt

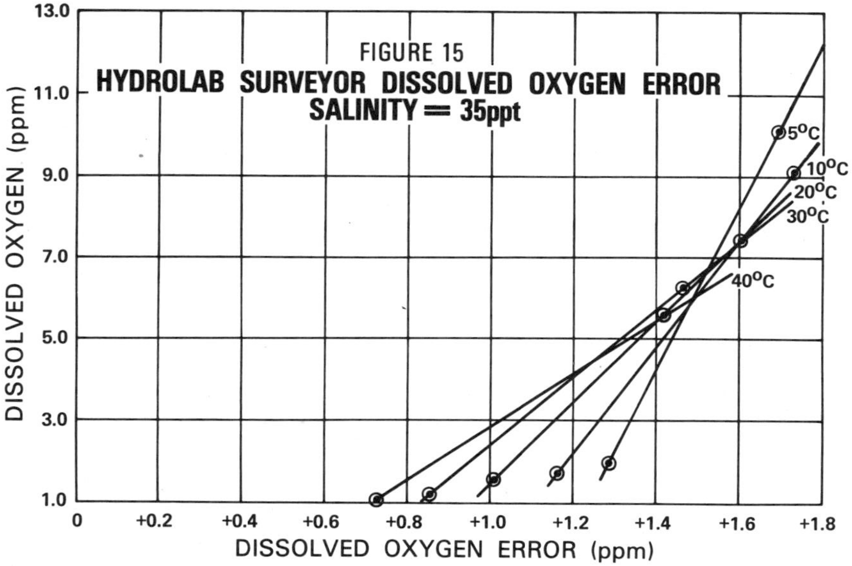

FIGURE 15
HYDROLAB SURVEYOR DISSOLVED OXYGEN ERROR
SALINITY = 35ppt

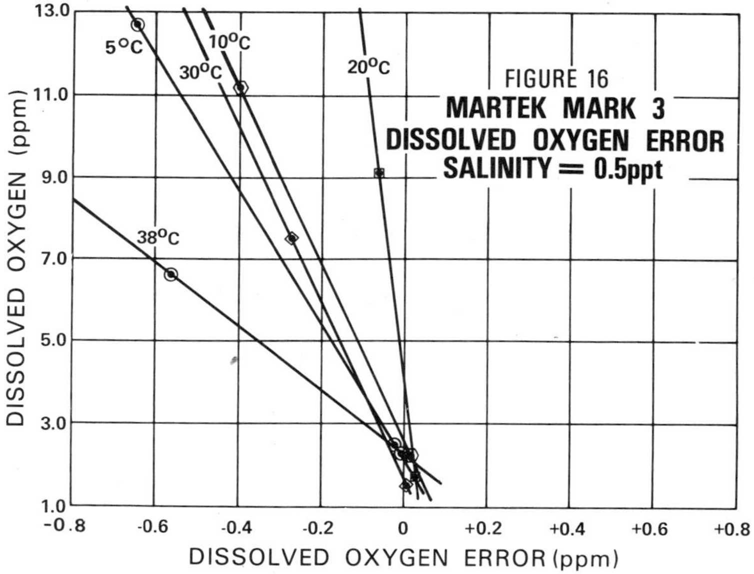

FIGURE 16
MARTEK MARK 3
DISSOLVED OXYGEN ERROR
SALINITY = 0.5ppt

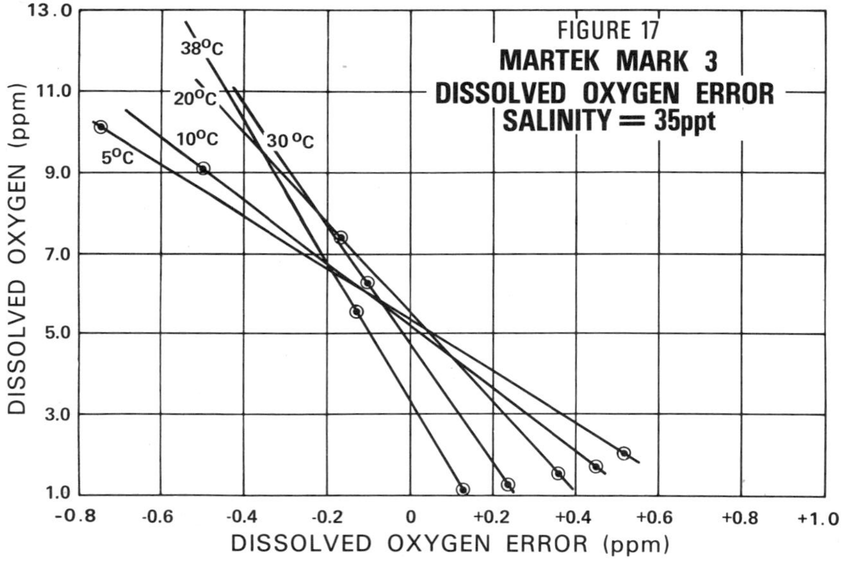

FIGURE 17
MARTEK MARK 3
DISSOLVED OXYGEN ERROR
SALINITY = 35ppt

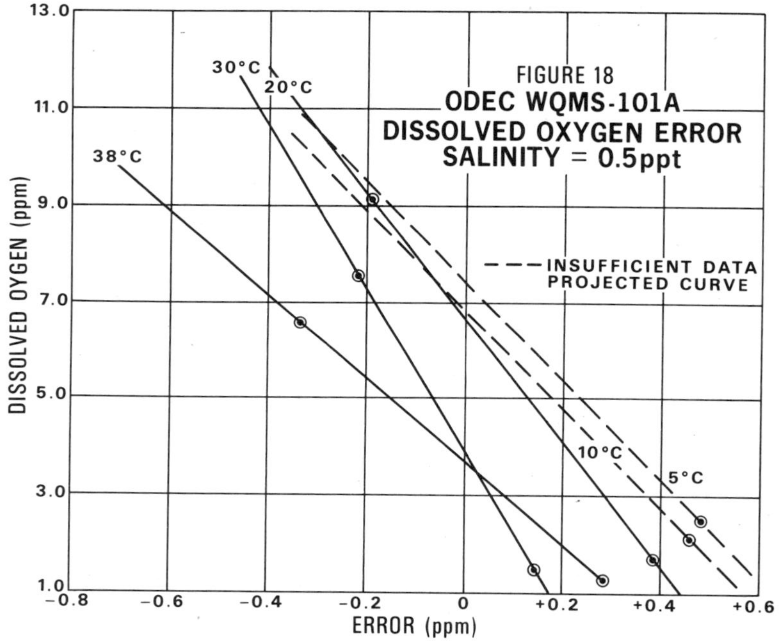

FIGURE 18
ODEC WQMS-101A
DISSOLVED OXYGEN ERROR
SALINITY = 0.5ppt

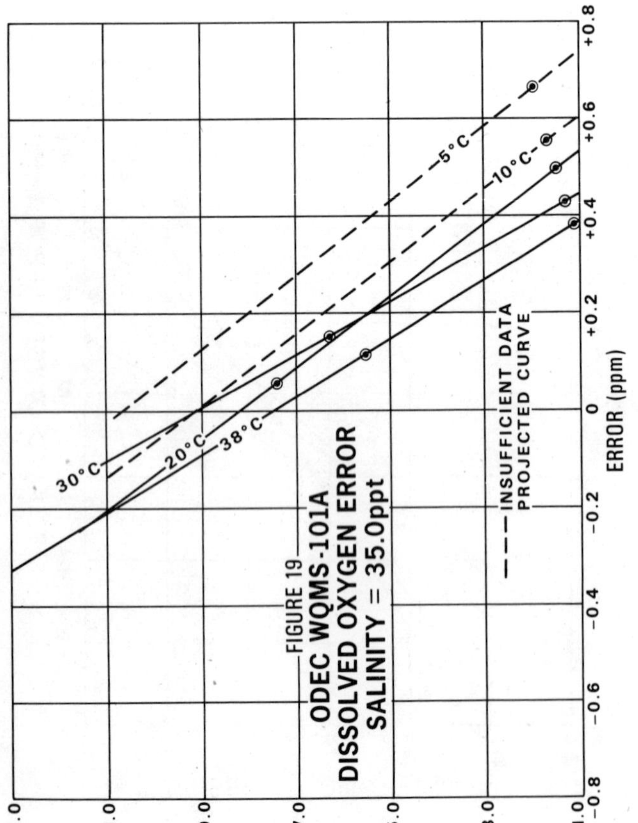

FIGURE 19
ODEC WQMS-101A
DISSOLVED OXYGEN ERROR
SALINITY = 35.0ppt

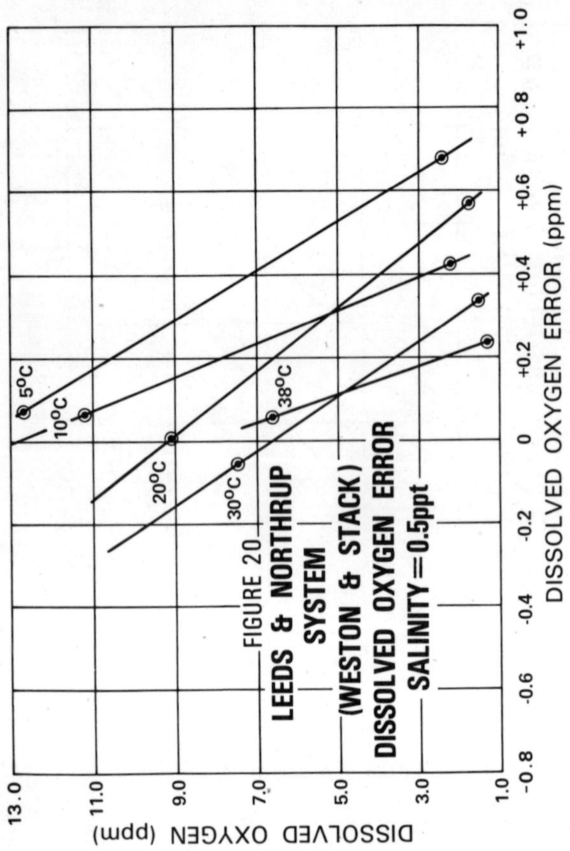

FIGURE 20
LEEDS & NORTHRUP SYSTEM (WESTON & STACK)
DISSOLVED OXYGEN ERROR
SALINITY = 0.5ppt

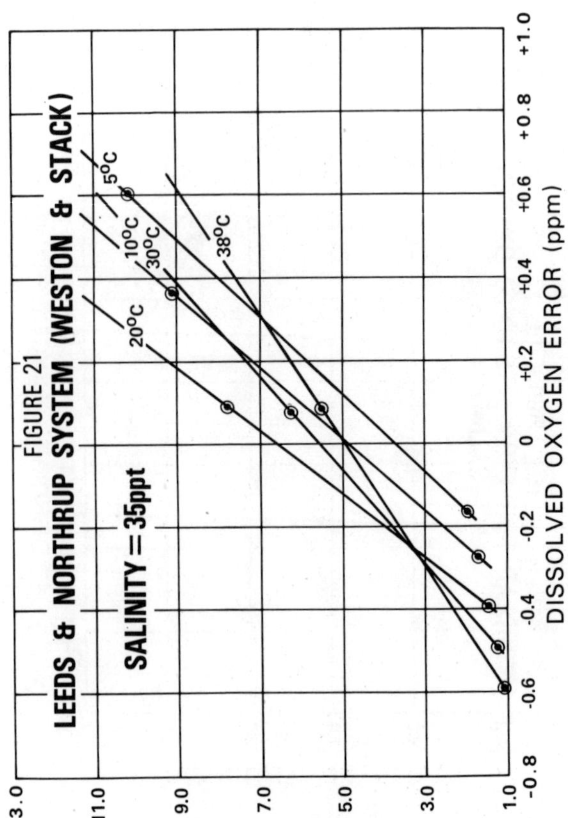

FIGURE 21
LEEDS & NORTHRUP SYSTEM (WESTON & STACK)

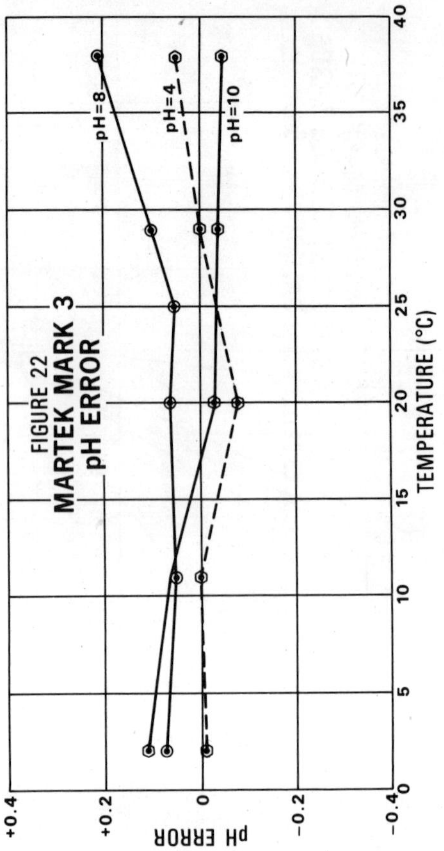

FIGURE 22
MARTEK MARK 3
pH ERROR

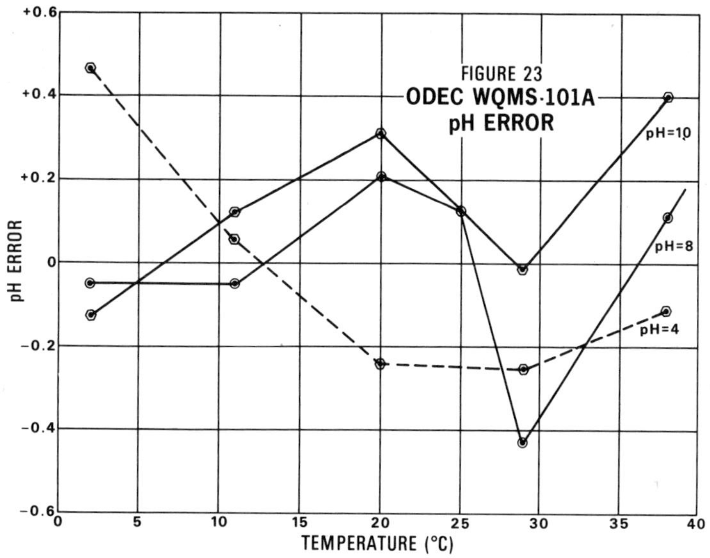

Table II

TEST RESULTS

Type	Hydrolab Surveyor	Martek Mark 3	ODEC WQMS-101A	Leeds & Northrup	Whitney* Mark II
Temperature					
Accuracy**	-0.26°C See fig. 5.	+0.09°C See fig. 6.	+0.13°C See fig. 7.	-0.40°C See fig. 8.	
Response Time (recorder only)	3.4 sec	1.2 sec	1.7 sec	18.1 sec	62.0 sec
Conductivity					
Accuracy**	See text & fig. 9.	-0.22 mmho/cm See fig. 10.	+0.20 mmho/cm See fig. 11.	-1.09 mmho/cm See fig. 12.	
Response Time	Recorder: 0.2 sec Meter: 1 sec	Recorder: 0.2 sec Meter: 1 sec	Recorder: 0.3 sec Meter: 1 sec	Recorder: 0.2 sec Meter: 1 sec	Recorder: 22.0 sec Meter: 10.7 sec
Dissolved Oxygen					
Accuracy	See figs. 14 & 15.	See figs. 16 & 17.	See figs. 18 & 19	See figs. 20 & 21.	
Response Time	← A function of temperature. See figure 3. →				
pH					
Accuracy***	At 2: +0.25 At 12: +0.02 See text.	At 1: -0.35 At 12: -0.39 See text & fig. 22.	At 1: +0.56 At 14: -0.36 See text & fig. 23.	See text. See text.	
Response Time	← A function of temperature. See figure 4. →				
Power Variation Effect (±15%)	See fig. 1. (DC only)	AC: No effect DC: No effect	AC: No effect DC: ±0.05 ppm D.O. ±0.05 pH	No effect (AC only)	AC: No effect DC: See fig. 2.
Effectiveness of D.O. Agitator	Adequate	Adequate	Adequate	Adequate	
Environmental Effects***					
Temperature	Temp: +0.7°C	Temp: +1.3°C pH: -0.1	Temp: +0.13°C pH: +0.5 Cond: ±0.1 mmho/cm	Temp: +0.9°C	
Humidity	pH: -0.4	pH: +0.4	None	Not available	
Dimensions & Weight					
Sensor (diameter & length)	Cylindrical 6.7 x 18.1 in. (17.0 x 46.0 cm) 14.8 lb (6.6 kg)	Irregularly Shaped 10.5 x 18.1 in. (26.7 x 46.0 cm) 12.2 lb (5.5 kg)	PVC Tubing Frame 9.0 x 45.2 in. (22.9 x 114.8 cm) 18.6 lb (8.4 kg)	Irregularly Shaped Dissolved Oxygen 7.6 x 16.0 in. (19.3 x 40.6 cm) 13.25 lb (6.0 kg) pH 3.7 x 11.25 in. (9.4 x 28.6 cm) 2.5 lb (1.1 kg) Conductivity 1.0 x 5.0 in. (2.5 x 12.7 cm) 3 oz (85 mg) Temperature 0.3 x 15.0 in. (0.8 x 3.8 cm) 3 oz (85 mg)	Cylindrical 4.4 x 2.18 in. (11.2 x 5.5 cm) 14.7 lb (6.7 kg)
Terminal Unit (D x L x H)	8.8 x 12.2 x 8.5 in. (22.4 x 31.0 x 21.6 cm) 12.0 lb (5.4 kg)	19.0 x 12.2 x 5.1 in. (48.3 x 31.0 x 13.0 cm) 25.2 lb (11.4 kg)	9.2 x 13.8 x 9.2 in. (23.4 x 35.1 x 23.4 cm) 28.0 lb (12.7 kg)	3 Units 19.0 x 7.0 x 7.0 in. (48.3 x 17.8 x 17.8 cm) 22.0 lb (10.0 kg) 1 Unit 19.0 x 8.5 x 5.0 in. (48.3 x 21.6 x 12.7 cm) 18.9 lb (8.6 kg)	9.6 x 16.0 x 9.6 in. (24.4 x 40.6 x 24.4 cm) 28.2 lb (12.8 kg)

*The Whitney Mark II system failed early in the testing program which accounts for the limited data presented. See text.
**These are worst-case average errors.
***These are worst-case errors. See text for details on environmental testing.

ments remained on the bottom for seven days and were monitored from the habitat by divers as various tests were performed.

Actual numerical data as obtained from the systems during the monitoring period were of secondary importance compared with the anticipated indentification of problem areas and potential failures. As expected, each system experienced problems not all of which have been resolved at this writing. The systems were transferred to the test site by a small work boat in 6- to 8-foot seas which subjected them to considerable vibration and shock. All terminal units were transferred to the habitat in dry containers. Although most of the cables were designed for underwater use, some connectors were not and required special preparation to keep them dry during the transfer. The preparation was not entirely successful and some effort was required to dry out the unsuccessful attempts. The Leeds and Northrup cables were not designed for submersion so these were fed through a rubber hose for protection.

All systems were operational and calibrated before transfer to the habitat. The following problems were encountered during underwater testing.

Hydrolab

The conductivity system was completely inoperative. This was later determined to be the result of the batteries being dislodged during transfer because retaining clips had been omitted. The magnetic stirrer stopped spinning and required physical manipulation to restart. However, it operates satisfactorily at surface pressure.

Martek

The pH system was inoperative, the result of water leaking under pressure into the sensor housing.

ODEC

The entire system was not operational. All output displays indicated unreasonable values. This was a result of broken wires in the cable probably induced by the weight of the cable itself causing strain on the connector and a faulty temperature probe. The D.O. stirrer was also nonfunctional, the result of a leak under pressure into the motor housing. Numerous mounting nuts and washers in the terminal unit were vibrated free during shipping and could have caused some of the electronic component failures pinpointed during post-testing examination.

Leeds & Northrup

The motor for the stirrer on the D.O. system was not operational in the habitat; however, it proved to work quite satisfactorily when returned to the surface and has been working ever since. Conductivity ouput remained constant at a value well below the actual level. The sensor appears to be faulty and the problem is still being investigated.

Pressure tests were carried out during the field work to a depth of 50 feet. Results indicate that the Leeds and Northrup system showed no effect of pressure on any parameters. The Hydrolab dissolved oxygen probe showed a negative pressure effect of 0.1 ppm/10 ft for the first 30 feet, then leveled off. The Martek dissolved oxygen system showed ±0.2 ppm instability over the pressure range tested.

GENERAL COMMENTS

After working with these systems simultaneously for almost a year, several qualitative and personal judgements have been formulated concerning the design and handling characteristics of the ideal *in-situ* system. These comments are offered to provide pertinent information to potential users and designers; however, it must be emphasized that they are personal views based on many months of operating this type of instrument and talking to users.

The ideal water quality system for *in-situ* use should be comprised of compact, easily portable units. The terminal unit should have direct readout in engineering units on simple, easily readable scales. Some systems have several scales superimposed on the same meter, resulting in confusion and erroneous interpretation. Calibration adjustments for each sensor should be easily accessible but protected from accidental movement. Lockwashers should be used on all screws and nuts to avoid vibration-induced loosening. It is often convenient and practical for the system to have several options for supplying power which can be selected by the user depending on his application such as internal d-c rechargeable batteries or external a-c or d-c power supplies. The system should also be designed to make readings impossible to obtain if the power supply is too low. This would obviate erroneous readings when sensors are not sufficiently powered.

The underwater unit should be compact and more or less symmetrical for convenient handling in deployment. It should be fabricated of lightweight, corrosion-resistant material and should not rely on the use of supplemental supports or braces which have a tendency to be misplaced over a period of time. Sensors should be exposed directly to the environment but protected from accidental bumping. They should be easily replaceable with underwater connectors if possible. Stirring or agitation must be provided for the dissolved oxygen sensor.

The cable should be single and flexible and have as small a diameter as possible to carry the weight and signals required. A single connector into the terminal unit is advisable to avoid any possible confusion in wiring. Connections to the underwater unit should be as simple as possible and easy to manipulate with cold, wet, uncoordinated hands. It is also helpful to be able to disconnect any one or more of the sensors from the cable; this requires that the underwater end of the cable be supplied with plugs. Connectors should be chosen so that mating the sensors improperly is not possible.

FUTURE PLANS

Plans for the immediate future include repeating pH testing on the systems not already completed and preparation of a series of Instrument Fact Sheets to be published by NOIC. The Whitney system will be completely reevaluated because of the extensive modifications made by the manufacturer since initial phases of testing.

In addition, it is anticipated that tests for calibration stability and battery life will be carried out and reported in the near future.

REFERENCES

(1) Pijanowski, B.S., "A Quantitative Evaluation of Dissolved Oxygen Instrumentation," Joint Conference on Sensing of Environmental Pollutants, Palo Alto, Calif., November 1971

(2) Pijanowski, B.S., "Salinity Corrections for Dissolved Oxygen Measurements," Environmental Science & Technology, October 1973, pp. 957-958

APPENDIX A

CONDUCTIVITY-TEMPERATURE-SALINITY RELATIONSHIPS

$$C_{S,T} = (C_{35,15})(C_{35,T}/C_{35,15})(C_{S,T}/C_{35,T})$$

$C_{S,T}$ = Conductivity of water sample at temperature T and Salinity S

T = Temperature of unknown water sample

T_1 = Temperature of unknown water sample in laboratory salinometer

R_{T1} = Conductivity ratio from laboratory salinometer $C_{S,T1}/C_{35,T1}$

$C_{35,15}$ = 42.896 mmho/cm International Units

$$C_{35,T}/C_{35,15} = 0.6764545 + 0.02012758T + 1.014000 \times 10^{-4}T^2 - 2.63840 \times 10^{-7}T^3 - 5.782014 \times 10^{-9}T^4 \quad (1)$$

$$C_{S,T}/C_{35,T} = R_{T1} + A(T_1) - A(T)$$

$$A(t) = 10^{-5}R(R - 1)(t - 15)[96.7 - 72R + 37.3R^2 - (0.63 + 0.21R^2)(t - 15)] \quad (2)$$

$$R_{15} = R_{T1} + A(T_1)$$

$$S = 0.08996 + 28.29720R_{15} + 12.80832R_{15}^2 - 10.67869R_{15}^3 + 5.98624R_{15}^4 - 1.32311R_{15}^5 \quad (3)$$

where

(1) is from Brown and Allentoft corrected for 1968 IPTS change in temperature scale

(2) and (3) are from the UNESCO Tables

© 1973, ISA JSP 6671

AN EVALUATION OF A HIGH-VOLUME CASCADE PARTICLE IMPACTOR SYSTEM

G. A. Sehmel
Research Associate
BATTELLE,
Pacific Northwest Laboratories,
Richland, Washington 99352

ABSTRACT

Commercially available 20 cfm cascade impactors were evaluated under field sampling conditions for particle sampling bias caused by interstage losses and by non-wind direction sampler orientation. An integrated sampler using an impactor and a wind-direction self-orienting cowl attachment decreased particle sampling bias.

INTRODUCTION

Accurate sampling is vital in determining airborne particulate concentrations, airborne particulate size distributions,(13) and particulate pollution composition by subsequent chemical analysis. An important application of sampling is in field studies of particle resuspension research. In this research, wind surface stresses cause soil particles to become airborne, which subsequently are sampled. When the host soil particles are contaminated by tracers such as radioactive material, the overall removal processes or particle resuspension rates need to be determined. Resuspended particles can be in the respirable size range (less than about 3 μm diameter) or can be much larger. This potential airborne hazard can be determined through sampling airborne material without regard for particle sizes, with the assumption that all of the contaminant is associated with the respirable size range. This sampling procedure does give a maximum airborne concentration and consequently the most conservative airborne concentration for radiological assessment. However, sampling is more realistic when airborne concentrations are determined as a function of particle diameter. In this case, the potential inhalation hazard is not only quantified, but information for larger particles is also obtained on the basic physics of resuspension. These physical concepts are needed to develop generalized models for predicting resuspension rate. Resuspension studies are underway at Battelle-Northwest.

Economic requirements often limit both the number of sampling sites as well as refinements in sampler design. Air samplers should be designed to sample isokinetically,(2,4-6,8-11) i.e., ambient air direction and velocity is not altered as air enters a sampler. When the sampler is a cascade particle impactor for determining airborne particle size distributions, the sampler is designed to operate at one constant air flow rate. Consequently, even if an impactor is directed into the wind, sampling with an impactor is nearly always anisokinetic unless an almost prohibitively expensive sampler inlet aperature were designed. Isokinetic sampling with the constant flow rate requires the sampler inlet aperature to be automatically controlled as a function of wind speed. Although anisokinetic sampling errors are often large when the wind speed is not matched sampling errors associated with not matching wind direction (4-6,8,15) can be equal or of greater importance. For instance, with small diameter sampling tubes oriented at right angles to the wind direction,(15) particle inertia can cause particles to by-pass the sample tube inlet. In this sampler orientation, fewer particles are collected in the sampler tube than were originally in air drawn into the sampler. This decreased collection produces an apparent particle concentration which can be much less than true airborne particle concentration. The larger the particle inertia, the smaller will be the apparent airborne particle concentration.

The purpose of this paper is to describe and evaluate a particle impactor system which minimizes directional sampling errors. Each sampling system was assembled from a high-volume, 20 cfm cascade impactor which was fitted with a wind-direction sensitive rotating cowl which directed the sampler into the wind.

EXPERIMENTAL

A sampler system was designed to separate airborne particles into aerodynamic particle diameters in the respirable size range and to simultaneously collect larger airborne particles. Each particle impactor system was assembled from a 20 cfm Andersen 2000 Inc.* high volume cascade impactor and a rotating cowl attached to the impactor. These impactors separate particles into nominal diameter ranges of 1.1, 2.0, 3.3, and ≥ 7 μm which are stage 50 percent cut-off diameters for unit density spheres. For

* Andersen 2000, Inc., model 65-100 High-Volume Sampler Head, P.O. Box 20769, AMF, Atlanta, Georgia 30320.

Superior numbers refer to similarly-numbered references at the end of this paper field test evaluation, these systems were mounted on towers at heights up to 30 m, at sites on the Atomic Energy Commission's Hanford reservation near Richland, Washington, and a site at the Rocky Flats plant near Boulder, Colorado.

An exploded view of the sampler system consisting principally of the impactor and rotating cowl are shown in Figure 1. The system support arm was bolted to the sampling tower. A variable transformer* attached to the support arm was used to control flow rate through the cascade impactor. The 8 × 10 in. filter in the high-volume sampler** and the cascade impactor inlet faced toward the ground. In order to attach the wind direction sensitive rotating cowl to the impactor, the factory supplied speedball handle at the impactor inlet was replaced with a threaded spindle extension. The cowl was attached to the spindle extension by a bolt passing through the spindle bearing housing. When the spindle extension was attached to the cowl, cowl rotation caused by the wind orientation tail fin was facilitated by the spindle bearing.

Particles entering the cylindrical sample inlet either settled on the cowl floor or were drawn up into the impactor. At the 20 cfm flow rate, only particles with an aerodynamic equivalent diameter of less than 63 μm in diameter should have been pulled up towards the impactor. If plug flow occurred in the cowl body, larger particles than this should have settled out.

Air and particles entering the impactor are shown schematically in Figure 2. Air flows through the multiple orifices in each support plate. Due to the particle inertia caused by each orifice air jet, successively smaller particles are impacted on the fiber glass filter collectors on each support plate. These collectors are perforated to match the orifice pattern of the succeeding stage. Particles are collected not only at each desired collection site on the four collectors and back-up filter, but also deposit on the support plate as shown by the interstage loss.

The cowl-impactor system is currently our best field system for sampling airborne particles in the respirable size range as well as for larger sized airborne particles. In other instances, impactors without cowls have been operated with the impactor inlet facing down, facing up, or facing horizontally.

Filter Collection

Cowl-impactor systems and impactors without cowls have been simultaneously evaluated in field tests by gravimetric techniques. The perforated collector media on each impactor stage were neutral pH Gelman Type A glass filter papers. Before initial weighings, filter papers were placed individually in open racks and allowed to equilibrate for several days in an air-conditioned laboratory. The relative humidity was nearly constant at 50 percent. After equilibration, filter papers were brushed to dislodge any loose filter remaining from manufacturing the perforated filter collectors and weighed. Filter weights could be reproduced to within 2 mg. After field sampling, the filters were again equilibrated prior to weighing.

Interstage Loss

As indicated earlier, particles deposit in the impactor not only at the designed collection sites on the perforated filter paper, but also principally on the back side of the metal collector support plates. Minor amounts were also deposited on gaskets and on exposed metal within the filter collector perforations. To determine interstage losses, each stage collector support plate was brushed to remove most adhering dirt particles. This dirt was collected and weighed to determine the interstage loss for each stage.

In order to determine the size distribution of the interstage loss particles, particles brushed from the interstage surfaces of twenty impactors used at the Rocky Flats site were combined for each stage. Subsequently, these combined particles were sieved through a series of 210, 177, 149, 125, 105, 86, 74, 63, 53, 37, 20, and 10 μm sieves and subsequently weighed. Sieving was done with an Allen Bradley Co., model L3P sonic sifter.*

DATA ANALYSIS

Interstage loss on each support plate was normalized to the weight on the impactor stage immediately downstream of the interstage loss site. That is, the loss between each two support plates was divided by the weight found on the next perforated stage collector. Similarly, loss is also reported for the support plate immediately in front of the 8 × 10 in. back-up filter.

The effect of impactor orientation on each impactor stage collection was determined at identical sampling heights for a cowl-impactor system and an impactor without cowl. For the impactor without cowl, the impactor inlet was faced vertically down toward the ground surface or was faced up away from the ground surface. Weight collected on each stage was normalized to the corresponding stage collection in the cowl-impactor system by taking the ratio of weight collected on a stage for the non-cowl system to that collected on the same stage for the system with cowl.

RESULTS AND DISCUSSION

All results were obtained under field sampling conditions. The atmosphere over the Hanford reservation was dry and essentially unpolluted except for airborne soils resuspended by winds. In contrast, the atmosphere at Rocky Flats was often wet due to local thunderstorms and was probably more chemically

---

* General Electric Co., model 9T92A87, Variable Transformer, Fort Wayne, Ind.

** General Metal Works, Inc., model GMWL-2000- hi-vol air sampler with filter holder, 8368 Bridgetown Road, Cleves, Ohio 45002.

* Allen Bradley Co., model L3P Sonic Sifter, Milwaukee, Wisconsin.

polluted due to near-by industrial areas. By visual inspection of the Rocky Flats filters, it is believed that some increased filter weights must have been caused by other than airborne soil. A significant filter weight gain often occurred for these neutral pH collectors(1) without a correspondingly apparent filter collector discoloration. Identification of the possible chemical pollutants and weights was considered beyond the scope of the present research.

Normalized Interstage Losses

Interstage losses were determined at two Hanford sampling sites and one Rocky Flats sampling site. For the first Hanford site, losses were determined for both dusty and non-dusty atmospheres. In Figure 3, losses for clear days are shown as a function of the mg of dirt collected on each stage. The different symbols indicate impactor 50 percent cut-off diameters for unit density spheres. In general, the line through the data shows an interstage loss increase with an increase in stage collection. This suggested increase might[7] be expacted if particles rebounded from a partial monolayer of particles covering the jet area on the collector.

There is some indication that interstage loss is a function of airborne soil concentration and particle size distribution. The same trend line is also shown as a broken line in Figure 4 for sampling at the identical site during periods of high airborne dust concentration caused by wind produced soil resuspension. In this higher airborne concentration[13] case, data points in general show greater interstage losses than suggested by the broken line. Apparently, interstage losses could be more significant when sampling from a higher airborne particle concentration than for clear days.

Since interstage losses may depend on airborne soil type and particle size distribution, interstage losses were also obtained at a second Hanford site and the Rocky Flats site for relatively non-dusty days. Impactors without cowls were faced into the prevailing wind direction for the Hanford test, while cowl-impactor systems were used at Rocky Flats. Interstage loss data for all these visibly non-dusty time periods are shown in Figure 5. In these cases, there is a much greater range in the weight collection on each impactor stage. These data confirm that there is no relationship between interstage loss and stage loading. These data show that for other than dust storm conditions, soil particle interstage losses can be expected to be from about 1 to 20 percent. These soil interstage losses are greater than the 10 percent or less interstage losses reported[3] for 1 μm diameter particles. Interstage losses can be expected to depend on particle diameter.

Interstage Loss Particle Size Distributions

Interstage soil loss particle size distributions were determined for the combined weight loss for twenty impactors used at Rocky Flats. Each sample collected from between two adjacent stages was sized by sieving. Size distributions are shown in Figure 6 for each impactor stage. The average particle size in the interstage loss at each stage is much greater than physical design would permit to appear at that stage. The average particle diameter for the first stage is about 40 μm while the average particle size for interstage losses adjacent to the back-up filter is about 90 μm. The brushing techniques used to remove interstage loss particles from support plates might have preferentially removed larger particles for subsequent size analysis. Nevertheless, large particles should not penetrate so deeply through the impactor. Apparently, if larger than respirable particles are not collected on the 7 μm perforated collector some will bounce and be partially re-entrained all the way through to the back-up filter.

It should also be emphasized these interstage loss particles are much larger than the particle diameter which should deposit on each perforated collector. There is no apparent relationship between interstage particle diameter and stage cut-off diameters. Interstage loss has been normalized to the collector weight at that stage rather than to collector plus interstage loss.

The weight associated with each particle size for the interstage loss was examined. In Figure 7, interstage losses are shown for each support plate as a function of particle diameter. Particle diameter in this case is the sonic sifter sieve size upon which the coresponding weight was collected. For 20 μm particles, interstage losses are greatest for the 7 μm (1st) stage and decreased through the impactor to the back-up filter. Although not determined, a reasonable assumption is that a proportional collection on corresponding stage collectors might also be expected. Similarly, 37 μm particle weights show a decrease from inlet stage to back-up filter. For even larger particles, interstage losses are more nearly independent of stage number. The rapid increase in interstage loss for 105 μm particle is in part due to metal particles, (source not apparent since places and cowls were cleaned before assembly), and material coming from between stage gaskets.

Sampler Orientation

Two purposes of the rotating cowl were: (1) to collect[13] by gravity settling within the cowl body the airborne particles greater than about 63 μm which enter through the cylindrical sampler inlet and, (2) to minimize or eliminate particle sampling bias caused by failure to direct the sampler inlet continually into the wind. Anisokinetic sampling errors, however, are not eliminated by this design. Since air flow rate is held constant at 20 cfm as required by impactor design, air entering the 6 in. diameter cylindrical sampler inlet corresponds to a wind speed of only 1.2 mph. At higher wind speeds ambient air would always be sampled subisokinetically. Consequently, particle inertia will consistently cause more particles to enter the sampler inlet than were initially in air drawn through the sampler inlet.[10] Airborne particle concentrations calculated from this increased particle collection will be greater than actual concentrations.

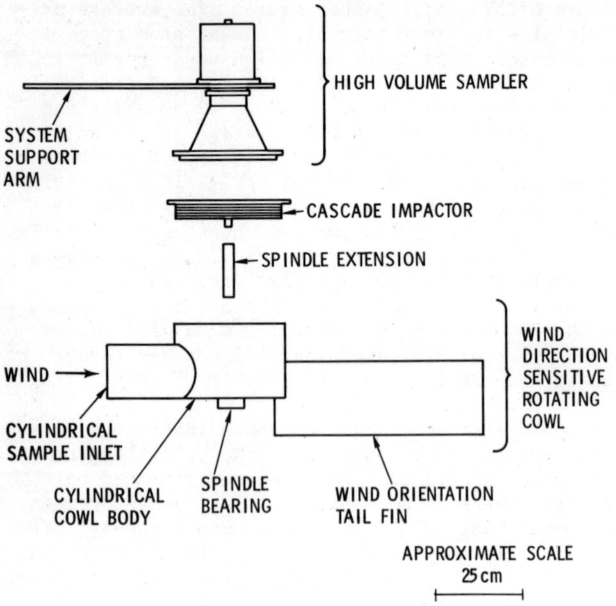

FIGURE 1. EXPLODED VIEW OF ROTATING COWL AND IMPACTOR

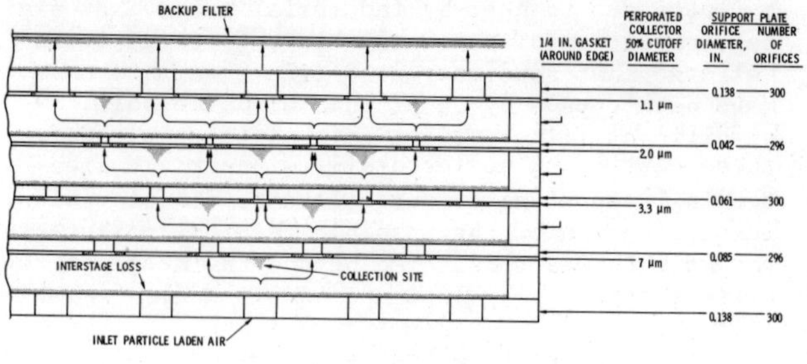

FIGURE 2. ORIFICE PLATE CASCADE IMPACTOR, PARTIAL SECTION

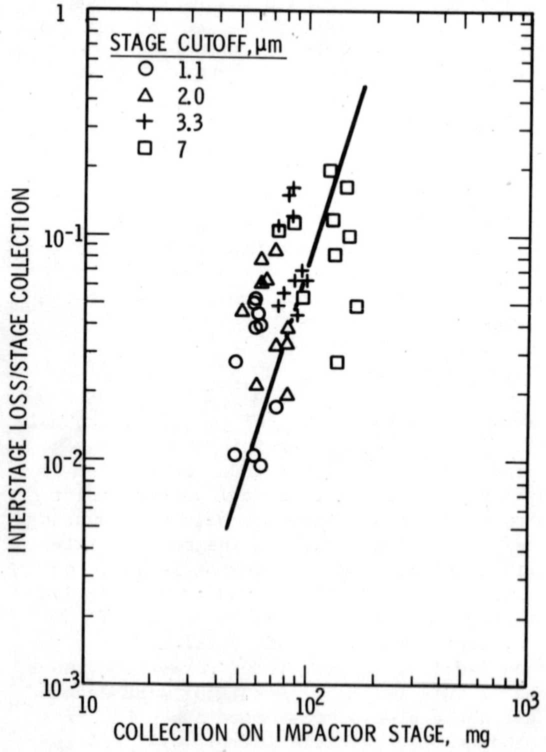

FIGURE 3. IMPACTOR INTERSTAGE LOSSES DURING A 430 HOUR SAMPLING PERIOD

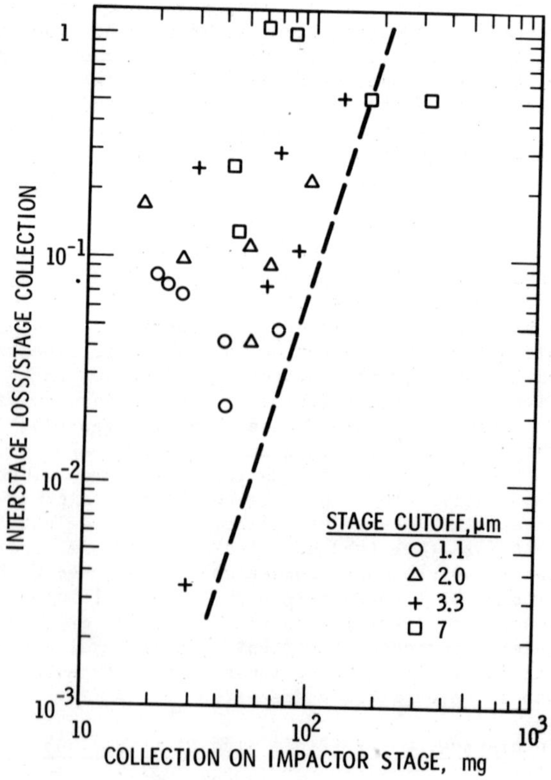

FIGURE 4. IMPACTOR INTERSTAGE LOSSES DURING A 38 HOUR SAMPLING PERIOD

Airborne particle soil collection on each impactor stage was determined at identical sampling heights for a cowl-impactor system and an impactor without cowl. The effect of impactor orientation was determined using a meteorological tower with impactors located at the 0.3, 1, 2, 3 and 10 m heights above ground in a semiarid Hanford region. Particles collected were wind-caused resuspended soil. Impactor orientation comparisons are shown in Figure 8 as a function of the 50 percent cut-off particle diameter at each impactor stage. Also shown is a comparison for particles less than 1 μm collected on the filter back-up. Curves show in general that: (1) normalized stage collection depends on particle diameter; (2) an impactor faced upward will collect, stage for stage, from 25 percent less to 50 percent more than a cowl-impactor system; (3) an impactor faced downward will collect up to about 40 percent less than a cowl-impactor system facing into the wind; and (4) normalized collection is also unexpectedly a minor function of sampling height. This minor function must be the result of different wind speeds and airborne particle concentrations as a function of sampling height. Although additional data would be required to completely explain these orientation effects, airborne particulates are most accurately sampled when a wind direction-sensitive inlet or self-orienting cowl is attached to an impactor.

CONCLUSIONS

These interstage loss and sampler orientation evaluations of impactors and cowl-impactor systems under atmospheric conditions in the field have pointed out some limitations in cascade particle impactors in practical usage. The limitations include: (1) interstage losses for ambient particles in this impactor usually range from 1 to 20 percent; (2) interstage deposited particles are not related in size to the 50 percent cut-off diameter for the stage; (3) particles larger than respirable size penetrate through the impactor to the back-up filter, and (4) the impactor should always be faced into the wind. These limitations, while severe, are also limitations which the field experimentalist must at present still accept. A particle sizing system better than the impactor concept is yet to be designed for both *in situ* sizing airborne particles and collecting particles for subsequent analysis.

Interstage losses for field experiments are greater than interstage losses for 1 μm particles under laboratory conditions. Nevertheless, realistic field measurements of resuspended radioactive particles can be made with this impactor system. Since laboratory experiments have shown that impactors operate satisfactorily for respirable sized particles, the important question then concerns the effect of larger airborne particles on airborne concentrations and size distributions. If resuspended radioactive particles are in the respirable size range, size distributions from field sampling with a large volume impactor will probably be comparable to results from laboratory measurements.

However, in application to field sampling, should radioactive contamination on large particles represent an appreciable fraction of the total airborne radioactive material, interstage losses would be very significant. The attendant penetration and deposition of larger particles on stage collectors beyond the first would make it appear that the air sampled had a larger number of particles in the respirable size range than actually present. This would always bias the sample conservatively toward assessing the potential respirable particle hazard as greater than actual. This study did not address the important question of the degree to which the stage collector itself collects these larger airborne particles; hence, the quantitative significance of errors from this source are as yet unknown, and will be a topic for study.

Those using or intending to use large volume impactors are thus cautioned that results may be biased when the impactors are to be used where large particles (> 20 μm) are in the stream sampled.

ACKNOWLEDGEMENT

This paper is based on work performed under U.S. Atomic Energy Commission Contract No. AT(45-1)-1830.

REFERENCES

(1) Burton, R.M., J.N. Howard, R.L. Penley, P.A. Ramsay, and T.A. Clark, "Field Evaluation of the High-Volume Particle Fractionating Cascade Impactor," J. Air Pollution Control Association, Vol 24 (1973), 277-281.

(2) Davies, C.N., "The Aspiration of Heavy Particles into a Point Sink," Proc. R. Soc. A. London, Vol 279 (1964), 413-419.

(3) Erickson, C., Andersen 2000 Inc., P.O. Box 20769 AMF, Atlanta, Ga., 30320, (1973) unpublished data.

(4) Glauberman, H., "The Directional Dependence of Air Samplers," Amer. Ind. Hyg. Assoc. J., Vol 23 (1962), 235-239.

(5) Green, H.L., W.R. Lane, Particulate Clouds: Dusts, Smokes, and Mists, 2nd Ed., D. Van Nostrand Co., Inc., Princeton, N.J. (1964), 271-275.

(6) Lundgren, D., S. Calvert, "Aerosol Sampling with a Side Port Probe," Amer. Ind. Hyg. Assoc. J., Vol 28 (1967), 208-215.

(7) Mercer, T.T. "Air Sampling Problems Associated with the Proposed Lung Model," CONF-661018, C.M. West, compiler, Proc. 12th Annual Bioassay and Analytical Chemistry Meeting, Gatlinburg, Tennessee, October 13-14, 1966, 87-97.

(8) Raynor, G.S., "Variation in Entrance Efficiency of a Filter Sampler with Air Speed Flow Rate, Angle, and Particle Size," Amer. Ind. Hyg. Assoc. J., Vol 31 (1970), 294-304.

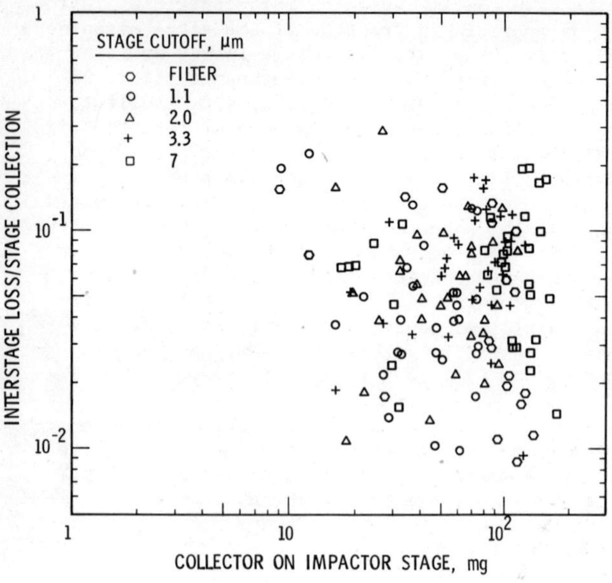

FIGURE 5. IMPACTOR INTERSTAGE LOSSES

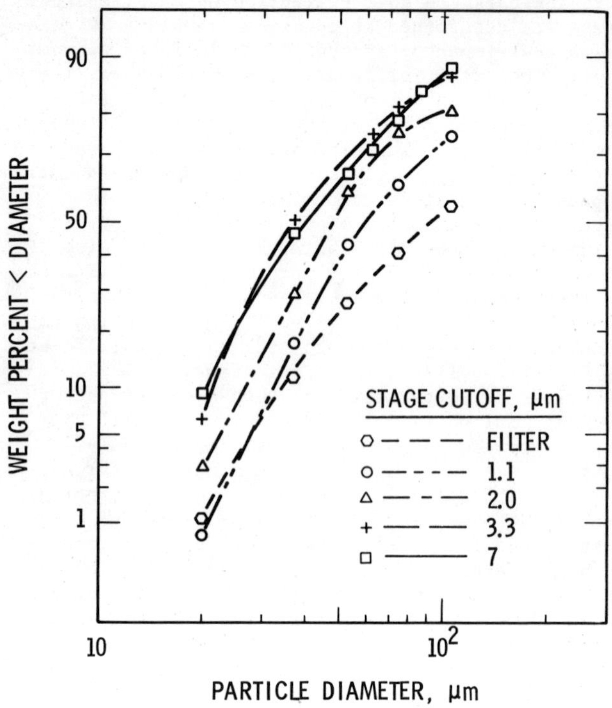

FIGURE 6. SIZE DISTRIBUTIONS FOR INTERSTAGE LOSS PARTICLES

STAGE WALL LOSSES

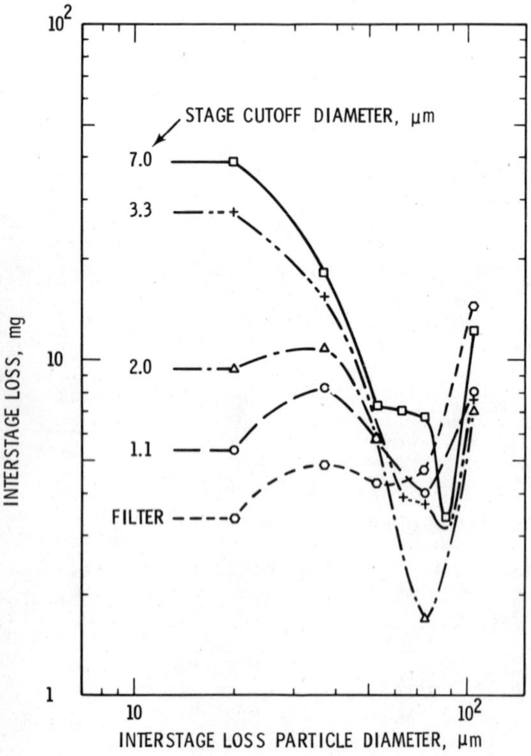

FIGURE 7. INTERSTAGE LOSSES

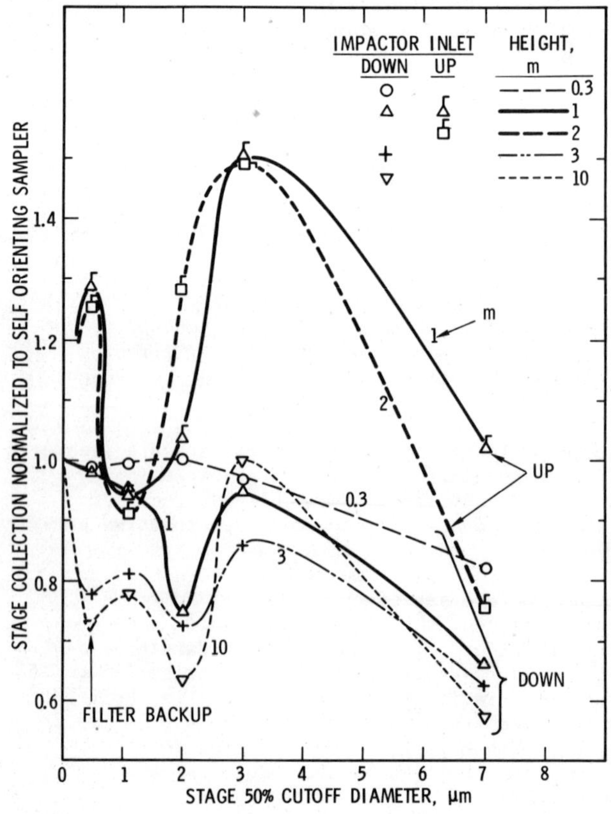

FIGURE 8. EFFECT OF IMPACTOR ORIENTATION ON STAGE COLLECTION

(9) Rüping, G., "The Importance of Isokinetic Suction in Dust Flow Measurement by Means of Sampling Probes," Staub-Reinhalt Luft, Vol 10 (1968), 1-11.

(10) Sehmel, G.A., "Estimation of Air Stream Concentration of Particulates from Subisokinetically Obtained Filter Samples," Amer. Ind. Hyg. Assoc. J., Vol 28 (1967), 243-253.

(11) Sehmel, G.A., "Particulate Sampling Bias Introduced by Anisokinetic Sampling and Deposition within the Sampling Line," Amer. Ind. Hyg. Assoc. J., Vol 31 (1970), 758-771.

(12) Sehmel, G.A., "Errors in the Subisokinetic Sampling of an Air Stream," Ann. Occup. Hyg., Vol 10 (1967), 73-78.

(13) Sehmel, G.A., Influence of Soil Erosion on the Airborne Particle Size Distribution Function, paper 73-162 presented at the 66th Annual Meeting of the APCA, Chicago, Ill., June 24-28, 1973.

(14) Vitols, V., "Theoretical Limits of Errors Due to Anisokineitc Sampling of Particulate Matter," J. Air Polluction Control Assoc., Vol 16 (1966), 79-84.

(15) Watson, H.H., "Errors Due to Anisokinetic Sampling of Aerosols," Amer. Ind. Hyg. Assoc. J., Vol 15 (1954), 21-25.

©1973, ISA JSP 6673

THE APPLICATION OF THE CORRELATION SPECTROMETER TO AMBIENT AIR QUALITY AND SOURCE EMISSIONS

Lee Langan
Environmental Measurements, Inc.
San Francisco, California

Andrew J. Moffat
Barringer Research Ltd.
Rexdale, Ontario

Abstract

The applications of the Correlation Spectrometer, designed as a remote sensor to measure the content of sulfur dioxide and/or nitrogen dioxide in the open air, have been expanded and compared to other measurement methods for verification of the results. Data accumulated between 1971 and 1973 are used to depict the use of this remote sensor for developing an emission inventory, for describing the distribution of gases and their relationship to ground-level concentrations and as an open-path ambient monitor.

Introduction

The Correlation Spectrometer is a spectrophotometer which utilizes a replica of the absorption spectra of the target gas as its means of identification. Since 1967, Barringer Research has produced instruments to measure two principal air pollutants by this method, sulfur dioxide (measured around 3150 angstroms) and nitrogen dioxide (measured around 4400 angstroms). As the United States licensee for this Barringer equipment, Environmental Measurements has conducted surveys since 1968.

The principle of operation of the instrument and displays of results of its use have been previously described.(1,2,3,4) Advances in the instrumentation are discussed in a companion paper at this conference. (5)

This paper reports on applications in which the Correlation Spectrometer (COSPEC) is compared with or used in conjunction with other instrumentation. In this manner, the uniqueness of the COSPEC data can be viewed in the extra dimension it adds to air quality sampling.

Most analyzers used in measuring gaseous air pollution measure the *concentration* of the gas in parts-per-million which is directly equivalent to mass-per-unit-volume. The remote sensor, which views gas in the open air over an undetermined pathlength, measures the *concentration-pathlength* product. The unit of measurement is parts-per-million meters. When the remote sensor is pointed vertically upward (a right angle front-surface mirror is used for convenience, Figure 1), this remote sensor measures *total burden*. The concentration-pathlength may then be stated in grams-per-square-meter (total overhead mass) as follows:

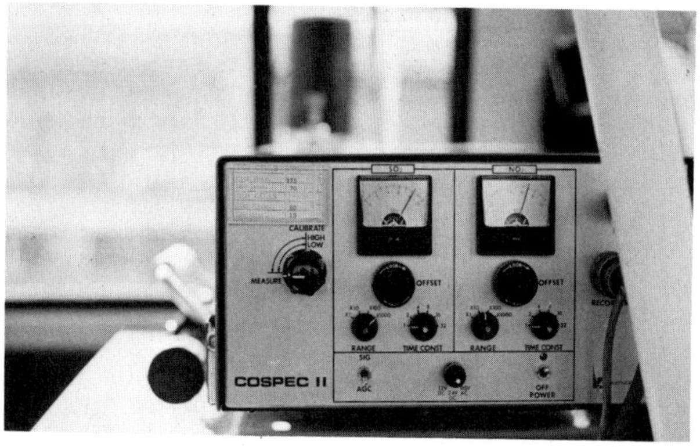

Figure 1. Correlation Spectrometer; dark cylinder is fastened to 45° mirror.

$$\text{ppmM} \times \frac{\text{molecular weight}}{22{,}400} =$$

$$\frac{\text{gram-meters}}{\text{cubic meter}} = \frac{\text{grams}}{\text{meter}^2}$$

Alternatively when the remote sensor is pointed at a light, the intervening distance (in meters, M) is used to convert concentration-pathlength directly to the average concentration in ppm:

$$\frac{\text{ppmM}}{\text{M}} = \text{ppm}$$

The following compendium of data are presented in three sections:

1) Results comparing the use of the COSPEC to measure emissions of a specific source are compared with alternative instrumentation, using two-gas remote sensor data alone, measurements to verify mathematical models, and those where no other technique could be used.

2) Measurements gathered on-the-move are used to demonstrate the use of the COSPEC in conjunction with rapid-response physical monitors which measure ground-level concentrations.

3) Results of continuing experiments using the remote sensor as an open-path monitor of ambient air.

Emission Measurements

The Total Burden data can be used to measure the rate of flow of the gas across a route of travel. Mass flow can be calculated for each segment as follows:

$$F = BvL\sin\emptyset$$

F is flow in grams per second; B is total burden (grams per square meter); v is wind speed in meters per second; L is length of the segment in meters; Ø is the road/wind angle.

Source Emission Comparison

In cooperation with the Kaiser Steel Corporation, in 1971, a one-hour comparison was made between source emissions measured by a wet-chemical in-stack sampler and two COSPEC instruments (6). The sensors were manufactured in separate lots in approximately March and September 1971 and were placed side by side in an open-roofed vehicle. A one-kilometer-long traverse route was selected to circumnavigate a sinter plant stack inside Kaiser's Fontana Steel Mill. Wind data were gathered from an adjacent anemometer mounted at about stack height on top of an adjacent water tower.

Between 1400 and 1500 hours, an average bubbler measurement was made in both breaches entering the stack, and eight encircling passes were made around the stack. The wind was blustery and ranged from .9 to 3.6 meters per second (2-8 MPH) over an arc of about 90°: wind speed is the greatest source of error in these measurements.

The results, shown in Table I show the average emission daily rate obtained with each COSPEC to be within 2.5 percent of each other over the hour. The standard deviation of the data was 24%.

Table I

Time 22 Oct 71 (PST)	Wind SPEED (MPH)	EPA COSPEC (MT/D)	EMI COSPEC (MT/D)
1406-1413	5	25.7	25.8
1413-1419	5	29.9	29.1
1419-1427	5	21.5	22.4
1427-1432	5	26.0	23.4
1432-1438	4	14.3	15.7
1438-1443	5	18.9	21.6
1443-1448	4	27.4	25.1
1450-1456	6	30.9	36.2
Averages:		24.32	24.91

The in-stack results measured 647 ppm at one input and 496 at the other. A sample average value of 577 ppm was used with a flow rate of 272,000 standard cubic feet per minute. (The flow was estimated from current readings which were compared to previous velocity relationships). These relationships are equivalent to a single average flow rate of 17.3 metric tons per day, 70% of the mean COSPEC results.

Emission Buildup

NO_2 is created by the oxidation of NO which derives from combustion sources. The rate of oxidation is affected by the presence of ozone and other chemicals which contribute to photo-chemical reactions. A set of data gathered in September 1972 near an oil refinery in Southern California have been used to demonstrate the buildup of NO_2 over a seven kilometer distance. A *flow map* was produced by a computer-driven plotter where each line, placed in the direction of wind flow, represents a specific number of grams per seconds (tons per day) flowing across the intersecting transect. Figure 2 shows the flow of NO_2 downwind of an extended, complex source of gas, the oil refinery. A similar map (not shown) was prepared for SO_2.

The average mass flow on seven adjacent traverses was 34.5 tons per day of sulfur dioxide; alternatively the NO_2 appeared to increase from approximately 15 tons per day, near the refinery, to approximately 45 tons per day at ten kilometers distance (about 30 minutes away). Because of the consistency of the SO_2 data, each SO_2 transect was used to normalize the NO_2 results, While the NO_2 and SO_2 did not come from exactly the same sources, within the complex, they did derive from the same refinery. As they dispersed downwind, the same meteorological effects controlled the dispersion of both gases and the normalization (depicting the ratio of NO_2 results to the SO_2 results) more clearly demonstrates the NO_2 buildup. The mass flow data are shown in the graphs of Figure 3.

Model Verifications

The emission rates established with the Correlation Spectrometer have been used as the input "Q" in Gaussian plume modeling. Subsequent calculations of ground-level concentrations can then be compared, or direct comparisons to alternative emission calculations can be made.

The comparison between values measured with the COSPEC and emissions calculated on the basis of fuel analyses has been made in the Maryland Power Plant Siting Program (4). A plot of the ratio of the measured-to-calculated SO_2 was prepared from 44 calculations taken at the Wagner Power Plant near Baltimore, Maryland, and 74 calculations taken at the Chalk Point Power Plant near Benedict, Maryland, Figure 3 is a plot of the standard deviation of the ratio about the means obtained. The mean values of the ratio are close to unity, while the standard deviation of these means are large (0.85 at Wagner and 0.7 at Chalk Point).

These data were gathered over a period of several weeks. Variations in the daily means were likely due to inaccurate wind speed measurements or variations in the sulfur content of the fuel. The variation gathered on any given day is much less, typically half that shown on the graph.

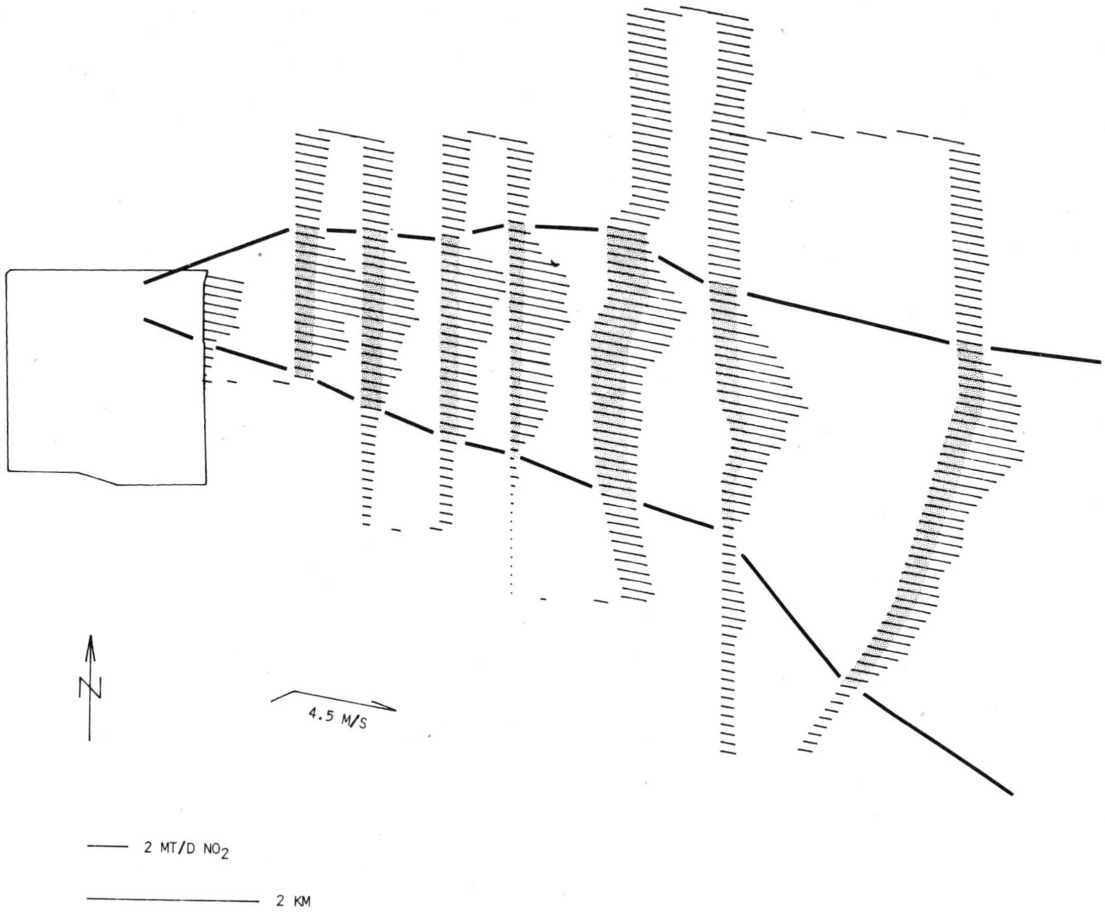

Figure 2. Flow of nitrogen dioxide as it builds up downwind of an oil refinery (located within the NE quandrant of the mapped roadways). Approximate boundary of plume is sketched; contribution of regional background is shaded.

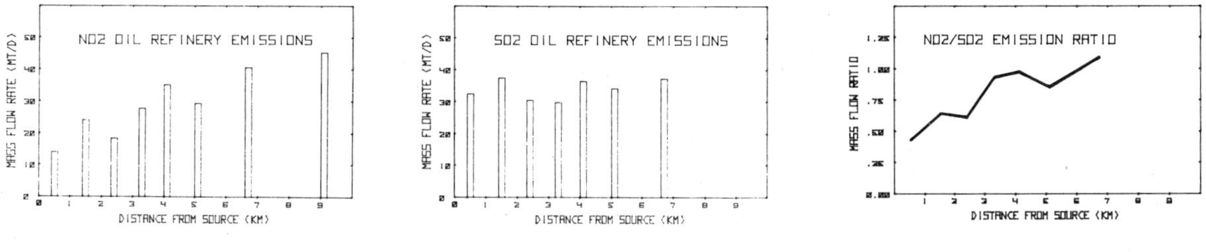

Figure 3

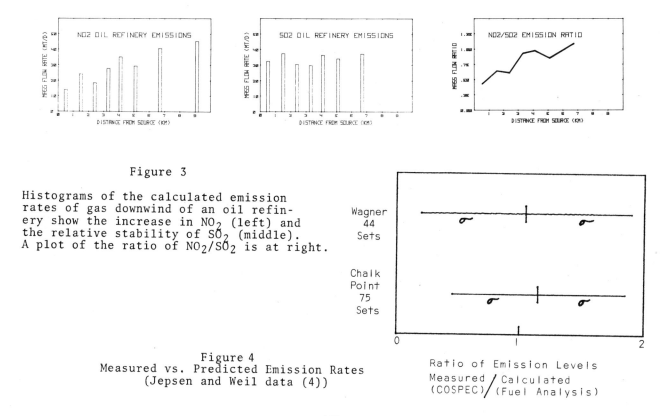

Histograms of the calculated emission rates of gas downwind of an oil refinery show the increase in NO_2 (left) and the relative stability of SO_2 (middle). A plot of the ratio of NO_2/SO_2 is at right.

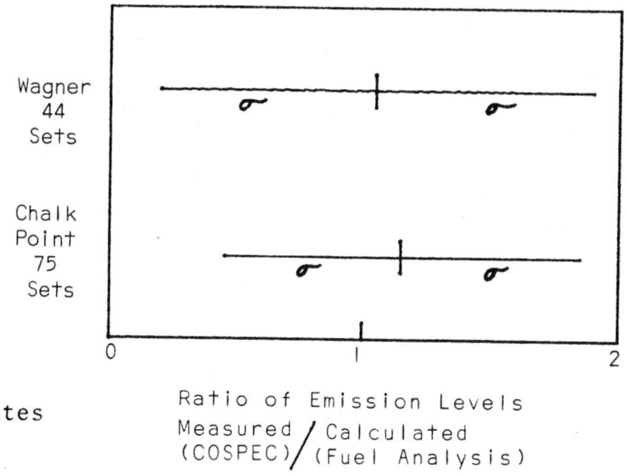

Figure 4
Measured vs. Predicted Emission Rates
(Jepsen and Weil data (4))

Natural Emissions

There are circumstances where the remote sensor appears to be the only reliable means of obtaining source-emission rates. Stoiber and Jepsen (6) reported on data gathered in 1972 which describes flows from each of the major volcanoes in Central America. Langan and Moffat (9) and Okita (10) reported in company publications measurement days in Japan where volcanic plumes were measured. Table II lists these reported levels of flow; the larger volcanoes rival power plants in their outflow of SO_2.

Table II

Volcano	SO_2 (MT/D)
Guatemala	
Santioguito	420
Fuego	40
Pacaya	260
Nicaragua	
San Cristobal	360
Telica	20
Momotombe	50
Masaya	180
Japan	
Mihara	345
Asama	142

Moving Measurements

When the COSPEC is placed together with more conventional air quality measuring equipment, as shown in Figure 5, concurrent measurements of the overhead distribution of sulfur dioxide and nitrogen dioxide can be compared with the simultaneous measurement of ground-level concentrations of these gases. It is the combination of both techniques which makes the data most useful.

Alone, the ground-level concentrations which relate to sulfur are difficult to distinguish, if more than one source are present, and are often difficult to interpret due to local micrometeorological effects.

Ground-level measurement of NO and NO_2 are more severely hampered than are sulfur results because of the contamination of automobiles. Considerable care is needed in excluding anomolies created by passing automobiles or traffic from that created from sources. Frequently it is necessary to drive the vehicle from the normal route of travel to an undisturbed area to get the few-moment measure of the plume's effect.

Three examples provide a variety of simultaneous measurements to demonstrate this usefulness.

Houston, Texas

An extensive survey of the Houston Ship Channel (3) is illustrative of an ability to identify the contribution to ground-level gases in an area where a number of different sulfur contributors exist (Figure 6). Because of the demonstration nature of these data, specific facilities were not identified along the route of travel; yet, individual plumes are identifiable. The two maps show the SO_2 mass flow across highways on both sides of the Channel and the local peaks of gas reaching the ground along these routes.

Three hours were utilized in traversing the 45 mile route during the demonstration, as several stops were made. In this composite record, four wind speeds were used to compute the mass flow. It is estimated that on the order of 1000 to 2000 metric tons per day were flowing out of the channel during the survey period. Several individual sources were contributing over 100 tons per day toward this total. The ground-level map shows the varying concentrations of sulfur gas measured along the same survey route. The highest total sulfur peak measured was 140 parts-per-billion (.14 ppm).

Figure 5

A self-contained moving laboratory, here in San Francisco, contains a flame photometric analyzer to measure total sulfur, a chemiluminescent monitor to measure nitrogen oxides, a remote sensor to measure total burdens and auxillary recording, calibration and power instrumentation.

St. Louis, Missouri

In an example of a rural set of concurrent measurements, St. Louis, Missoui, was treated as a single source. A travel route, extending 120-kilometers to the North of the Gateway Arch, is shown in Figure 7 together with the sulfur dioxide burden, ground-level sulfur concentration and the NO_2 burden. The City at this distance appears as a source in itself. In addition, the plume on the left probably results from a major minehead power plant in Southern Illinois. The peak values of ground-level sulfur measured were 80 parts-per-billion in the vicinity of Carrollton, Illinois. The total flow of sulfur was measured as 5353 metric tons per day (8).

Figure 6

The Houston, Texas, Ship Channel map on the right depicts the composite flow of SO_2 during a three-hour period in March 1973. Wind arrows show the direction during the time of each traverse segment. Ground levels of total sulfur are shown in a similar map (below) as lines plotted normal to the route of travel.

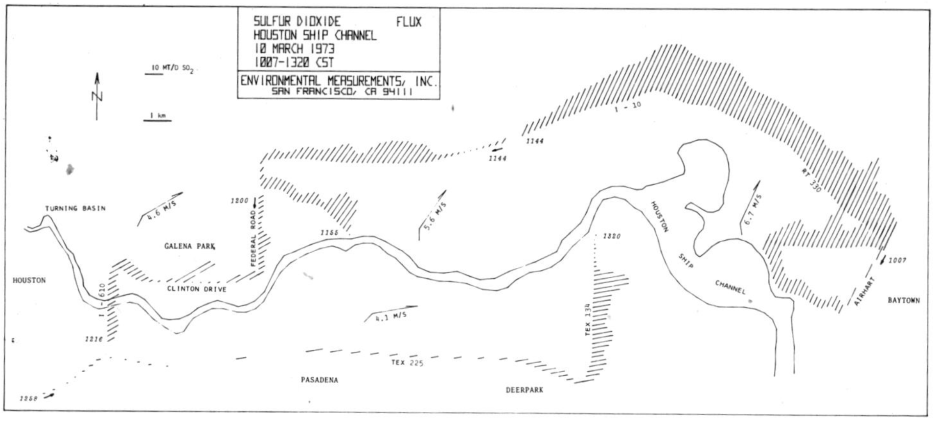

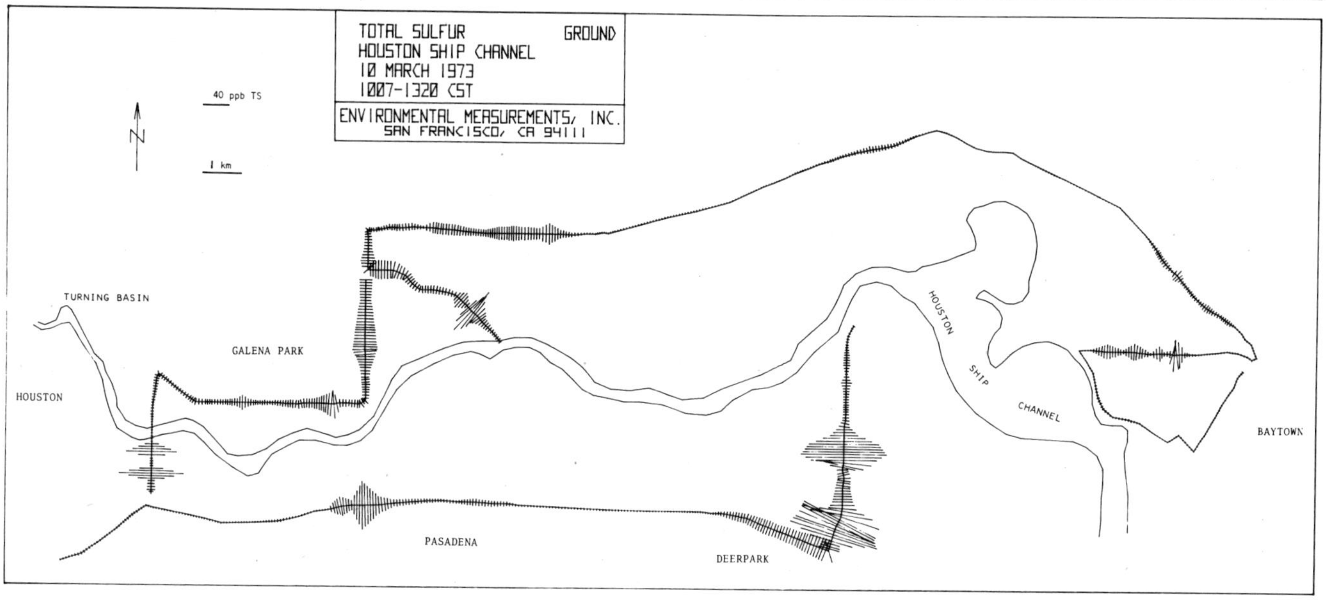

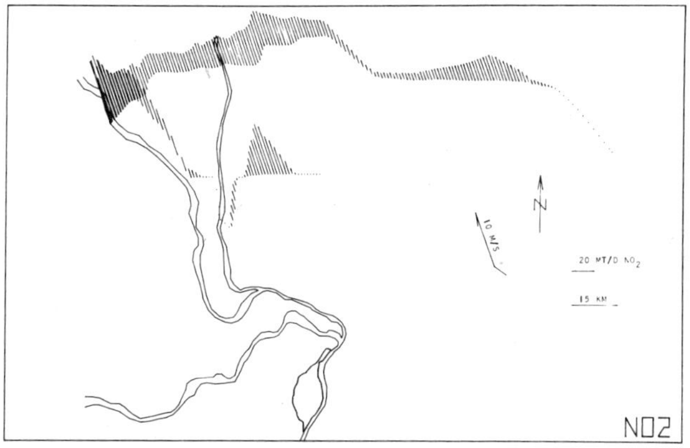

Figure 7

Three regional maps depict pollutant flows (left) and ground-level gas (below) for 14 April 1973. The Mississippi, Missouri, and Illinois Rivers, and the city of St. Louis (shaded area) provide geographical references. Scales of flow lines are noted; transport wind speed labels wind arrows.

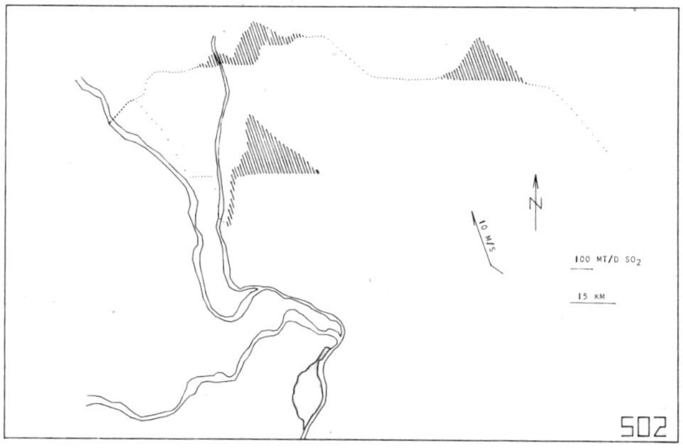

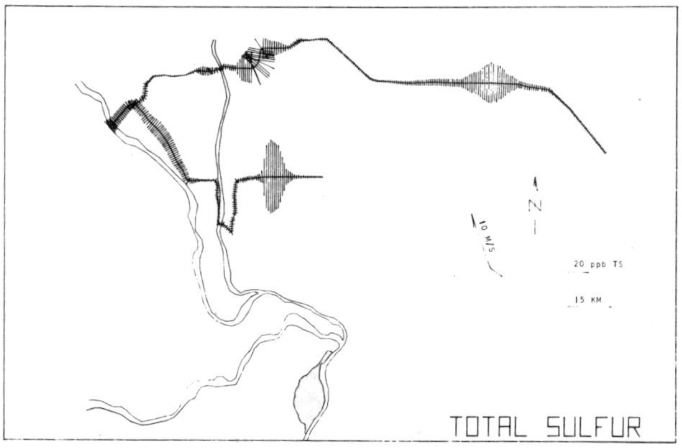

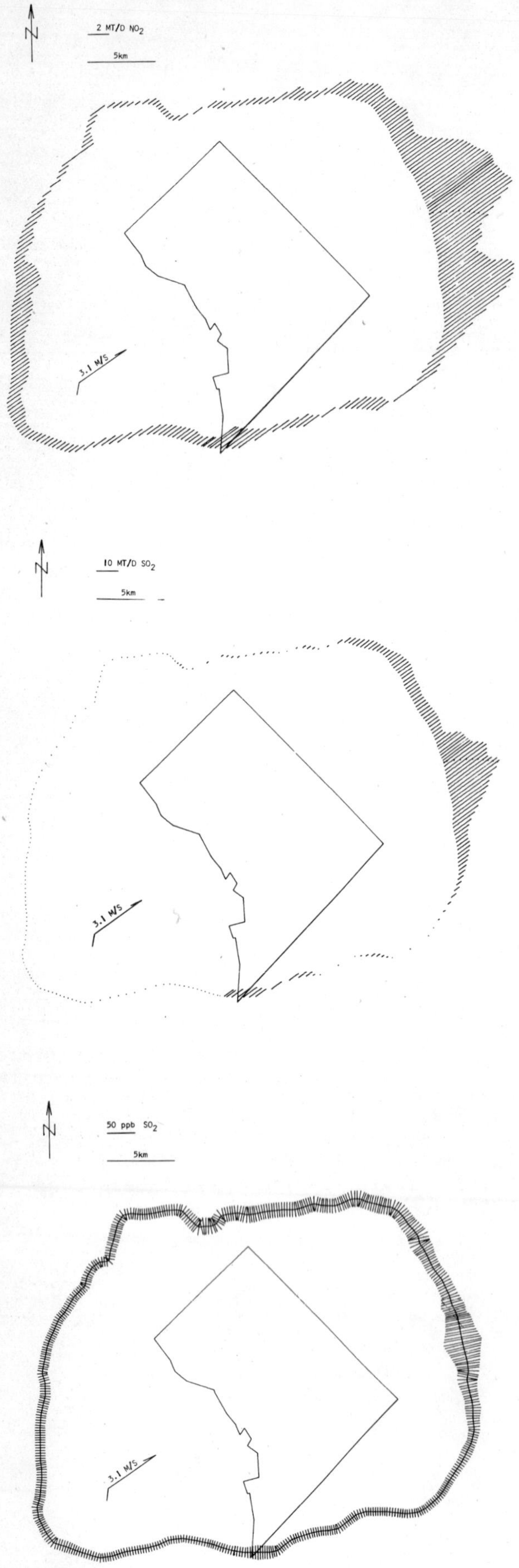

Figure 8

Flow and ground-level maps for 30 August 1973 (left) plotted from data collected 1245-1350 EDT. The traverse on I-495 began and ended at the lower center (6 o'clock) point on the beltway. The District of Columbia is outlined for reference.

Washington D.C.

As an example of a traverse route which encircles an entire metropolitan area data have been selected from Washington D.C. The three maps shown in Figure 8 clearly define the flow of gases from the area within the beltway and show the ground-level concentrations along this roadway. Four power plants contribute to the sulfur dioxide flow; three are within the traverse route and one is a few miles south of it. The total sulfur dioxide flow from the area, on the order of 700 metric tons per day, probably results from a combination of these specific sources and that contribution of sulfur dioxide made by automobile traffic.

Longline Ambient Monitoring

The remote sensing Correlation Spectrometer can be used as an ambient air quality monitor by directing it at a suitable light source. To establish longer paths, over which the average concentration of SO_2 or NO_2 can be measured, a xenon light source has been produced as a COSPEC accessory. In this application, the instrument measures the absorption of each gas on the broader xenon emission spectra. The pathlength between the light and the COSPEC is used to determine the sensitivity of the system; typically a few parts-per-billion is established over a kilometer long line.

In Beltsville, Maryland, in June, 1973, experiments were conducted advancing the development of this open-path technique. In cooperation with the Environmental Protection Agency (11) a kilometer-long line was established parallel to a suburban roadway; a chemiluminescent monitor was mounted on an electrical cart to traverse along the path to measure variations in nitrogen dioxide. Without contributing to the levels of NO_2 in the ambient air, contamination in the ambient air, along the path, could be detected.

The COSPEC system is modified internally to adapt for 24-hour, long-line measurements. A phase-lock detector is included in the electronics to enable it to identify the absorption only from the electronically modulated xenon source; thus, the potential interference of absorption of daylight is removed. Figure 9 is a photograph of the well-collimated xenon light beam showing in the early morning mist. Figure 10 shows the EPA monitoring cart prepared for a parallel traverse.

Figure 9

Xenon light beam produces path for ambient pollutant monitoring. Tent protects equipment from the weather.

Figure 10

Parallel monitoring along the light path using a mobile cart provides comparative data.

The levels of NO_2 experienced over a five-day period of experimentation remained unexpectedly low. Nevertheless, under ambient levels of NO_2 frequently less than 25 parts-per-billion, two chemiluminescent monitors and the COSPEC agreed often within 10% and always by a factor of two.

An opportunity for a longer and more extensive use of the longline presented itself later in the year, when data were derived as input data being gathered for regional model studies in the San Francisco area (12). Figure 11 is a graph of the results obtained at San Jose, California, on September 29, over a 1080-meter line for a fifteen-hour period. The ambient levels of NO_2 were measured with both the longline and a chemiluminescent monitor located at one end of the line. Concentrations varied from 160 ppb to 50 ppb (0.16 to 0.05 ppm) and were observed on both systems. Noise levels of both instruments, at this urban site, are seen in the data.

This application remains in its developmental stages and efforts are continuing to evaluate the remote sensor as an alternative means of monitoring the open air in a fashion which does not perturb the substance being monitored and allows free passage of this substance over an area which is representative of the region under study. Data which represent kilometer-long or mile-long cells may be most useful as inputs to and verification of regional air pollution models.

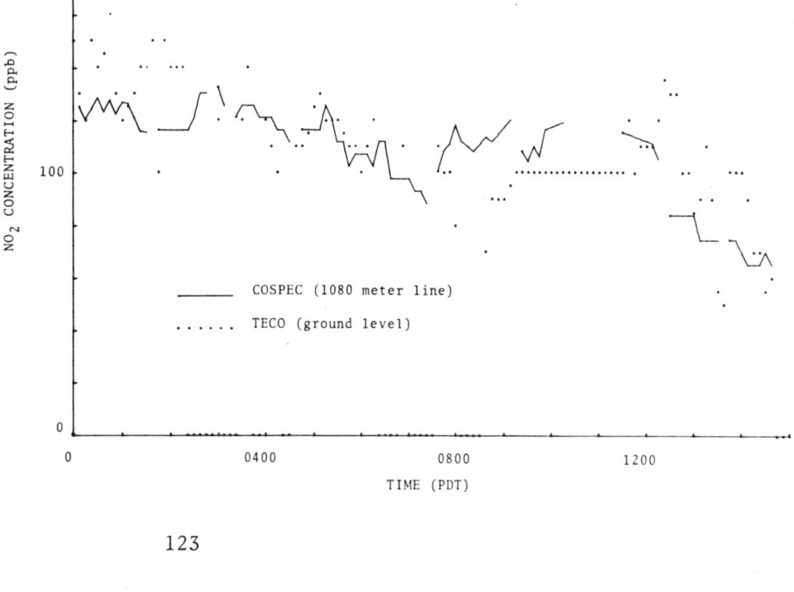

Figure 11

NO_2 concentrations as measured in San Jose, California are compared between a Barringer Correlation Spectrometer (lines) and a Thermo Electron Chemiluminescent Monitor (dots). Gaps in the longline data represent times when the system was zeroed; TECO results are six minute averages.

Conclusions

These data have been presented as representative of the extensive use now made of the correlation spectrometer remote sensor as a complementary means of measuring air quality data. The routine use of the COSPEC, now common in several projects, adds an extra dimension to the understanding of the distribution of gases and the inter-relationships of adjacent sources.

An efficient and rapid means of producing an emissions inventory for the key gases, sulfur dioxide and nitrogen dioxide, is now available. The alternate use of the same type of instrumentation for measuring the ambient air directly adds an additional dimension to its value.

References

1. Langan, Lee, "Remote Sensing Measurements of Regional Gaseous Pollution", 1st Joint Conference on Sensing of Environmental Pollutants, Palo Alto, California, AIAA Paper No. 71-1060, November, 1971.

2. Langan, Lee, "Gaseous Pollution Measurements Along Paths of Light", 65th Annual Meeting of the Air Pollution Control Association, Miami Beach, Florida, Paper No. 72-28, June 1972.

3. Sperling, Roger B, "In-transit Air Quality Measurements", 1st Annual Symposium on Air Pollution Control in the Southwest, College Station, Texas, November, 1973.

4. Jepsen, Anders F. and Weil, Jeffrey C., "Maryland Power Plant Air Monitoring Program Preliminary Results", 66th Annual Meeting of the Air Pollution Control Association, Chicago, Illinois, June, 1973.

5. Barringer, A.R. and Davies, J.H., "Further Developments in Correlation Spectroscopy for Remote Sensing Air Pollution", 2nd Joint conference on Sensing of Environmental Pollutants, Washington, D.C., December, 1973.

7. These data were originally reported to the Environmental Protection Agency in December, 1971, and were partially sponsored under their Contract No. 6H-02-0124 (Task A).

6. Stoiber, Richard E. and Jepsen, Anders, "Sulfur Dioxide Contributions to the Atmosphere by Volcanoes", Science, Vol. 182, p. 577-578, Nov., 1973. References therein: Okita, Jasco Report 8; EMI, AirNote Volcanic SO_2.

8. The rural data north of St. Louis were gathered in cooperation with the of Air Pollution Study team of the Fate Natural Center for Atmosphere Research. Further details of this program, describing NCAR results from 1972 (when EMI also provided regional COSPEC data) are recently published: Breeding, R.J. et al.,, "Background Trace Gas Concentrations in the Central United States", J. Geophys. Res., 78, No. 30, October, 1973.

9. Langan, Lee and Moffat, A.J., Barringer Research, Vol. 5, No. 1, Spring, 1974.

10. Okita, Toshiichi, "Detection of SO_2 and NO_2 Gas in the Atmosphere by Barringer Spectrometer", JASCO Report, Vol, 8, No. 7, July, 1971.

11. Longline data obtained in June and July, 1973, were measured in cooperation with the Environmental Protection Agency and partially funded by EPA. Equipment, encouragement and assistance were provided by Dr. William McClenny, the senior EPA representative from the National Environmental Research Center, Research Triangle Park, North Carolina.

12. Figure 11 is a portion of the data delivered by EMI as subcontractor to a National Science Foundation RANN Grant No. GI-36390. EMI provided data to the Bay Area Air Pollution Control District which shares this NSF grant with the Lawrence Livermore Laboratories and the NASA Ames Research Center. Appreciation is given for permission to release these data.

©1973, ISA JSP 6674

REMOTE ACOUSTIC WIND SENSING

Martin Balser
Xonics, Inc.

The use of coherent acoustic radars to measure winds both in aircraft trailing vortices and in the ambient atmosphere has been under intensive investigation for nearly four years at Xonics. These efforts have led to the development of two systems. One of these is a fully engineered system that has been demonstrated to display in real time vortex tracks generated by heavy aircraft in routine commercial traffic at a major airport. The other is a system, actually a family of systems, for measuring ambient wind and wind profiles for meteorological, environmental and related applications. The latter are currently being manufactured for delivery in the near future. A test of such a wind sensor was conducted very recently to provide a comparative test against conventional sensors. This brief and rather informal paper describes those tests.

Figure 1 shows a typical wind-measuring configuration using two receivers. In this configuration, the assumption is made that the wind is horizontal and is therefore completely defined by two components. Figure 2 shows a configuration with three receivers. This system is capable of measuring three components, and can therefore measure the full vector wind velocity with no restrictive assumptions. In either case, the principle of wind measurement is the same, and is illustrated in Figure 3. The transmitter illuminates a vertical column of air. At a given height, some of the energy is scattered through an angle χ and picked up at the receiver. The receiver signal is found to be shifted in frequency from the transmitted frequency. The magnitude of this frequency, or doppler, shift is directly related to the component of the wind along the direction $\vec{q}$ which bisects the angle at the scattering volume between the rays to the transmitter and receiver. The components thus found in each of the (two or three) wind-measuring elements can be combined into a full wind vector and transformed to any desired co-ordinate system.

A key to the successful operation of such a system (or any acoustic system) is the use of appropriate techniques to suppress sidelobes in the antenna patterns so as to minimize annoying and interfering radiation from the transmitter and also interfering noise and direct signals at the receiver. Figures 4 and 5 are examples of actual measured antenna patterns of the transmitting and receiving antennas respectively that were used in the tests reported in this paper.

The prototype, two-component, wind-measuring system whose antenna patterns were just shown was recently transported to the White Sands Missile Range in New Mexico where the Atmospheric Sciences Laboratory of the U.S. Army Electronics Command had kindly made meteorological and support facilities available for a demonstration to them of the system's capability. The system was set up with the transmitter near the base of a meteorological tower and the two receivers 300 ft. away roughly south and west of the transmitter. The receiver beams were aimed to intersect the transmitter beam at an altitude of 300 ft. This common scattering volume was approximately 50 ft. removed from a cup anemometer and vane (Beckman and Whitley 170 - 41) at the 300 ft. level of the tower. While it would have been desirable to compare winds at the very same point rather than separated by 50 ft., the presence of the tower in the beam caused such strong scattering that the wind-scattered signals were frequently masked.

The processing in the acoustic system was carried out largely by a minicomputer in real time. The signals from the receiving antennas were first digitized, read into the computer, frequency analyzed and then examined by computer algorithms to determine the doppler shift and finally the wind velocity. The final wind velocity values thus determined were printed out on a teletypewriter. These values were compared with the strip-chart output of the tower anemometer.

Two examples of data obtained in this manner on 5 and 6 December 1973 are shown in Figures 6 and 7. The first, labeled Run 1, was taken under relatively low-wind conditions, the second, Run 11, with higher wind (principally in one component). In each case, the continuous curve is the output of the anemometer, while the curve made up of straight-line segments is the result from the acoustic system. The actual data points from the acoustic system occur once every two seconds, which is the spectral processing interval, and are connected by straight lines. As can be seen in the two figures, the agreement between the two data sources is generally good, differences being mostly in the 1-2 mph range. Such differences seem quite reasonable considering the separation in location between the two measurement points. Even the delay between the two curves near the start of Figure 7 is consistent with the time required for the wind to cover the distance between the two measurement points.

In summary, a remote doppler-acoustic wind sensing system has been successfully demonstrated against a standard meteorological wind sensor mounted on a tower at 300 ft. altitude. The small differences between the two sets of measurements are reasonable and in some cases explainable in view of the separation between measurement points. Several configurations of remote wind-measuring systems using the same acoustic radar approach are now available to meet a wide variety of meteorological and environmental requirements.

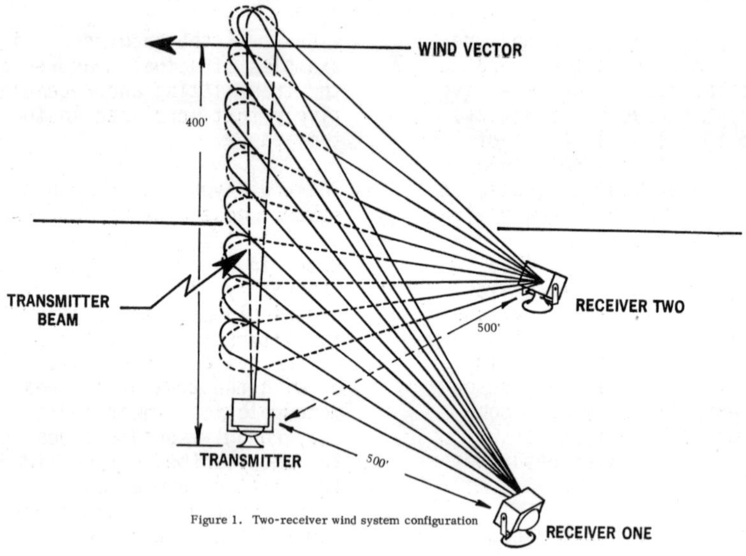

Figure 1. Two-receiver wind system configuration

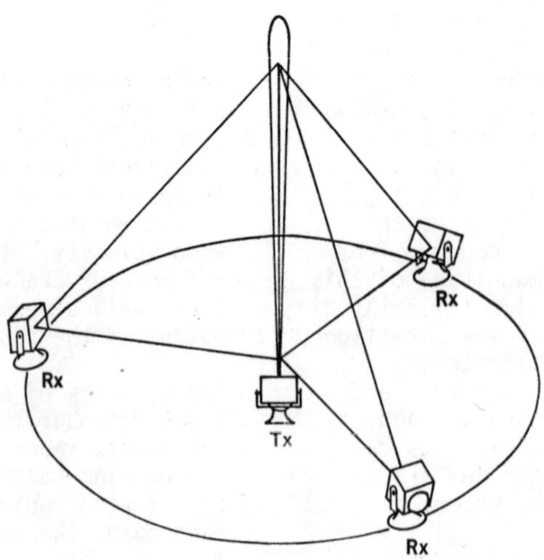

Figure 2. Three-receiver wind system configuration

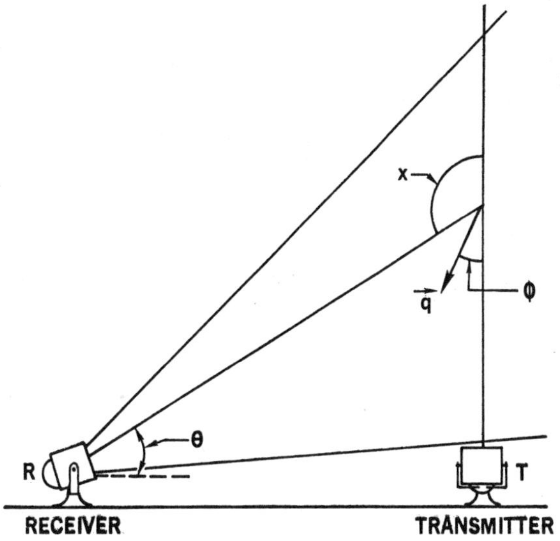

Figure 3. Basic wind-measuring element

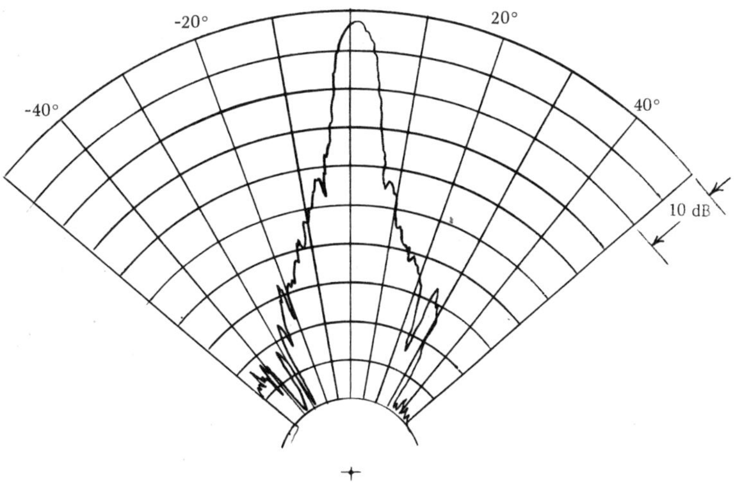

Figure 4. Transmitting Antenna Pattern.

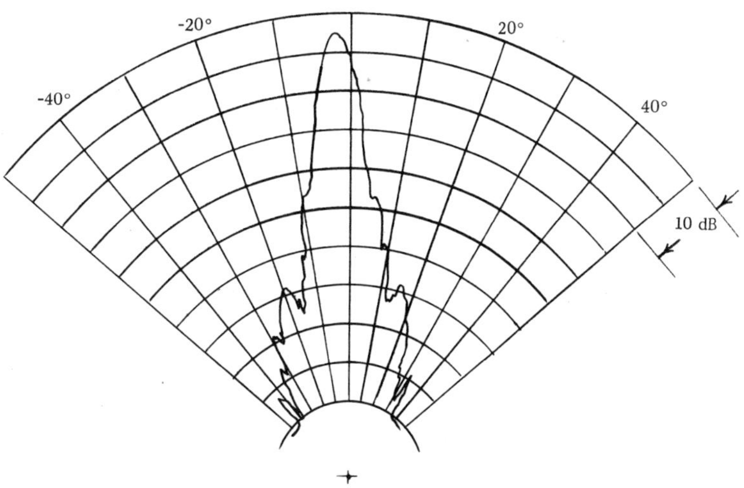

Figure 5. Receiving Antenna Pattern.

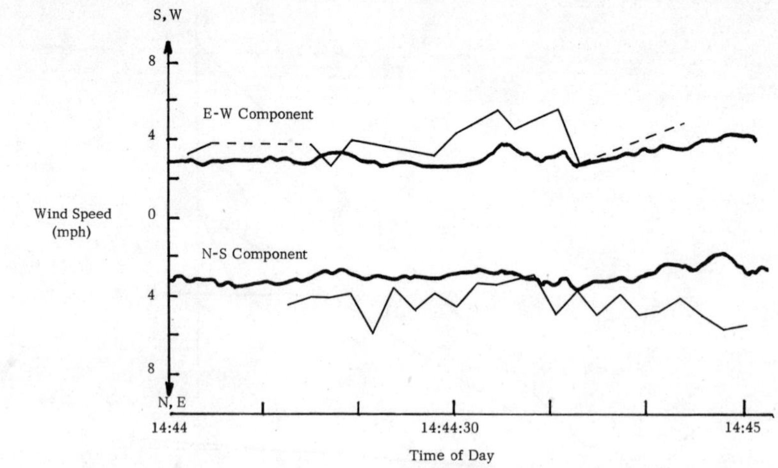

Figure 6. Comparison of acoustic wind measurements (segmented curves) with cup anemometer measurements (continuous curves) - Run 1.

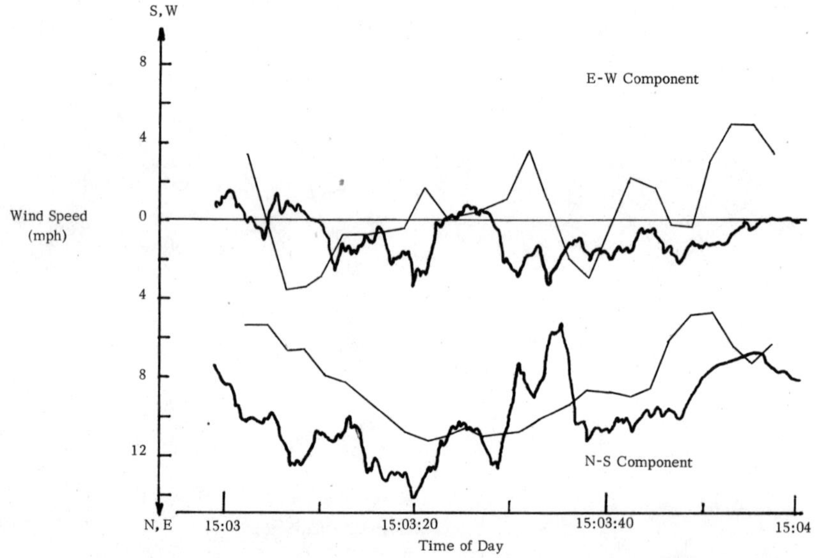

Figure 7. Comparison of acoustic wind measurements (segmented curves) with cup anemometer measurements (continuous curves) - Run 11.

© 1973, ISA JSP 6675

ANALYSIS OF LASER DIFFERENTIAL ABSORPTION REMOTE SENSING USING DIFFUSE REFLECTION FROM THE EARTH

R. K. Seals, Jr.,*† and Clayton H. Bair*
NASA Langley Research Center
Hamtpon, Virginia

ABSTRACT

A computer model analysis of an infrared laser differential absorption remote sensing technique (DARS) is presented. An infrared laser source operating at two or more wavelengths and a heterodyne detection system are considered to be mounted on either an aircraft or satellite platform to monitor the differential absorption between the laser source and the earth surface. The capability of this technique for measuring gas concentrations in the lower 5 kilometers of the atmosphere is emphasized. Numerical results are presented simulating measurements of vertical burdens of CO, NO, CH_4, CO_2, and O_3. These results indicate that measurements of expected concentrations of these gases can be made with greater than 80% accuracy using realistic laser powers and system parameters.

INTRODUCTION

In areas such as pollution monitoring and atmospheric modeling, a knowledge of the regional or global variation of the total burden of pollutants or trace gases in the lower atmosphere is of great importance.(1) Passive radiometric measurements depend on thermal emissions from the monitored gas and are of limited use below 3 kilometers altitude due to interfering gas effects and a lack of temperature discrimination between the earth and the lower atmosphere.(2,3) However, direct absorption techniques can provide relatively simple and sensitive measurements of gas concentrations in this lower region of the atmosphere. Techniques involving differential absorption of infrared laser radiation using natural objects as diffuse retroreflectors have been mentioned(4,5) as showing great promise for measuring small gas concentrations. Such differential absorption techniques do not depend on a temperature discrimination between the gas and the background and allow background and interference effects to be largely canceled out.

This paper presents a computer model study of potential applications of an infrared laser differential absorption remote sensing technique (DARS) for monitoring the lower atmosphere vertical burdens of such gases as CO, NO, and CH_4 and the total vertical burdens of CO_2 and O_3. A laser source, either tunable or fixed frequency, is considered to be mounted on either a 5-km-altitude aircraft or a 200-km-altitude satellite or space shuttle looking downward at a diffusely reflecting earth. The two-way absorption between the laser platform and the earth is assumed to be monitored by a high spectral resolution (10^6 Hz) heterodyne detector(4,6) mounted on the laser platform. Differential absorption is provided by operating with two or more closely spaced laser wavelengths where one wavelength serves as a background and interference reference and the others represent different degrees of absorption by the gas of interest. In this way any background or interference effects which are not highly wavelength dependent can be eliminated. For an aircraft instrument, the results presented here emphasize tunable lasers as both the laser source and the heterodyne local oscillator in order to provide maximum sensitivity and specificity. Calculations indicate that adequate signal-to-noise ratio would be available with state-of-the-art or near state-of-the-art diode laser powers. For satellite or space shuttle use, the higher powers of fixed frequency lasers would presently be required. Results are also presented illustrating the possibility of obtaining limited altitude profiling, or layering, using a tunable laser on a low altitude aircraft.

THEORY

The system analyzed in this paper would consist of one or more laser sources located in an aircraft or satellite and pointed downward toward the earth as illustrated in Figure 1. A heterodyne detector located with the laser source would monitor a portion of the laser radiation reflected from the earth's surface. For a laser of power P_o pointed vertically downward from an altitude L, the power reflected from a unit area of a diffusely reflecting earth into a unit solid angle is given by

$$P_r = \sigma_A \sigma_R \tau P_o/(\pi^2 d^2/4) \qquad (1)$$

where σ_A is the attenuation due to nonabsorption mechanisms such as aerosol scattering, σ_R is the

*Aero-Space Technologists, Environmental and Space Sciences Division.
†Associate Member, AIAA.

Superior numbers refer to similarly numbered references at the end of this paper.

L-9028

reflectivity of the earth for the wavelength of the laser, τ is the one-way transmittance of the atmospheric path due to molecular absorption, and d is the diameter of the laser spot on the earth. For a heterodyne detection system with receiving optics which are matched with the transmitting optics and the laser spot, d is given by

$$d = 4\lambda L/\pi D \qquad (2)$$

where λ is the laser wavelength and D is the diameter of the receiving optics. The coherent reflected power seen by the heterodyne detector is then given by

$$P = P_r\left(\frac{\pi d^2}{4}\right)\left(\frac{\pi D^2}{4L^2}\right)\tau = \frac{\sigma_A \sigma_R D^2 P_o \tau^2}{4L^2} \qquad (3)$$

The goal of a differential absorption measurement is to increase the sensitivity and specificity by eliminating effects other than molecular absorption by the gas of interest. This can be done by using two laser wavelengths at which these interference effects are equal. One wavelength serves as a reference while the second wavelength is chosen coincident with a strong absorption line of the target gas. In the work presented here, two closely spaced laser wavelengths (less than 0.1 cm^{-1} separation for most cases) are chosen with the result that

$$\sigma_{A_1} = \sigma_{A_2} \qquad (4)$$
$$\sigma_{R_1} = \sigma_{R_2}$$

Thus, the ratio of the difference signal $P_1 - P_2$ to the reference signal P_1 can be related to the transmittances due only to molecular absorption by

$$I_2 = \frac{P_1 - P_2}{P_1} = 1 - \frac{\tau_2^2}{\tau_1^2} \qquad (5)$$

where it is assumed that laser source powers at the two wavelengths are equal. The transmittance τ_i is the product of a transmittance $\tau_{i,t}$ due to absorption by the pollutant or trace gas t and a transmittance $\tau_{i,f}$ due to absorption by interfering gases. For the closely spaced wavelengths of this analysis, the assumption can be made that

$$\tau_{1,f} = \tau_{2,f} \qquad (6)$$

provided the wavelengths are selected carefully. The transmittance $\tau_{i,t}$ over a vertical path between the earth surface and an altitude L is given by

$$\tau_{i,t} = e^{-cM_t \int_{p_L}^{p_o} k_i x_t \, dp} \qquad (7)$$

where c is a constant, M_t is the molecular weight of gas t, k_i is the absorption coefficient for gas t at wavelength λ_i and pressure p, x_t is the volume mixing ratio $(x_t = p_t/p)$ for gas t at pressure p, and p_o and p_L are the atmospheric pressures at ground level and altitude L, respectively.

The mixing ratio x_t can be a function of pressure (altitude) and may be written as

$$x_t = \bar{x}_t y_t \qquad (8)$$

where $\bar{x}_t$ is a constant average mixing ratio and y_t represents the pressure dependence of x_t. Combining Equations (5) - (8) gives $\bar{x}_t$ in terms of the signal ratio I as

$$\bar{x}_t = \left[-ln(1 - I)^{1/2}\right] \Bigg/ \left[cM_t \int_{p_L}^{p_o} (k_2 - k_1) y_t \, dp\right] \qquad (9)$$

The total burden U_t of gas t in units of atm-cm is given by

$$U_t = \int_o^L (p_t/p_o) dh = \frac{\bar{x}_t}{p_o} \int_o^L y_t p \, dh \qquad (10)$$

If the function y_t is known or can be approximated for a particular gas, Equations (9) and (10) can be used to calculate the total burden of the gas between the earth surface and altitude L from the signal ratio I. Knowledge of the absorption coefficients k_1 and k_2 and of the pressure and temperature profiles of the atmosphere are required for unfolding the data contained in I. Exact data on the temperature profile is not as essential as in thermal radiance measurements since for a differential absorption measurement the temperature enters only as a secondary effect in calculation of the absorption coefficients. If the function y_t is completely unknown, an average attenuation mixing ratio $\bar{x}_{A,t}$ can be determined by setting y_t equal to unity in Equation (9). The quantity $\bar{x}_{A,t}$ is the average mixing ratio which results in the same absorption over the path as the actual mixing ratio x_t. The two quantities $\bar{x}_t$ and $\bar{x}_{A,t}$ are related by

$$\bar{x}_t = \left[\frac{\int_p^{p_o} (k_2 - k_1) dp}{\int_p^{p_o} (k_2 - k_1) y_t dp}\right] \bar{x}_{A,t} \qquad (11)$$

If y_t is unity, as is the case for uniform mixing, $\bar{x}_t$ and $\bar{x}_{A,t}$ are identical.

Remote sensing for gas profiles of the two types illustrated in Figure 2 is considered in detail here. For the Case I profile, the mixing ratio x_t is constant with altitude and y_t is given by

$$y_t = 1 \qquad o \leq h \leq L \qquad (12)$$

This type of profile would be typical of gases such as CO, NO, CH_4, and CO_2 in a well-mixed lower atmosphere. Case II profiles would correspond to an atmosphere with an inversion layer at altitude h_1 with y_t given by

$$y_t = \begin{cases} 1 & o \leq h \leq h_1 \\ C_1 & h > h_1 \end{cases} \qquad (13)$$

For this type of profile, which might be typical of gases such as CO and NO in a polluted atmosphere, the concentration below the inversion layer has built up over a period of time and can be considerably greater than the concentration above the inversion layer.

The number of wavelengths required for measurement of the total burden U_t is determined by the number of unknowns in the expression for x_t given by Equation (8). If there are N unknowns, N + 1 wavelengths are required. Thus, for Case I profiles only two wavelengths are required, while for Case II profiles three wavelengths are necessary. For Case II profiles, one wavelength is again used as a reference while the second is chosen coincident with an absorption line of gas t. The third wavelength is chosen between the first and second wavelengths in the wings of the absorption line. Then the unknown quantities $\bar{x}_t$ and $C_1\bar{x}_t$ are the solutions of the matrix equation

$$\begin{bmatrix} A_{21} & A_{22} \\ A_{31} & A_{32} \end{bmatrix} \begin{bmatrix} \bar{x}_t \\ C_1\bar{x}_t \end{bmatrix} = \begin{bmatrix} y_2 \\ y_3 \end{bmatrix} \tag{14}$$

where

$$A_{i1} = cM_t \int_{p_1}^{p_o} (k_i - k_1)dp$$

$$A_{i2} = cM_t \int_{p_L}^{p_1} (k_i - k_1)dp \tag{15}$$

$$y_i = -ln(1 - I_i)^{1/2}$$

$$I_i = \frac{P_1 - P_i}{P_1}$$

For Case I profiles the total burden up to altitude L is given by Equation (10) while for Case II profiles, it is given by

$$U_t = \frac{\bar{x}_t}{p_o}\left(\int_o^{h_1} p\ dh + C_1 \int_{h_1}^{L} p\ dh\right) \tag{16}$$

For Case II type profiles, determination of the total burden below the inversion layer is extremely important for pollution monitoring. This information can be obtained by a layering process involving multiple wavelengths. For Case II profiles, three wavelengths are required, and the total burden below the inversion layer is given by

$$U_{t,h_1} = \frac{\bar{x}_t}{p_o} \int_o^{h_1} p\ dh \tag{17}$$

where $\bar{x}_t$ is part of the solution of Equation (14).

CALCULATIONS

To demonstrate the potential of the DARS technique, calculations have been made simulating measurements of CO, NO, CH_4, CO_2, and O_3 gases in an atmosphere corresponding to the 45°N July model of the 1966 U.S. Standard Atmosphere Supplements.(7) Interfering absorptions of the measured gases and of H_2O vapor, N_2O, and SO_2 have been included. Measurements of the vertical burden below 5 km altitude for Cases I and II profiles of CO and NO and for Case I profiles of CH_4 have been analyzed for an aircraft instrument using tunable diode lasers. Measurements of the total vertical burden for Case I profiles of CO_2 have been simulated for a 200-km-altitude satellite or space shuttle instrument using fixed frequency CO_2 lasers. In addition, a measurement of the average attenuation mixing ratio $\bar{x}_{A,t}$ for a typical O_3 profile has been analyzed for a 200-km-altitude instrument using CO_2 lasers.

Detailed line-by-line calculations of the atmospheric absorption over a vertical path from the earth surface to the instrument altitude have been made to select appropriate wavelengths for detection of the various gases discussed and to estimate the laser powers required. Calculations of the absorption coefficients have been made using a Voight profile(8) line shape near absorption line centers for atmospheric pressures less than 0.250 atmospheres with a Lorentz profile line shape being used otherwise. Absorption line parameters for H_2O vapor, CO_2, O_3, N_2O, CO, and CH_4 have been obtained from the listing of McClatchey et al.(9) Estimates of the line parameters for SO_2 and NO have been obtained from various sources. Concentration profiles varying with altitude have been used for H_2O vapor(7) and O_3.(10) In the calculations of interference absorption due to CO_2, N_2O, CO, CH_4, and SO_2, constant mixing ratio profiles with magnitudes of 330 ppm, 280 ppb, 75 ppb, 1.6 ppm, and 200 ppb, respectively, have been used.

Signal ratios I_i have been calculated using the wavelengths selected for each target gas t along with Equation (5) for a range of total burdens distributed according to the two profile types shown in Figure 2. For the Case I profiles, the total burden is varied by changing the constant mixing ratio. For the Case II profiles, the total burden is varied by changing the mixing ratio below the inversion layer while leaving the upper layer mixing ratio unchanged. With the simulated signal ratios I_i and calculated absorption coefficients for the target gas, either Equation (9) or Equation (14) is used along with Equation (10) or Equation (16) to determine a retrieval total burden $\bar{U}_t$. This value is then compared with the actual U_t used in calculating the signal ratio to determine the accuracy of the procedure. Expected ranges of retrieval error are determined by including the effects of finite signal-to-noise ratios in the retrieval procedure. For the two-wavelength case, the signal-to-noise ratios of importance are

$$\frac{P_1 - P_2}{N} = \frac{\Delta S_2}{N} = \frac{\eta P_o \sigma_A \sigma_R D^2}{16hcL^2}\left(\frac{\tau_1^2}{\bar{\nu}_1} - \frac{\tau_2^2}{\bar{\nu}_2}\right)\left(\frac{T}{\beta}\right)^{1/2}$$

$$\frac{P_1}{N} = \frac{S}{N} = \frac{\eta P_o \sigma_A \sigma_R D^2}{16hcL^2}\left(\frac{\tau_1^2}{\bar{\nu}_1}\right)\left(\frac{T}{\beta}\right)^{1/2} \tag{18}$$

where η is the quantum efficiency of the detector, $\bar{\nu}_i$ is the wave number of wavelength i, P_o is the laser source power, T is the postdetection integration time, and β is the heterodyne detection bandwidth of 10^6 Hz. A factor of 0.5 has been included to allow for chopping the incoming radiation. Signal-to-noise effects are to be included by replacing I_i in Equation (9) with

$$\bar{I}_i = I_i \left[\frac{1 \pm 1/(\Delta S_i/N)}{1 \mp 1/(S/N)} \right] \tag{19}$$

For the three-wavelength case an additional signal-to-noise expression of the form shown in Equation (18) is required for the difference signal between the third wavelength and the reference wavelength. The signal-to-noise effects are included by replacing y_i in Equation (15) with

$$\bar{y}_i = -ln(1 - \bar{I}_i)^{1/2} \tag{20}$$

RESULTS AND DISCUSSION

Table 1 shows the wavenumbers selected for each of the five gases considered and the required laser powers and potential ranges for measurements of the total burdens of Case I profiles. Also shown in parentheses are the third wavelengths selected for measurement of Case II profiles of CO and NO. For CO, NO, and CH_4, the results are for measuring the lower 5-km vertical burden from an aircraft or balloon using tunable diode lasers. The CO_2 results are for measuring the total vertical burden of a Case I profile from 200 km altitude using the P(18) line of the (00^01)-(10^00) emission band of $C^{12}O_2^{18}$ and the R(8) line of the (00^01)-(02^00) emission band of $C^{12}O_2^{16}$. The O_3 results are for measuring the average attenuation mixing ratio $\bar{x}_{A,O_3}$ of a typical O_3 profile shape from 200 km altitude. The wavenumbers correspond to the P(24) and P(20) lines of the (00^01)-(10^00) emission band of $C^{12}O_2^{18}$. For all of these calculations a quantum efficiency η of 0.5, a telescope diameter D of 25 cm, and a detection bandpass β of 10^6 Hz have been used. Postdetection integration times T of 0.1 sec for CO_2 and O_3, 1.0 sec for CO and CH_4, and 10 sec for NO have been assumed. An average surface reflectivity[10] σ_R has been used for each wavelength ranging from 0.044 for the O_3 wavelengths to 0.074 for the CO wavelengths. The quantity σ_A has been assumed to be unity corresponding to clear conditions in the field of view.

Figure 3 indicates that the DARS measurement from 5 km altitude would be accurate to within 20% for total burdens of Case I profiles of CO ranging from 0.012 to 0.16 atm-cm. This corresponds to constant mixing ratios ranging from 32 to 429 ppb. The solid line in Figure 3 shows the perfect detection result, while the dotted lines indicate the maximum error limits due to finite signal to noise as predicted by Equations (9) and (19). For total burdens less than 0.012 atm-cm, the failure of the assumption in Equation (6) becomes dominant. The difference in the interfering absorption at the two signal wavelengths appears as a signal of 0.0019 atm-cm of CO. For total burdens greater than 0.16 atm-cm, the rapid increase in error is caused by insufficient signal return due to the increased absorption of CO itself.

Figures 4 and 5 show results for monitoring Case I profiles of NO and CH_4 from 5 km altitude. For NO the range of total burdens for which less than 20% error has been calculated is 0.002 to 0.075 atm-cm, corresponding to constant mixing ratios of 5.3 to 200 ppb. The error for low concentrations of NO is due primarily to a difference in interfering absorption at the signal wavelengths which appears as a signal of -0.00017 atm-cm of NO. For CH_4, the range of total burdens resulting in less than 20% error is 0.021 to 2.24 atm-cm, corresponding to constant mixing ratios of 56 ppb to 6 ppm. In this case the constant error due to interfering absorption differences appears as a signal of 0.0024 atm-cm of CH_4. For both NO and CH_4, the large error at higher total burden values is due to decreased signal return.

Figures 6 and 7 show results of the calculations simulating measurements of Case II profiles of CO and NO. The inversion level height h_1 has been taken as 1 km with the instrument altitude L being 5 km. In both cases, the mixing ratio above h_1 has been kept constant, with values of 50 ppb for CO and 2.5 ppb for NO, while the mixing ratio below h_1 varies. Figure 6 indicates that measurements of total burdens ranging from 0.0186 to 0.20 atm-cm, corresponding to mixing ratios below the inversion layer of 50 ppb to 2 ppm, would be within 20% error limits. The lower limit on this measurement is set by the 50 ppb concentration chosen above the inversion layer, while the upper limit is due to insufficient signal return. Figure 7 shows that measurements of total burdens of NO of from 0.01 to 0.085 atm-cm, corresponding to 100 to 910 ppb concentrations below the inversion layer, have calculated error limits of less than 20%. The lower limit on the range of this measurement is set predominantly by interfering absorption while the upper limit again is set by the signal return.

Results similar to those in Figures 3, 4, and 5 have been obtained for measuring Case I profiles of CO_2 from a 200-km altitude instrument. These calculations indicate that measurements with less than 20% error could be made for total burdens of CO_2 ranging from 7.6 to 530 atm-cm. This would correspond to constant mixing ratios of 10 to 700 ppm as compared to an ambient concentration of approximately 330 ppm. The error for a measurement of this ambient concentration has been calculated to be less than 2.2%.

Results have also been obtained for monitoring the average attenuation mixing ratio $\bar{x}_{A,O_3}$ of a typical O_3 profile[10] from a 200-km altitude. As shown in Table 1, the two laser wavelengths are separated by greater than 3 cm^{-1}, and, as a result, additional error due to differences in background and absorption interference can be expected. However, results indicate that $\bar{x}_{A,O_3}$ can be measured to within 10% accuracy. In order to relate $\bar{x}_{A,O_3}$ to $\bar{x}_{O_3}$ and finally to the total burden U_{O_3}, the functional form of the altitude dependence of the mixing ratio must be known.

As has been mentioned previously, determination of the total burden below the inversion layer can be important for Case II profiles. Figures 8 and 9 show results of simulations of such measurements for CO and NO for a 1-km inversion height and a 5-km altitude instrument. This type of measurement is more sensitive to both interference effects and the effects of finite signal to noise. As a result, greater laser powers are required to improve the accuracy. Figure 8 shows the calculated error for measurements of the lower km burden of CO with the solid line indicating the perfect detection case and the large dashed line and the small dashed line indicating the error limits for laser powers of 6 mW and 10 mW, respectively. For the 10-mW case, the results show that measurements could be made with less than 20% error for burdens below the inversion layer of from 0.04 atm-cm to 0.18 atm-cm, corresponding to mixing ratios of 430 ppb to 1.95 ppm. In all cases the CO mixing ratio above the inversion has been kept at 50 ppb. For NO, Figure 9 shows the calculated errors for similar measurements of NO assuming laser powers of 30 mW and 50 mW. For the 50-mW case, NO burdens of from 0.014 to 0.09 atm-cm in the lower km, corresponding to mixing ratios of 150 ppb to 970 ppb, could be measured with less than 20% error. The NO concentration above the inversion has been set at 2.5 ppb in these examples.

SUMMARY AND CONCLUSIONS

A detailed computer analysis of applications of an infrared laser differential absorption remote sensing technique (DARS) has been presented. Numerical simulations of measurements of the lower atmosphere burdens of typical profiles of CO, NO, and CH_4 from a 5-km altitude and of the total vertical burdens of CO_2 and O_3 from a 200-km altitude have demonstrated the potential of this technique with reasonable laser powers and system parameters. Results have also been presented for a limited amount of altitude layering using multiple wavelengths by simulating the determination of the lower 1-km burdens of CO and NO in a polluted atmosphere from a 5-km altitude.

Results indicate that measurements of CO and NO burdens below 5 km can be made with less than 20% error for constant mixing ratios ranging from 32 to 429 ppb for CO and from 5.3 to 200 ppb for NO. The CH_4 and CO_2 results indicate accurate measurements are possible over a wide range encompassing the ambient concentrations. Accurate measurements of the average attentuation burden of O_3 are possible and can be related to the actual total burden if the altitude dependence of the mixing ratio is known. For profiles typical of polluted atmospheres, results have shown that accurate measurements of the lower 5-km burdens of CO and NO can be made for concentrations below a 1-km inversion layer ranging from 50 ppb to 2 ppm for CO and from 100 to 910 ppb for NO. Laser powers of 2 mW for the CO and CH_4 measurements, 10 mW for the NO measurements, and 25 W for the CO_2 and O_3 measurements have been assumed. With these laser powers and realistic system parameters, it has been shown that adequate signal-to-noise ratios are attainable to allow measurements with less than 20% error over the range of concentrations listed above.

All of these calculations are based on line-by-line absorption calculations using the best available absorption line parameters. Such parameters as exact line positions, line widths, and line strengths should be confirmed experimentally for the selected sensing channels before these results are extended to actual instruments. Such a program is presently underway at our laboratory using tunable diode lasers.

REFERENCES

(1) "Remote Measurement of Pollution," NASA SP-285 (Scientific and Technical Information Office, NASA, Washington, D.C., 1971).

(2) Seals, R. K., Jr., "Theoretical Analysis of Improvements in Remote Sensing of Atmospheric and Pollutant Gases Through High Resolution Detection of Individual Infrared Emission Lines," AIAA Paper 73-703 (1973).

(3) Ludwig, C. B., R. Bartle, and M. Griggs, "Study of Air Pollutant Detection by Remote Sensors," NASA CR-1380, 1969.

(4) Menzies, R. J., and M. S. Shumate, "Usefulness of the Infrared Heterodyne Radiometer in Remote Sensing of Atmospheric Pollutants," AIAA Paper 71-1083 (1971).

(5) Jacobs, G. B., and L. R. Snowman, "Laser Techniques for Air Pollution Measurement," IEEE Journal of Quantum Electronics, QE-3, 603 (1967).

(6) Melngailis, Ivars, "The Use of Lasers in Pollution Monitoring," IEEE Transations on Geoscience Engineering GE-10, 7 (1972).

(7) U.S. Standard Atmosphere Supplements, 1966. U.S. Government Printing Office, Washington, D.C. (1966).

(8) Young, C., "Calculation of the Absorption Coefficient for Lines with Combined Doppler and Lorentz Broadening," Journal of Quantitative Spectroscopy and Radiative Transfer 5, 549 (1965).

(9) McClatchey, R. A., W. S. Benedict, S. A. Clough, D. E. Burch, R. F. Calfee, K. Fox, L. S. Rothman, and J. S. Garing, "AFCRL Atmospheric Absorption Line Parameters Compilation, Environmental Research Paper 434, AFCRL-TR-73-0096 (January 1973).

(10) McClatchey, R. A., R. W. Fenn, J. E. A. Selby, F. E. Volz, and J. S. Garing, "Optical Properties of the Atmosphere," Environmental Research Paper 354. AFCRL-71-0279 (May 1971).

TABLE 1. SELECTED WAVENUMBERS, REQUIRED LASER POWERS, AND TOTAL BURDEN RANGES CALCULATED FOR DARS MEASUREMENTS

Gas	$\bar{\nu}_i$ (cm^{-1})	P_o(W)	Case I U_{min}, U_{max} (atm-cm)
CO	2068.749, 2068.849 (2068.812)	0.002	0.012, 0.160
NO	1900.23, 1900.13 (1900.168)	0.010	0.002, 0.075
CH_4	1231.03, 1230.93	0.002	0.021, 2.24
CO_2	1070.507, 1070.458	25	7.6, 530
O_3	1065.756, 1068.942	25	0.03, 0.68

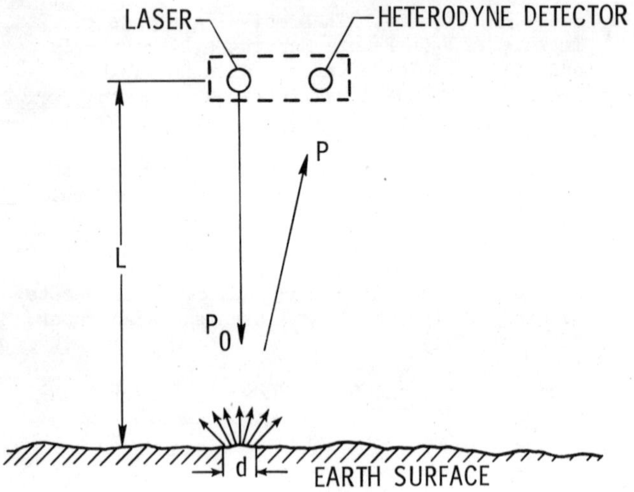

Figure 1. Schematic of DARS measurement of atmospheric gases.

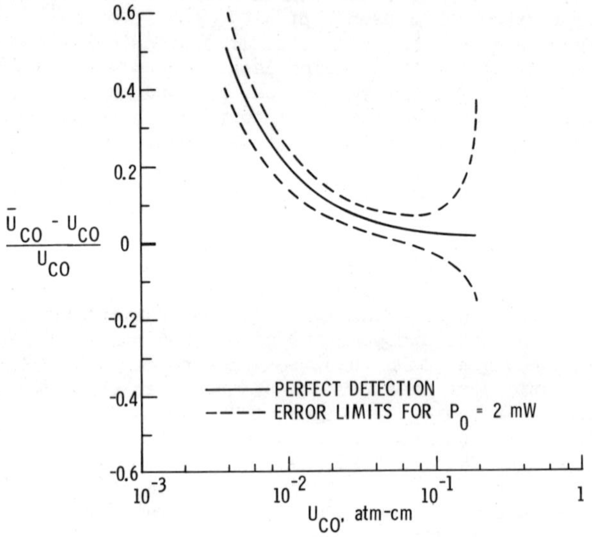

Figure 3. Calculated error for measurement of lower 5-km burden of Case I CO profiles.

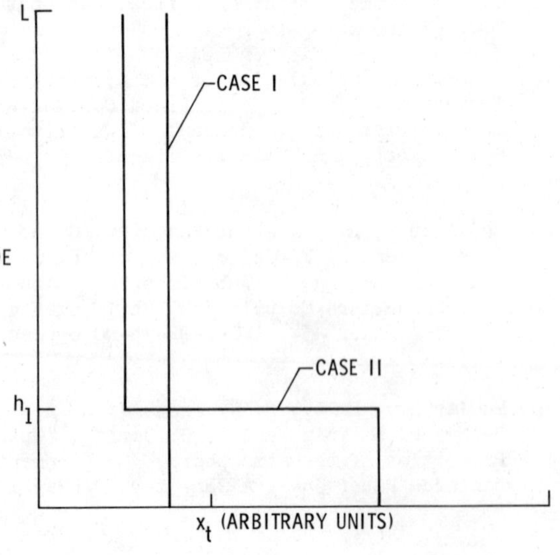

Figure 2. Altitude variation of mixing ratio for two types of profiles.

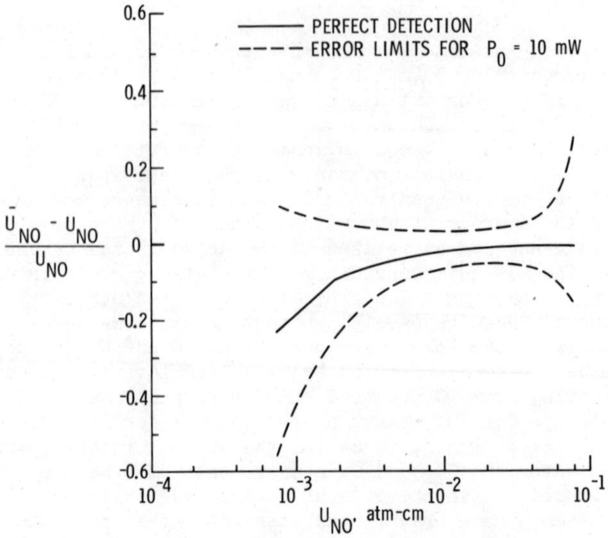

Figure 4. Calculated error for measurement of lower 5-km burden of Case I NO profiles.

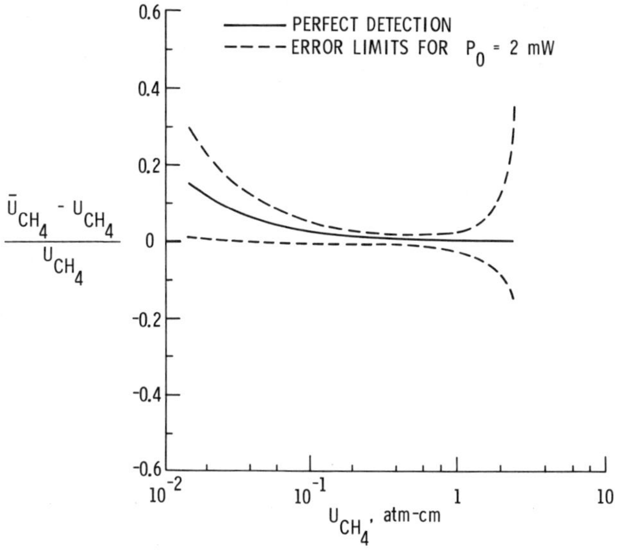

Figure 5. Calculated error for measurements of lower 5-km burden of Case I CH_4 profiles.

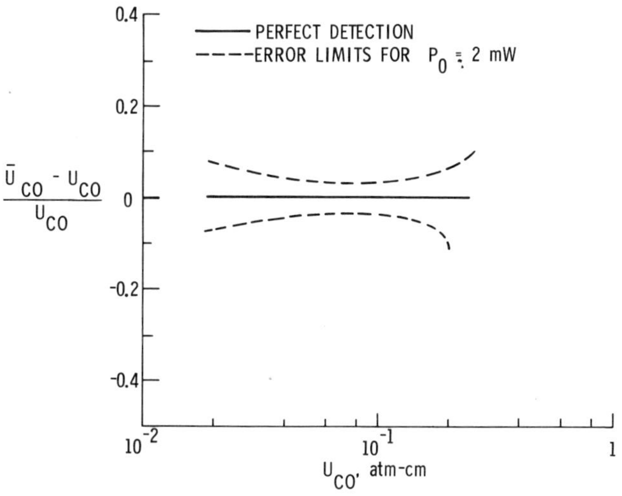

Figure 6. Calculated error for measurements of lower 5-km burden of Case II CO profiles.

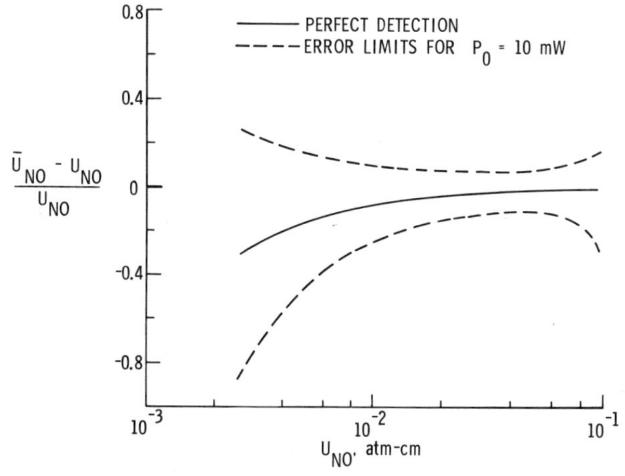

Figure 7. Calculated error for measurements of lower 5-km burden of Case II NO profiles.

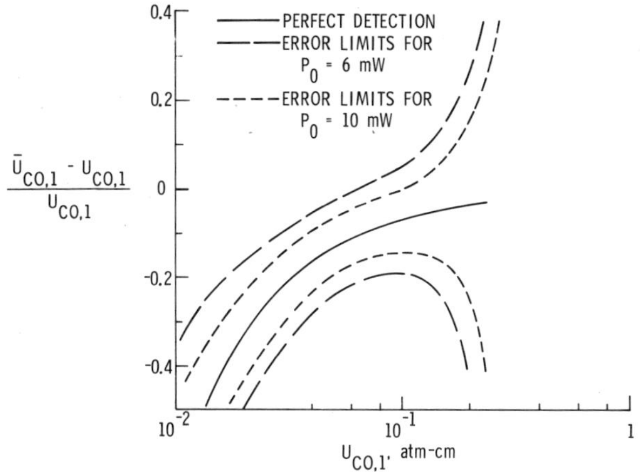

Figure 8. Calculated error for measurements of lower 1-km burden of Case II CO profiles.

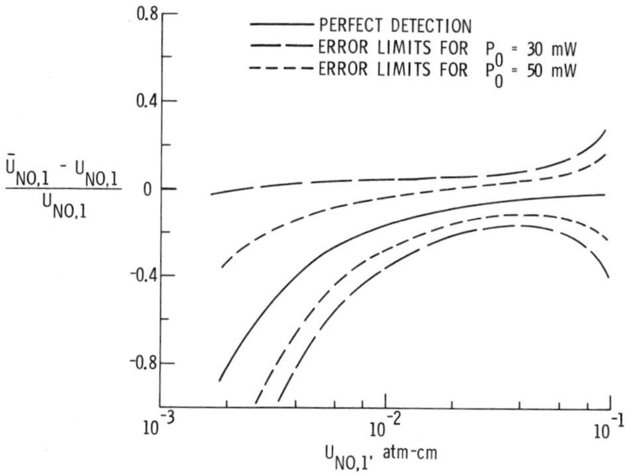

Figure 9. Calculated error for measurements of lower 1-km burden of Case II NO profiles.

© 1973, ISA JSP 6677

USE OF A MONOSTATIC ACOUSTIC SOUNDER IN AIR POLLUTION DIFFUSION ESTIMATES

Ivar Tombach, Director of Advanced Development
Paul B. MacCready, Jr., President
Lal Baboolal, Research Scientist
AeroVironment, Inc., Pasadena, California

ABSTRACT

An acoustic sounder has been developed to study atmospheric structure by transmitting a pulse of sound upward and then receiving the energy which is scattered from turbulent fluctuations of atmospheric temperature. This instrument has been applied to the analysis and modeling of the diffusion of atmospheric pollutants. Information obtained or derived from its records in these studies has included the character of the stability of the atmosphere, the height of elevated inversions, the depth of the mixed layer, and estimates of the turbulent eddy scale in thermal plumes. A quantitative technique, using the acoustic sounder and one low-level anemometer, has been developed for the estimation of diffusion parameters in the lowest part of the atmospheric boundary layer.

INTRODUCTION

Acoustic energy propagating through the atmosphere is scattered by turbulent fluctuations of temperature and wind velocity. Since the initial successes by McAllister, et al[(1)*] in 1968, several investigators have applied this principle to the remote sensing of atmospheric structure and winds by transmitting acoustic waves into the atmosphere and detecting some of the scattered energy.**

In the special case when only the backscattered energy is received, i. e., when the source of acoustic energy and the receiver are co-located, only scattering due to thermal fluctuations is detected since wind fluctuations do not produce scatter in the backward direction[(3)]. An instrument which transmits an acoustic pulse and receives and processes the backscattered energy returned by the atmosphere is called a monostatic acoustic sounder. Because of the analogy between its operation with acoustic waves and the operation of radar with electromagnetic waves, other descriptive terms for an acoustic sounder are "acoustic radar" and "sodar" (for Sound Detection and Ranging).

* Superior numbers refer to similarly numbered references at the end of this paper.

**Reference 2 is a recent compilation of research results in this field.

Here we will describe one such system and discuss its application to air pollution meteorology.

SYSTEM DESCRIPTION

A photograph of a monostatic acoustic sounder appears in Figure 1. The acoustic energy is generated by the white transducer above the circular antenna, is transmitted downward through the cone and is then reflected into the atmosphere in the form of a narrow beam by the paraboloidal antenna dish. The path of the acoustic energy is reversed for receiving the echoes of scattered sound, with the transducer now acting as a microphone.

The rectangular box in Figure 1 contains the necessary electronics and a recorder which displays the received signal. A simplified block diagram of the electronic configuration for this unit is shown in Figure 2 and will be discussed below. Some operating specifications will also be given to illustrate representative values of system variables.

The transmitted tone burst for this system can have a length of 50 to 200 ms, at a frequency of 1600 Hz and an electrical power input to the transducer of 35 to 140 w. During the transmission of the pulse the receiver is disconnected from the transducer, and several portions of the highly sensitive receiver circuit (overall gain $\sim 10^8$) are clamped at a fixed level. Subsequent to the transmission there is a delay (typically 60 ms) before the receiver is coupled to the transducer, in order to allow mechanical and electrical ringing time to damp out.

When the receiver is turned on, the echo signals from the transducer are amplified to reasonable levels and are then further amplified by a multiplier circuit, one input to which is a linearly increasing signal which starts at zero at the time of the transmitted pulse. The purpose of this multiplication by a ramp is to compensate for the r^{-2} attenuation of acoustic energy due to spherical spreading. Since the system responds to acoustic pressure, which is proportional to the square root of the energy, a correction for $1/r$

decay is needed, which is provided by the ramp. (The correction is only valid when scatterers which intercept the entire transmitted beam are considered, which is the usual case. Compensation varying as t^2 would be required for scatterers of fixed dimension smaller than the beam width.) No attempt is made to compensate for attenuation due to atmospheric molecular absorption, the magnitude of which varies with pressure and humidity.

The signal then passes through a narrow filter, centered on the transmitter frequency, which rejects much of the ambient noise. A compromise is necessary here, however, since the best noise rejection would be obtained by a very narrow filter, but the atmosphere is continually in motion and the consequent doppler shift of the signal frequency could shift the echo out of the passband of a too-narrow filter. A tunable filter with a bandwidth of 10 to 50 Hz at -3dB is thus used in this system. With such narrow bandwidths it is necessary to insure that receiver filter and transmitter oscillator drift could not also cause the received signal to be outside the receiver passband. To remedy this problem, the filter is caused to oscillate and thus serve as the transmitter oscillator, thereby insuring that the transmitted pulse frequency is always exactly in the middle of the receiver passband.

The filtered and amplified signal is then demodulated and amplified further for recording. The recorder contains a stylus which is drawn across a special paper and which makes a mark whose intensity is related to the magnitude of the received signal. The stylus starts at the bottom of the paper when the pulse is transmitted and reaches the top when echoes from the desired full scale range (typically 500m or 1 km) have been received. The paper moves laterally through the recorder at slow speed so that each pulse record is slightly offset from the previous one, resulting in a recording of regions of dark and light where one axis is that of time of day and the other that of height of the scatterer above the antenna (or, more precisely, the round-trip travel time of an acoustic wave between the antenna and the scattering region.) Several such records appear later in this paper.

Although the beam radiated and received by the parabolic antenna is well focused, typical ambient noise levels are so much greater than the minute echo strength and consequently noise reception through the antenna side lobes can affect signal quality. In addition, the sound levels in the side lobe of the transmitted signal can be significant enough to create a "noise pollution" problem in quieter inhabited surroundings. To further suppress the side lobes of the antenna an acoustical enclosure such as that shown in Figure 3 is normally used. This enclosure, which is fabricated from 1/2' plywood (for structural rigidity and sound level attenuation), 1/64" lead sheet (for further sound attenuation), and convoluted urethane foam (to damp internal reflections) attenuates horizontally traveling signals by more than 20 dB. The lead/plywood combination used has a mass of about 11 kg/m^2 and, because of the elastic properties of the lead, has significantly better attenuation properties than an equivalent mass of plywood alone. With the enclosure installed the peak transmitted sound pressure level at 10 meters from the enclosure is less than 60 dB(A). Simmons, et al[4] discuss the limitations these figures place on the locations where the sounder can be used. In practice ambient sound levels in excess of 70 dB(A) noticeably affect signal readability at ranges in excess of about 500m, and operation of the sounder closer than about 100m to a residence in a typical residential neighborhood will disturb the sleep of some individuals.

THEORY

As was mentioned above, a monostatic sounder can detect only scattering due to thermal fluctuations, since air velocity fluctuations do not scatter sound in the backward direction. This follows from the equations for the scattering of acoustic radiation described by Kallistratova[5], which are discussed by Little[3] and McAllister, et al[1].

For the specific case of backward scattering the scattering cross section η per unit volume of turbulent atmosphere is

$$\eta = 2\pi^2 k^4 \frac{\varphi(2k)}{T^2} \tag{1}$$

where k is the wave number of the incident wave ($k = 2\pi\lambda^{-1}$), T is the temperature, and φ is the three-dimensional power spectral density of the temperature fluctuations. Since only the spectral density at wave number 2k contributes to φ, only thermal variability over a scale of $\lambda/2$ contributes substantially to the scattered power. For the 1600 Hz system discussed above $\lambda/2$ is approximately 0.2m. Obviously only variability along the radial vector from the transmitter/receiver to the scattering region can affect the return.

Atmospheric turbulence with eddies of this scale is often nearly isotropic and homogeneous and lies within the inertial subrange[6], so that Kolmogorov's similarity hypothesis holds. This turbulence spectrum can be easily treated analytically and thus it is informative to consider it in order to obtain some quantitative information on the nature of the scattered signal. For the inertial subrange

$$\varphi(2k) = 0.033\, C_T^2 (2k)^{-11/3} \tag{2}$$

and thus Eqn. (1) becomes

$$\eta = 0.004\, \pi^2 \left(\frac{C_T}{T}\right)^2 (2k)^{1/3}$$

$$= 0.1\, \lambda^{-1/3} \left(\frac{C_T}{T}\right)^2 \quad . \tag{3}$$

C_T is a thermal structure function defined as

$$C_T^2 r^{2/3} = \overline{[T(x) - T(x+r)]^2} \tag{4}$$

where T(x) is the instantaneous temperature at a point x along the radial vector, T(x+r) is the instantaneous temperature at a point which is removed from x by a distance r along the same radial vector, and the bar denotes an average over time. Invoking the hypothesis that spatial averages are interchangeable with temporal averages in a homogeneous turbulent field, C_T^2 can be interpreted as the mean square difference in temperature between two points separated by unit distance.

In summary, a monostatic acoustic sounder detects energy scattered from turbulent fluctuations of temperature of scale $\lambda/2$. The scattering is greatest when λ is the smallest, i.e., for the highest transmitted frequency, although absorption due to humidity (which increases with frequency[7]) limits usable frequencies to less than 5 kHz if a range in excess of 100-200 m is desired. Although the sounder in this configuration does not detect turbulence, per se, turbulence is necessary for thermal fluctuations of the proper scale to be created. For different eddies to have different temperatures, and hence a non-zero C_T, it is also necessary (for the vertically pointing sounder) that the atmospheric stratification be other than neutral, i.e., statically stable or unstable, or that heat be injected by other means (e.g., heat of condensation or a heated plume from a smokestack) into the turbulent medium.

Any mechanism which produces a significant temperature change in a distance of order $\lambda/2$ will cause an acoustic echo. Normal temperature inversions probably do not have sufficiently strong temperature gradients to provide significant echoes directly, and thus one must rely on turbulent fluctuations within the inversion to produce the scattering. This is not invariably so, however, and reflective echoes have been reported from extremely strong inversions in very cold climates. Also, acoustic measurements in stable conditions have frequently indicated a stronger return than that which would be predicted from directly measured values of C_T, with specular reflection being a possible explanation for the excess return[8].

INTERPRETATION OF SOUNDER RECORDS

Figure 4 is an acoustic sounding record which illustrates many of the phenomena which can be observed by such a system.* This particular record, which is actually a composite of three facsimile recordings, shows the transition from stable nocturnal conditions to unstable daytime convection.

At the earliest time shown (0500) the record shows three echo layers centered at about 40 m, 100 m, and 130 m. These layers are in statically stable air in which turbulence is generated by kinetic energy extracted from the mean wind flow. The clear regions in between the layers, and the minute clear area below the bottom layer before 0520, correspond to more adiabatic conditions. Two more layers at higher altitudes become visible as the lower layers descend and merge into a surface based turbulent layer. This formation of a surface based stable turbulent layer is characteristic of the lower atmosphere on a clear night as the surface temperature falls due to radiation and turbulent convection transmits this cooling to the lowest levels of the atmosphere. Its thickness may vary during the night and would increase if the wind speed remained constant. If, as the stratification becomes more stable the wind dies down, as occurs near dawn in Figure 4, the layer thickness will decrease as the shear-generated turbulence decreases.

The stable layer is rather rapidly pushed up from the surface soon after sunrise as a shallow adiabatic region forms beneath (as indicated by the small clear area) at 0705. The upper layers also move upward by the same amount and the region below the now-elevated lowest layer fills with a thickened stable turbulent surface layer (which heretofore was in the blanked region below 25 on the record), indicating that an increase in wind speed has taken place. The high frequency wavy structure in the upper layers and the sawtooth nature of the top of the rising layer are characteristic of what are probably breaking Kelvin-Helmholtz waves resulting from wind shear at the stable interface at the top of each layer[9].

A longer period wave motion is apparent also from about 0700 to 0900 in both the layer aloft and the bottom layer, with obvious coupling between the motions of the two layers. These waves are probably gravity waves generated on the highly stable nocturnal radiation inversion layer[10, 11].

Solar heating succeeds in destroying the surface based inversion at about 0920 and the vertically rising structures which are indicative of thermal plumes in an unstable atmosphere are present on the chart after that time. Earlier manifestations of limited thermal activity appeared at 0830 and attempted to penetrate the stable inversion above them but were not successful in doing so until about 0905, following which time both of the remaining inversion layers broke up. Various levels of thermal activity are visible throughout the rest of the day with the echoes from some plumes rising up to 400 m. Such dark regions correspond to warm air, typically found to be 0.2 to 0.6°C warmer than the surrounding atmosphere, with the weakening and eventual loss of echo returns with height suggesting a decreasing vertical temperature gradient within the plume[1]. The regions between the plumes, marked by an absence of echoes, have descending air[12] with near neutral stability.

Another record of a different type of occurrence is shown in Figure 5. The layer at 160 m at 0830 is the top of a layer of fog, which was verified by direct observation from nearby elevated terrain. Solar heating eventually raises temperatures enough to dissipate the fog, leaving only a thin

*This record was made by Messrs. Edward Miller and Robert Bartz at Oregon State University with a sounder of their own design. Their permission for its use is appreciated.

Figure 1. Monostatic Acoustic Sounder System

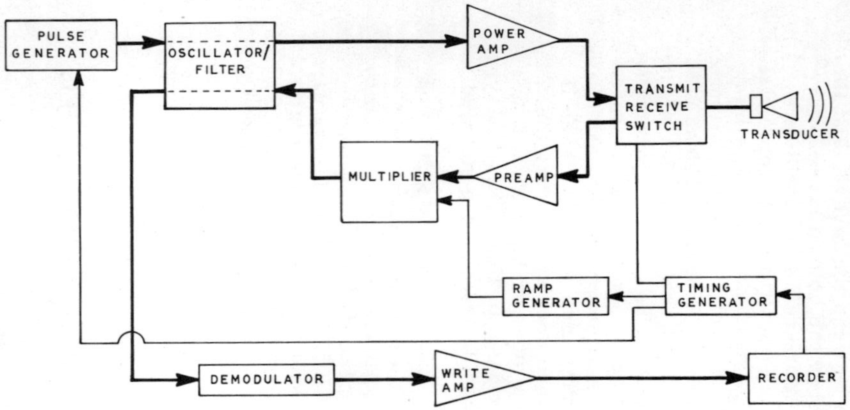

Figure 2. Block Diagram of Acoustic Sounder. The signal path is indicated by the heavy line.

Figure 3. Acoustic Enclosure, with one of the five sides removed to show the antenna inside.

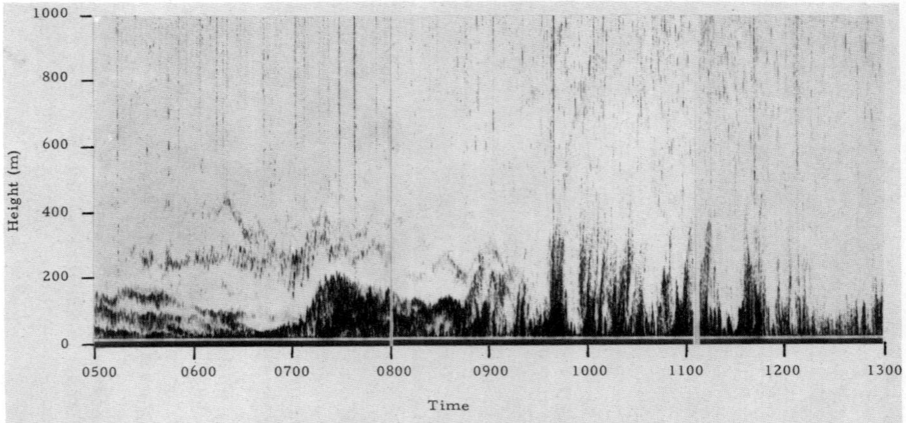

Figure 4. Acoustic Sounder Record at Corvallis, Oregon, on 16 June 1972. The dark band across the bottom represents the transmitted signal, the narrow clear space is a delay before the receiver is turned on, and the darker areas above represent echo returns. Stable atmospheric conditions are portrayed before 0830, while unstable conditions prevail after 0930.

layer of stratus by 0930. The subsequent breakthrough of the sun results in rapid formation of thermal plumes which drive the stable layer upward to 450 m at 1230. Breakup of the layer begins at 1300 and is visible at the upper right of the figure. As a matter of operational interest, ambient noise levels during the making of the record were around 70 dB(A), accounting for the noise darkening of the top of the record by the linear increase of gain with height. The electrical input to the transducer was 140 watts.

As was noted earlier, either static stability or instability of the atmosphere is one prerequisite for the presence of acoustic echoes for a monostatic sounder. Determination of which of these two conditions prevails is obvious at a glance. In Figure 4, for example, the horizontal structure of the echoes before 0830 indicates a stable stratification, the vertical structure after 0930 indicates unstable stratification, and the period from 0830 to 0930 is a period of transition with unstable conditions developing upward from the surface.

In summary, there are four significant meteorological situations (and a fifth unusual one) which result in thermal structure of small enough scale to result in echoes.

1) The base regions ("roots") of rising thermal plumes;
2) The stable, surface layer of the planetary boundary layer;
3) Stably stratified turbulent shear layers aloft;
4) Penetrations of a stable layer by thermal plumes (thereby indicating the top of the mixing layer).

These are categorized in the table below by whether they are ground based or aloft, and by the atmospheric stability which the echo region marks. All of them contain small scale turbulence.

	Ground Based	Aloft
Unstable	Thermal roots.	
Stable	Surface Layer.	Shear. Layer. Top of mixing layer.

The fifth situation, that of a non-turbulent extremely strong inversion, was discussed earlier and is rather uncommon.

APPLICATION TO AIR POLLUTION METEOROLOGY

The above discussion suggests some of the ways in which acoustic sounding can assist the air pollution meteorologist. From a monostatic sounder record he can determine, for example:

a) The time of onset of unstable mixing conditions in the morning and the period for which these conditions prevail;
b) The time at which stable conditions appear in the evening and their period of persistence;
c) The minimum height to which thermal plumes will carry surface-generated emissions;
d) The thickness of the surface based turbulent layer in stable conditions; and
e) The heights of stable layers and inversions aloft, and the depth of the mixing layer.

Further inferences can be made using an acoustic sounder in conjunction with in-situ meteorological instruments such as anemometers, rawinsondes, and sigma (r. m. s. turbulence) meters[13]. Examples of some applications of acoustic sounding to real air pollution situations will be presented below. Later in this paper a quantitative technique for the prediction of low-level dispersion using an acoustic sounder, an anemometer, and a sigma meter will be discussed. Other authors have also addressed themselves to the use of acoustic sounding in air pollution meteorology[4, 14, 15].

During the summer of 1973 the acoustic sounder described earlier was operated adjacent to a power plant site in a desert region of the southwestern United States. Anemometers were mounted on a nearby smokestack, and were 240 m above ground level. Data from these instruments was used to fill in gaps of knowledge concerning the mechanism and timing of stability buildup at night and breakdown during the day, for the purpose of assessing the dispersive characteristics of the atmosphere some 200-500 m above the surface.

Figure 6 gives typical results for three consecutive 24-hour periods in late summer at this location. The most distince feature is the buildup and decay of convection during the day, starting about 0900 and ending about 1730 each day. The roots of the convective cells can be seen dimly up to 200 to 250 m. The dark areas (such as at 1015 on the second day) descending from the top of the record are noise, primarily noise from wind blowing over the acoustic enclosure, where the increase in darkness with height from steady noise is a consequence of the instrument's automatic gain increase with range.

The acoustic sounder shows only the convective roots of thermal plumes and verifies strong mixing at least through the depth of the returning echoes. Convection in the summer at this location can be expected to go considerably higher than this and the anemometer records indicate turbulence appearing at 240 m whenever any thermal echoes can be seen, no matter how weak. The absence of echoes does not necessarily imply the absence of turbulence, however. Especially as convection diminishes in the late afternoon, turbulence and mixing (though not thermals) may still be occuring but ground cooling may be establishing a neutral lapse rate and thus removing the sharp temperature fluctuations needed for the presence of echoes. Thus, if there are echoes there is atmospheric mixing; if there are no echoes, there may or may not be mixing.

Interpretation of the situation at night is somewhat different. As the ground cools appreciably, any low-level mixing will be accompanied by echoes. This is observed starting about 2030 on September

17 and lasting until 0100 on the 18th. The strong, ground-connected mixing may go up only the distance shown by echoes. The stack instruments do not show strong turbulence until 2300, although some very slight turbulence is detectable starting at 2030. Turbulence then continues at a low, varied, but finite level until 0830 when it increases because of convection. Some light turbulence persists prior to midnight of the 19th, after a brief lull at 1900. Turbulence on the 20th is very small until 0930. The acoustic sounder echoes at this time are strong but very shallow, suggesting a shallow drainage flow condition.

A significant feature of the nocturnal acoustic sounder records is the lack of any visible layering, such as is regularly seen at night in many other locations, which appears to be a consequence of significant drainage flows from nearby high mountains. Thus at this location the filling up of the lower layers with cooler air as a consequence of drainage evidently involves, at some stage, enough small scale mixing to smooth out temperature gradients (which is confirmed by aircraft temperature soundings). The absence of any visible structure at all on the night 18-19 September is also rather unusual (though there may be some in the bottom 15-20 meters which are blanked on the record).

In summary, then, the records in Figure 6, coupled with the anemometer inputs, show that during sunlight the thermals whose roots are visible on the radar contribute to turbulence at stack top height. At night, deep echoes correlate with turbulence at stack top while shallow drainage flow echoes do not. Also, layering aloft is not observed, suggesting that small scale mixing in the lower 500 m keeps sharp gradients from forming.

In another study concerned with the impact on air quality of the proposed construction of a major airport in the Mojave Desert of California, the monostatic sounder was used in conjunction with a sensitive, low-level anemometer. In this study, where there was a lack of satisfactory historical data on wind or atmospheric structure, these instruments were used to define the time of day and period of the year of greatest pollution potential for inert contaminants. Appropriate mathematical models were then used to estimate emissions and dispersion during these "worst case" meteorological conditions.

Records such as those in Figure 7 were analyzed to help define the meteorological conditions of interest. In this figure the period from late afternoon through the night until late morning is shown for three consecutive days. Each day's record begins with thermal activity until about 1800. The dark regions at the tops of the records during this time period are caused by noise from rustling dry leaves in a large tree nearby. Each afternoon was windier than the one before, as can be seen by the greater amount of wind noise on each successive record, reaching a peak wind speed of 40 km/hr from 1600 to 1700 the last afternoon. Because of the greater wind speed, the thermal activity that afternoon died down about an hour earlier than on the preceding two days, and the echoes from the thermals are narrower (being transported past the antenna in a shorter time), more numerous, and weaker (in terms of echo darkness).

The nocturnal structure is quite different on the three nights also. The first night shows the formation of a surface turbulent layer about 80-100 m thick, which thickens to a multiple layered region about 150 m thick at 0700. Thermal plume activity begins eroding the stable layer at about 0715 and has totally eliminated it by 0945. The second night was windier until about 0500, as the noise at the top of the record shows. The anemometer indicated wind speeds in excess of 10 km/hr until 0600 on this morning. The surface layer was considerably thinner this night, except for the evening hours, but was topped by two stable layers aloft at 130 m and 240 m for many hours of the night. On this morning, thermal activity penetrated the surface layer at 0845 and became quite vigorous soon after 1000. The higher winds during the third night prevented formation of a steady stable layer until around 0300. Once formed, this turbulent layer thickened up to 200 m at 0700 and then rapidly dissipated after 0830 in the presence of convective activity and increasing wind.

Study of the acoustic radar records thus indicates that, in general, thermal activity from around 0830 until about 1900 will insure good dispersion of air pollutants during this period. The rather thin stable layer which forms during the night could result in trapping of pollutants near the surface from around midnight until the layer is dissipated around 0830. The anemometer records show, however, the presence of reasonable drainage winds during this same period until about 0600, after which there is a calm period of several hours. Thus the period from about 0600 until 0830 should be one of poorest dispersion because of the absence of wind, the presence of a strong surface-based inversion resulting in a thin surface layer, and the absence of thermal activity. Since this time interval coincides with one of significant emissions due to morning auto and air traffic, it is likely that this defines the "worst case" period of the day for this part of the year for non-reacting pollutants.

One rather interesting record from the same study (but from a different sounder site nearer to an airport, with rustling leaves replaced by low-flying jet aircraft) is shown in Figure 8, and runs from one afternoon until the following morning. Here again the anemometer assists in the interpretation of the record. Thermal plume echoes extending up to 150 m during the afternoon die down as the sun sets so that by 1800 there is no activity left. The sudden burst of activity at 1840 is a consequence of a 180° wind reversal during the preceding hour, which results in well-heated desert air being advected back over the antenna in place of the normal daytime flow of marine air entering the desert through a nearby mountain pass. Considerable wind shear is present behind this front,

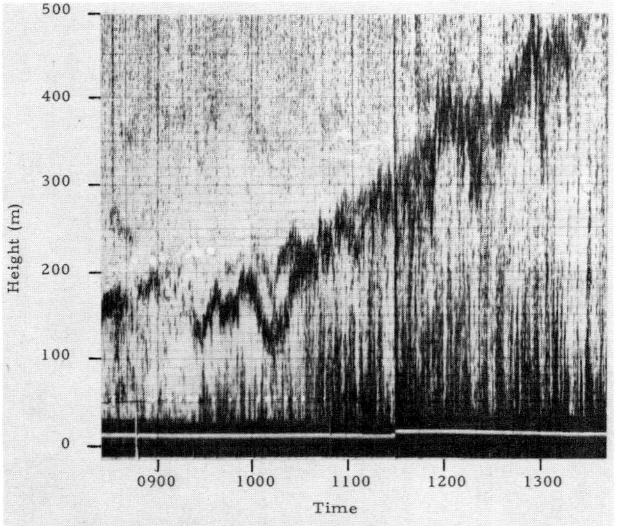

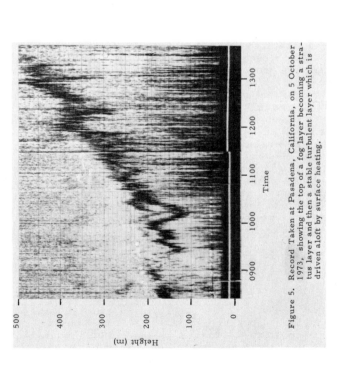

Figure 5. Record Taken at Pasadena, California, on 5 October 1973, showing the top of a fog layer becoming a stratus layer and then a stable turbulent layer which is driven aloft by surface heating.

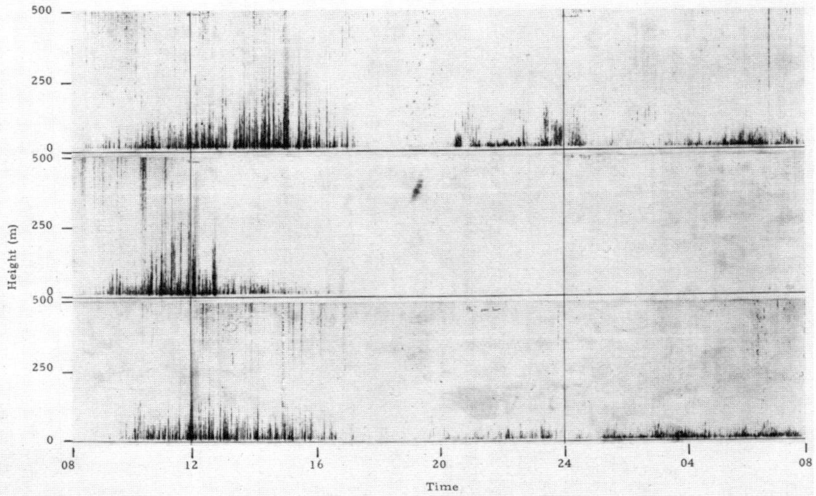

Figure 6. Sounding of atmospheric structure for 72 hours from 0800 on 17 September to 0800 on 20 September 1973 in a desert region in the southwestern United States. Note the complete lack of echoes during the night of the 18th-19th.

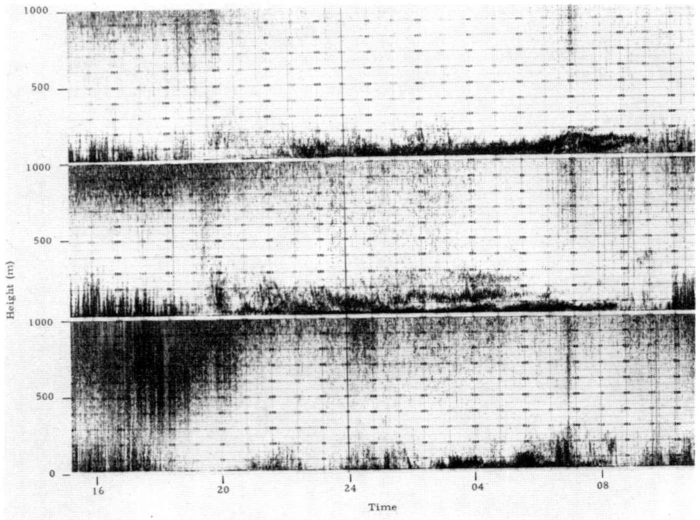

Figure 7. Records for three consecutive late afternoon to early morning periods in the Mojave Desert in the spring of 1973, showing three different forms of nocturnal structure. The dark regions at the tops of the records are caused by wind rustling of dry leaves in a nearby tree.

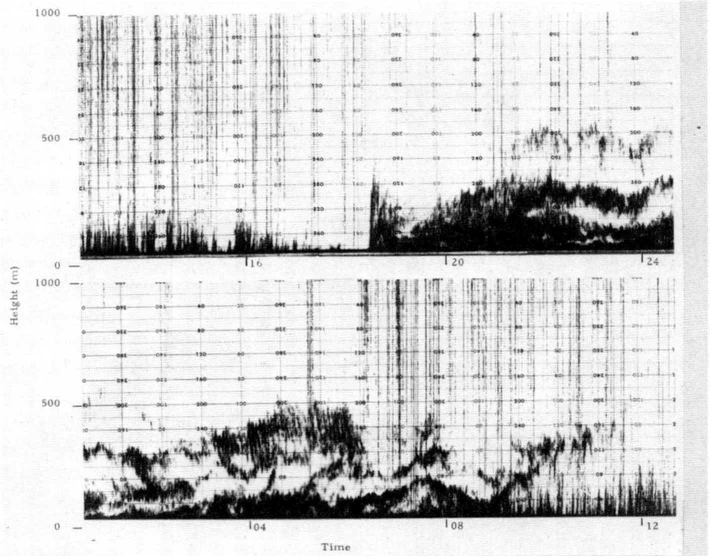

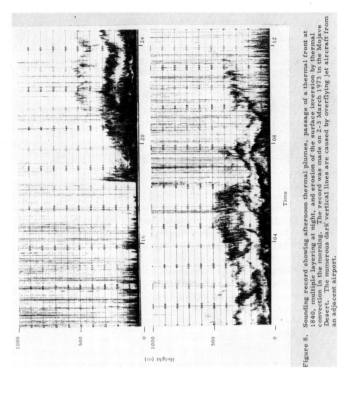

Figure 8. Sounding record showing afternoon thermal plumes, passage of a thermal front at 1840, multiple layering at night, and erosion of the surface inversion by thermal convection in the morning. The record was made on 2-3 March 1973 in the Mojave Desert. The numerous dark vertical lines are caused by overflying jet aircraft from an adjacent airport.

as can be seen from the waves in the layered structure from 1940 to 1910.

The airflow returns to normal by 1900 and the usual nocturnal layering of stable strata takes place. The record around 2300 shows nicely the strong echo of the 50 m thick surface layer, topped by the less dense records of echoes from two merging layers aloft at 100 m, a thicker layer aloft at 250 m, and a weak layer at 450-500 m. Note that all layers rise and descend together to the tune called by the wind over the surface. Also, all layers except the surface-based one show a continuous structure of waves of a few minutes period and 10-20 m amplitude, indicating the presence of significant wind shear across them. The surface layer deepens considerably to 130 m by 0600, with occasional neutral "holes" in the lowest portion of the record as different structure is transported past the antenna by the wind.

Sunrise brings an increase in wind (and airplane noise), which results in the surface layer and layers above it being forced aloft as the increased turbulence and solar radiation begin to make the atmosphere below them more neutral and echo-free. The wind becomes calm again about 0715 and the layers again subside. The same process of erosion of stability by surface-based thermal activity as was shown in Figures 4 and 5 takes place here after 0830 and the inversion is totally eliminated by 1130.

AN APPROACH TO QUANTITATIVE ASSESSMENT OF DISPERSION

Figure 9 shows in detail two nighttime records from the same study as that in Figures 7 and 8. The top record indicates a shallow (< 50 m) strongly stable surface layer after midnight which is topped by less intense echoes indicating further stable structure up to 300 m at times. One layer appears aloft at 600 m at 0400, rises rapidly to 900 m, and then vanishes. The surface layer is still strong at 0800.

The succeeding night's record shows even less of a strong-echo surface layer, but a turbulent and relatively stable region extends now up to 600 m. The clear regions which separate several layers before midnight are more neutral in stability, and the wavy structure at the uppermost portion of the echo region indicates that there is significant wind shear in the stable region.

The depth of the surface-based region, and its apparent strength in terms of echo intensity indicate the extent of coupling between the flow and turbulence at the surface and that aloft. Its depth is an indication of the thickness of this coupled region. Strong echoes indicate significant mixing in a stable layer, while an echo-free region between two layers may indicate a lesser degree of vertical turbulent transfer of information and thus a degree of independence in the turbulent flows in the two layers.

One can use such information about the structure of the lowest layers of the planetary boundary layer in combination with appropriate low-level <u>in-situ</u> measurements of wind and turbulence to arrive at quantitative values for dispersion parameters in the bottom 20 to 200 meters of the atmosphere. The approach to be used is that the parameters of interest, typically wind speed U, wind direction φ, and r.m.s. vertical turbulence σ_w, are measured near the ground and appropriate extrapolation formulas are used to determine values higher aloft, with the acoustic sounder providing information about the height to which these extrapolation formulas can be applied and suggesting the proper formula to use.

A quantitative theory for the structure of the turbulence in the layer near the ground was developed by Monin and Obukhov(16), and will be briefly summarized here. Considering only the turbulent transport of heat and momentum in a steady-state, horizontally homogeneous atmosphere, the equations of motion in the boundary layer are

$$\overline{u'w'} = -u_*^2$$

$$\overline{w'T'} = -u_* T_* \qquad (5)$$

$$\frac{\partial}{\partial z}\left(\frac{\overline{P}}{\overline{\rho}}\right) + \frac{g}{\overline{T}}\,T' - \frac{\partial}{\partial z}\left(\overline{w'^2}\right) = 0 \quad ,$$

where u' is the fluctuating component of wind in the direction of the mean wind, w' is the fluctuating component of vertical wind, $\overline{T}$ and T' are the mean (time average) and fluctuating temperatures, $\overline{P}$ and $\overline{\rho}$ are the mean pressure and density respectively, z is the height, and the parameters u_* and T_* are defined by the first two equations.

The Monin-Obukhov similarity theory assumes that the structure of the turbulence in the layer near the ground is uniquely determined by the parameters in Equations (5), viz., the constant vertical fluxes u_* and T_*, the buoyancy parameter $g/\overline{T}$, and the height z (where it is assumed that $z \gg z_o$, where z_o is the roughness scale, but z is small enough to allow the assumption that u_* and T_* are not functions of height). Any characteristic, such as $\partial U/\partial z$, σ_w (the standard deviation of w) or Φ_w (the spectrum of w), of the turbulence should then be a universal function of these parameters. Since only two independent non-dimensional numbers can be derived from these parameters, one finds that

$$\frac{\kappa z}{u_*}\,\chi = S\left(\frac{z}{L}\right) \qquad (6)$$

where χ represents the turbulence characteristic of interest, S is a universal function for each characteristic, $z/L = \kappa z g T_*/(u_*^2 \overline{T})$ with L called the Monin-Obukhov length, and κ is the von Karman constant (κ = 0.4). L is a parameter related to the stability (lapse rate) of the atmosphere, with $z/L \approx 0$ corresponding to neutral conditions, $z/L > 0$ to stable stratification, and $z/L < 0$ (i.e., $L < 0$) to unstable conditions.

The forms of the universal functions S for $\chi = \partial U/\partial z$, σ_w , or Φ_w must be determined experimentally or by appropriate theoretical inference. For our discussion here the main interest is in σ_w , the r. m. s. value of the vertical velocity fluctuations, which is a major factor in the vertical dispersion of pollutants (see, for example, reference 17). It has been found experimentally that σ_w is relatively invariant with height and stability in the surface boundary layer where, indeed, the vertical fluxes are constant and thus meet the assumptions of the Monin-Obukhov theory. One recent experimental study[18] finds $\sigma_w/u_* \approx 1.55 \pm .15$ for z/L ranging from -.75 to +10.0. Other experimenters have found mean values of σ_w/u_* between 1.2 and 1.4. The exact value is not of importance here, however, but its relative constancy (± 10%) over a wide range of z/L is, since this means that a measurement of σ_w at one height can be applicable also to other heights.

The discussion so far has been confined to conditions where turbulent mixing arises as a consequence of the normal generation processes in the shearing flow in the boundary layer. For more unstable conditions, free convection begins to dominate and the scaling changes. In free convection conditions (say, $z/L \lesssim -.5$ [18]) Lumley and Panofsky[19] have predicted theoretically that $\sigma_w/u_* \sim |z/L|^{1/3}$, i. e., it increases very gradually with height.

Similar extrapolation is possible for other flow characteristics. For example, for forced convection, the wind profile is given by

$$U = \frac{u_*}{\kappa}\left[\ln\frac{z}{z_o} - \Psi\left(\frac{z}{L}\right)\right] + Cz \qquad (19, 20) \qquad (7)$$

where the linear increase Cz due to the Coriolis force is negligible below z = 30 m or so and $\Psi(z/L)$ is zero for neutral conditions, negative for stable conditions and positive for unstable conditions. Values for Ψ and C are given in references 19 and 20, respectively. In highly unstable conditions with free convection and moderate wind speeds mixing length theory gives $U \sim z^{1/3}$.

Another quantity of interest to air pollution meteorologists is the turbulence dissipation rate, ϵ , for turbulence in the inertial subrange. Here one finds that[21] $\epsilon^{1/3} \sim U^{2/3}\,\sigma_\varphi$ where σ_φ is the r. m. s. variation of wind vane angle φ for wavelengths in the inertial subrange. There is no theory, however, which relates the lateral dispersion σ_φ to z/L and there are questions as to whether such a relation should exist. For example, experimental data in reference 17 shows large scatter in a plot of $\sigma_\varphi U/u_*$ versus z/L . For a wide range of atmospheric stabilities, experiments have shown, though, that $\epsilon^{1/3} \sim z^{-1/3}$. (22)*

Based on the above discussion, it is possible to define an approach for evaluating vertical dispersion from an acoustic sounder record and the output from a sigma meter connected to a propeller or vane for measuring vertical velocities. This anemometer should be on a tower at least 7 meters or so above the ground, and it is desirable (though not absolutely necessary) that the acoustic sounder record be able to display echoes down to the anemometer level. A standard wind speed and direction anemometer is also necessary for complete diffusion evaluation and, if ϵ is of interest, a special sigma meter tailored for the inertial subrange and connected to the vane is needed.

These sensors supplement each other to allow derivation of the total low-level diffusion situation:

1. The acoustic sounder shows qualitatively what is happening, and how far up turbulent mixing extends;
2. The tower mounted instruments provide quantitative information on airflow and turbulence at their level;
3. The surface boundary layer equations discussed above provide a technique for extending the flow and turbulence measurements vertically, and for interpreting appropriate diffusion coefficients from them;
4. The acoustic sounder suggests how far vertically these extrapolations can be extended; and
5. The tower measurements help clarify the interpretation of the acoustic sounder record.

Assume, for example, that the acoustic sounder shows a dark, signal-rich region up to 25 m and no signal above. If the meteorological sensors show turbulence existing at their level, then the sounder record suggests the turbulent region extends to 25 m and no higher. If there were turbulence above 25 m, the associated temperature fluctuations would be visible. Even if the air above had exactly a neutrally stable lapse rate, echoes would still be seen if turbulence were present to transport small eddies upward from the non-adiabatic region below. Thus, in this case, the presence of an echo implies turbulence throughout the echo region and the scaling laws presented previously are applicable (at least as a first approximation) up to the top of the echo.

It is rather unlikely that an echo could be obtained without the anemometer showing turbulence, unless the atmospheric stability were extreme enough to cause significant signal reflection. In this case vertical mixing would be nil and the extrapolation formulas would not apply. This condition would be one of extremely poor pollution dispersion.

If now the surface air was neutrally stable and the ground also fit into the neutral profile, then there would not be an acoustic signal close to the surface. The tower measurements would show here if turbulence existed. If they show no turbulence, then there is no diffusion through their level, and so whatever the sounder shows at higher levels is

*A useful discussion of the behavior of many atmospheric properties in the lower part of the planetary boundary layer appears in Chapter 6 of the report by Lissaman, et al[23]. Although slanted toward the aircraft wake turbulence problem, the concepts and formulas presented therein are equally applicable to air pollution diffusion.

unimportant for diffusion of material from the surface. If turbulence is shown, then the same extrapolation formulas apply as for the echo-rich case with turbulence, except that the height to which they can be applied is unknown. If a layer is shown aloft, the extrapolation is definitely not valid above the top of that layer (for the same reasons as in the echo-with-turbulence case), but it may not even be valid up to the base of the layer.

If the echoes have the vertical structure characteristic of highly unstable stratification, then the free-convection extrapolation formulas given earlier may be used at least up to the vicinity to which most of the echoes ascend.

The table below summarizes the various possibilities. The choice of appropriate extrapolation formulas depends on whether the echo structure is horizontal (stable, forced convection) or vertical (unstable, free convection).

Surface Based Echo Region	Turbulence at Anemometer	Extrapolation Formulas Apply
Yes	Yes	Yes, up to top of echo region.
Yes	No	No. Extremely stable case, no diffusion.
No	Yes	Yes. Maximum height not well defined.
No	No	No. No diffusion.

CONCLUSIONS

This paper has shown some of the applications of the new technique of acoustic sounding to the solution of pollution dispersion problems. The approches which were discussed were ones which indicated the utility of even a quick study of a sounding record in providing an improved understanding of the mixing properties of the atmosphere. Quantitative techniques, such as that presented in the last section, are still in their infancy of development. Consequently, the increasing use of acoustic sounders by air pollution meteorologists will surely lead to more powerful ways in which acoustic sounding can be utilized.

(1) McAllister, L. G., J. R. Pollard, A. R. Mahoney, and P. Shaw, "Acoustic Sounding - A New Approach to the Study of Atmospheric Structure," PROC. OF IEEE, Vol 57 (1969), 579-587.

(2) Proceedings of Symposium of Inter-Union Commission on Radio Meteorology, 5-15 June 1972. BOUNDARY LAYER METEOROLOGY, Vol 4 (1973), 3-523 and Vol 5 (1973), 3-256.

(3) Little, C. G., "Acoustic Methods for the Remote Probing of the Lower Atmosphere," PROC. OF IEEE, Vol 57 (1969), 571-578.

(4) Simmons, W. R., J. W. Wescott, and F. F. Hall, Jr., "Acoustic Echo Sounding as Related to Air Pollution in Urban Environments," National Oceanic and Atmospheric Administration Technical Report ERL 216-WPL 17, May 1971.

(5) Kallistratova, M. A., "Experimental Investigation of Sound Wave Scattering in the Atmosphere," TRUDY AKAD. NAUK. SSSR, INST. FIZ. ATMOSFERY, No. 4 (1961), 203-256.

(6) MacCready, P. B., Jr., "The Inertial Subrange of Atmospheric Turbulence," J. GEOPHYSICAL RESEARCH, Vol 67 (1962), 1051-1059.

(7) Harris, C. M., "Absorption of Sound in Air Versus Humidity and Temperature," J. ACOUST. SOC. AM., Vol 40 (1966), 148-159.

(8) Beran, D. W., W. H. Hooke, and S. F. Clifford, "Acoustic Echo-Sounding Techniques and Their Application to Gravity-Wave, Turbulence and Stability Studies," BOUNDARY-LAYER METEOROLOGY, Vol 4 (1973), 133-153.

(9) Emmanuel, C. B., B. R. Bean, L. G. McAllister, and J. R. Pollard, "Observations of Helmholtz Waves in the Lower Atmosphere with an Acoustic Sounder," J. ATMOS. SCI., Vol 29 (1972), 886-892.

(10) Hooke, W. H., J. M. Young, and D. W. Beran, "Atmospheric Waves Observed in the Planetary Boundary Layer Using an Acoustic Sounder and a Microbarograph Array," BOUNDARY-LAYER MET., Vol 2 (1972), 371-380.

(11) Ottersten, H., K. R. Hardy, and C. G. Little, "Radar and Sodar Probing of Waves and Turbulence in Statically Stable Clear-Air Layers," BOUNDARY-LAYER MET., Vol 4 (1973), 47-89.

(12) Staff, Wave Propagation Laboratory, "Potential Capabilities of Four Lower Atmosphere Remote Sensing Techniques," National Oceanic and Atmospheric Administration Technical Report ERL 227-WPL 20, December 1971, p. 23.

(13) Jones, J. I. P., and F. Pasquill, "An Experimental System for Directly Recording Statistics of the Intensity of Atmospheric Turbulence," QUART. J. ROYAL METEOROLOGICAL SOCIETY, Vol 85 (1959), 225-236.

(14) Beran, D. W., F. F. Hall, Jr., J. W. Wescott, and W. D. Neff, "Application of an Acoustic Sounder to Air Pollution Monitoring," paper presented at Air Pollution Turbulence and Diffusion Symposium, December 7-10, 1971, New Mexico State University, Las Cruces, N. M.

(15) Wyckoff, R. J., D. W. Beran, and F. F. Hall, Jr., "A Comparison of the Low Level Radiosonde and the Acoustic Echo Sounder for Monitoring Atmospheric Stability," J. APPLIED METEOROLOGY, Vol 12 (1973), to be published.

(16) Monin, A. S., and A. M. Obukhov, "Basic Regularity in Turbulent Mixing in the Surface Layer of the Atmosphere," TRUDY GEOFIZ. INST. AKAD. NAUK. SSSR, Vol 24 (1954), 163-187.

(17) Lissaman, P. B. S. L., "A Simple Unsteady Concentration Model Explicitly Incorporating Ground Roughness and Heat Flux," Air Pollution Control Association paper 73-129 (1973). (AeroVironment Inc., Pasadena, California, Technical Paper AV TP 311).

(18) McBean, G. A., "The Variations of the Statistics of Wind, Temperature and Humidity Fluctuations with Stability," BOUNDARY LAYER METEOROLOGY, Vol 1 (1971), 438-457.

(19) Lumley, J. L., and H. A. Panofsky, THE STRUCTURE OF ATMOSPHERIC TURBULENCE, Interscience, New York, 231 pages.

(20) Fiedler, F., and H. A. Panofsky, "The Geostrophic Drag Coefficient and the 'Effective' Roughness Length," QUART. J. ROYAL METEOROLOGICAL SOC., Vol 98 (1972), 213-220.

(21) MacCready, P. B., Jr., and H. R. Jex, "Turbulent Energy Measurements by Vanes," QUART. J. ROYAL METEOROLOGICAL SOC., Vol 90 (1964), 198-203.

(22) Pasquill, F., "Some Aspects of Boundary Layer Description," QUART. J. ROYAL METEOROLOGICAL SOC., Vol 98 (1972), 469-494.

(23) Lissaman, P. B. S., S. C. Crow, P. B. MacCready, Jr., I. H. Tombach, and E. R. Bate, Jr., "Aircraft Vortex Wake Descent and Decay Under Real Atmospheric Effects," Department of Transportation, Federal Aviation Administration report FAA-RD-73-120, (1973).

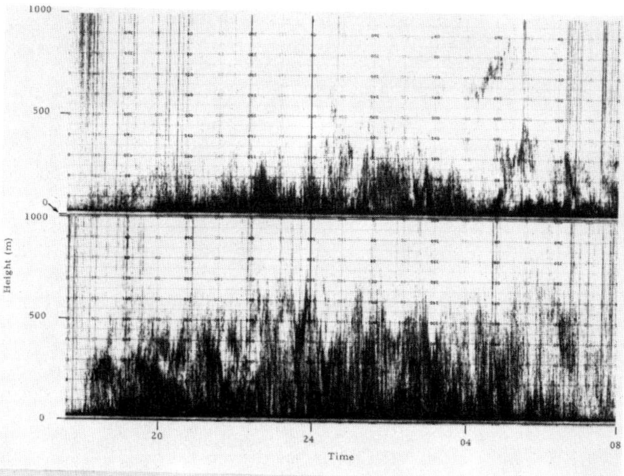

Figure 9. Nocturnal structure for two consecutive nights in the Mojave Desert in March 1973.

© 1973, ISA JSP 6679

A REMOTE SENSOR OF ATMOSPHERIC TEMPERATURE PROFILES USEABLE UNDER ALL AIR POLLUTION CONDITIONS

H. Dean Parry

INTRODUCTION

Confucius reportedly said that one picture is worth 10,000 words. A logical extension of this thought is that one quantitative measurement is worth 10,000 pictures. The system about which I shall talk today continuously measures the temperature profile of the atmospheric layer extending from near the ground to about one kilometer. It does this, from the ground, inexpensively, without benefit of balloons, sky hooks, towers, kites, carrier pigeons, satellites or space ships. The system is called RASS - for Radio Acoustic Sounding System. It penetrates clouds, heavy pollution, rain, snow, and low level inversions to measure the temperature profile reliably. In fact, the only meteorological element which severely restricts the height to which the system will work is strong wind at or above the ground. Since significant air pollution does not occur with strong winds near the ground, the system is unusually well adapted to monitoring of weather conditions for air pollution control.

As this group is well aware, an understanding of weather mechanisms which trap pollutants in, or remove pollutants from the atmosphere is an essential part of any air pollution control program. In fact, Session 9 of this conference is entitled "Measurement of Meteorological Parameters that Impact on Atmospheric Pollutants". The temperature profile which RASS measures is one of the most important meteorological parameters, if not the most important in the study and control of air pollution.

RASS, then, works best under light wind conditions when air pollution is most severe and measures that meteorological element which, along with wind, controls the rate of accumulation of pollutants in the atmosphere.

THE RASS CONCEPT

The fundamental concept employed in RASS is the tracking of sound pulses by means of electromagnetic radar. This is possible because the compression and rarefaction of air by a sound wave alters the dielectric constant slightly, thereby producing a partial, but very weak, reflection of the radar's electromagnetic energy. The intensity of the reflection is enhanced by two effects which long have been known and understood. The first of these effects occurs when the CW radar antennae are colocated with the acoustic antenna, Figure 1. Parts of acoustic shells having a common center at the sound source and delimited by the beamwidth of the acoustic antenna will then reflect the radar energy and focus it on the radar receiving antenna. The second effect is achieved by selecting the radar wavelength so that it is twice the acoustic wavelength. This condition for maximizing first order Bragg scattering, phases the radar energy reflected from successive shells.

These principles were well known to the experimenters, who, in the 1960's, tried to measure wind by the RASS concept which was then called EMAC. The essential difference between that work and the work here reported, done at Stanford University under NOAA-National Weather Service sponsorship, is in the frequencies employed. The EMAC seemed to be limited in range to a few hundred feet. It has been possible, using an acoustic frequency of 85 Hz ($\lambda a = 4$ meters) and a matching radar frequency of about 36.8 MHz to get returns from heights of 1.5 km frequently, if not regularly. The reason for better performance at the lower acoustic frequencies is not perfectly understood. The lower acoustic attenuation at the lower frequency and the fact that long acoustic waves are not distorted so much by turbulence undoubtedly constitutes at least part of the reason.

The temperature profile is derived by measuring the speed of the sound pulse as it moves upward through the atmosphere. This is done by measuring the Doppler shift in the returns of the electromagnetic radar. The speed of sound through air is given by:

$$S = w + K\sqrt{T}$$

or

$$T = \frac{S^2 - 2Sw + w^2}{K^2}$$

where S is the speed of sound propagation, w is the speed of vertical movement of the air through which the sound is passing, K is nearly constant, but varies slightly with humidity, and T is the absolute temperature in degrees Kelvin of the air through which the sound is passing. Temperature can be computed from the second form of this equation provided the vertical motion can be averaged out and the variation in speed, due to changes in humidity, are small enough to make the humidity induced error acceptably small. The effect of vertical speed and humidity are fully analyzed in an article by the Stanford developers and the author, published in the September issue of the American Meteorological Society Bulletin. It is shown that for the light wind conditions and stable lapse rates typical of air pollution episodes, an RMS temperature error of about .5° C is easily possible with RASS.

THE PHYSICAL SYSTEM

The roof was removed from an old house to provide shielding walls for the acoustic transmitter. Twenty-five meters on either side of this house two Yagi antennae were installed, one to serve the transmitter and the other the receiver of the CW electromagnetic radar. Inside the house is a 3 by 3 array of horns. These horns are made of 3/4 inch plywood and each has a square mouth about 5 ft by 5 ft. The horns face downward toward the house's concrete floor which serves as a reflector, to send a beam of sound vertically upward. The horns are acoustically driven by 4 woofers located within the boxlike type of each horn. The woofers are driven by a tunable oscillator through a 1 kw power amplifier.

The electronic equipment is unsophisticated and relatively inexpensive. The radar is only 5 watts CW and uses a homodyne receiver which produces an output frequency equal to the acoustic frequency. The data processing part of the system consists of solid state counters which measure the interval between zero-crossings to determine the instantaneous period of the Doppler signal in real time.

SAMPLE TEMPERATURE PROFILES MEASURED BY THE SYSTEM

Figure 2 -- This is a very early result. Multiple measurements were made at each level and each dot represents an instantaneous measurement. The scatter of these dots is due to the net effect of all errors inherent in the system, plus a number of errors due to inexperience of the operator. At the very least, this early result showed the necessity for averaging a number of measurements at each level to obtain the temperature at that level. A need to improve the overall accuracy of individual measurements is also strongly indicated. At one point, the author believed -- perhaps hoped is more accurate -- that the dispersion of points at each level represented a measure of the vertical motions due to turbulence at that level. The fact that the dispersion was minimum in the inversion and greatest in the most unstable layer gave rise to this idea. The acoustic beamwidth is 24 degrees and the pulse length converted to distance is 120 meters. At 1 km the volume sampled by RASS is roughly a cylinder about 200 meters in radius and 120 meters high. Variations within this volume would tend to average each other out and result in Doppler broadening. Furthermore, there is an error associated with the signal to noise ratio which in turn is a function of the match between acoustic and electromagnetic wavelengths. If that match had been best at the level of inversion, the small dispersion at that level would be explained thereby. The possibility of obtaining a temporal spectrum of vertical motion for each level may still exist and the question of the practicality of doing this should be investigated further. Figure 2 which suggests the feasibility of obtaining such a spectrum is subject to misinterpretation and should be used with extreme caution.

An NCAR type tethered balloon was used to carry an in situ temperature probe to heights of about 650 meters. Temperatures measured by this probe were used to check the measurements made by RASS. Since RASS profiles usually extended above 650 meters, the San Jose EMSU radiosonde was used as a check for the higher levels. Figure 3 shows one set of the three measurements just described. The solid line is the RASS profile. This was calculated and plotted before the check data were entered or consulted. The dotted line is the average of the profiles made by the tethered balloon as it went up and came down. This balloon was operated from a point only a few meters away from the RASS acoustic transmitter. The dashed line is the San Jose radiosonde which was released within a few minutes of the time of the RASS measurement, but at a point some 30 km away. The tethered balloon trace is not as straight a line as the radiosonde. Tethered balloon measurements were made every 33 meters. The radiosonde, on the other hand, has an artificial linearity built into both the sounding instrument and to the manner in which the sounding is worked up. On the basis of this tethered balloon trace, it seems fair to state that the atmosphere does not know its own temperature to more than a degree. Agreement between the RASS profile and the tethered balloon profile is excellent between about 300 meters and the top of the tethered balloon run.

There is a significant difference between the San Jose measurement and the balloon measurement near the ground and this difference persists up to near the height of the top of the tethered balloon measurement. It is a reasonable meteorological assumption that this local difference between the RASS site and the San Jose measurement would smooth out with increasing height above the ground. If this is indeed the fact, there is excellent agreement between the RASS and the radiosonde from 650 meters to the top of the RASS profile. The misbehavior of the RASS profile below 300 meters is probably due to the geometry of the system. For air pollution studies, this limitation of the present system is quite serious. It is highly probable that the design can be modified to overcome this difficulty, thereby making the RASS even more useful in air pollution studies. Some ideas for system modification are now being tested by the Stanford developers. One idea proposed to correct the low level problem is to bring the Yagi antennae in close to the walls of the acoustic shelter. Doing this will decrease the isolation between the transmit and receive antennae which will in turn decrease the signal to noise and consequently the probable maximum range. The RASS was envisioned as a sounding system for all meso scale work of the National Weather Service. If indeed the low level problem can not be overcome without penalty of reduced range it would appear quite feasible to use one pair of antennae to sound up to 500 meters and a second pair to sound above that level. The system could be either automatically or manually switched to first one pair then the other to obtain a complete sounding of the first one or two kilometers of the troposphere. A pulsed radar using a single antenna located just on top of the acoustic transmitter and a CW acoustic tone have been considered. Such a system would generate considerably more noise pollution and RFI than the present mode. Achieving a very short minimum range in an unsophisticated radar might well also present a difficult problem.

The profile of Figure 3 extending to $1\frac{1}{2}$ kilometers is one of the higher levels to which comparisons with in situ sensors have been obtained. All comparisons made to this time have indicated that the RASS can produce accurate temperature profiles from 300 meters to a height determined by the winds, both surface and above the ground. Strong winds at any level will prevent the present system from making accurate measurements above that level. Within the vertical range of RASS, presently from 300 meters to something over a kilometer for most wind conditions, the temperature error is much less than 1° C as judged by a rather large number of comparisons with in situ measurements.

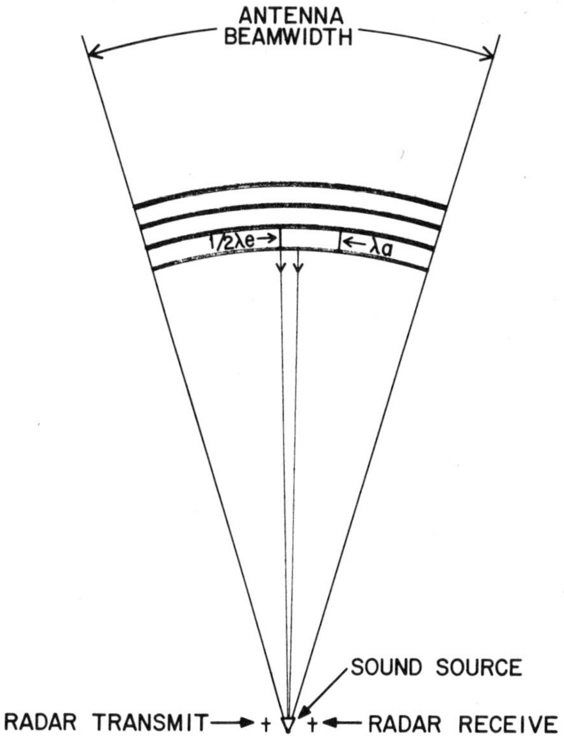

Fig. 1

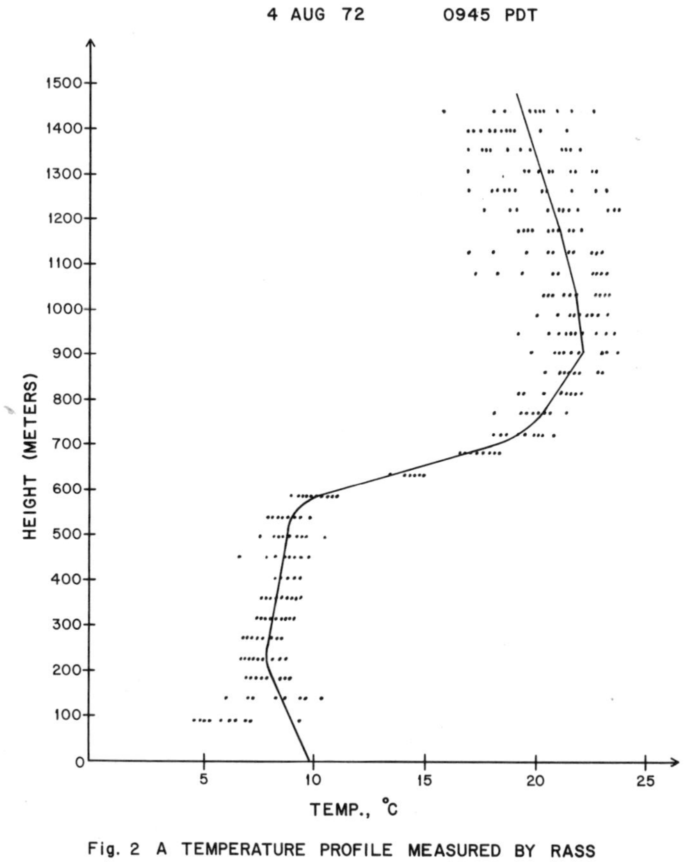

Fig. 2 A TEMPERATURE PROFILE MEASURED BY RASS

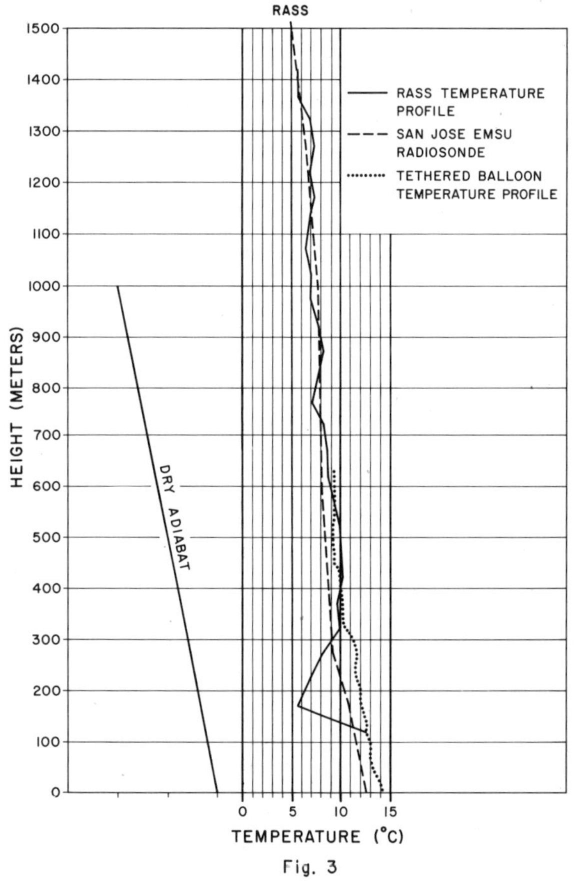

Fig. 3

© 1973, ISA JSP 6682

THE APPLICATION OF ELECTRO-OPTICAL TECHNIQUES TO SENSING OF STATIONARY SOURCE POLLUTANTS

William F. Herget
Chemistry and Physics Laboratory
National Environmental Research Center
Environmental Protection Agency
Research Triangle Park, NC 27711

INTRODUCTION

The basic function of the Stationary Source Measurement Research Section of the Chemistry and Physics Laboratory is to conduct research on and development and evaluation of methods and instrumentation for the identification and measurement of particulate and gaseous pollutants in emissions from stationary sources. The types of methods and instrumentation being studied in this effort may be divided into two main categories: extractive (the pollutant sample is transported from within the stack to the analyzer outside the stack,) and non-extractive (where the pollutant sample is analysed within the stack by an electro-optical technique). This paper will concern itself with non-extractive electro-optical techniques and emphasize those techniques applicable to gaseous pollutant analysis.

The electro-optical techniques are generally applied to stationary source emissions in the three operational modes: in-situ, long-path, and remote. In the in-situ mode the pollutant under study is flowing through a duct or stack and the measurement (usually absorption as a function of wave length) is made "across-the-stack" with an energy source on one side of the stack and a receiver unit on the other side; in some applications the source and receiver are on the same side of the stack. In the long-path mode a double-ended absorption measurement is made in the vicinity of an extended area stationary source, such as oil refinery; the path length might vary from 100 to 2000 meters. In the remote mode, measurements are generally single-ended observations of infrared emission or laser-stimulated scattering from pollutants; range information can be obtained in some of the scattering techniques.

The operational modes defined above are meant to be general, and the techniques are often applicable to other than stationary source measurements. Long-path and remote techniques can be used also in ambient air measurements. In-situ techniques are generally applicable to mobile source emission measurements (such as jet engine or auto exhausts) since these sources are usually stationary when being studied. At this writing, the in-situ mode must be considered the most advanced of the three modes since there are several commercial instruments for particulate opacity and gas concentration measurements on the market now, and the number will certainly increase substantially during 1974. The technology for long-path instruments is only slightly behind that for in-situ instruments, but few instruments are commercially available; a number of research prototype instruments are either in use or nearly so. The remote sensing techniques are the least developed of the three techniques. However, because remote techniques can cover large areas from a single location and are the only method of effectively measuring the net pollutant flow from extended area sources, these techniques are sure to develop rapidly during the next few years.

To efficiently develop and evaluate the techniques described above, it is necessary to know the spectral signatures of both the gaseous pollutants and potential interferences encountered in various stationary source emissions. Toward this aim, an infrared telespectrometer is being used for both long-path and remote measurements to spectrally characterize various source emissions. The remainder of this paper will discuss the use of this instrument and review the status of the electro-optical techniques currently under study at EPA.

INFRARED TELESPECTROMETER SYSTEM

The EPA infrared telespectrometer system, which bears the acronym ROSE (Remote Optical Sensing of Emissions) was designed and fabricated under contract with General Dynamics/Convair[1,2]. A schematic of the major optical components is shown in Fig. 1. For absorption measurements the source is a 1500°K blackbody chopped at 570 hz. Radiation is collimated and focused with Dall-Kirkham telescopic optics using 60 cm diameter primary mirrors. For remote emission measurements chopping occurs at 330 hz in the receiver section. Wavelength dispersion is obtained with a 1/4-meter, linear-wave-number-drive monochromator. Two gratings and two long-pass filters are used to cover the 3-5.5 and 7-13μ spectral ranges, respectively. A closed cycle cryogenic cooling system maintains the Ge:Hg detector at 28°K. The detector output is amplified and processed by conventional electronics and displayed on a strip chart recorder. Data may also be recorded in digital form on either magnetic tape or printed paper tape. Wave number information is recorded simultaneously with spectral data. Figure 2 shows the ROSE receiver section, and Fig. 3 shows the two electronics consoles. To date, long-path absorption measurements have been made with path lengths of 0.5, 1.5, and 3 km(2). Remote emission measurements have been made at gas-fired power plants(2), coal-fired power plants(3), and oil-fired power plants.

In the measurements referred to above, the ROSE system was transported via rental truck with hydraulic tail-gate, and considerable time was spent loading, moving, and unloading equipment on field trips. To aleviate this problem, a mobile laboratory has been designed and fabricated to house the ROSE system on a quasi-permanent basis. The receiver unit and associated components will be located as indicated in Fig. 4. Special features of the mobile lab include: stabilizing jacks attached to the vehicle frame, provision for complete isolation of the receiver unit from vehicle motions, an external mirror assembly to provide various viewing elevations, and provision to lower the generator unit to the ground to eliminate vibration effects. For single-ended measurements only the external mirror assembly need be moved into position. For double-ended absorption measurements the source unit and its console, which will be transported in the rear of the lab, will be removed for use at the desired remote location.

Remote Emission Measurements

Figures 5, 6, and 7 summarize recent data on power plant emissions(3). In Fig. 5 the lower spectrum shows atmospheric gases in emission against the cold background of outer space for the spectral range 7 to 13μ. This spectrum was obtained by sighting the ROSE receiver off to the side of the stack plume. The upper spectrum of this figure shows the coal-burning power plant plume emission spectrum with the sky spectrum as background. This plume has had over 99 per cent of the particulate matter removed, so that only the gaseous species contribute to the spectrum. Figure 6 shows similar type spectra for the same plume from 3.5 to 6.5μ. Probably the most important feature of these spectra is that although the 4μ SO_2 band is the weakest of the three bands shown, it is probably the most suitable for remote measurements since there is essentially no interference from other gaseous species. The lower two spectra in Fig. 7 are similar to Fig. 5 except that the sky background is obscurred by plumes from adjacent stacks that were heavily laden with particulate matter; these dispersed dirty plumes formed an essentially opaque screen at ambient temperature behind the clean plume. Using a plume temperature of 400°K (measured 15m below the stack exit) and a blackbody intensity calibration, the SO_2 concentration (determined 8.6μ, Fig. 7, middle curve) was calculated to be 650 ppm. Extractive instrumentation was measuring between 675 and 750 ppm at the same time. The upper curve in Fig. 7 shows an emission spectrum from an oil-fired turbine (jet engine) exhaust plume. There is more CO_2 and H_2O than in the coal-fired plume but essentially no SO_2. Figure 8 shows the emission spectrum from an oil-fired power plant plume along with the lower two spectra from Fig. 7. Comparison shows that the spectral features are nearly identical; in fact, the only observable differences are due to higher particulate concentrations in the oil-fired plume than in the coal-fired plume.

A major goal of the above measurements was to evaluate the capabilities of the ROSE system for remote measurements. It was found that for a relatively high concentration pollutant such as SO_2 with minimal H_2O interference, a reasonable concentration measurement could be made if the plume temperature was known independently. It would have been very difficult to determine the plume temperature from the ROSE data. Because of the low plume temperature (~400°K) the spectrometer had to be operated at low spectral resolution (~4 cm^{-1} at 8μ and ~16 cm^{-1} at 4μ) to give a reasonable signal-to-noise ratio. For this reason it was not possible to observe species such as NO and CO. The data obtained did emphasize the need for an accurate temperature measurement. At the observed plume temperatures, a 5 per cent temperature would produce a 10 per cent concentration error. The greatest value of the ROSE system in remote measurements is sure to be in species identification. However, it will be necessairy to increase the spectral resolution by an order of magnitude to fully exploit this capability.

Long-path Absorption Measurements

Long-path measurements were made with the ROSE system during contractor evaluations(2). Over a 3 km path it was possible to measure hourly ozone concentrations increasing from 0.05 ppm in the morning to almost 0.2 ppm at noon and then decreasing to about 0.05 ppm by late evening. These measurements were made in the Pomona Valley east of Los Angeles. In another experiment CO concentrations on the order of 10 ppm were measured along a 0.5 km path parallel to a highway. The spectral resolutions obtainable at the 3 km path were 3 and 1 cm^{-1} at 4 and 8μ, respectively. At the 0.5 km path resolutions of 1.5 and 0.5 cm^{-1} were obtained. Additional measurements have been delayed pending the completion of the mobile lab described previously and a computerized data reduction program.

The computerized data reduction program is being developed for EPA under an interagency agreement with NOAA, Environmental Research Labs, under the direction of Dr. V. E. Derr. A report detailing this effort is in preparation(4). The input data to this computer program will be the spectral data recorded in digital form on magnetic tape using the ROSE system. Initial computer processing of the tapes will place the data in a format containing per cent transmission as a function of wave number at preselected intervals (variable from 0.1 to 2.0 cm^{-1}). The main computer program contains a library of spectra of various gases at several concentrations. This library was compiled from existing experimental data and some additional laboratory measurements. Figure 9 shows the gases and their principal regions of absorption, and Fig. 10 is a selection of the spectra of the more important species. When in operation, the computer program varies the concentration of the species in the library and calculates the resulting spectra; this procedure is iterated until a best fit between the calculated and input spectra is obtained. Only those gases suspected of being in a given spectrum need be included in a particular calculation (to save computer time) and

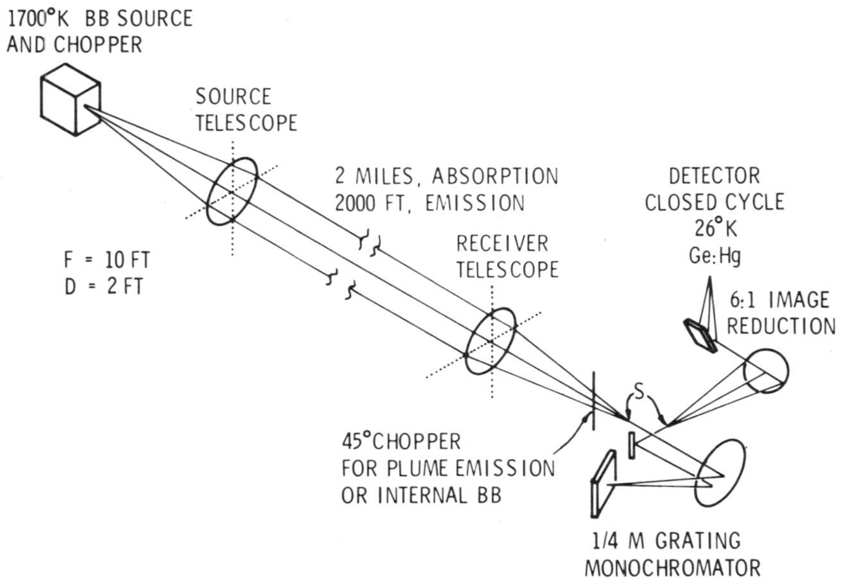

Figure 1. ROSE System Optical Schematic.

Figure 2. ROSE System Receiver Unit.

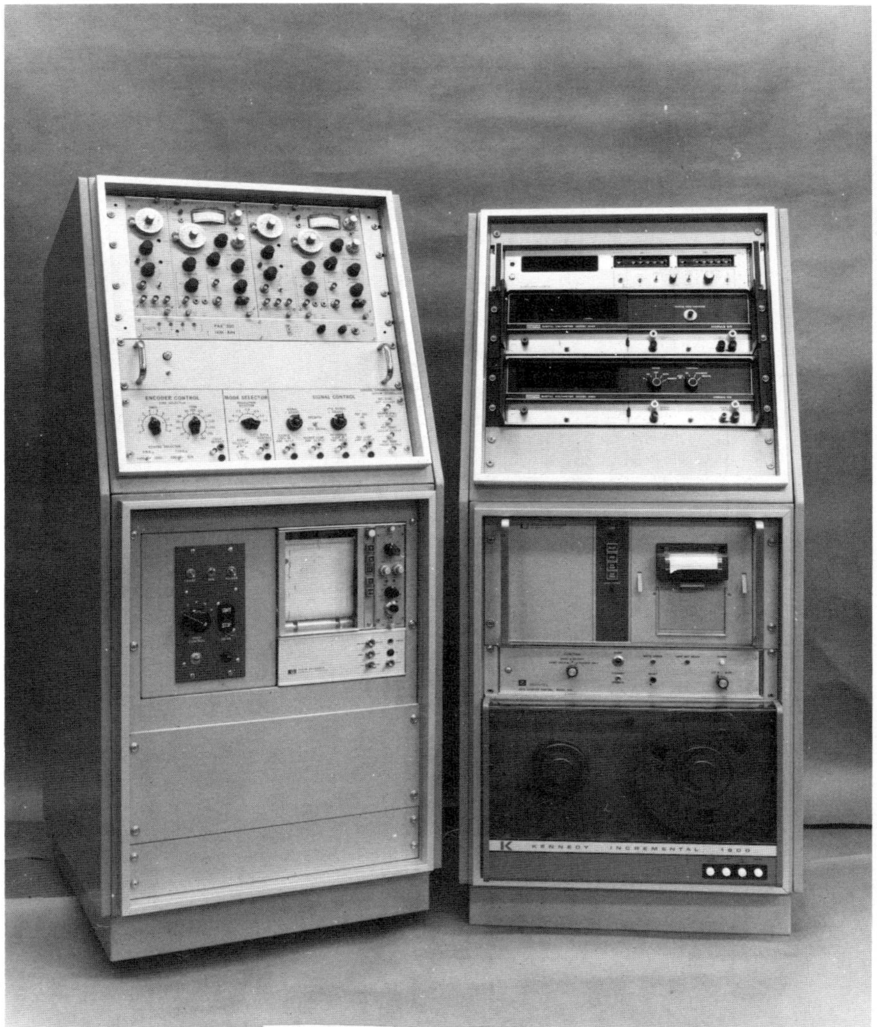

Figure 3. ROSE System Electronic Consoles.

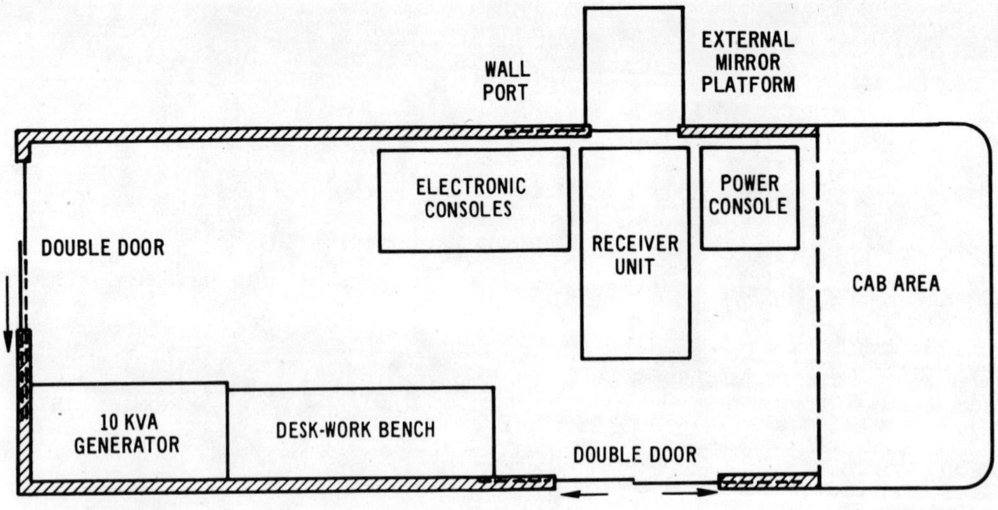

Figure 4. Mobil Laboratory Schematic.

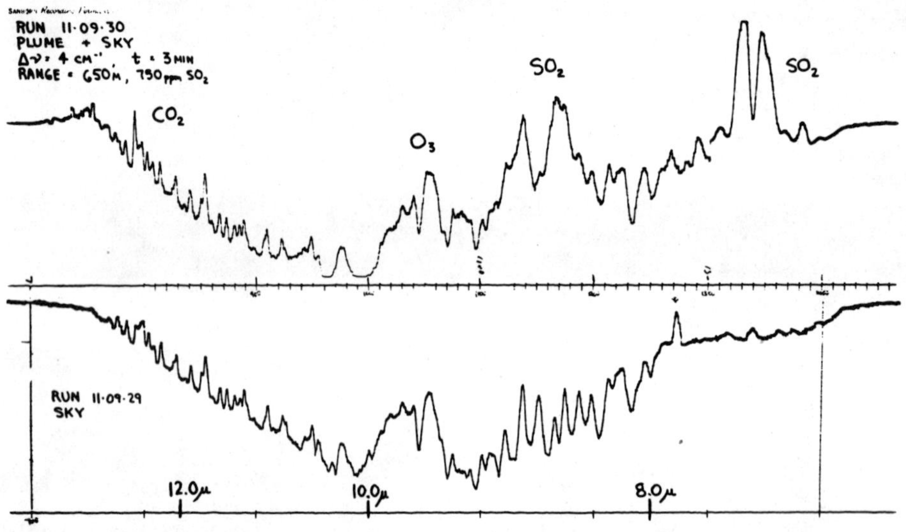

Figure 5. Clean Coal-Fired Plume and Sky Background, 7 to 13μ.

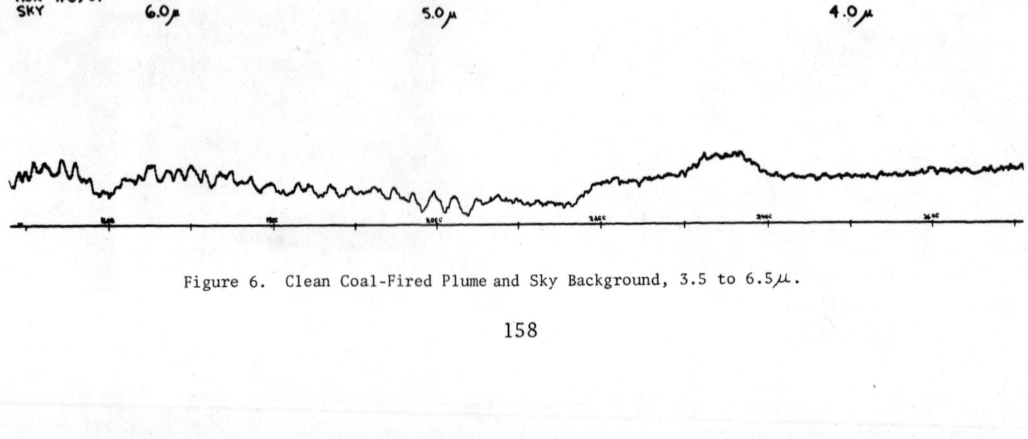

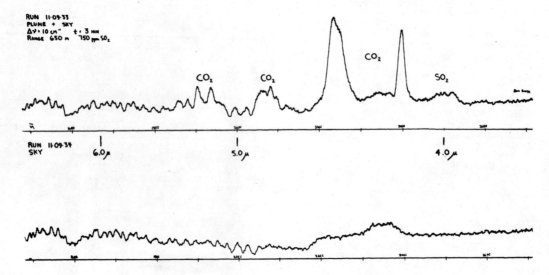

Figure 6. Clean Coal-Fired Plume and Sky Background, 3.5 to 6.5μ.

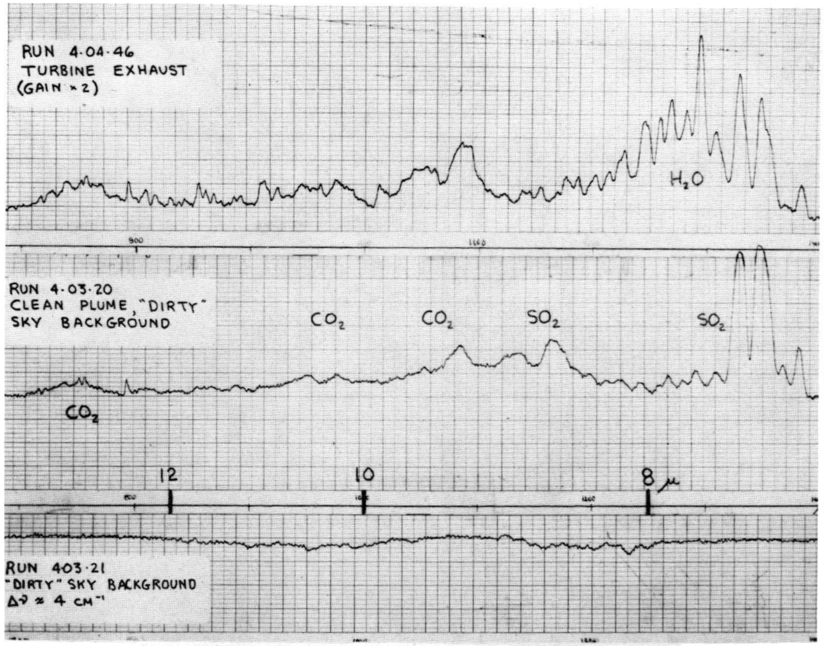

Figure 7. Coal-Fired Plume with Opaque Background and Turbine Exhaust Plume.

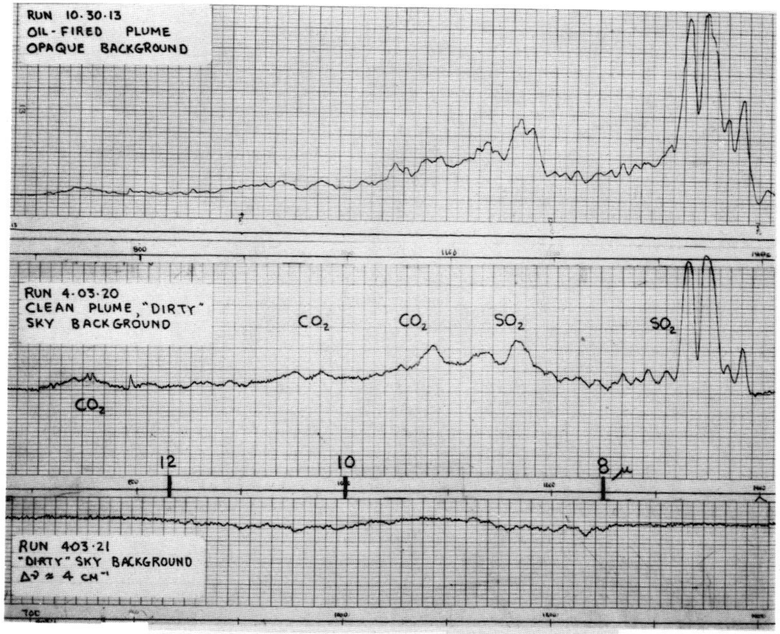

Figure 8. Coal-Fired and Oil-Fired Plumes with Opaque Background.

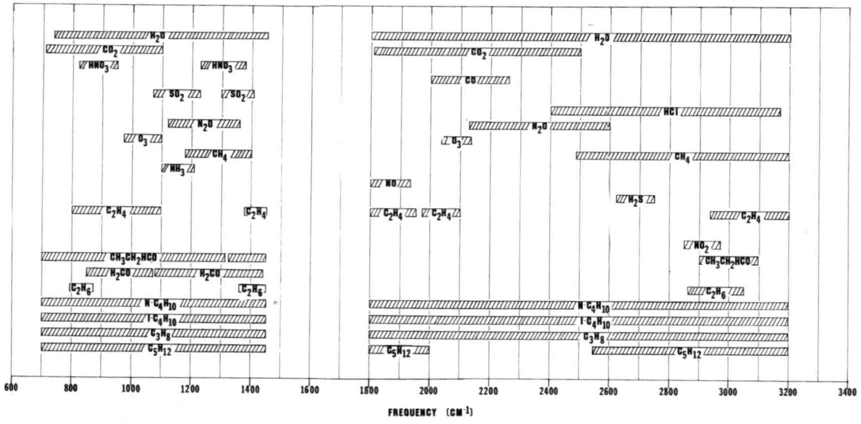

Figure 9. Principal Regions of Absorption of Selected Normal and Pollutant Atmospheric Constituents.

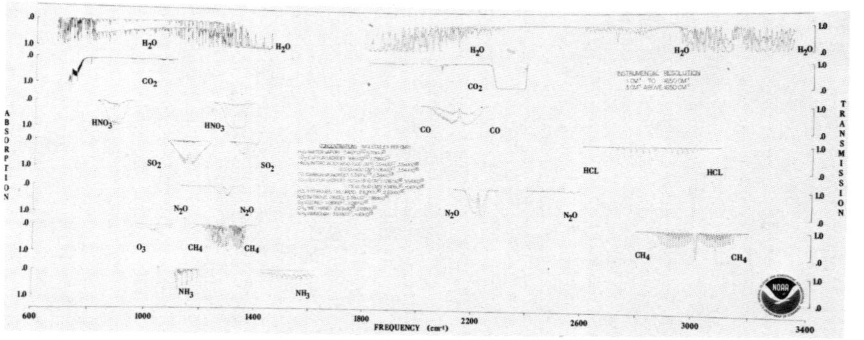

Figure 10. Infrared Spectra of Selected Gases.

the wave number interval to be covered may be varied as desired.

At present, spectral absorption data obtained with the ROSE system in the laboratory using the gases CO, SO_2, and CH_4 at known concentrations in a 10 cm cell are being used as input to the computer program to check out the overall data handling and reduction procedure. Results to date show that the entire procedure is operable and that it is easy to obtain concentration measurements accurate to within 25 per cent. For the general purpose of spectrally characterizing pollutant emissions from various sources, this type accuracy would probably be acceptable. However, since higher accuracy is certainly expected, it is being pursued.

In addition to uncertainties in knowing the concentrations of the gases in the 10 cm cell, the two difficulties that appear to contribute most to the above error are determination of true per cent transmission data and match-up of instrument resolution with that used in the calculation of the spectra which are contained in the computer program library. Two methods are being used to overcome these difficulties: one is to ratio-record using a second light source and path which simulates the remote source and path but contains no pollutants; the second is to ratio spectra obtained at different times under clean and polluted conditions. Both methods introduce uncertainties into the data. The spectral resolution problem occurs because the library spectra were calculated with a constant spectral resolution (3 cm^{-1} for the 3 to 7μ range and 1 cm^{-1} for the 7 to 13μ range), while the ROSE grating monochromator resolution is a function of wavelength. Methods of overcoming this second problem are to change the monochromator slit width at intervals during a scan, change the library spectra resolution to match the monochromator, or replace the monochromator with an instrument (such as a Fourier-transform spectrometer) whose resolution is not a function of wavelength. These methods are all under study. It has been verified that the computer program itself does not introduce error into the calculations. This was done by using data from the library itself as input data to the computer program.

ELECTRO-OPTICAL TECHNIQUES UNDER STUDY

The purpose of this section is to briefly summarize those activities of the Chemistry and Physics Laboratory related to the application of electro-optical techniques to the sensing of stationary source emissions. Table I lists those techniques which are currently under study, have been studied recently, or are planned for study during FY'74. For those techniques related to gaseous concentration measurements, the principal species under study are given; of course, most such techniques are applicable to a wide variety of gases. The type of study concerning a particular technique is catagorized as follows: "Laboratory Feasibility Study" implies that existing non-optimized components have been assembled into a system to demonstrate in the laboratory the potential capabilities of the technique; "Field Evaluation" implies that an existing instrumentation system (not necessarily developed for air pollution measurements) is being evaluated in the field for stationary source measurements; and, "Instrument Developement" implies that a protype instrument is being fabricated for field evaluation. The diode and gas laser techniques listed are currently being evaluated for ambient air measurements.

In addition to studies of individual electro-optical techniques, considerable effort is being expended in comparing results obtained with in-situ and extractive instruments in use on the same power plant stack. This work has been reported by J. B. Homolya[3]. Future comparison experiments will concern in-situ and remote measurements and also long-path and point monitor measurements.

SUMMARY

A general review has been presented of Chemistry and Physics Laboratory activities concerning the application of electro-optical techniques to the measurement of pollutants emitted by stationary sources. This review included a description of the program under way to characterize the infrared spectra of various pollutants using long-path absorption spectroscopy and remote emission spectroscopy. A brief summary was given of activities related to the application of particular electro-optical techniques to the measurement of specific pollutant properties.

ACKNOWLEDGEMENT

Appreciation is expressed to the Duke Power Co. and the Carolina Power and Light Co. for permission to make the remote measurements at their facilities and for their assistance and cooperation.

REFERENCES

(1) Streiff, M. L., and C. L. Claysmith, "Design and Construction of a System for Remote Optical Sensing of Emissions," General Dynamics/Convair, San Diego, October 1972; EPA Report No. R2-72-052.

(2) Streiff, M. L., and C. B. Ludwig, "Remote Sensing of Pollutants in Urban Areas," General Dynamics/ Convair, San Diego, August 1973; EPA Report No. 650/2-73-026.

(3) Stevens, R. K., and W. F. Herget, editors, "Analytical Methods Applied to Air Pollution Measurement", Ann Arbor Science Publishers Inc. (in publication).

(4) Derr, V. E., "Remote Sensing of Pollutants", final report on EPA Contract No. IAG-077(D), (in preparation).

Table I. Electro-Optical Techniques Under Study.

Technique	Species or Property Measured	Mode[a]	Type Study[b]	Investigator	Status[c]	Spectral Region
Fluorescence	SO_2	I, R	LF	United Aircraft	Current	UV
Fluorescence	Particulates	I	LF	SRI	R2-73-219	UV
Vibrational Raman	NO, SO_2	R	FE	NASA/Langley	Current	UV
Rotational Raman	NO	R	LF	Princeton U.	Current	UV
Diff. Absorption LIDAR	NO, SO_2	R	LF		Future	UV
LIDAR	Opacity	R	FE	EPA	Current	VIS
Transmissometry	Opacity	I	FE	EPA	Current	VIS
LIDAR, CWFM	Opacity	R	LF	SRI	Current	VIS
Transmissometry	Opacity/Mass Ratio	I	LF	Philco-Ford	Current	VIS, IR
Optical	Mass	I	LF		Future	
Doppler Velocimetry	Velocity	R	FE		Future	IR
Interference Fringes	Size Distribution	I	LF	ARO/AEDC	Current	VIS
Dispersive Correlation	NO_2, SO_2	P, R	FE	EPA	Current	VIS, UV
Gas Filter Correlation	NO, SO_2, HF,	I	ID	Philco-Ford	Current	IR
Gas Filter Correlation	HCl, CO	P	FE	EPA	Future	IR
Gas Filter Correlation	NO, H_2S	R	LF	Philco-Ford	Current	IR
Gas Filter Correlation	SO_2	R	ID	JRB Assoc.	Current	IR
Diode Laser	SO_2	I	ID	Lincoln Labs	R3-73-218	IR
Diode Laser	SO_2, CO	P	FE	Lincoln Labs	Current	IR
Gas Laser	O_3, NH_3, CO_2, etc.	P	FE	General Electric	Current	IR

a. I = In-situ, P = Path, R = Remote
b. LF = Laboratory Feasibility Study, FE = Field Evaluation, ID = Instrument Developement
c. EPA report number given for completed work

© 1973, ISA JSP 6684

APPLICATION OF THE TRIBOELECTRIC EFFECT TO THE MEASUREMENT OF AIRBORNE PARTICLES

Arnold H. Gruber and E. Karl Bastress*

IKOR Incorporated

Burlington, Massachusetts

ABSTRACT

The electrical charge transfer process, or "triboelectric effect," has been developed as the operating principle for a series of instruments used to measure mass concentration of particulate material under a variety of emission conditions. The charge transfer occurs between particles suspended in a moving gas stream and solid surfaces bounding the stream, and hence, the process can be used to detect the presence of particles in the stream.

The instruments which have been developed by IKOR Incorporated combine an electronic sensing system based on the charge transfer process with an extractive sampling train to provide truly continuous monitoring of particulate mass flow. The monitor can be calibrated by gravimetric analysis obtained with an integral filter assembly, or by parallel testing with a comparison sampling system. The IKOR instruments have proven to be highly effective in sampling and monitoring of particulate materials. The instruments also provide a convenient and highly viable technique for maintenance of continuous dust collection equipment.

INTRODUCTION

Requirements exist in various industries for methods of continuously monitoring concentrations or flow rates of particulate materials in gaseous flow streams. The instruments developed by IKOR Incorporated† utilize the charge transfer process, and consist of extractive sampling systems into which particle flow monitoring systems have been incorporated. Collisions between particles in the sampling stream and a sensing probe result in the flow of an electric current which is processed electronically to produce an output voltage signal proportional to the charge transfer current. This output voltage, which is thus indicative of the particle-sensor collision on an instantaneous basis, provides a continuous indication of the rate of flow of particulates in the sample stream.

Two versions of the IKOR Air Quality Monitor (AQM) are available:

1. A permanent installation featuring automatic isokinetic sampling, direct readout in grains/standard cubic foot (or other desired units), and high/low alarm functions.

2. A portable, light weight unit which can be readily utilized for stack and occupational monitoring.

The portable unit is the most commonly used version at the present time. This paper describes the portable monitor, the nature of the charge transfer process, and the performance characteristics of the monitor.

MONITOR DESCRIPTION

Portable Monitor Configuration

The portable model of the IKOR AQM, shown in Figure 1, consists of three components: a probe, a sensor unit and a control unit. In stack monitoring, the probe is inserted in the duct or stack to be monitored where it simultaneously extracts a continuous sample and measures stack flow conditions.** A modified probe is available for ambient or occupational air quality monitoring. The sensor unit houses the electronic sensor, gravimetric filter (directly behind the sensor) and blower. The control unit houses the instrumentation and controls regulating and monitoring sample flow conditions, and also houses the electronic readout instrumentation for the particulate mass flow sensor. Each of the components is portable and light weight so that the entire AQM can be transported readily to any test site, assembled and placed into operation quickly.

Accessory components are normally used for recording and integrating the electronic particulate mass flow output signal. A strip chart recorder provides a permanent record of output signal with time. An integrator is used to automatically totalize the area

---

† Covered by U. S. and foreign patents

*Present Affiliation - Arthur D. Little, Inc., Cambridge, Massachusetts

**A. H. Gruber and E. Karl Bastress, "A Comparison of a New Continuous Electric Charge Particulate Monitor with the EPA Sampling Train," Proceedings of the TAPPI Engineering Conference, Boston, Mass., p. 205 - 216, October 8 - 11, 1973.

under the output curve during the specific period of operation. The totalized electronic signal can be compared to the filter weight gain during the identical period of operation.

To operate the AQM, only electrical service is required. The AQM can be operated by one man.

Electronic Sensing System

The electronic sensing system consists of an electrically conductive sensing element mounted coaxially in the sampling train and connected to a high-gain amplification circuit. Any of a variety of materials and configurations may be used for the sensing element, depending upon the nature of the particles to be monitored and their gaseous environment. The amplification circuit produces an output voltage which is proportional to the rate of transfer of electrical charge between the sensing element and the particles in the sample stream. The amplification ratio can be varied over five decade ranges of sensitivity to provide a particulate mass flow sensing range of over 100,000 to 1.

The particle sensing mechanism consists of the transfer of electrical charge between the particles in the sample stream and the sensing element. The charge transfer process, or "triboelectric effect," occurs between dissimilar solid materials when they come into physical contact. Charge transfer occurs between particles suspended in a moving gas stream and solid surfaces bounding the stream, and, hence, the process can be utilized to detect the presence of particles in the stream. The particles need not be precharged for the charge transfer process to occur. Nor are the particles captured as a result of the charge transfer process. The extracted sample passes through the sensing unit unchanged except for particle charge.

The charge transfer process has been observed under a wide variety of circumstances and has been studied by a number of researchers. Two primary (but not mutually exclusive) theories have been formulated to explain the process. The simpler theory proposes that during instantaneous contact of two solid bodies of different work functions, the equilibrium state requires that their Fermi levels coincide with the transfer of electrons from the surface having the lower work function to the one with the higher work function. This gives rise to a contact potential difference. When contact is suddenly broken (as in a continuous flow condition), the body having a lower work function becomes negatively charged. In this way, a metal sensor subjected to collision by a particulate cloud can give an electric current as a signal due to redistribution of charges from particulate impact. After impact, the exiting particle cloud differs only in electron inventory.

The double layer theory proposes that each particle is surrounded by an adsorbed surface layer of air molecules which participate in the charge transfer process. The air layer is in a charged state since it is in work-function equilibrium with the body to which it is attached. When the particles impinge on the surface of a compact body (such as a metallic sensor), the ions of the loosely bounded air layer are detached by inertial forces. Thus, the sensor becomes charged by the ions separated from the double layer of the dust particle, with the dust particle entraining the charges of the more strongly bonded lower level of the double layer.

The charge transfer process results in a flow of electrical current from the sensor element to the system electronics, and ultimately to a virtual ground. A signal equal in magnitude to the current generated by charge transfer is produced, amplified, and converted into an output voltage precisely related to particulate charge transfer. The magnitude of this signal is thus dependent upon the characteristics which affect the charge transfer process - primarily composition and surface characteristics of the particles and sensor - and the sample stream, including the flow characteristics of the sample stream which determine the probability of contact between the particles and the sensor element. As a result of these interdependencies between stream, particle, and sensor characteristics, effective use of the monitor is dependent upon knowledge of factors affecting the relationship of electronic signal and particulate mass flow.

Sampling System

The sampling train incorporated in the IKOR AQM is shown schematically in Figure 2. The sampling train includes the following components:

1. Stainless steel sampling tube with sharp-edged inlet nozzle and 0.93 inch internal diameter.

2. Venturi incorporated in the sampling tube providing a pressure differential indicative of the sampling rate.

3. Pitot tube mounted in parallel with the sampling tube providing a pressure differential indicative of the flow velocity in the duct or stack.

4. Thermocouple mounted near the sampling nozzle to indicate stack gas temperature.

5. Flexible hose to convey the sample from the probe to the sensor unit.

6. Thermocouple mounted at the sensor element to indicate sample temperature in the sensor.

7. The electronic sensor.

8. Filter assembly mounted downstream from the sensor element for collecting particulate samples for calibration purposes.

9. Variable speed centrifugal blower for maintaining sampling rate through the train.

The sampling tube, pitot tube, and stack thermocouple are installed in a stainless steel sheath which forms a rigid probe structure which is readily installed in standard stack port fittings. An adjustable collar is provided on the sampling probe for simplified stack traversing. Both the probe and the flexible hose are electrically heated to provide control of sample temperature. Sample temperature can be adjusted and

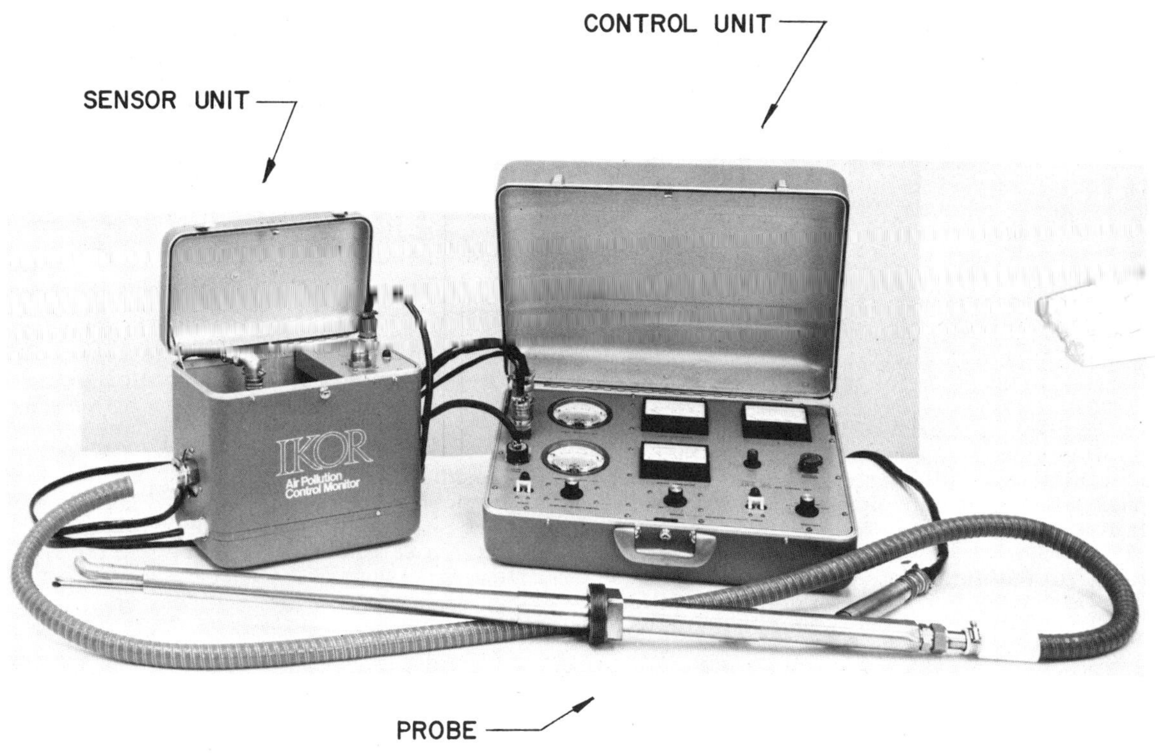

FIGURE I. IKOR AIR QUALITY MONITOR

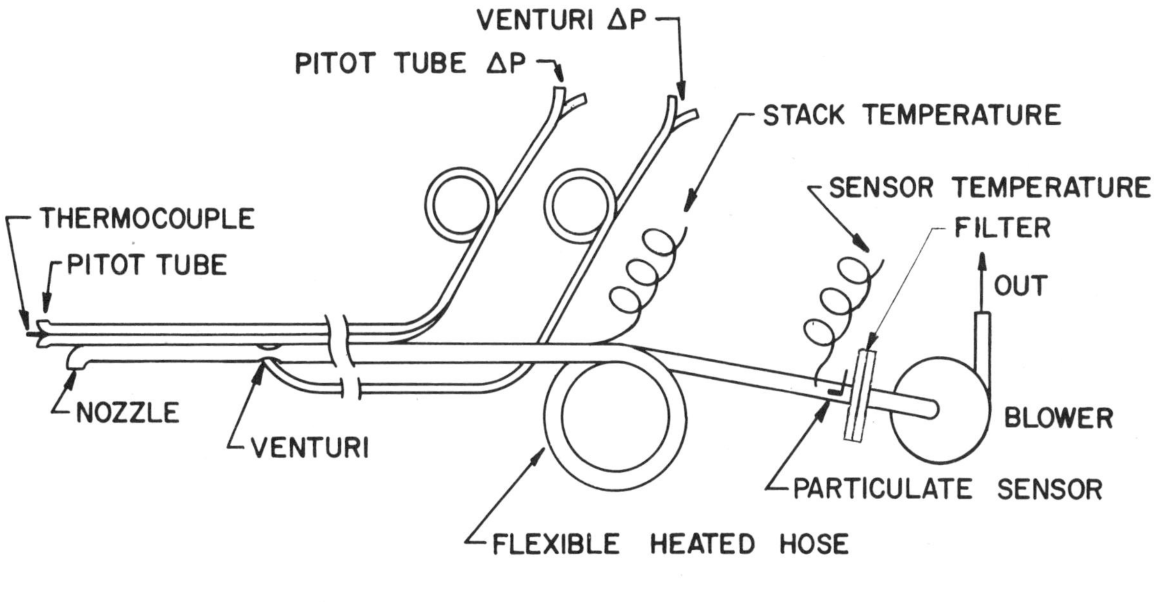

Figure 2. Conceptual view of the sample handling system for the IKOR Air Quality Monitor.

maintained at or above the stack temperature to prevent condensation of vapors in the sampling train.

The venturi used for monitoring sampling rate is located near the inlet of the sampling tube so that it is immersed in the stack and operates at stack temperature. The venturi also is designed to operate with the same flow coefficient as the pitot probe. With these design features, isokinetic sampling - that is, sampling tube inlet velocity equal to stack velocity is achieved by matching the venturi pressure differential to the pitot tube pressure differential. This characteristic of the sampling train simplifies operation of the portable AQM and also provides for simple automation of isokinetic sampling whenever automatic operation is required.

The sampling rate is controlled by manual variation of the blower speed. At any speed setting, the blower is controlled by a solid-state speed control circuit which maintains a constant blower speed with variations in line voltage.

The filter assembly includes a cartridge-type filter holder in which a 142-mm diameter glass fiber filter is mounted. The filter holder can be inserted into and removed from the filter assembly quickly, and filter samples can be obtained over precisely controlled time intervals without interrupting the sample flow. Whenever a filter is not installed, the filter holder is replaced with a blank which allows the sample to flow directly into the blower without filtration. The sample stream leaving the blower may be vented to the atmosphere or reinjected into the stack.

MONITOR APPLICATION

Operating Modes

The IKOR AQM can be utilized in four different modes of operation, as follows:

1. Event recording and process tuning (continuous electronic analysis)

2. Independent monitoring mode (continuous electronic analysis)

3. Dependent monitoring mode (continuous electronic analysis)

4. Sampling mode (intermittent gravimetric analysis)

In the event recording and process tuning mode, the AQM is operated in conjunction with a continuous strip chart recorder. Instantaneous particulate events taking place, such as leaky bags in a pulsed bag collector or the effect of electrostatic precipitator raps are observed and recorded with a permanent real-time record. In addition, dust leaks can be located by probing and observing areas of maximum excursion of electronic signal.

In the independent monitoring mode, a calibration factor is determined by gravimetric analysis of one or more filter samples obtained with the monitor operating in the sampling mode. Subsequently, the electronic sensing system is used to convert the sensor output signal to particulate mass flow rate.

In the dependent operating mode, the calibration factor is determined by parallel test with a separate sampling system, and thereafter the electronic sensing system is used for continuous monitoring as in the independent monitoring mode. This operating mode must be used when particulate mass concentrations in the stack exceed the operating range of the AQM in the sampling mode.*

In the sampling mode, the AQM is operated in a manner similar to the EPA train and other particulate sampling systems. Particulate samples are collected by the filter assembly and particulate mass concentrations are determined by gravimetric analysis of the filters. In this operating mode, the electronic sensing system is not utilized.

Data Reduction

Processing of data with the IKOR Monitor is a simple procedure. The first data processing step involves determination of a calibration factor by relating the sample mass collected on the filter to the integrated sensor mass flow. These quantities are related by the following expression:

$$m = BI \tag{1}$$

where m = sample mass determined gravimetrically

I = integrated sensor mass flow

B = calibration factor at constant sampling velocity.

The integrated sensor mass flow (I) actually consists of three factors, as follows:

$$I = M_i K/S \tag{2}$$

where M_i = actual integrator scale reading

K = integrator time constant

S = sensor sensitivity setting

The factors K and S correspond to instrument adjustments which provide for wide ranges of operating conditions but do not enter into the interpretation of the test results.

After the calibration factor (B) is obtained from eq. 1, it can be used to determine the instantaneous particulate mass flow rate ($\dot{m}$) or particulate mass concentration (C), as follows:

$$\dot{m} = B\,\dot{M}_s/S \tag{3}$$

*The AQM cannot be operated over as great a range of particulate mass concentration in the sampling mode as in the continuous monitoring modes. This range limitation is due to the inability of the sampling train to maintain isokinetic sampling rates at high particulate concentrations when the filter assembly is used. This limitation affects the operation of the AQM only in the sampling mode and at very high particulate concentrations.

$$C = B \dot{M}_s / Q_{std} S \qquad (4)$$

where $\dot{M}_s$ = sensor mass flow reading

$\dot{m}$ = instantaneous particulate mass flow rate

Q_{std} = volumetric sampling rate calculated to STP

As noted, the B factor under isokinetic sampling conditions is constant at constant sampling velocities. Similarly, the B factor is also constant but slightly different under isokinetic sampling conditions and constant velocities if the temperature conditions of stack and sensor are constant, but different. This is due to the fact that the velocity of the stack and sampling line are matched in the probe, while the temperature of the stack is determined at the probe and that of the sensor after 10 feet of preheater line beyond the probe.

Where variations in velocity are encountered, the velocity sensitivity correction factor (B*) can be determined as follows from any of several data bases, provided a consistent data base is always used:

$$B^* = (Q_{sensor,\ cal}/Q_{sensor})^n ,$$

$$= (Q_{stack,\ cal}/Q_{stack})^m . \qquad (5)$$

A velocity adjusted correlation factor (B^1) is obtained from the relationship

$$B = B^1 B^* \qquad (6)$$

where n, m = velocity sensitivity indices

Q = volumetric sampling rate

sensor = conditions at the sensor

stack = conditions at the stack

cal = conditions existing during calibration.

Note: n or m can be obtained from a slope of a log-log plot of Q ratio versus B.

Thus, equations (3) and (4) can be augmented under varying velocity conditions to

$$\dot{m} = B^1 B^* \dot{M}_s / S \qquad (7)$$

and

$$C = B^1 B^* \dot{M}_s / Q_{std} S . \qquad (8)$$

MONITOR PERFORMANCE

Sensitivity

The IKOR AQM responds in its electronic sensing mode to all types of suspended particulate material with which it has been tested to date, including liquid droplets. Tests have been conducted both in the IKOR laboratory and in industrial facilities involving coal ash, oil ash, incinerator ash, metals, metal oxides and inorganic salts, carbonaceous combustion products, and automobile exhaust particulates. Tests also have been conducted involving monitoring of indoor and outdoor ambient air environments where the nature of the particulates was not known. In all of these tests, the monitor has demonstrated an electronic response amplitude which will provide monitoring capability over a wide range of particulate concentrations.

As mentioned above, the electronic amplification circuitry provides a sensitivity range of five decades, or 100,000 to 1. In a coal-burning combustion system, this sensitivity range corresponds to a particulate concentration range of approximately 0.0005 to 50 grains/scf at typical stack velocities. This sensitivity range cannot be matched by conventional filter sampling systems, including the filter assembly incorporated in the IKOR AQM. Because of this limitation of the useful range of filter samplers, the useful range of the IKOR AQM is occasionally limited by the availability of suitable calibration methods.

The lower sensitivity limit of the IKOR AQM has not as yet been fully explored. In an experimental evaluation of the monitor for ambient particulate monitoring, satisfactory detection was obtained at particulate mass concentrations of approximately 0.00001 grains/scf using the standard sensor element configuration. Further increases in sensitivity are possible with variations in the design of the sensor element.

Particle Size Effects

In the discussion of the charge transfer particle detection mechanism it was noted that the processes involved are sensitive to the characteristics of particles being detected. The charge transfer rate from particles making contact with the sensor element is surface-area dependent so that the sensitivity to particle mass flow is inversely proportional to particle size. On the other hand, the probability of particle-sensor contact increases with particle size due to inertial separation effects. The net result of these counteracting effects is a particle size sensitivity which is small, except at the extremes of the particle size range of interest. Small changes in particle size during operation of the monitor have a negligible effect on its response. However, if substantial changes in particle size occur, it is necessary to recalibrate the instrument.

No lower limit in particle size sensitivity has been observed as yet with the monitor. Tests with monodisperse, spherical particles of 1 μm diameter revealed a high sensitivity to particles of this size. Additionally, tests of particles known to be in the sub-micron range, such as tobacco smoke and diesel and gas turbine exhaust particulates, have indicated similar monitoring sensitivities. Consequently, variations in particle size and operations with particles of extremely small size can be accommodated with the IKOR AQM.

Particle Composition Effects

As discussed above, the IKOR AQM is sensitive to all types of suspended particulate material. However, its response does vary with the composition of the particulates. This variation is observed to cover a range of approximately 5 to 1. The monitor calibration factor defined above, is found generally to range between 0.1 and 0.5 (grains/min)/scale-unit. The monitor does respond to mixtures of suspended particulate materials, and if the composition of a mixture does not vary, the monitor will provide an accurate measurement of particulate mass flow. If the particulate mixture composition does vary, an error will be introduced into the mass flow measurement, but the performance of the monitor in indicating variations in mass flow will not be affected. Furthermore, if infrequent changes in particulate composition occur and the operator is aware of the changes, accuracy of mass flow measurement can be preserved by recalibration of the monitor.

Sensor Surface Effects

The reproducibility of electronic signal obtained by particle impingement is also dependent upon the nature of the sensor surface. Corrosion of the sensor will affect the surface and also provide a current as a result of the electrochemical reaction. For this reason IKOR manufactures sensors in highly corrosion resistant metals and alloys, such as tantalum, various stainless steels and others, as required by process conditions. Smearing of the surface of the sensor, and particulate buildup as a result of deposition of wet, oily or waxy particulates will also affect the reproducibility of the signal. This can be minimized by maintaining the sample temperature at or above the stack temperature. In normal operation with discrete, dry particulates, a slight dusting of the sensor takes place; however, equilibrium conditions are rapidly established (normally within 10 minutes), and the attainment of equilibrium conditions can be observed by the leveling of the signal on the continuous strip chart recorder. As a general rule, all dry and discrete particles which might be collected in a pulsed bag collector can be suitably measured in the IKOR monitor. Other applications have to be demonstrated.

MONITOR PERFORMANCE DEMONSTRATIONS

It has been anticipated that the greatest demand for the IKOR AQM would be for use in monitoring particulate emissions from stationary combustion sources and dry dust collectors. Consequently, an extensive amount of developmental and demonstration testing has been conducted on effluent ducts and stacks from utility boilers, large combustion systems and dust collector operations.

An extensive series of parallel tests was conducted with an IKOR AQM and an EPA train at a coal burning facility, and the results are summarized in Table I. The results of these tests indicated that the performance of the IKOR AQM is equivalent to that of the EPA train with the IKOR AQM operated in either its sampling or electronic monitoring modes.

Performance tests of the IKOR AQM have also been conducted at oil-fire combustion facilities, boilers operating on coal-bark mixtures and Kraft recovery boilers. Consistent and reliable electronic and mechanical performance was demonstrated by the monitor at these facilities.

A variety of other tests have been conducted with the IKOR AQM in monitoring effluents from non-combustion facilities. Of particular pertinence is the effectiveness of the IKOR AQM in evaluating the performance of emission control systems with unsteady state operating characteristics. Figure 3 presents strip chart data (1 in = 5 min) obtained in particulate mass flow measurement of a dust collector outlet at an aluminum reduction facility. In this operation, the highly efficient baghouse dust collectors were completely automated, and normally pulse every 10 - 11 seconds. In operation, however, the specific dust collector pulsed for six consecutive 10.5 sec pulses, followed by a 51 sec no-pulse cycle. The results are vividly demonstrated and highly reproducible. Replicate runs revealed an excellent correlation of electronic signal to filter weight gain with a standard deviation of $\pm$ 3% of the calibration factor.

Figure 4 presents strip chart data from a pulsed dust collector operation at another aluminum reduction facility. In this operation, the baghouse dust collectors pulse every 65 seconds, with eight sequential pulses per cycle. Each cycle number is superimposed on the strip chart. From the data it can be observed that cycles 2 and 1 are consistently higher than the other peaks, and indicative of poor bag performance. The peaks, as well as the time between peaks, provide a characteristic fingerprint of that specific baghouse during the period of test. A series of replicate runs provided an electronic signal to filter weight ratio with a one sigma standard deviation of $\pm$ 7.6% of the calibration factor.

Limited Monitor Applications

Not all attempted applications have been unqualified successes. Facilities with particulate emissions of highly variable composition, such as incinerators, present monitoring difficulties due to the sensitivity of the IKOR AQM to particulate composition. The monitor will provide a continuous emission record with such facilities, but the actual emission rates indicated will be inaccurate as a result of the variable particulate composition.

Another monitoring problem area is associated with the tendency of some particulate materials to deposit on the walls of the sampling train. Because of this phenomenon, special precautions must be taken in monitoring of emissions from certain types of sources, such as combustion systems fired with residual oil, to minimize particle deposition. This problem is not unique to the IKOR AQM, but is common to all measurement systems requiring extractive sampling. The IKOR staff currently is engaged in development programs which will lead to improved sampling systems free of the particle deposition problem.

CONCLUSION

The IKOR Air Quality Monitor has been demonstrated to be an effective instrument for continuous measurement of particulate mass emissions from a wide variety of industrial facilities. Its potential utility in providing real-time emission data for air pollution control

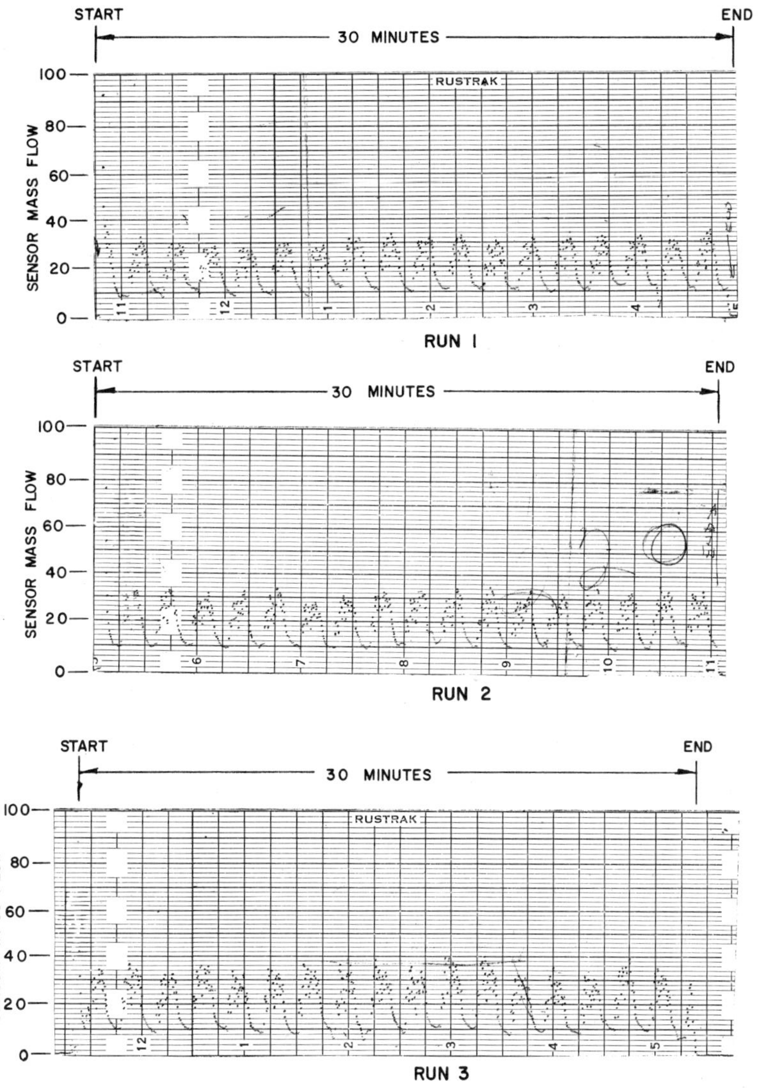

FIGURE 3. CONTINUOUS STRIP CHART TRACES - AL_2O_3 DUST COLLECTOR TESTS

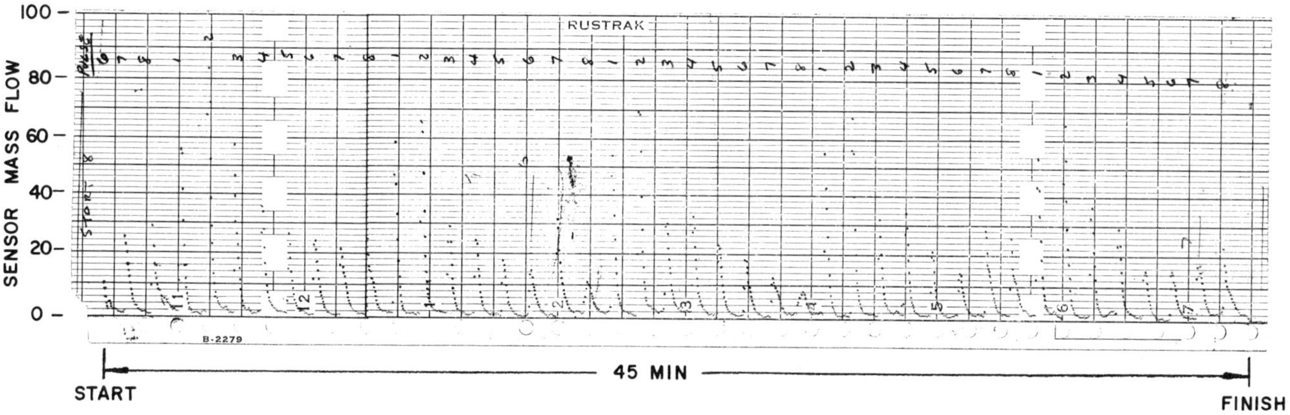

FIGURE 4. CONTINUOUS STRIP CHART TRACE - PULSED DUST COLLECTOR, DIFFERENT Al_2O_3 FACILITY

records or for process control and maintenance applications has been demonstrated for utility boilers, manufacturing facilities, metallurgical processing plants, and other types of industrial facilities. The natural variability of the electrical charge transfer process with variations in particle characteristics has been accommodated partly by the design of the particle detection system and partly by the provision of a convenient calibration method. As a result of these monitor features, the IKOR AQM provides accurate and reliable particulate mass emission data collection capabilities in a variety of operating modes.

TABLE I

COMPARISON OF IKOR AQM AND EPA TRAIN DATA

IKOR AQM					EPA TRAIN		
Test	Port	Part. Mass. Conc. (gr/SCF) Sampling Mode (1)	Part. Mass. Conc. (gr/SCF) Independent Monitoring Mode (2)	Part. Mass. Conc. (gr/SCF) Dependent Monitoring Mode (3)	Test	Port	Part. Mass. Conc. (gr/SCF)
4	3	.006	.007	.007	6	3	.009
5	5	.016	.021	.020	4	5	.014
6	4	.013	.011	.010	5	4	.014
Avg. Tests 4 - 6		.012	.013	.012			.012
7	4	.014	.014	.014	8	4	.006
8	5	.019	.015	.014	9	5	.024
9	3	.008	.011	.010	7	3	.015
Avg. Tests 7 - 9		.014	.013	.013			.015
Avg. All Tests		.013	.013	.013			.014

NOTES:

(1) *IKOR AQM data derived from gravimetric analysis of IKOR filter samples

(2) *IKOR AQM data derived from IKOR electronic sensor data and calibration factor based on IKOR filter sample (Test 7)

(3) *IKOR AQM data derived from IKOR electronic sensor data and calibration factor based on EPA Train data (average for Tests 4 - 6)

© 1973, ISA JSP 6685

AIR QUALITY STANDARDS-SETTING

D. S. Barth, Ph.D
Director
NERC-LV

F. G. Hueter, Ph.D., Director
Special Studies Staff; Office of the
Director, NERC-RTP

J. Padgett, Director
Strategies & Air Standards Division
Office of Air Quality Planning & Standards

ABSTRACT

A brief summary of the provisions of the Clean Air Amendments of 1970 relating to regulatory authorities will be presented. Following this introduction, considerable detail will be given concerning the scientific documents required to serve as a foundation for air quality standards-setting. An overview of the EPA research program, both intramural and extramural, underway to provide the necessary documentation will be provided.

Particular attention will be given to the difficulties inherent in translating state-of-the-art scientific knowledge into regulatory standards. Problems associated with defining the most sensitive population at risk and estimating an adequate safety margin to assure public health protection will be addressed. The roles of the National Air Quality Criteria Advisory Committee in assembling the required supporting data for standards-setting will be discussed.

Finally, the underlying importance of verified measurement accuracy to all of the above will be emphasized.

INTRODUCTION

It is not possible in a paper of limited length to present both a comprehensive and an in-depth analysis of all of the complex factors involved in air quality standards-setting. In the following we shall attempt to make our presentation as comprehensive as possible at the expense of sacrificing some of the in-depth coverage. In order to put air quality standards into perspective we shall provide a brief overview of other types of standard-setting authorities granted to the U. S. Environmental Protection Agency by the Clean Air Amendments of 1970. This is considered to be necessary to emphasize that the preferred method for adequate control of air pollution is not always the promulgation of air quality standards.

We feel it necessary to present an overview of the philosophy of research in EPA since clearly any air quality standard must be based on adequate research data. To support air quality standards, adequate information for the pollutant of concern must be available on effects on health and welfare, sources, measurement technology, existing air quality, available control technology and models linking source emissions to subsequent air quality.

A discussion will be given of the procedures used to develop our existing Air Quality Criteria Documents on Sulfur Oxides, Particulates, Carbon Monoxide, Hydrocarbons, Oxidants and Nitrogen Oxides. The extensive review process used, involving the National Air Quality Criteria Advisory Committee as well as many other external and internal reviewers, insured that an objective and comprehensive review of the documents' contents was obtained prior to publication.

Current activities, including both research and state-of-the-art reviews, directed to the possible development of new or the revision of existing Air Quality Criteria Documents will be presented and discussed. In addition to the factual information content of each Air Quality Criteria Document many additional research needs are identified. A major portion of our research program is oriented to accomplish these research needs and thus fill the gaps in our existing state of knowledge. We also must be continually evaluating selected new or newly identified air pollutants to determine whether there is a need for control to abate adverse effects on health or welfare. If such a need is identified and the pollutant can best be controlled through the promulgation of Air Quality Standards, then a new Criteria Document must be prepared and issued.

Perhaps the most difficult task of all is the final conversion of research information contained in Criteria Documents into Air Quality Standards. The most pertinent information relative to known adverse effects on health and welfare must be identified. A reference measurement method must be specified.

An appropriate safety margin must be determined. Present procedures used to accomplish the above will be presented and discussed.

REGULATORY AUTHORITIES OF THE CLEAN AIR AMENDMENTS OF 1970

AIR POLLUTION STANDARDS

It is important to bear in mind that Federal policy for air pollution control is based on the need to protect the public from the adverse effects of pollutants on health and welfare and to enhance the quality of the total environment.

A brief discussion of each authority will be given.

CRITERIA AND CONTROL TECHNIQUES DOCUMENTS

The EPA Administrator is directed by the Clean Air Act, as amended, to publish air quality criteria and control technique documents for those air pollutants

"(A) which in his judgement have an adverse effect on public health and welfare; "and

"(B) the presence of which in the ambient air results from numerous or diverse mobile or stationary sources."

Air quality criteria for an air pollutant "shall accurately reflect the latest scientific knowledge useful in indicating the kind and extent of all identifiable effects on public health or welfare which may be expected from the presence of such pollutant in the ambient air, in varying quantities." Information on control techniques "shall include data relating to the technology and alternative methods of prevention and control of air pollution." The Act states that "effects on welfare include, but are not limited to, effects on soils, water, crops, vegetation, man-made materials, animals, wildlife, weather, visibility, and climate, damage to and deterioration of property, and hazards to transportation, as well as effects on economic values and on personal comfort and well-being."

NATIONAL AMBIENT AIR QUALITY STANDARDS

For those materials for which criteria and control technique documents are issued the Administrator is directed to prescribe national primary and secondary ambient air quality standards which are defined as follows:

1. "National primary ambient air quality standards shall be ambient air quality standards the attainment and maintenance of which in the judgement of the Administrator, based on such criteria and allowing an adequate margin of safety, are requisite to protect the public health."

2. "National secondary ambient air quality standards shall specify a level of air quality the attainment and maintenance of which in the judgement of the Administrator, based on such criteria, is requisite to protect the public welfare from any known or anticipated adverse effects associated with the presence of such air pollutant in the ambient air."

To date criteria and control technique documents have been issued and National Primary and Secondary Ambient Air Quality Standards have been promulgated for six common air pollutants: sulfur dioxide, particulate matter, carbon monoxide, photochemical oxidants, hydrocarbons and nitrogen dioxide. These national standards are being implemented, maintained, and enforced by the States in accordance with implementation plans developed by the States which must follow specified requirements and are subject to review and approval by EPA.

STANDARDS OF PERFORMANCE FOR NEW STATIONARY SOURCES

"The term 'standard of performance' means a standard for emissions of air pollutants which reflects the degree of emission limitation achievable through the application of the best system of emission reduction which (taking into account the cost of achieving such reduction) the Administrator determines has been adequately demonstrated."

The Administrator is required to designate stationary source categories and then publish standards for new sources that in his judgement "may contribute significantly to air pollution which causes or contributes to the endangerment of public health or welfare."

NATIONAL EMISSION STANDARDS FOR STATIONARY SOURCES OF HAZARDOUS AIR POLLUTANTS

"The term 'hazardous air pollutant' means an air pollutant to which no ambient air quality standard is applicable and which in the judgement of the Administrator may cause, or contribute to, an increase in mortality or an increase in serious irreversible, or incapacitating reversible, illness." The Act requires the Administrator to publish a list and then, subsequently, to promulgate national emission standards for those air pollutants deemed hazardous. These emission standards must provide "an ample margin of safety" to protect the public health. They are applicable to both new and existing stationary sources.

NATIONAL EMISSION STANDARDS FOR MOTOR VEHICLES AND AIRCRAFT AND REGULATORY AUTHORITY FOR FUELS AND FUEL ADDITIVES

Emission standards are required for any air pollutant from any class or classes of new motor vehicles or new motor vehicle engines which in the judgement of the Administrator "causes or contributes to, or is likely to cause or contribute to, air pollution which endangers the public health or welfare." Emission standards are also required for similarly harmful air pollutants from any class or classes of aircraft or aircraft engines.

The effective date for emission standards must be reasonable. It must allow time to develop and apply the requisite technology. The cost of compliance within the set time period must be given appropriate consideration.

Issuance of proposed aircraft emission standards must be preceded, i.a., by a study of the effects of aircraft emissions on air quality, and of the technological feasibility of controlling emissions. The Secretary of Transportation must be consulted.

The manufacture or sale of any motor vehicle fuel or fuel additive may be controlled or prohibited if the emission product of such fuel or fuel additive

"(A) ...will endanger the public health or welfare," or

"(B) ...will impair to a significant degree the performance of any emission control device or system" which is or may be expected to be in general use.

Before controlling or prohibiting the manufacture or sale of a fuel or fuel additive, the Administrator must consider an array of scientific, medical, economic, and technological data relevant to determining the need for regulation and to making findings on the technological and economic feasibility and consequences of such regulations. This includes, for example, an appraisal of adverse effects of substitutions if the manufacture or sale of a fuel or fuel additive is to be prohibited or, as another example, an analysis of alternate emission control devices if the manufacture or sale of a fuel or fuel additive is to be controlled or prohibited because emission products will impair the performance of emission control devices or systems.

EPA RESEARCH PHILOSOPHY

The U.S. Environmental Protection Agency is first and foremost a regulatory agency. Briefly stated the mission of the EPA is to control environmental pollution to acceptable levels. This simple statement, upon careful analysis, turns out not to be simple at all. In fact the development and maintenance of an adequate regulatory program is an incredibly complex matter. As we shall see, a major research and development program is needed as an integral part of EPA to provide the necessary scientific information base to serve as the foundation for EPA environmental standards and regulatory programs.

To determine the need for and required extent of a regulatory program to control a given environmental pollutant, answers to the two following fundamental research questions must be obtained:

1. As the given environmental pollutant is presently introduced into the environment and subsequently cycled and recycled through the biosphere, is it or any of its byproducts harmful or objectionable to a significant portion of either the human population or the environment?

2. If the given environmental pollutant or any of its byproducts is determined to be harmful or objectionable, what minimum degree of control is necessary to abate the adverse effects to acceptable levels, and in what manner should that control be instituted in order to be most efficient in the sense of having a minimum impact on the total U.S. economy?

Let us now examine some of the required scientific information which must be accumulated in order to answer the above questions for a wide variety of environmental pollutants and, in turn, to develop adequate environmental standards and regulatory programs.

First, we must precisely define the environmental pollutants of concern as well as the adverse effects caused to man or his environment by various levels of the pollutant. Since the pollutant must usually be controlled at its source, we must have an accurate procedure for determining the degree of source control which will be required to reduce the environmental levels sufficiently to adequately abate the adverse effects. This also implies that we have adequate measurement methods and sufficient knowledge of control technology to clearly indicate how the necessary source control is to be accomplished. In summary then, we must know, or develop an adequate research program to find out substantial information required in the following subject areas:

1. Effects

2. Measurement Methods

3. Existing Environmental Levels

4. Sources

5. Predictive Models Linking Source Emissions to Subsequent Environmental Levels

6. Control Technology for Significant Sources

Once a sufficient amount of information is assembled in each of the six key subject areas and published in appropriate documents, we are then in a position to select the best available control measures to abate the adverse effects on health or welfare of a given environmental pollutant or combination of environmental pollutants. Principal considerations entering into this selection include:

1. Severity of observed adverse effect on health or welfare and size of the critical population at risk.

2. Distribution of significant sources.

3. Time by which controls can be effective in abating adverse effects.

4. Cost-effectiveness of different available control measures.

5. Efficacy of available control techniques for significant sources.

In almost all instances this is an extremely difficult decision to make. Economic factors of major importance are invariably involved. Thus, a careful analysis must always be made of cost control vs. benefits to be achieved from control. The decision to take regulatory action or not is rarely an easy one. But it must be made.

DEVELOPMENT OF EXISTING AIR QUALITY CRITERIA DOCUMENTS

In a previous section we have defined the term Air Quality Criteria Document. We have also discussed in a previous section the research information needs which are required to serve as a foundation for Air Quality Standards. In this relatively short section we will now discuss briefly the minimum requirements of a Criteria Document and the procedures which were utilized to develop our existing Air Quality Criteria Documents.

Air quality criteria are compilations of the latest available scientific information on the sources, prevalence, and manifestations of recognized air pollutants. Most important, they describe what effects have been associated with, or may be expected from, an air pollutant level in excess of a specific concentration for a specific time period. Such effects generally involve visibility reduction, damage to materials, economic costs, vegetation damage, nuisance aspects, and adverse effects on the health and well-being of man and animals.

Air quality criteria documents are planned to provide comprehensive information in the following major areas whenever pertinent:

I. Environmental Appraisal

A. Origin (natural and man-made) and fate

B. Physical and chemical properties

C. Spatial and temporal distribution in the atmosphere

D. Atmospheric alterations

1. Chemical transformations
2. Meteorological influences

E. Environmental Cycling

1. Air
2. Water
3. Soil
4. Food

F. Measurement technology

II. Effects on Human Health (including laboratory animal studies)

A. Toxicologic appraisal

1. Behavioral and sensory responses
2. Biochemical and physiological mechanisms and responses

B. Epidemiological and clinical appraisal

1. Field studies
2. Clinical studies

III. Other (welfare) Effects

A. Biological and Physical effects (including significant microfauna and flora)

1. Plant life (natural and cultivated)
2. Domestic and wild animal life
3. Materials

B. Ecological considerations

C. Social and aesthetic effects

D. Economic impact (includes that resulting from all other effects listed on the previous page plus effects on specific economic parameters such as real property values, etc.)

IV. Gap Areas in Current Knowledge

Criteria production begins by the development of a state-of-the-art document which pulls together all that is known regarding the different important subject areas defined earlier. This is usually done by contract. Once this document is in hand we then begin a long and tedious review process which ultimately results in a completed Air Quality Criteria Document.

The state-of-the-art document is first extensively reviewed intramurally to assure that all required subject areas have been covered in adequate depth. Additional material is prepared and incorporated as deemed appropriate. The end product of this effort is then titled as a Draft Air Quality Criteria Document. At this point the document is transmitted to 50 to 100 outside experts in the field for their review and comments. Simultaneously the Draft is transmitted to the National Air Quality Criteria Advisory Committee for its review. This Committee is composed of technically qualified experts from outside EPA primarily from the academic community with some representation from industry and State or local governments. This Committee is authorized by the Clean Air Amendments of 1970.

All review comments are then incorporated or rebutted and a revised Draft Criteria Document is prepared. This revised document is then referred back to the National Air Quality Criteria Advisory Committee for any additional comments which they may care to make. After incorporating these comments, the Criteria Document is then forwarded to higher headquarters for final approval prior to publication.

The entire procession of events outlined above usually takes about one year or longer but we feel that the required quality of the product warrants proceeding in this manner. The extensive review obtained has proven to be of inestimable value.

REVIEW OF CRITERIA DOCUMENTS

Perhaps the most intensive review of the criteria since they were published, is now going on in conjunction with Congressional "Oversight Hearings on the Clean Air Act." The Senate Subcommittee on Air and Water Pollution has contracted with the National Academy of Sciences (NAS) for an evaluation and review of the health basis for the primary ambient air quality standards. One report has been made by NAS in the form of a conference held October 3-5, 1973. The arrangement with NAS calls for a one-year study to be completed July 1974.

An in-depth review of the Criteria for Sulfur Oxides by a Department of Health, Education and Welfare committee was also carried out early this fall. Their report indicates more research is needed on SO_2 and its products. They conclude that current information indicates that the primary standard should stay as it stands.

REVIEW, BACKGROUND, STATE-OF-THE-ART DOCUMENTS

For some years EPA or its predecessor organization has contracted with outside contractors to prepare background documents that summarize the current state of knowledge regarding the effects of specific pollutants. Information on sources, emissions, environmental transport and measurement methodology is also included. These documents have served as one major input into Agency documents that serve as a basis for regulatory action.

Though a number of centers of expertise have provided such reports, the NAS is currently most relied upon. The NAS mode of operation in preparing these documents is to appoint a panel of widely recognized experts for each pollutant or class of pollutants for which a report is desired. The panel develops the report which is reviewed by the NAS committee on the Biological Effects of Air Pollution, by NAS selected independent reviewers, and by EPA. A formal document (contract report) is submitted to EPA. NAS is permitted to and has been publishing its own documents on these reports.

As part of EPA's program to review and update Criteria Documents, NAS was asked at the beginning of FY 1974 to review and prepare documents on carbon monoxide, hydrocarbons, photochemical oxidants, and nitrogen oxides. Since these pollutants are also being studied for Congress, NAS will coordinate these two activities and provide EPA with its report earlier than with the documents prepared by NAS under its usual procedure. The documents prepared by NAS for EPA in the past have required about 18 months for completion.

In addition to the 4 pollutants given above, NAS is also beginning reviews of arsenic and the platinum group of heavy metals.

NAS reviews currently underway and their approximate submission dates to EPA are:

Nickel	March 1974
Vapor Phase Organic Matter	March 1974
Fine Particulate	March 1974
Zinc	April 1974
Chlorine-Hydrogen Chloride	April 1974
Selenium	April 1974
Copper	August 1974

Documents on manganese, vanadium, and chromium have recently been completed by NAS and submitted to EPA.

Documents on asbestos, fluorides, lead, and particulate polycyclic organic matter have been submitted to EPA and published by NAS.

Two review documents have been prepared for EPA by the Karolinska Institute in Sweden. These are entitled "Mercury in the Environment" (1971) and "Cadmium in the Environment" (1971 and 1973).

To assist other programs of EPA, internal summary documents, usually written as "Scientific Summary Reports," are prepared to further condense information provided in contractor prepared review documents. Added to these is information on EPA projects that are expected to close gaps in the available information.

RESEARCH UNDERWAY TO FILL GAPS IDENTIFIED IN CRITERIA DOCUMENTS

CHESS STUDIES

Each criteria document indicated gaps in epidemiological information. The CHESS research program is helping to fill some of these gaps. CHESS is an acronym for "Community Health and Environmental Surveillance System." It is a national research effort which relates community health to changing environmental quality. It consists of a series of standardized epidemiologic studies designed to measure simultaneously environmental quality and sensitive health indicators in sets of communities representing exposure gradients to common air pollutants. The program is conducted by the Environmental Protection Agency in cooperation with local health agencies, universities, and private research institutes. The purpose of the CHESS program is: to evaluate existing environmental standards; obtain health intelligence for new standards, and document health benefits of air pollution control. CHESS became fully operational for air pollution effects in 1973 and will be operational for multimedia toxic substances in 1975.

CO IN MAN

Carbon Monoxide (CO) effects in man are well documented at high concentrations. At levels apt to be found in ambient air, there are conflicting data. To help fill this gap, EPA is carrying on clinical research to measure effects on individuals with chronic heart disease who are exposed to measured concentrations of CO.

OXIDANT TOXICOLOGY

It is felt that oxidant exposure and other stresses may potentiate each other. Therefore, animal toxicologic studies are being carried out that are designed to provide information on the interrelationships between exposure to photochemical oxidants (essentially ozone) and stresses such as those of climate or exercise. Combination effects with other pollutants such as sulfur dioxide and nitrogen dioxide are also being studied. The effects of oxidant on pulmonary defense mechanisms against biological agents have been studied in more detail since early findings were reported in criteria documents.

MEASUREMENT OF NITROGEN OXIDES

Problems with existing measurement techniques for NO_2 were pointed out in the NOx criteria document. The primary methods being used were re-evaluated and suspected short-comings of the Jacobs-Hochheiser method were confirmed. Currently, at least four methods are being evaluated by EPA with the expectation that at least one will be precise, accurate and reliable enough to be designated as the reference method.

ATMOSPHERIC PHOTOCHEMISTRY

The criteria documents on oxidants, hydrocarbons, and nitrogen oxides deal with pollutants involved with photochemical reactions. Many research needs were noted concerning these pollutants. EPA is trying to fill these research needs through internal and external work. This research includes support of grants studying mechanisms of photooxidation; photolytic oxidation in the presence of NO_2; gas phase ozonolysis rates; reaction of oxyradicals with NO, NO_2, SO_2, and CO; photochemistry of SO_2 and reactions of electronically excited SO_2.

ATMOSPHERIC DISPERSION AND FATE OF POLLUTANTS

The need for mathematical modeling of the atmospheric dispersion of photochemical air pollutants was pointed out in the oxidant and NO_x criteria documents. These documents relied on an observational model to relate early morning hydrocarbon levels and NO_2 with afternoon oxidant levels. Work on a mathematical model for photochemical air pollution has been supported for several years with positive results.

EPA investigators have also studied the atmospheric fate of several pollutants of great interest. One approach has been field measurement of precipitation scavenging of SO_2, NOx, sulfates, nitrates, and hydrogen ions from coal fired power plants and other urban sources.

PROCEEDING FROM AN AIR QUALITY CRITERIA DOCUMENT TO STANDARDS

The standard-setting process begins with the gathering of all available data on the health and welfare effects of a pollutant. This information comes from EPA's own research studies and from throughout the scientific community. The information is studied and evaluated by EPA's own scientific and technical experts. Technical advisory committees and outside contractors may be called upon for assistance. EPA's Office of Air Quality Planning and Standards participates in the development and assessment of this criteria information, and is the office ultimately responsible for proposing a standard. As a part of this evaluation process a decision must first be made on the appropriateness of regulatory action, and specifically which of the several regulatory alternatives is best suited for control of the candidate pollutant. This evaluation procedure is identified as a preferred standards path analysis.

The preferred standards path analysis is one of the major milestones in guiding the standard-setting process and includes an assessment of at least the following factors:

1. An objective summary of the existing problem, such as known adverse effects, ambient concentrations; measurement techniques; sources of the pollutant; and feasible control measures.

2. A summary of existing regulatory efforts, i.e., what State and local regulations are in effect and any voluntary controls.

3. A consideration of all possible options under the Act, previously described.

If the candidate pollutant has an adverse effect on public health or welfare, and ambient levels result from numerous or diverse mobile or stationary sources, National Ambient Air Quality Standards will probably be the recommended regulatory option.

The Clean Air Act requires that primary ambient air quality standards be set to fully protect both specifically susceptible sub-groups and healthy members of the population. The Act excludes persons who require an artificial environment. In theory at least, exacerbation of existing disease in hospitalized or institutionalized patients with severe pre-existing illnesses, might not constitute an appropriate adverse effect upon which to base an ambient air quality standard. In practice, the implications of this restriction have not been emphasized. On the other hand, possible adverse effects on a large number of relatively small susceptible segments of the population have not been specifically and separately considered in setting standards.

After an assessment of available data has been completed and a tentative standard developed, an extensive review process takes place both within EPA and with other government agencies. A first draft of a regulation setting out the proposed standard is prepared by the Office of Air Quality Planning and Standards. The draft is circulated within EPA for independent review by other Agency divisions, with the Office of Planning and Management coordinating the process through its steering committee and special working groups.

EPA's Office of Research and Development reviews the scientific basis for the standard, as well as the surveillance and monitoring implications. The Office of Enforcement and General Counsel then reviews the proposal to make sure all legal requirements are satisfied.

In addition to overseeing the internal coordination process, the Office of Planning and Management also reviews the policy implications of the standard, the cost-effectiveness of alternative ways of achieving the standard, and the standard's potential impact on other pollution control programs. (For instance, will the new air pollution control standard affect energy and fuels programs?)

All branches of EPA that can contribute to the final product are involved throughout this initial process. Questions are asked, positions are challenged, changes may be proposed. The objective is the fullest possible inquiry and consideration.

The product that emerges from this process thus represents EPA's best judgement on what is required, workable and supportable on scientific, technical, legal and policy grounds, to protect public health and the environment.

The standard-setting process then moves outside of EPA. What effect would the standard have on the goals of other Federal agencies such as Commerce, Defense, Interior, Transportation and on the general economy? To achieve this objective the proposal is circulated among other Federal agencies, including the Council on Environmental Quality, for review and comment. The views of State agencies and interested nongovernmental organizations are also solicited.

After the comments of other groups are considered, EPA publishes the standard in the Federal Register as a proposed regulation. The views of the general public -- interested individuals and organizations -- are solicited, with at least 30 days usually provided for comments. Following this process, a public hearing may be appropriate, depending on the response to the proposed regulation.

Concurrent with the proposal of a National Ambient Air Quality Standard a reference method must be specified for evaluating the ambient concentrations of the pollutant. An alternative to the reference method is an equivalent technique which can be demonstrated, to the Administrator's satisfaction, to have a consistent relationship to the reference method. This measurement method is an extremely important part of the standard-setting process and provides the basis for data accumulation and subsequent abatement strategies. A sound method must be developed, tested and proven prior to proposal of the standard.

After receiving comments on the proposed regulation, EPA, in effect, begins its internal process again. A summary of the comments is prepared and circulated to all concerned within EPA, along with any revisions suggested.

The decision is then made. The standard as EPA intends to issue it is approved by the EPA Administrator, and is sent to other appropriate Federal agencies for final review. This accomplished, the regulation is then, promulgated by EPA and published in the <u>Federal Register.</u>

When finally issued by EPA, an air quality standard is, therefore, the product of EPA's own scientific expertise, with due consideration given to the views of other Federal agencies, State agencies, interested organizations in the private sector, including scientific, technical, industrial, environmental groups, and concerned individuals. The milestones for this process are presented below as summary of the events included in the development of a standard:

	Issuance of air quality criteria and proposed primary and secondary standards.
w/in 3 mos.	Promulgation of standards.
w/in 9 more mos. *1	Submittal of State plans.
w/in 4 more mos.	Approval/disapproval of State plans. If disapproved, "prompt" proposal of EPA plan. *2
w/in 2 more mos.	Promulgation of EPA plan if State plan remains inadequate.
w/in 3 more years *3, *4	Attainment of primary *5 standard.

*1 - Deadline for submittal of plan for secondary standard may be extended by the Administrator for additional 18 months.

*2 - If the Administrator notifies a State of the need to revise a plan which is already in effect, EPA revisions are to be proposed "promptly" if the State fails to revise within 60 days or a longer period.

*3 - "As expeditiously as practicable" but within three years.

*4 - If necessary technology unavailable, extension of up to two years for attainment of primary standard may be granted by the Administrator. If necessary technology unavailable and continued operation of a particular source is essential to national security or to public health or welfare, postponement of compliance with a requirement of a state plan for one year may be granted to the source.

*5 - Secondary standards must be attained at a "reasonable time" specified by the plan.

© 1973, ISA JSP 6688

ULTRASONIC TECHNIQUES TO MEASURE WATER POLLUTANTS

Dr. Kalwant S. Seklon
Staff Engineer
Hughes Aircraft Company
Fullerton, California

and

Dr. Raymond C. Binder
Professor
University of
Southern California
Los Angeles, California

ABSTRACT

Ultrasonic velocity and absorption of sound energy were measured to determine the degree of pollution in fresh and sea water. An improved sing-around technique was employed to measure sonic velocity in the media. The accuracy of the test equipment was checked by measuring sonic velocity in distilled water. The data was compared with that of National Bureau of Standards. The mean deviation between the author's data and National Bureau of Standards data was less than 0.2 percent. Sonic velocity and percent attenuation were measured in a number of test samples collected from various locations between Los Angeles Harbor and San Diego Beach. The results are presented.

Three commonly encountered contaminants were selected to prove the feasibility of ultrasonic technique. These were crude oil, diesel oil and chlorine solution. Velocity and absorption measurements were made by varying percent concentration of these contaminants, over a wide range of temperature. The parameters sonic velocity, velocity gradient with respect to temperature, velocity gradient with respect to pollution concentration and percent attenuation were used to determine water pollution qualitatively and quantitatively. It has been shown that these parameters provide information concerning the pollutance of water.

INTRODUCTION

Increasing development in technology has been accompanied by increasing pollution of our environment. Very little attention has been given to the prevention of the pollution of our air and water except in the last decade. Most of the technological efforts have been concentrated in the aerospace industry. With the achievement of our goal to land a man on the moon, some of the technological efforts are now directed toward ecology. However, a vast amount of work is needed to clean our air and water. This study is related to the detection of water pollution using ultrasonic techniques.

The United States was using fresh water at the rate of 360 billion gallons per day in 1969. Probably over 400 billion gallons per day now are consumed. Pollution of the fresh water supplies will make it unfit for human and for industrial uses. Similarly the pollution of the sea is causing a devastation of marine life. Water pollution causes several biological ill effects. Polluted water is conducive to undesirable plant and animal growth and enhances the spread of disease. It generates putrid water and odors. It kills desirable aquatic and other wildlife which provide food and recreation.

The oil pollution is the major cause of the sea pollution. The oil pollution is caused by two sources: crude oil transport tankers and offshore drilling. Since 1954, some 10,000 wells have been drilled with 8 resulting in oil blowouts and 17 in gas blowouts. It has been predicted that if offshore drilling continues at the present rate, there will be 3,000 to 5,000 offshore wells drilled each year by 1980. This will increase the number of oil and gas blowouts proportionately unless some technological breakthrough is made to prevent these blowouts. Similarly, there were 9,700 reported oil spills during last year according to the United States Environmental Protection Agency. The unfortunate aspect of the oil pollution is that most of these oil blowouts and oil spills could have been controlled if they were detected in time.

The pollution of the sea either by dumping of oil or by leakage of crude oil pipe lines is not easily traced to the source of pollution on a real time basis with the present instrumentation available. The present water pollution detection methods are divided into two categories: the physical methods and the chemical methods. These methods involve the measurements of temperature, pH values and colors, electrical conductivity, radioactivity, chromatography, absorption spectroscopy, emission spectrography, flame photometry, polarography and physical titration. Most of these methods are quite involved and do not yield results in real time.

The purpose of this study is to investigate the feasibility of a new physical method to detect pollution in fresh or sea water by using the ultrasonic technique. The ultrasonic method consists of measuring the sonic velocity and attenuation through the polluted water. It has been observed that the sound velocity and attenuation are proportional to the degree of pollutance in water. Sonic velocity and absorption were measured, by contaminating sea water with crude oil and diesel oil, over a wide range of temperature. Similar tests were conducted by adulterating fresh water with chlorine solution. The results illustrate that the sonic velocity and attenuation can be utilized as parameters to measure water pollution. Besides sonic velocity and attenuation, the velocity gradient with respect to temperature, at constant pollution and pressure, was also determined.

Even though the tests were conducted with three types of contaminants: crude oil, diesel oil and chlorine, the ultrasonic technique can be extended to other types of pollution like pesticides, organic and inorganic contaminants from municipal waste waters and detergents. It requires sonic calibration for any kind of pollutance which is miscible in water.

The ultrasonic method is very precise and fast. The sonic velocity can be measured within ±0.01 percent accuracy. The test data can be translated into meaningful results instantaneously. The ultrasonic method cannot replace the detailed and involved chemical analysis but it can be used to monitor the waste discharge from various pollution sources on a real time basis. The chemical testing methods are time consuming and costly. It is not practical to maintain such devices around the clock to monitor pollution. The ultrasonic method can be made automatic with minor modifications.

THEORY

When an ultrasonic wave travels through a compressible medium, the temperature, pressure and density of the medium change periodically with respect to time. If we assume an ideal fluid and the sound travels through the fluid isentropically then there is no attenuation of the sound energy in the medium. The velocity of the sound wave is given by [19]

$$C = \frac{\omega}{k} = \left(\frac{dP}{d\rho}\right)^{1/2} \tag{1}$$

where

C = wave velocity

ω = angular frequency

$k = 2\pi/\lambda$ = propagation constant

P = pressure of the medium in the absence of sound waves

dP = change in pressure due to sound waves

ρ = density of the medium in the absence of sound waves

$d\rho$ = change in density of the medium due to sound waves

The equation (1) is not very useful to measure sound velocity accurately in the medium. The small periodic changes in pressure and density cannot be measured accurately. However, the equation (1) can be written in terms of compressibility of the medium using the thermodynamic equation of state. The thermodynamic equation of state of a fluid involves three parameters – pressure, temperature, and density. Two of these parameters are independent. In order to calculate the changes in pressure and density, we need to know the changes in temperature during compression. It is a well established phenomenon that no heat leaves or enters the fluid during compression and rarefaction. Therefore the process is adiabatic. Assuming the process is also reversible, then we can write

$$\left(\frac{\partial P}{\partial \rho}\right)_s = \beta_s/\rho = \frac{1}{\rho K_s} \tag{2}$$

where

s = entropy

β_s = adiabatic bulk modulus

K_s = adiabatic compressibility

substituting equation (2) in (1) we have

$$C = \sqrt{\frac{1}{\rho K_s}} \tag{3}$$

The variation of sound velocity in sea water with respect to pollution can be expressed in terms of changes in density and adiabatic compressibility as follows:

$$\left(\frac{dC}{dF}\right)_T = \frac{C}{2}\left[\frac{1}{\rho}\left(\frac{\partial \rho}{\partial F}\right) + \frac{1}{K_s}\left(\frac{\partial K_s}{\partial F}\right)\right]_T \tag{4}$$

where

F = pollutance concentration by volume

It is clear from equation (4) that the change in velocity with pollutance depends on the density gradient and compressibility gradient with respect to pollution. Both of these dependent variables have the opposite signs. Therefore the change in sound velocity with pollutance depends on the density as well as on the adiabatic compressibility of the contaminants.

Similarly the variation of sound velocity with respect to constant pollutance was also studied. The change of sound velocity with temperature at constant F can be written as

$$\left(\frac{dC}{dT}\right)_F = \frac{C}{2}\left[\frac{1}{\rho}\left(\frac{\partial \rho}{\partial T}\right) + \frac{1}{K_s}\left(\frac{\partial K_s}{\partial T}\right)\right]_F \tag{5}$$

where

T = temperature of the fluid in the absence of sound pulse

Here again the change in velocity depends on two parameters:

$$\left(\frac{\partial \rho}{\partial T}\right)_F \text{ and } \left(\frac{\partial K_s}{\partial T}\right)_F$$

Ultrasonic velocity measurements have been carried out, in three types of pollutance in sea and fresh water, over wide ranges of concentration and temperature. The study of these velocity measurements strongly indicates the feasibility of ultrasonic technique to measure water pollution.

Since the actual process of sound propagation in polluted water is not isentropic, some energy is dissipated due to irreversibility. Test data were taken by maintaining the polluted water in a constant temperature bath. Therefore the process was isothermal. To be precise, isothermal compressibility should be used in equations (2),

(3), (4), and (5). The isothermal compressibility can be calculated from adiabatic compressibility by using the well known thermodynamic relationship:

$$K_T = K_s \left(\frac{C_P}{C_V}\right) = K_s + 0.02391\ (L^2T)/C_P\rho \qquad (6)$$

where

K_T = isothermal compressibility

K_s = adiabatic compressibility

C_p = specific heat at constant pressure

C_v = specific heat at constant volume

$L = \left(\frac{\partial V}{\partial T}\right)_p \quad V$

L = volume coefficient of thermal expansion

V = volume

T = temperature

ρ = density

For aqueous solutions at room temperature and pressure, the difference between K_s and K_T is approximately 1 percent.

Since polluted water is hardly an ideal fluid and sound propagation is not isentropic, sound waves are attenuated due to viscous, heat conduction and molecular exchange losses in the medium. The attenuation of the sound energy was also measured. The major difficulty associated with attenuation measurements in polluted water is that the attenuation is too small at frequencies below 1 megahertz to measure in the laboratory. The sound path length required is too large to be accommodated in the relatively small confines of a laboratory. The frequency utilized was 3.6 megahertz and a reverberation technique was employed to increase path length. The attenuation was still too low to detect the degree of pollutance at low concentration levels.

TEST APPARATUS AND PROCEDURE

Test Apparatus

A number of commercially available instruments were investigated to measure the sound velocity and attenuation in liquids. Most of the available instruments for direct measurement of sound velocity in liquids are based on the "Sing-Around" technique developed by Greenspan and Tschiegg in 1957. An instrument known as Sonic Solution Analyzer, developed by Nusonics Inc., was selected for this study. This instrument is also based on the modified "Sing-Around" method. The functional schematic diagram of the test setup is shown in Figure 1. It consists of

1. Pulse Generator
2. Transducer Assembly
3. Amplifier
4. Threshold Detector
5. Frequency Counter
6. Constant Temperature Bath

The pulse generator, the amplifier, the threshold detector and the other necessary electronics to form the "Sing-Around" circuit are packaged in one cabinet. We shall refer to this cabinet as Sonic Solution Analyzer. The transmitting transducer, the receiving transducer and the two reflectors are mounted on one circular stainless steel ring assembly. We shall refer to it as the Transducer Assembly henceforth.

The Sonic Solution Analyzer is equipped with a control panel. The control panel incorporates an attenuation meter which includes the attenuation range at 0 to 10 percent and 0 to 100 percent of full range. The proper attenuation range is selected by an attenuation control switch. Similarly the control panel is equipped with four velocity range selection switches. The overall velocity selection range varies from 500 to 2500 meters per second. The control panel also incorporates one red and one green lamp, indicating out-of-velocity-range and within-velocity-range respectively.

The Transducer Assembly consists of one transmitting transducer, one receiving transducer and two reflectors. The front surface of the sensing element of the transducer is mounted on a stainless steel container and the back surface is sealed by epoxy. The space between the back surface of the crystal and the epoxy is filled with silicone grease to insure a proper seal and to allow the crystal to expand and contract freely. The reflectors are made of 316 stainless steel with the front face machined at an angle to reflect sound pulses from the first reflector to the second reflector and then to the receiving transducer. The transducers employed in the Transducer Assembly have a resonance frequency of 3.6 megahertz.

Test Procedure

The liquid under test was placed in a 500 milliliter beaker with sufficient liquid used to immerse the transducers completely. A liquid sample of 100 milliliter was sufficient; however, the amount actually used varied from 150 to 250 milliliters. This insured complete immersion of the transducers during the stirring process. The beaker was placed in a constant temperature bath and the transducer assembly was placed in the beaker. The temperature of the liquid in the beaker was monitored by means of a copper-constantan thermocouple.

The tests were conducted by keeping the sound path, temperature and time delay of the electronic circuit constant. Thus the effect of various degrees of concentration of contaminants were studied. The sound velocity through the test sample was calculated as follows:

$$C = \frac{1\ (1 + \alpha T)}{t - D} \qquad (7)$$

where

C = sound velocity

1 = sound path length

α = coefficient of thermal expansion of transducer assembly

T = temperature

t = time

D = electronic time delay

The time t used in equation (7) is the reciprocal of frequency repetition rate. It was more convenient to measure time to the eleventh decimal place of a second than the frequency repetition rate. The attenuation of each test sample was also measured.

Preparation of Test Samples

The samples obtained from the Pacific Ocean were taken in clean colorless glass bottles. The size of each bottle was one quart. This was sufficient to make at least ten test samples. Each bottle was rinsed thoroughly with the sample water before taking the sample. Every sample bottle was labelled as to the type of water collected, the source, and the position of the sampling point and the date.

To mix the oil with the ocean sample water, a specific method was adopted. This method was established after a number of trials. Test samples of 10, 20, 30, 40 and 50 percent oil by volume and 90, 80, 70, 60 and 50 percent of sea water by volume, respectively were prepared for two types of oils. The two types of oil used were cracked crude oil of sixth cut and diesel oil of fourth cut. The two oils were selected from a number of available oil samples because these oils gave a very stable mixture with sea water. These oils also represented the most commonly transported oils through the ocean either by tankers or by offshore drilling through the pipe lines.

The desired proportion of oil and water were put in a 500 milliliter beaker. Five to ten drops of benzol trimethyl ammonium hydroxide (triton-X-100), depending on oil proportion, were added to the mixture to insure a stable mixture. Then the mixture was stirred by a mechanical stirrer at a constant speed of fifty thousand revolutions per minute for fifteen to twenty minutes. The mixture was passed through an homogenizer one time. This provided a very stable mixture of oil and sea water. The test sample was stirred before taking a reading to insure a uniform mixture and temperature during testing.

DISCUSSION OF DATA AND RESULTS

The sonic velocity was measured in a number of test samples of sea water collected from various locations in the Pacific Ocean. These locations included Huntington Beach, San Clemente Beach, Fox's Snug Harbor (inland), Oceanside Beach and Los Angeles Harbor. The purpose of this investigation was to measure the relative difference in sonic velocity through these samples and consequently correlate these sonic velocity differences with pollutance. The results for sonic velocity temperature for these sea water samples are shown in Figure 2. It may be observed from Figure 2 that sonic velocity for all these samples varied less than 0.1 percent with the exception of the Los Angeles Harbor. The sonic velocity in the Los Angeles Harbor sample was consistently lower, by 8 to 10 meters per second, than the other sea water samples. It was also observed that the Los Angeles Harbor sample was not as clean as the other samples. It also had a foul odor to it. The lower sonic velocity in the Los Angeles Harbor sample was attributed to its higher degree of pollutance. It was further proved that the pollutance was due to oily substances of specific gravity lower than that of the sea water. Figure 2 also illustrates sonic velocity as a function of temperature in fresh water. The sonic velocity in fresh water is 35 to 40 meters per second lower than that of the sea water. The higher sonic velocity in sea water is attributed to its salinity and higher specific gravity than that of fresh water.

Sonic velocity and attenuation measurements were made by adding two commonly encountered contaminants in the sea water. These were crude oil and diesel oil. The effect of crude oil pollution in sea water on sonic velocity at various concentrations and temperatures are illustrated in Figure 3. An interesting phenomenon was observed. Unlike water, the sonic velocity in crude oil decreased with increase in temperature. This enables one to distinguish between the clean water and the oil polluted water more readily. It may be further observed from Figure 3 that the slope of change in sonic velocity, in oil polluted water, with respect to temperature is proportional to its degree of pollutance. The sonic velocity for oil polluted samples was approximately the same at 12°C. But as the temperature was increased the difference in sonic velocities became larger due to the variation in slopes. For example, the difference in sonic velocities between 10 percent and 20 percent oil polluted sea is 0.5 meters per second at 15°C. This difference in sonic velocity increases to 18.6 meters per second at 35°C. Similarly the difference in sonic velocities is also a function of oil concentration. The difference between 10% and 20% oil concentration is 18.6 meters per second at 35°C. This difference increases to 65.4 meters per second between 10% and 50% oil concentration.

Another parameter investigated to study water pollution was percent attenuation in ultrasonic energy as it travelled through it. The frequency used was 3.6 megahertz and the distance travelled through the water was 0.087121 meter. The percent attenuation at low concentration levels was too low to be measured accurately by using Sonic Solution Analyzer. This was attributed to short path length. However for higher concentrations the relative differences in percent attenuations were detected. The results are plotted in Figure 4 and Figure 5. Figure 4 illustrates the change in percent attenuation with respect to temperature at various concentration levels. The percent attenuation decreases with the increase in temperature. For example the percent attenuation for 50 percent oil concentration decreases from 7.80 to 3.90 as the temperature is increased from 15°C to 40°C. However the changes in percent attenuation are less pronounced at lower concentration levels. For example the percent attenuation decreases from 1.75 to 1.70 at 10 percent concentration level for the same temperature range. This percent attenuation in the oil polluted water is due to viscous losses. The viscosity of crude oil decreases significantly with the increase in temperature. On the other hand the change in viscosity of water with temperature is relatively small as compared with crude oil. The viscosity of water decreases from 1.307 centipoises to 0.653 centipoises when its temperature is increased from 10°C to 40°C while the viscosity of crude oil decreases from 147.5 centipoises to 36.3 centipoises for the same change in temperature. The change is so large that the viscosity of crude oil near 0°C was so high that the agitator, inside the 50 percent oil sample, could not be moved. Figure 5 exhibits the relationship between the percent attenuation of ultrasonic energy and crude oil concentration in the sea water at 15°C and 40°C.

Similarly, tests were conducted for diesel oil as a pollutance instead of crude oil. The purpose was to investigate the feasibility of the ultrasonic technique to study oil pollution qualitatively.

Figure 6 represents the relationship between the sonic velocity and temperature in 10, 30 and 50 percent diesel oil polluted sea water. A comparison between Figure 3, for crude oil, and Figure 6, for diesel oil clearly shows variations in absolute sonic velocities and velocity gradients between the two contaminants. As an example the absolute sonic velocities for 50 percent crude oil and diesel oil at 15°C are 1486.8 and 1451.3 meters per second respectively. Similarly velocity gradients for each pollution level for both contaminants are different. This illustrates that ultrasonic technique can be utilized to measure water pollution not only quantitatively and qualitatively also.

The velocity gradient with respect to temperature and the velocity coefficient for diesel oil at various temperatures was also explored. The relationship between the velocity coefficient and the temperature at various diesel oil concentration levels are represented by Figure 7. It may be observed that the velocity coefficient decreases with an increase in temperature. The velocity coefficient also decreases with an increase in diesel oil pollution. The later results are attributed to the negative velocity coefficient for diesel oil. As an illustration the velocity coefficient decreases from 1.62 to 0.62 at 20°C as the diesel oil concentration is increased from 10 to 30 percent.

The crude and diesel oil both have negative velocity coefficients. A third commonly encountered contaminant, chlorine, in fresh water as well as in sea water of positive velocity coefficient was investigated. The chlorine solution consisted of 12.5 percent sodium hypochloride and 87.5 percent other inert ingredients. This chlorine solution is commonly used in swimming pools. The sonic velocity, in chlorine polluted fresh water, versus temperature are plotted in Figure 8. The velocity coefficient for chlorine solution is positive. It was interesting to observe that a small addition of chlorine solution in fresh water caused a significant increase in sonic velocity. As an illustration the sonic velocity at 15°C increases from 1470.42 to 1514.34 meters per second with the addition of 10 percent chlorine solution. It must be noted that 10 percent chlorine solution contains 1.25 percent sodium chloride. The sonic velocity versus chlorine solution concentration at 15°C, 25°C and 40°C are illustrated in Figure 9. The difference in sonic velocities due to increase in temperature decreases with the increase in concentration. As an example the difference between sonic velocities, at 10 percent chlorine concentration, between 15°C and 40°C, is 54.28 meters per second. This difference reduces to 24.69 meters per second at 50 percent chlorine concentration.

CONCLUDING REMARKS

It has been shown that ultrasonic velocity, absorption coefficient and velocity gradient with respect to temperature can be used to measure water pollution quantitatively and qualitatively. These parameters provide information concerning the composition of polluted water. Even though ultrasonic techniques cannot provide detail information about all the ingredients in the polluted water as obtained by chemical analysis, the ultrasonic technique is quick and can be used on a real time basis to monitor rivers, lakes, offshore drilling and industrial wastes. The present methods involve chemical analysis of test samples which are not only expensive but in most cases require days to get the results. The delay in obtaining these results may be disastrous to our waters in some cases. The selection of the chemical test samples is at present made on a random basis. The ultrasonic technique can be utilized to provide a more efficient selection of test samples to be analysed chemically.

A simple procedure is required to detect water pollution by using the ultrasonic technique. A liquid sample is calibrated by adding a known amount of contaminants at three concentration levels at least. The relationship between sonic velocity and concentration in the operating temperature range is determined. A similar relationship between sound absorption and concentration in the operating temperature range is established. The other parameters like velocity gradients with respect to temperature and concentration can be established from the sonic velocity versus temperature at various concentration level relationships.

It has been shown that sonic velocity in polluted water is very sensitive to its temperature. So much so that in some cases the temperature effect on the sonic velocity is more pronounced than the pollution effect. This property of polluted water can be used favorably to detect pollution more accurately by studying velocity gradients. As an illustration, the velocity versus temperature curves, at different pollution levels, diverge as the temperature is increased. This makes it more convenient to measure velocity differences, hence pollution levels at higher temperatures. Instruments are commercially available which can measure sound velocity and temperature simultaneously.

The ultrasonic absorption coefficient can also be used to measure pollution level at higher concentration levels. These tests were conducted by employing a transducer assembly of one frequency only. It is recommended that at least three frequencies should be used to establish the relationship between frequency and percent attenuation at various temperatures and concentrations. The frequencies selected should be of the order of megahertz to measure attenuation at lower pollution levels. The sound path length should be increased and it should be a variable.

GENERAL REFERENCES

1. Beranek, L. L., "Acoustic Measurements," John Wiley and Sons, Inc., 1959.

2. Bergmann, L., "Ultrasonics and Their Scientific Applications," John Wiley and Sons, Inc., 1938.

3. Bergmann, P. G., "Physics of Sound in the Sea," Gordon and Beach Science Publishers, 1969.

4. Camp, L., "Underwater Acoustics," Wily-Interscience, 1970.

5. Carlin, B., "Ultrasonics," McGraw-Hill Book Company, Inc., 1949.

6. Goodman, G. T., Edwards, R. W. and Lambert, J. M., "Ecology and the Industrial Society," John Wiley and Sons, Inc., 1964.

7. Grossman, C. C., Joyner, C., Holmes, J. H. and Purnell, E. W., "Diagnostic Ultrasound," Proceedings of the First International Conference, 11965.

8. Klein, L., "River Pollution," Butterworths, 1959.

9. Libby, H. L., "Introduction to Electromagnetic Nondestructive Test Methods," Wiley-Interscience, 1970.

10. Southgate, B. A., "Advances in Water Pollution," The Macmillan Company, 1964.

11. Krautkramer, H. and Krautkramer, J., "Ultrasonic Testing of Materials," Springer-Verlag New York Inc., 1969.

12. Blitz, J., "Fundamentals of Ultrasonics," New York Plenum Press, 1967.

13. Hueter, T. F. and Bolt, R. H., "Sonics," John Wiley and Sons, Inc., 1955.

14. Gooberman, G. L., "Ultrasonics, Theory and Applications," Hart Publishing Company, Inc., New York, 1968.

15. Bhatia, A. B., "Ultrasonic Absorption," The Clarendon Press, 1967.

16. Mason, W. P., "Physical Acoustics," Academic Press, 1965.

17. Kinsler, L. E. and Frey, A. R., "Fundamentals of Acoustics," John Wiley and Sons, Inc., 1967.

18. Albers, V. M., "Underwater Acoustics," Plenum Press 1961.

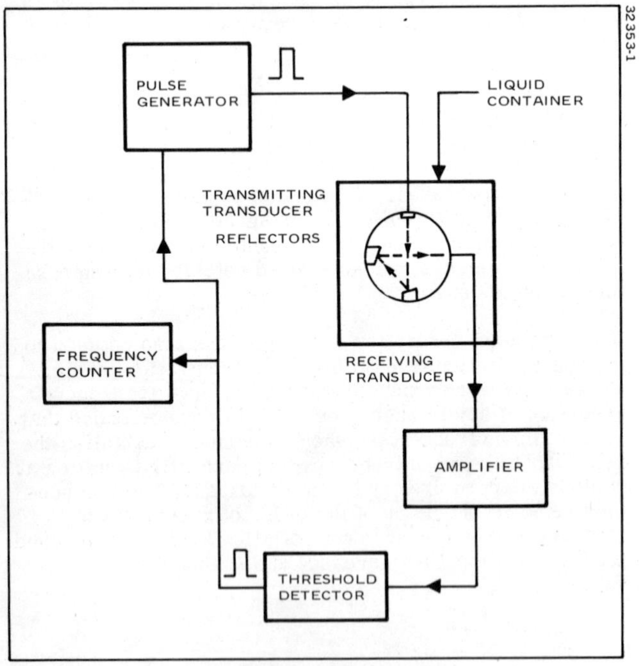

Figure 1. Functional Schematic Diagram Using Sonic Solution Analyzer

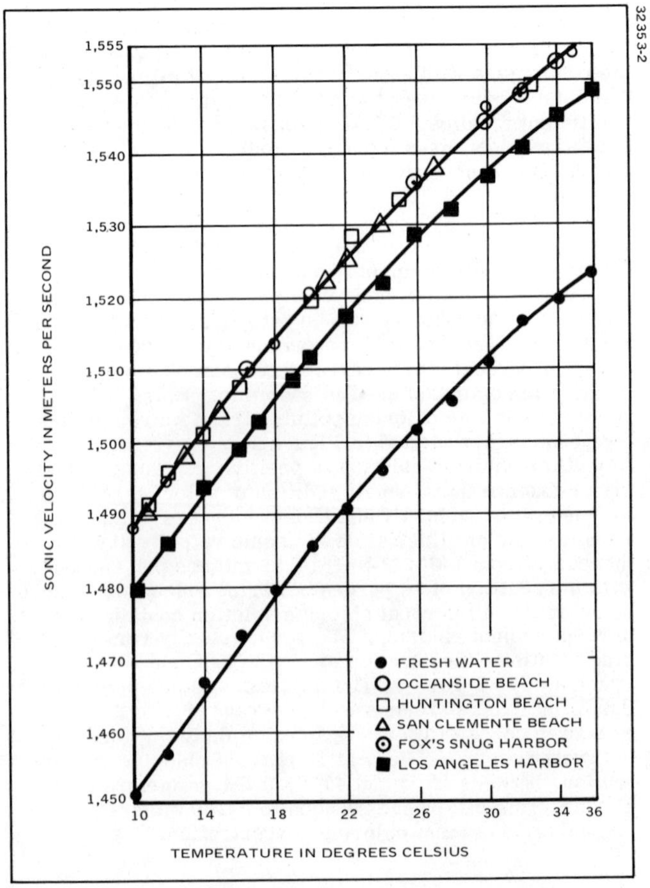

Figure 2. Sonic Velocity Versus Temperature for Fresh and Sea Water

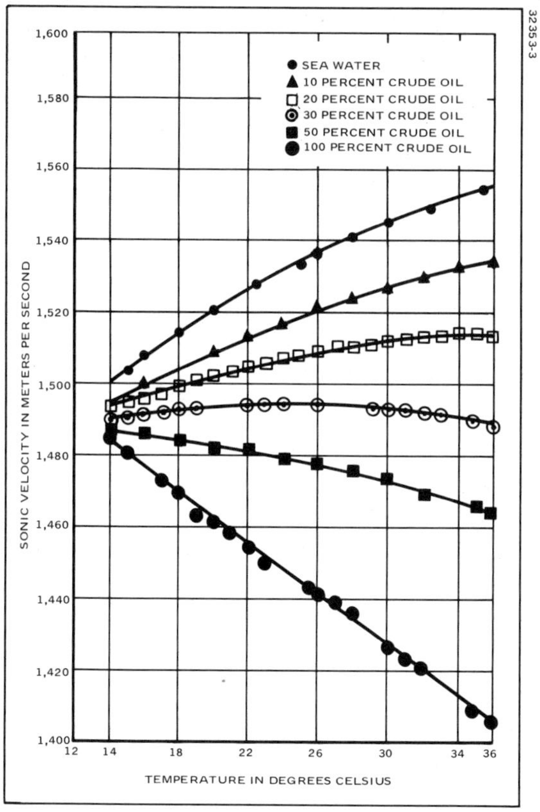

Figure 3. Sonic Velocity Versus Temperature for Crude Oil Polluted Water

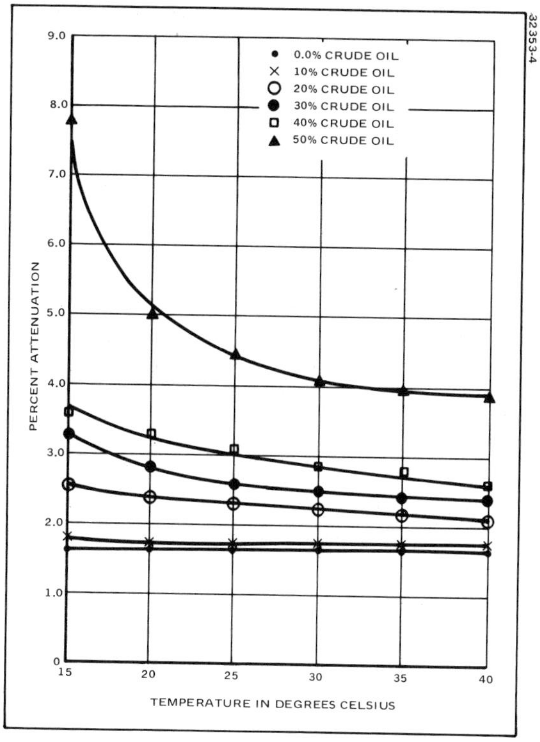

Figure 4. Ultrasonic Energy (f = 3.6 MHz) Percent Attenuation Versus Temperature for Crude Oil Polluted Sea Water

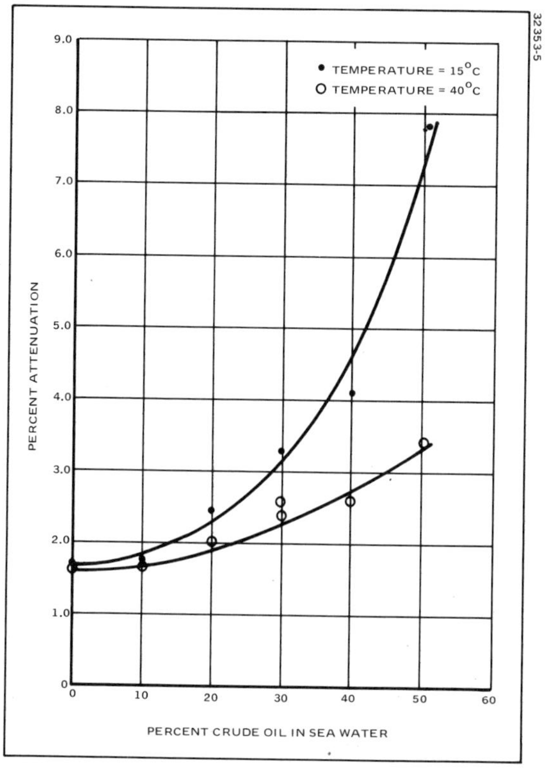

Figure 5. Ultrasonic Energy (f = 3.6 MHz) Percent Attenuation Versus Oil Concentration in Sea Water at Various Temperatures

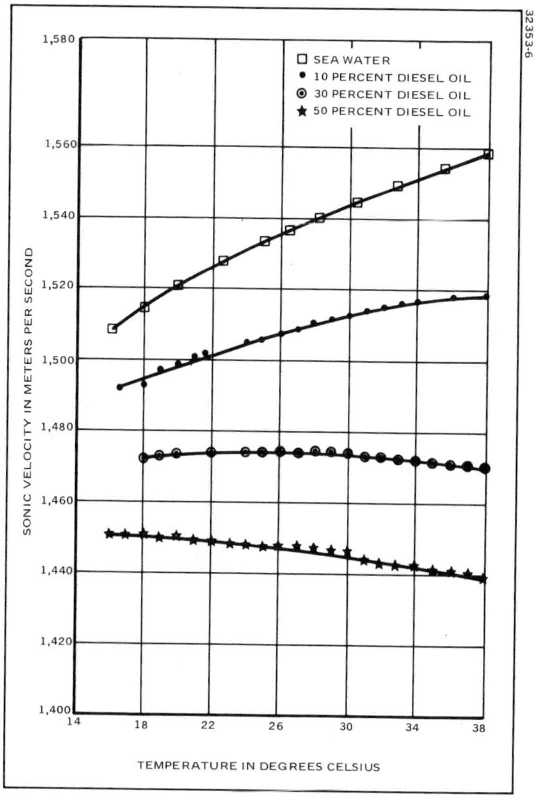

Figure 6. Sonic Velocity Versus Temperature for Diesel Oil Polluted Water

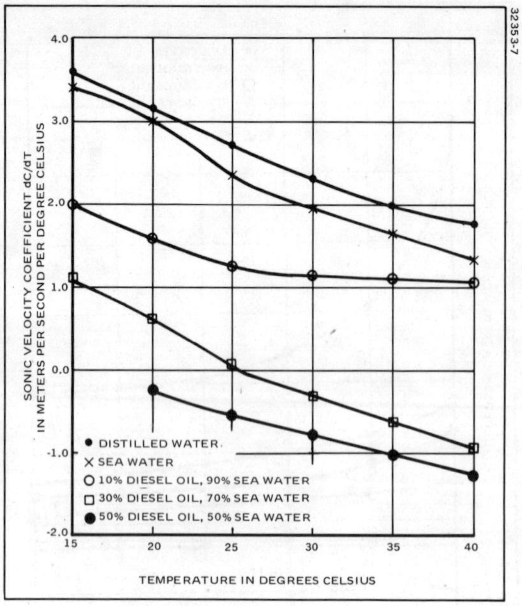

Figure 7. Sonic Velocity Coefficient Versus Temperature for Diesel Oil Polluted Water

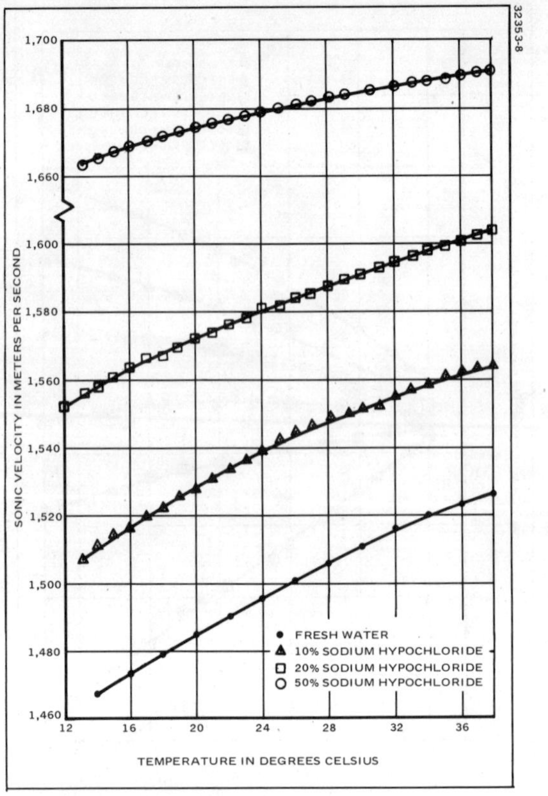

Figure 8. Sonic Velocity in Sodium Hypochloride Polluted Water Versus Temperature

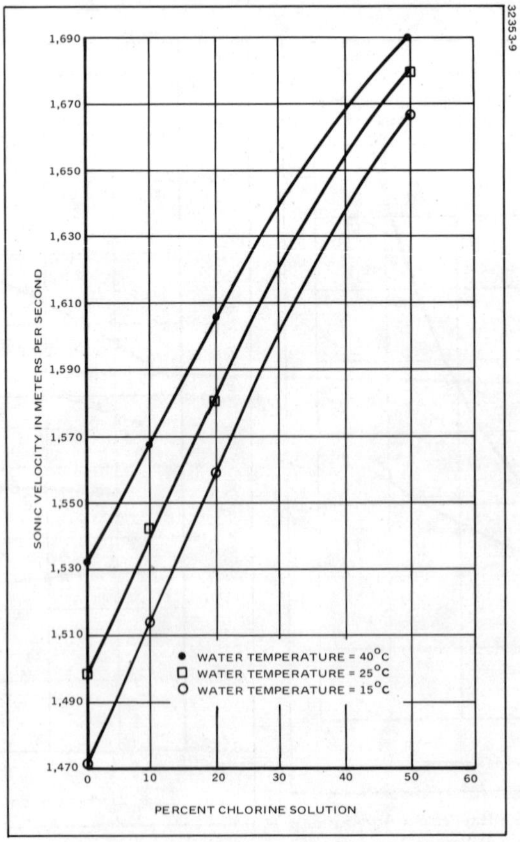

Figure 9. Sonic Velocity Versus Percent Chlorine in Fresh Water

© 1973, ISA JSP 6691

PHOTON ACTIVATION ANALYSIS

W. H. Zoller
Department of Chemistry
University of Maryland
College Park, Maryland 20742

ABSTRACT

Instrumental photon activation analysis (IPAA) using bremsstrahlung from an electron linac has been used to measure the concentrations of 19 elements in atmospheric particulate samples. The technique is sensitive, nondestructive and capable of analysis of fuels such as gasoline, fuel oil, residual oil and coal in addition to filter and cascade-impactor samples. The list of elements that are easily measurable down to the sub microgram range includes such important elements as Pb, Ni, As, Zn, Sb, Br and I. The sensitivity of IPAA is intermediate between neutron activation and atomic absorption.

INTRODUCTION

The analysis of minute quantities of atmospheric particulate material for trace elements using nondestructive nuclear techniques has been used extensively during the last few years. These techniques have gained popularity due to the extreme sensitivity for many elements, allowing one to analyze very small samples. In studies of atmospheric particulate material one would like to identify the sources that are responsible for the major fractions of particulates in a given area. Air pollution control personnel would like to identify the main sources of pollutant particles so that clean-up strategies can be applied to specfic sources, e.g. power plants, municipal incinerators, industrial plants, or heavy vehicular traffic.

In the cases of some of the sources noted above, there are trace elements that are indicative of emissions from these specific sources of particulates. For example, lead from gasoline combustion is converted to lead halogen aerosols upon combustion. This source of lead probably accounts for almost all of the lead in urban atmospheres and can be used as a chemical tag on automobile emissions. Likewise, vanadium is a good tracer for the combustion of residual fuels as shown by Zoller et al.[1] for Boston. Many scientists, for example Gordon et al.[2] and Friedlander[3] have attempted to account for the particulate material in the atmosphere from different sources. For these calculations a portion or all of the specific trace elements from a source (i.e. Pb from automobiles, Na from the sea, etc.) is used to estimate the quantities of other elements emitted from that source. The contributions from the different sources can then be summed up to predict how much of an element can be accounted for from the known sources. For many elements, wind blown crustal material or fly ash from coal account for most of what is observed. But in the case of Zn, Sb, Cu, Se, As, etc., we can account only for a small portion of what is observed (Gordon et al.[4]. To account for these elements, all of which are relatively volatile, new studies are needed of the trace element emissions from sources of urban particulates. The analytical techniques used must be sensitive for these and as many other elements as possible to estimate the importance of these sources for the anomalous trace elements that have been observed in urban areas. The source fuels, for example, coal and oil, as well as ash samples are of importance and also must be analyzed to give a complete picture of the trace metals emitted to the atmosphere.

Since the most important aspect of atmospheric particulate studies is the collection of a meaningful sample, the analytical techniques used for the chemical analysis must be suitable to many different types of samples. No matter how good a specific analytical technique is for obtaining precise elemental analyses, the data are meaningless unless the sample was collected carefully and something is known about the meteorological conditions during collection. If an analytical method is

Superior numbers refer to similarly-numbered references at the end of this paper.

sensitive enough, then short term, or low-volume samples can be collected on filters such as 47 mm Nuclepore or Millipore. If larger sample sizes are required then 10-cm diam or 20 x 25 cm rectangular filters of either polystyrene (Delgab) or cellulose (Whatman 41) can be used. In addition to the analysis of total filters the analytical technique should be capable of analyzing samples collected by cascade impactors.

The use of cascade impactors to measure the size distribution of atmospheric particulates has been demonstrated elsewhere (Lee[5], Gladney, et al.[6]). The samples from cascade impactors are usually deposited on some thin collection surface such as polyethylene, polycarbonate or Teflon coated with a sticky surface to prevent bounce-off of particles as discussed by Gladney et al.[6] These samples are usually quite small, making the chemical analysis difficult unless the analytical technique used is sensitive for the elements of interest.

ANALYTICAL TECHNIQUES

Of the possible analytical techniques for particulate analysis, nondestructive techniques are best due to the difficulty of dissolving certain portions of the sample by common techniques. This is especially true for the source materials such as coal, oil and other fuels. Many aluminosilicate minerals are difficult to dissolve completely and the use of reagents such as hydrofluoric acid may result in the loss of some elements such as the rare earths. Probably the worst problem with destructive analysis is that some volatile elements may be lost during heating, or the blank may be large to contamination or poor reagents used in the dissolution. The latter problem is especially serious for very small samples such as those collected by a cascade impactor.

Of the nondestructive techniques, x-ray fluorescence is very good for most of the first row transition elements and lead[7], but not capable of measuring elements such as Si, Ni, As, Se and Sb in air filters. The analysis of coal, oil and other fuels is also quite difficult and makes the use of x-ray technique of limited value for a wide range of studies.

Instrumental neutron activation analysis (INAA) can analyze all of the types of samples of interest for atmospheric studies, fuels, air filters, and cascade impactor for 30 to 40 trace elements.[8,9] The sensitivity of INAA is also adequate to allow the analysis of very small samples with great precision and accuracy. Unfortunately the important elements Pb, Ni and Si cannot be measured when thermal neutron activation is used.

INSTRUMENTAL PHOTON ACTIVATION ANALYSIS (IPAA)

This nondestructive nuclear technique uses the interaction of high energy γ-rays with matter to produce radionuclides. An electron linac delivers an intense beam (~100 μA) of about 35-MeV electrons to a heavy metal converter. The converter stops the electron, and produces an intense beam of bremsstrahlung radiation which hits the sample. The technique has been extensively utilized by Lutz[10] and Hislop[11] for the analysis of numerous elements in different matrices. Aras et al.[12] have discussed the use of this technique for the analysis of filters, Zoller et al.[13] for cascade impactor samples, and Cahill[14] for oil and fuel samples. If the electrons are stopped completely in the water-cooled converter, the samples are not subjected to much heating, allowing the analysis of volatile fuels. Samples are packaged with suitable elemental monitors for each element to be determined with flux monitor to correct for flux variation from one location in the irradiation facility to another. The types of nuclear reactions observed in photon irradiated air filters are given below in Table I.

TABLE I

Nuclear Reactions Observed

(γ,n)	(γ,p)	(γ,α)	(n,γ)
$(\gamma,2n)$	(γ,pn)	$(\gamma,\alpha n)$	

The product radionuclides are determined by observing the emitted γ-rays with Ge(Li) detectors coupled to 4096-channel analyzers. The γ-ray energy spectra obtained are then analyzed by high-speed computer and, after comparison with the standards, elemental concentrations are calculated. A typical γ-ray spectrum of a photon-irradiated filter sample is shown in Figure 1. The most intense lines observed are due to Pb, Br and Na, although the lines of many other isotopes are also observed.

The elements routinely measured and others that are observed are given in Table II. There are several cases in which more than one radionuclide can be used for the determination of a given element.

TABLE II

Elements and Radionuclides Observed by IPAA in Air Samples

Most Useful Activities

Element	Nuclide
Na	^{22}Na ^{24}Na
Si	^{29}Al
Cl	^{34}Cl ^{38}Cl
Ca	^{43}K ^{47}Ca
Ti	^{46}Sc ^{47}Sc ^{48}Sc
Cr	^{51}Cr
Ni	^{58}Ni
Zn	^{62}Zn ^{63}Zn ^{65}Zn ^{67}Cu
As	^{74}As
Se	^{75}Se
Br	^{77}Br ^{82}Br
Rb	^{84}Rb
Y	^{88}Y
Zr	^{89}Zr
Sb	^{120m}Sb ^{122}Sb
I	^{126}I
Cs	^{132}Cs
Ce	^{139}Ce

TABLE II (continued)

Other Observed Activities

Element	Nuclide
C	^{7}Be
K	^{42}K
Fe, Mn	^{52}Mn, ^{54}Mn, ^{56}Mn
Ni, Co & Cu	^{57}Co, ^{58}Co
As, Se	^{76}As
Ag	^{110m}Ag
La	^{140}La

There are many cases in IPAA where several nuclear reactions or different target nuclei can produce the same isotope. Care must be taken to be sure that the radionuclide measured comes from the reaction one is interested in using. This work has been done by Aras *et al.*[12] by irradiating pure samples of each element of interest, and observing the ratio of isotopes produced.

Table III lists comparative results for three elements determined by more than one isotope. The agreement between the

TABLE III

Comparison of Elemental Determinations Using Different Isotopes

Element	Isotope	Conc. $\mu g/m^3$
Ca	^{43}K	2.6
	^{47}Ca	2.4
Ti	^{46}Sc	0.28
	^{47}Sc	0.30
	^{48}Sc	0.29
Pb	^{203}Pb	2.4
	^{204m}Pb	2.9

different isotopes shows that they are all produced from the parent element that is being measured.

The limits of detectability of IPAA are

worse for many elements than INAA or flame methods, but for air samples the technique can easily be used for the determination of the elements listed in Table II. A detailed discussion of the sensitivities for analyzing atmospheric particulates has been given by Aras et al.[12] IPAA has been used successfully to analyze cascade-impactor samples for 19 elements and to obtain size distributions as reported elsewhere.[13]

IPAA is very useful for the analysis of Pb and the halogens Cl, Br, and I simultaneously, without destruction of the samples. This is important in studies of automobile exhaust particulates since they consist in large part of Pb, Br, Cl, $PbBr_2$ and $PbCl_2$ when initially emitted to the atmosphere. Chemical changes of these particles can release Br and Cl gaseous species and allow the residual lead aerosols to possibly react with such gaseous species such as I_2 to produce freezing nuclei as suggested by Schaefer[15]. It is very convenient to be able to measure all of the elements important for these reactions by the use of a single nondestructive technique.

In addition IPAA has been used by Cahill[14] to analyze hydrocarbon fuels for trace metal impurities. The usefulness of IPAA to various types of samples has been discussed by Hislop[11], and shows great promise for biological tissues and water samples, in addition to atmospheric particulates and fuels.

SUMMATION

IPAA is a very useful analytical technique for environmental analysis, but cannot totally replace INAA, atomic absorption on x-ray fluorescence. This technique is best as a supplemental nondestructive method for determination of a variety of those elements that are either difficult or impossible to measure by the other techniques. In conjunction with the INAA technique, IPAA offers the greatest possibilities for analysis of a wide range of sample types for a large number of trace elements. Some of the most useful aspects of IPAA are listed below:

- Many samples can be irradiated simultaneously due to the long penetrating range of γ-rays which allows one to analyze large numbers of samples with only a few hours of linear accelerator time.

- Coupled with INAA, about 45 elements can be determined on one sample of only a few mg weight without destroying the sample so it can be used for further work at a later time.

- IPAA is capable of analyzing tissue, hydrocarbon fuels, coal, air filters, and cascade impactor samples nondestructively.

REFERENCES

(1) Zoller, W. H., G. E. Gordon, E. S. Gladney and A. G. Jones, The Sources and Distribution of Vanadium in the Atmosphere, ADVANCES IN CHEMISTRY SERIES #123, TRACE ELEMENTS IN THE ENVIRONMENT, (1973), 31-47.

(2) Gordon, G. E., E. S. Gladney, A. G. Jones, P. K. Hopke and W. H. Zoller, The Size and Chemical Composition of Atmospheric Particulates in Boston, presented at the Amer. Chem. Soc. Nat'l. Meeting, Boston, Mass., April, 1972.

(3) Friedlander, S. K., Chemical Element Balances and Identification of Air Pollution Sources, Environ. Sci. and Tech. Vol. 7, (1973), 235-240.

(4) Gordon, G. E., W. H. Zoller and E. S. Gladney, Abnormally Enriched Trace Elements in the Atmosphere, Presented at the Seventh Annual Conference on Trace Substances in Environmental Health, Univ. of Missouri, Columbia, June 1973.

(5) Lee, R. E., Jr., The Size of Suspended Particulate Matter in Air, Science Vol. 178, (1972) 567-575.

(6) Gladney, E. S., W. H. Zoller, A. G. Jones and G. E. Gordon, Composition and Size Distributions of Atmospheric Particulate Matter in the Boston Area, Environ. Sci. and Tech. (1974), to be published.

(7) Giauque, R. D., F. S. Goulding, J. M. Jaklevic and R. H. Pehl, Trace Element Determination with Semiconductor Detector x-ray Spectrometers, Anal. Chem. Vol. 45 (1973) 671-681.

(8) Zoller, W. H. and G. E. Gordon, Instrumental Neutron Activation Analysis of Atmospheric Pollutants Utilizing Ge(Li) γ-Ray Detectors, Anal. Chem. Vol. 42 (1970) 257-265.

(9) Dams, R., J. A. Robbins, K. A. Rahn, and J. W. Winchester, Nondestructive Neutron Activation Analysis of Air Pollution Particulates, Anal. Chem. Vol. 42 (1970) 861-867.

(10) Lutz, G. J., Photon Action Analysis--A Review, Anal. Chem. Vol. 43 (1971) 93-103.

(11) Hislop, J. S., Photon Activation Analysis of Biological and Environmental Samples: A Review, Proceedings International Conference on Photonuclear Reactions and Applications, March 1973, 1159-1176.

(12) Aras, N. K., W. H. Zoller, G. E. Gordon and G. J. Lutz, Instrumental Photon Activation Analysis of Atmospheric Particulate Material, Anal. Chem. Vol. 45, (1973) 1481-1490.

(13) Zoller, W. H., N. K. Aras, E. S. Gladney and G. J. Lutz, Studies of Atmospheric Particulate Size Distributions Using Instrumental Photon Activation Analysis, Proceedings International Conference on Photonuclear Reactions and Applications, March 1973, 1007-1008.

(14) Cahill, R. C., Neutron Activation Analysis of Petroleum and Petroleum Fract Fractions, M.S. Thesis, Univ. of Maryland, 1974.

(15) Schaefer, V. J., Ice Nuclei from Automobile Exhaust and Iodine Vapors, Science Vol. 154, (1966) 155-157.

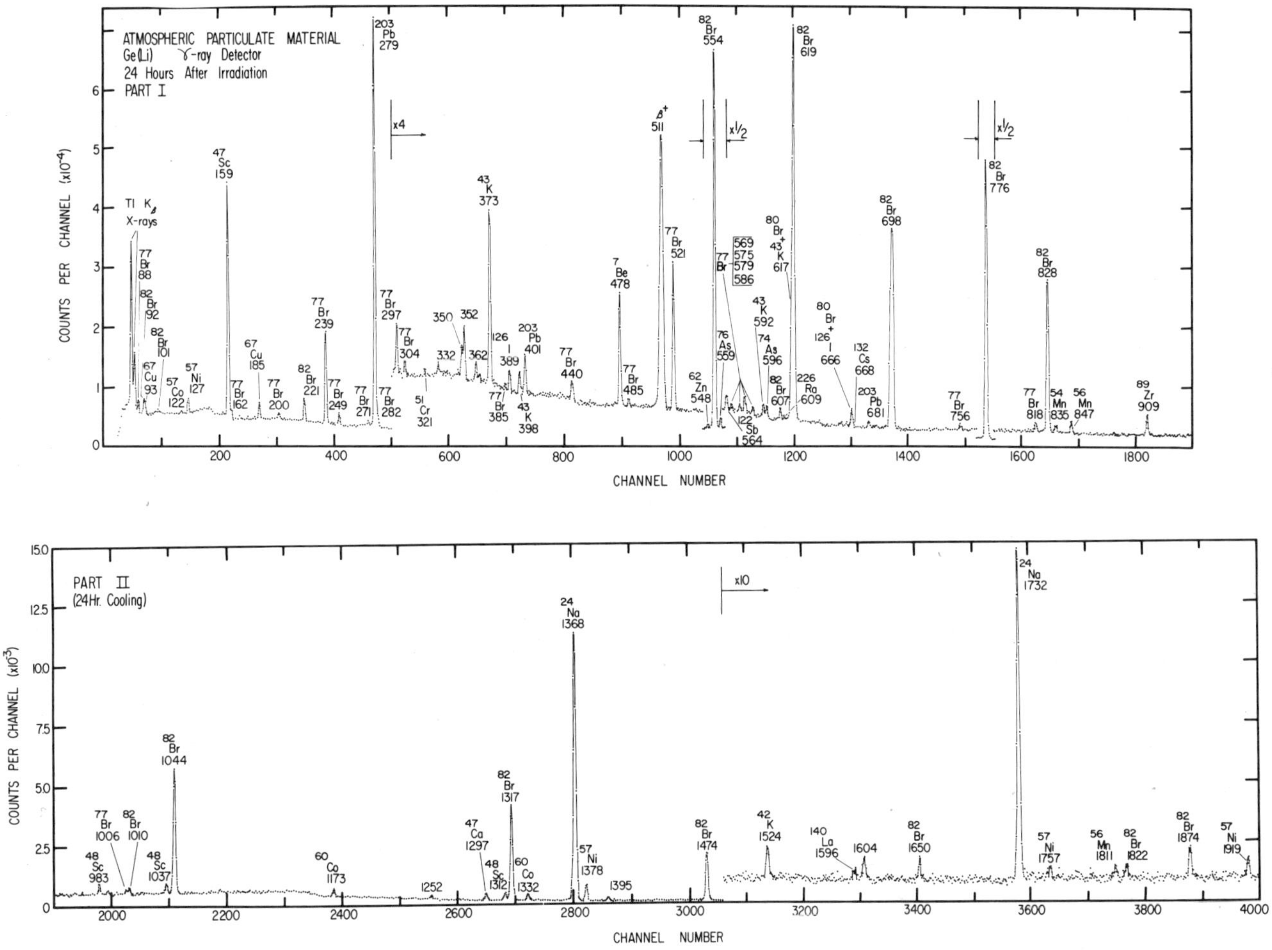

©1973, ISA JSP 6692

AMBIENT AIR AEROSOL SAMPLING

Jerome J. Wesolowski, Ph.D.
Chief
Air and Industrial Hygiene Laboratory
Laboratory Services Program
State of California Department of Health

ABSTRACT

The characterization of ambient air aerosol to obtain information on the sources and evolution of the aerosol requires a complex sampling and analytical scheme. This paper will discuss the measurements of the variations of the concentrations of specific chemicals in California air as a function of time, space and particle size. Particular emphasis will be placed on the sampling problems associated with the collection of the aerosol.

INTRODUCTION

This paper reviews the measurements of the chemistry of the aerosol in California ambient air. This is presented in three parts: the reasons for the measurements; examples of representative data; and a quantitative discussion of the efforts to validate the data, especially that obtained with size-segregating instruments.

The basic objective of our efforts in aerosol chemistry is to determine what control program to use to reduce the deleterious effects of air pollution on visibility and public health. These are closely related since the size range which scatters light and therefore reduces visibility is also respirable and has a large mass concentration, thus presenting a large potential health problem.

Two pieces of information required to meet this objective are the functional dependence of visibility on chemical species and particle size, and the sources of these particles. These are not simple problems, particularly in Los Angeles where 50% of the aerosol can be secondary aerosol formed in the atmosphere as a result of reactions between gases or between gases and particles. Therefore, the function we seek must include the contribution from secondary as well as primary aerosol. To adopt a control program which eliminates all primary aerosol and all gaseous precursors of secondary aerosol is impossible since it would effectively stop all human activity. Therefore, we wish to determine the functional dependence in order to maximize the effectiveness of any realistic control program.

To meet this complex objective we obtain in our aerosol research programs gas and meteorological data, and the concentrations of many chemical species as a function of time, space, and particle size.

REPRESENTATIVE DATA

Table 1 shows some of the specific objectives which can be reached by these measurements. We will give examples of how diurnal patterns aid in identifying specific sources, how spatial distributions can be used to determine the extent to which an aerosol sample collected is representative of an ambient air mass, and how particle size distributions aid in determining mechanisms of aerosol formation.

The data in Figure 1 is an example of the identification of a source from multi-element diurnal patterns obtained from ambient air. Pb was obtained by X-ray fluoresence analysis and the other elements by neutron activation analysis. The aerosol was collected on total filters. The excellent correspondence among these curves during the morning hours of December 10 indicates a common source of Pb, Zn, Hg, Sb, As, In, Ba, Ag, and Se. Many other elements, e.g., Na, Cl, Al, had different diurnal patterns. Indeed, a lead smelter located two miles upwind of the sampling unit which obtained these results, was polluting the atmosphere.(1)

Table 2 demonstrates how spatial distributions are used to determine if the aerosol sampling

NOTE: Superior numbers refer to similarly-numbered references at the end of this paper.

device is so located that the aerosol collected is representative of the ambient air one wishes to study. Measurements were made of the concentration of about 30 elements and the mass from aerosol collected on 24-hour high-volume samplers located at nine district air pollution stations distributed throughout the San Francisco Bay Area.[2] If all nine hi-vol samplers were measuring atmospheric aerosol representative of the community in which they were located we expected a high degree of correlation for many pairs of elements due to their common origin, e.g., wind-blown dust. The first column gives the correlation coefficients for some of the possible pairs of elements for all nine stations. The correlations are poor. The next column shows typical results when one station is removed from the analysis. In this case, station no. 9 was the one excluded. There is little difference between column one and column two. The last column shows the result of an eight-station analysis when the station removed was station no. 5. The correlation coefficients are now close to unity. A careful inspection of station no. 5 revealed a small vent upwind of the hi-vol sampler which led to a machine shop on the floor below. Thus the sampler at that particular local air pollution control district station was not measuring the ambient air, but rather machine shop air. This is an example of how spatial distributions, when combined with statistical analysis can be useful in assessing the suitability of specific sampler locations.

Figure 2 demonstrates how size distributions aid in determining the mechanism of aerosol formation. The slide shows the vanadium size distributions obtained simultaneously at nine stations in the Bay Area using Andersen impactors.[3] The distributions are quite similar and the shapes imply two separate mechanisms for the formation of the aerosol; a mechanical process, such as wind erosion, to account for the large particle component and a small particle production mode, such as combustion, to account for the small particle component. Aluminum distributions were also obtained and showed predominantly a large particle component consistent with a soil erosion mechanism for that element. Figure 3 shows the V/Al ratio as a function of particle size for the nine stations. The V/Al ratio for large particles is very similar to what one would obtain from soil whereas the ratio for small particles is more than an order of magnitude greater. We conclude that the large particles are predominently from wind-blown dust and the small, from a combustion source, such as the burning of residual fuels.

VALIDATION OF DATA

We have given examples of the type of information obtained from the measurement of the concentration of elements as a function of time, space, and particle size. Clearly the reliability of the information will depend on the accuracy of the measurements. The last part of this paper will discuss one part of our program to determine the accuracy of aerosol chemistry data, in particular, a study of the errors introduced by the use of non-sticky collection surfaces in size-segregating impaction instruments. The need for such a study arises from the fact that many of the sensitive analytical techniques we wish to use to obtain the concentration of chemical species require, for maximum specificity and sensitivity, that the aerosol be collected on smooth, dry impaction surfaces such as Teflon or Mylar films. The study consists of comparing the particle size distributions of specific elements obtained from two identical Lundgren impactors which simultaneously sampled air from a common manifold, one with sticky and one with non-sticky collection surfaces on the stages.

A useful feature of the Lundgren impactor (see Fig. 4) is that the collection surfaces are placed on metal cylinders which rotate with time. After a run the surfaces can be cut into sections, each representing a given time interval. Particle size distributions for time intervals as small as two hours can be obtained. The need for an impactor which yields such information prompted our choice of the Lundgren for our initial efforts to determine the differences in the size distributions obtained with sticky and non-sticky collection surfaces.

Figure 5 shows 24-hour size distributions obtained in Berkeley for four elements using neutron activation, for two impactors run in parallel into a common manifold, one with 0.0025 cm thick Teflon substrate and one with .0025 cm thick polyethylene covered with a .004 cm thick sticky resin. The two Na distributions are very similar. According to Junge 90% of the mass of sea-salt particles is carried by jet sea-spray particles, i.e. particles greater than 1 μ which originate upon the bursting of small air bubbles produced by breaking waves and only 10% by film particles, i.e. particles less than 1 μ which form from the bursting of the bubble film. Therefore the Na distribution is consistent with a sea source, as expected from Berkeley's near coastal location. Thus it is not surprising that the hygroscopic sea aerosol should give similar distributions on the sticky and dry surfaces. The Br and V distributions are both heavily weighted towards small particles consistent with a combustion production mechanism. Although the distributions for sticky and dry surfaces exhibit the same trends, the collection efficiency for the latter is lower, especially for large particles. This could be caused by the bounce-off of particles which impacted on the surface but did not stick to it.

The Al sticky distribution indicates most of the concentration in the large size fraction, consistent with a soil source. The dry distribution is completely different and in fact it appears that some of the large particles which bounce-off cross over to the after filter.

Figure 6 gives the results of a 24-hour experiment in San Jose, California, another near coastal location. One of the impactors had its cylinders covered with two collection surfaces, i.e. half was covered with Teflon and half covered with 0.0006 cm thick Mylar. The data shows the collection efficiency for these two non-sticky surfaces are very similar. The differences in the sticky and non-sticky distributions are similar to those found in the Berkeley experiment.

Figure 7 shows the results for Fresno, which is located in the central valley of California. Because of its location we expect a greater contribution from the soil. This is demonstrated by the increased concentrations for the large particle sizes for V, Al and Na. The Fresno Na distributions are different from those in the near coastal locations and are indicative of a large soil contribution. Also the Fresno Na distributions show large differences between sticky and non-sticky surfaces whereas the opposite was true for the near coastal locations. The greater soil contribution is also shown by the V/Al ratios for the three locations as shown in Table 3. The ratios for large particles are approximately the same as for soil for all three locations but the small particle ratios deviate considerably from the soil ratio. The smallest deviation is for Fresno, indicating the dominance of soil to the V distribution in this dry, inland location.

Therefore the collection efficiency of a surface is related to the source of the aerosol containing the element in question rather than to the element itself, precluding the possibility of determining a normalization function which could be used to correct the size distributions obtained from non-sticky collection surfaces.

Comparisons were also made between the sum of the concentrations on each of the stages (including the after filter) and the concentration on a total filter which was run in parallel with the impactors in the common manifold. This was done to estimate the wall losses within the impactor. The results shown in Table 4, which are average values for a number of experiments, indicate that Mylar and Teflon impactor surfaces yield similar wall losses, the greatest being for the soil-derived element, Al. When using sticky surfaces the wall losses are significantly reduced.

SUMMARY AND CONCLUSION

We have delineated the major reasons for studying the chemistry of the urban aerosol in California. Examples of measurements of the concentrations of specific elements as a function of time, space and particle size were related to specific objectives; in particular the identification of an industrial polluter, the determination of the representativeness of sampler location, and the determination of the production mechanism for specific elements in particulate matter. The accuracy required for such measurements depends on the objective. The last part of the paper presented experimental evidence that the use of non-sticky collection surfaces on size-segregating impactors may cause sufficient inaccuracies to make the determination of production mechanisms based on the size distributions measured invalid. This demonstrates the need to have a validation program as an integral part of any atmospheric research project.

ACKNOWLEDGEMENTS

The various experimental results cited above were obtained over a number of years with contributions from many of my colleagues at the Air and Industrial Hygiene Laboratory as well as the Lawrence Livermore Laboratory. Particularly significant contributions were made by Drs. B. R. Appel, W. John, C. Martens, P. K. Mueller, R. Ragaini; Messrs. A. Alcocer, R. Kaifer, and S. Wall, and by SuzAnne Twiss. Their help is gratefully acknowledged.

REFERENCES

(1) Wesolowski, J.J., W. John, R. Kaifer, "Lead Source Analysis by Multi-Element Analysis of Diurnal Samples in Ambient Air", Trace Elements in the Environment, ACS Series, 1973.

(2) John, W., R. Kaifer, K. Rahn, J.J. Wesolowski, Trace Element Concentrations in Aerosol from the San Francisco Bay Area, Atmospheric Environment, Vol. 7, 1973.

(3) Martens, C.S., J.J. Wesolowski, R. Kaifer, W. John, "Sources of Vanadium in Puerto Rican and San Francisco Bay Area Aerosols", Environmental Science and Technology, Vol 7, 817 (1973).

TABLE 1

TYPES OF INFORMATION OBTAINED FROM Δt, Δx, and ΔDp MEASUREMENTS

Δt (Diurnal Patterns)

1. Specific sources
2. Age of aerosol
3. Trajectory information
4. Secondary aerosol information

Δx (Spatial Distributions)

1. Specific sources
2. General sources (sea, soil)
3. Trajectory information
4. Representativeness of sample collected

ΔDp (Particle Diameter Distributions)

1. Mechanisms of aerosol production (combustion, grinding, soil erosion)
2. General sources (sea, soil)
3. Secondary aerosol formation

TABLE 2

EXAMPLES OF THE EFFECT ON THE CORRELATION COEFFICIENTS RESULTING FROM THE DELETION OF STATION 5

Elements	Correlation coefficients (× 100)		
	All 9 Stations	8 Stations (without 9)	8 Stations (without 5)
Sb–Total	38	37	99
V–Total	63	63	91
Co–Sb	67	66	94
Ce–Sb	55	56	96
Th–Sb	45	45	97
Al–Sb	56	54	94
Mn–Sb	69	66	91
Sm–Sb	65	64	97
V–Ce	77	77	92
V–Cr	57	56	70
Sc–V	67	67	83
V–Th	69	68	91

TABLE 3

V/Al RATIOS

	<0.5 μ	> 8 μ
Fresno 8/31/72	0.006	0.0015
Berkeley 7/28/72	0.663	0.0031
San Jose 8/25/72	0.037	0.0016

TABLE 4

SUM OF LUNDGREN IMPACTOR STAGES AND AFTER FILTER AS PERCENT OF TOTAL FILTER

Impaction Surface	Element V	Br	Na	Al
Dry Teflon	75	83	83	50 %
Dry Mylar	61	86	79	30
Sticky Polyethylene	80	84	86	72

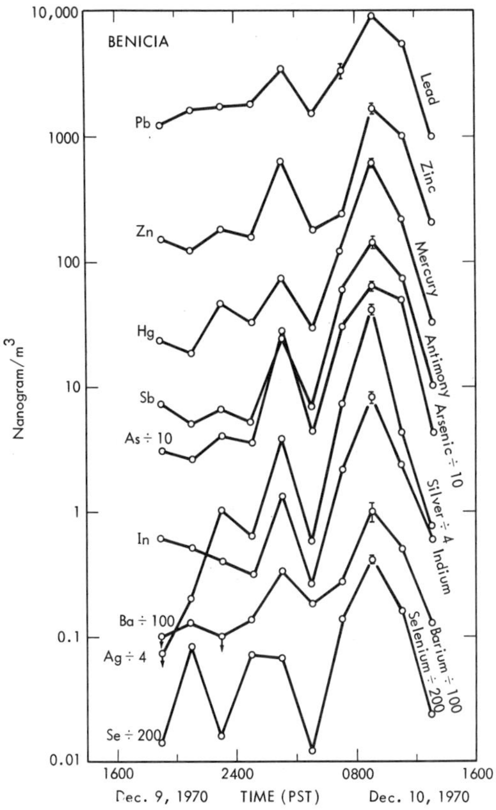

FIGURE 1

DIURNAL VARIATION OF CORRELATED ELEMENTS
IN BENICIA, CALIFORNIA

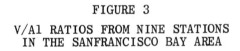

FIGURE 2

VANADIUM SIZE DISTRIBUTION
IN AEROSOLS COLLECTED ON ANDERSEN IMPACTORS
AT NINE STATIONS IN THE SAN FRANCISCO BAY AREA

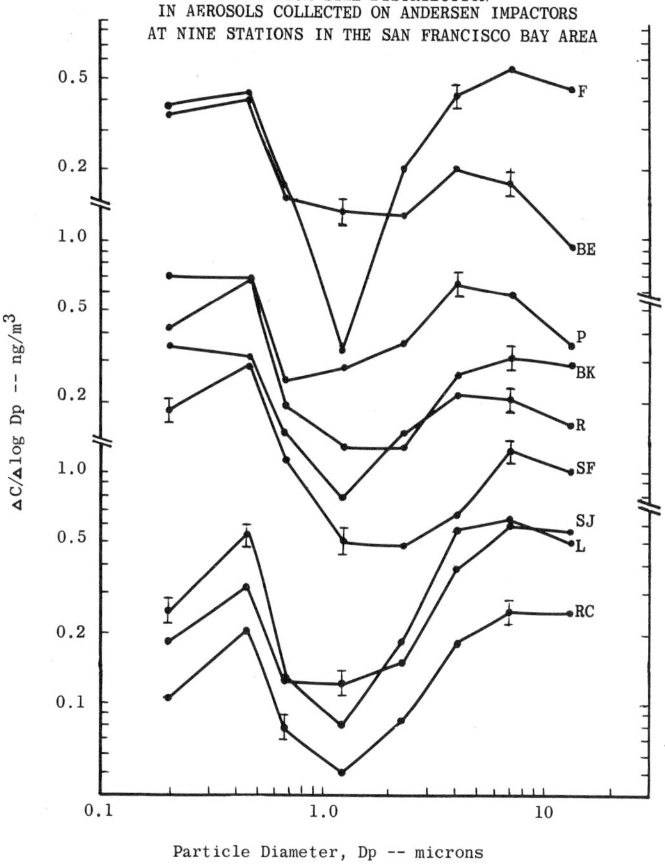

FIGURE 3

V/Al RATIOS FROM NINE STATIONS
IN THE SANFRANCISCO BAY AREA

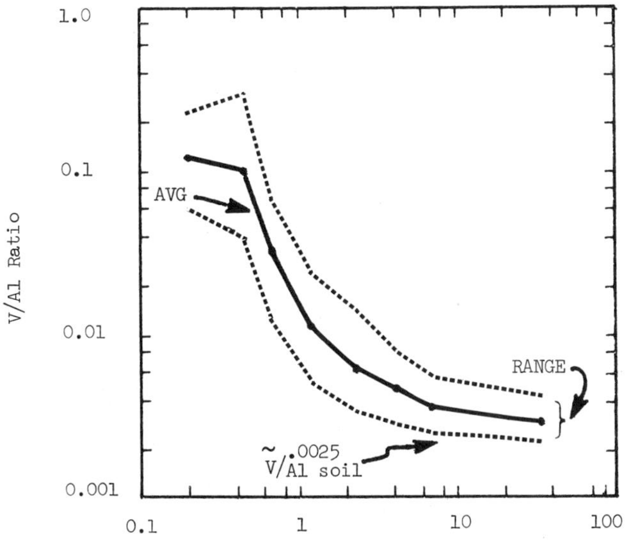

FIGURE 4

LUNDGREN CASCADE IMPACTOR

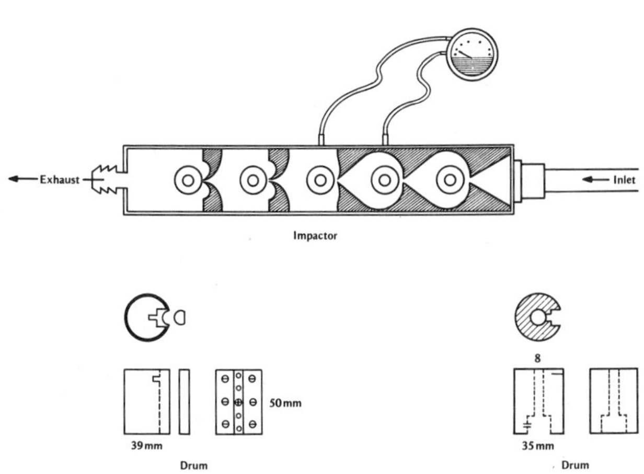

FIGURE 5
COMPARISON OF CHEMICAL DETERMINATIONS
FROM AEROSOLS COLLECTED ON VARIOUS LUNDGREN IMPACTOR SURFACES

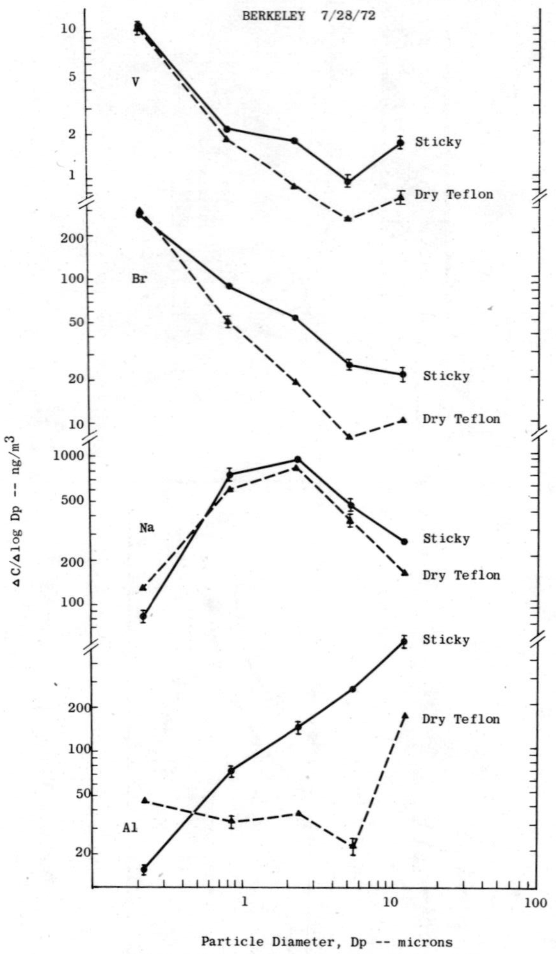

FIGURE 6
COMPARISON OF CHEMICAL DETERMINATIONS
FROM AEROSOLS COLLECTED ON VARIOUS LUNDGREN IMPACTOR SURFACES

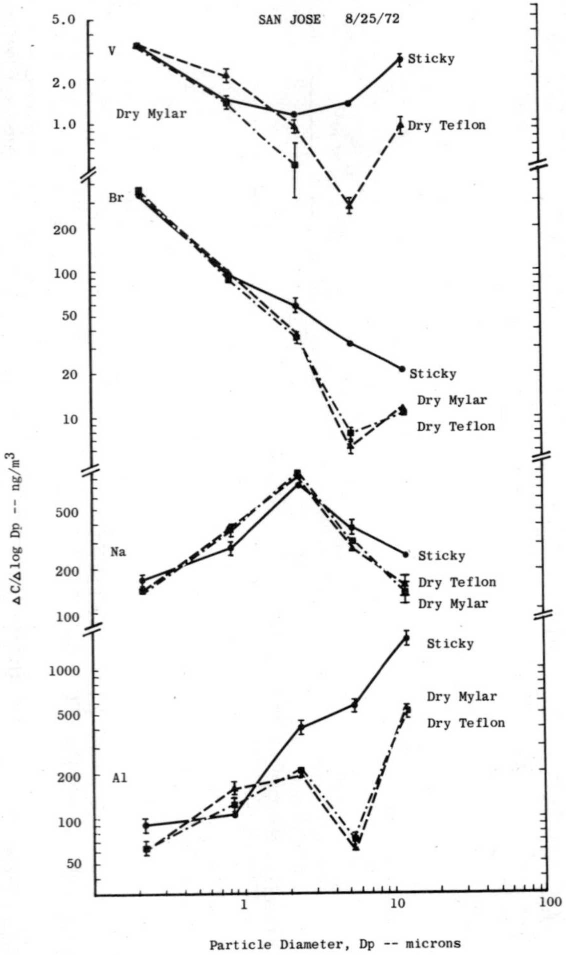

FIGURE 7
COMPARISON OF CHEMICAL DETERMINATIONS
FROM AEROSOLS COLLECTED ON VARIOUS LUNDGREN IMPACTOR SURFACES

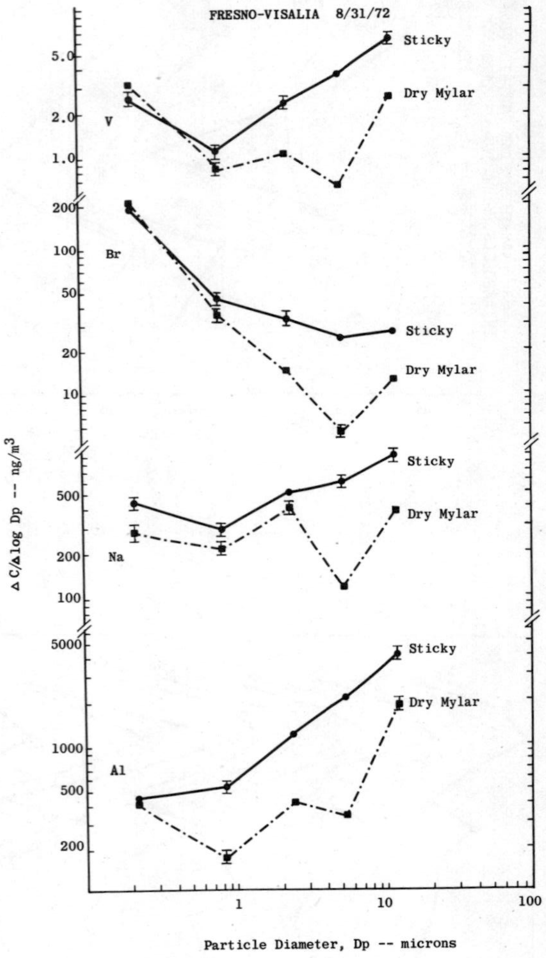

© 1973, ISA JSP 6693

CHEMICAL CHARACTERIZATION OF ATMOSPHERIC POLLUTION PARTICULATES BY PHOTOELECTRON SPECTROSCOPY

T. Novakov
Lawrence Berkeley Laboratory
University of California
Berkeley, California 94720

ABSTRACT

This paper addresses itself to the application of X-ray photoelectron spectroscopy (also known as ESCA or XPS) to the chemical characterization of pollution particulates, in particular to the determination of concentrations and chemical states of sulfur, nitrogen and carbon. In addition to the discussion of the capabilities of ESCA as an analytical tool, a short summary of the application of the method to a more comprehensive study of the formation and behavior of urban sulfates will be presented.

INTRODUCTION

The method has been shown to be a convenient and useful technique for the determination of chemical states of atoms in molecules[1]. It has also been demonstrated that this method has a potentially significant application to the chemical characterization of pollution aerosols. For example, Novakov et al.[2] have applied this method for the determination of lead and the determination of the chemical states of sulfur and nitrogen in smog particles as a function of particle size and time of day. Two oxidized states of sulfur were identified in their work; the more oxidized form was assigned to the sulfate species, while it was suggested that the other corresponds to an oxidation state of 4^+, i.e. sulfite or adsorbed sulfur dioxide.

Hulett et al.[3] used ESCA to study sulfur compounds in fly ash and smoke particles. They reported three chemical states of sulfur present on coal smoke particles. These included a single reduced state, which was assigned to hydrogen sulfide or a mercaptan, and two species in higher oxidation states corresponding to sulfate and sulfite. The sulfur in the fly ash was tentatively identified as a sulfate.

In the course of our work on the photoelectron spectroscopy of ambient aerosols we have carefully examined the spectra of about four hundred samples, containing different particle size classes, collected at various sites in California.[4] We have also made extensive use of this technique in our studies of source emission and laboratory produced particles.

Photoelectron Spectroscopy

Extensive reviews of X-ray photoelectron spectroscopy have been given in the literature[1] and therefore only a brief description of the method will be given here. ESCA or XPS consists of the measurement of kinetic energies of photoelectrons expelled from a sample irradiated with monoenergetic x-rays. The kinetic energy of a photoelectron E_{kin}, expelled from a subshell i, is given by $E_{kin} = h\nu - E_i$ where $h\nu$ is the x-ray photon energy and E_i is the binding energy of an electron in that subshell. If the photon energy is known the determination of the kinetic energy of the photoelectron peak provides a direct measurement of the electron binding energy, which is the main observable in this type of spectroscopy.

Because of the low energy of photoelectrons produced by Mg or Al Kα x-rays, which are most commonly used as photon sources, the effective escape depth for their emission without suffering inelastic scattering is small. Recent studies[5] have given an electron escape depth of 15 - 40 Å for electron kinetic energies between 1000 and 2000 eV. This renders the ESCA method especially sensitive to the surface conditions of solids.

The electron binding energies are characteristic for each element, which enables the method to be used for elemental analysis. The binding energies are not, however, absolutely constant, but they are modified by the valence electron distribution, so that the binding energy of an electron subshell in a given atom varies when the atom is in different chemical environments. These differences in the electron binding energies are known as the "chemical shift".

In the early stages of photoelectron spectroscopy it was realized that the chemical shifts can be related to the oxidation state of the element studied. Subsequent studies have shown that the electron binding energy shifts are correlated to a high degree with the effective charge which the atom possess in the molecules. Therein lies the usefulness of chemical shifts in the analysis of unknown molecular structures. The chemical shift can be adequately described by using a simple electrostatic model in which the charges are idealized as point charges on

[1] Superior numbers refer to similarly-numbered references at the end of this paper.

atoms in a molecule and the electron binding-energy shift relative to the neutral atom is equal to the change in the electrostatic potential, as experienced by the atomic core under consideration, resulting from all charges in the molecule. This model predicts a practically linear relationship between the binding energy shift and the effective charge. In short, the binding energies will be greater than the ones for the neutral configuration for positive effective charges, i.e. for oxidized species. Similarly the binding energies will show a negative shift for reduced species.

Good correlations between the estimated charges, based on the electrostatic model, and measured binding energy shifts have been obtained for a large number of compounds. The existence of this type of theoretical and experimental background facilitates our task in identifying the species in aerosols.

The ESCA technique can therefore be used for both elemental analysis and for determination of chemical states. We describe here the factors that govern the sensitivity of the method, for analysis of both homogeneous and heterogeneous solid samples.

The number of photoelectrons emitted per unit time from a homogeneous solid sample is

$$n = N \cdot S \cdot \ell \cdot \phi \cdot s$$

where N = the volume concentration of the given element (cm^{-3})

S = the sample area irradiated by the x-ray beam (cm^2)

ℓ = average thickness from which photoelectrons are emitted (cm)

ϕ = x-ray flux ($cm^{-2}sec^{-1}$)

s = photoelectric cross-section for given element and shell (cm^2)

The number of electrons detected, n_d, is less than n, by the fraction f :

$$n_d = f \cdot n$$

and $f = t \cdot \omega$

where t = transmission (solid-angle) factor of spectrometer

ω = factor describing the non-isotropic (energy dependent) angular distribution of photoelectrons with respect to the x-ray beam direction.

Some of these factors (S, ϕ, t) are constants of the apparatus and others (s, ω, ℓ) are functions of energy and Z, and thus depend on the details of the particular experiment.

Grouping these we write

$$n_d = K.F$$

where K = constant of the apparatus, and

F = elemental sensitivity.

Elemental sensitivities for all elements of interest can be determined with use of compounds of known stoichiometry.[6] In this way relative atomic concentrations of various elements in the sample are easily obtained. For practical purposes in the analysis of ambient samples the relative atomic concentrations are combined with actual (absolute) Pb concentration determined by, for example, X-ray fluorescence (XRF), gravimetric factors for the elements involved, flow, rates etc., yielding the concentrations of elements and chemical species in usual units of mass/volume of air. The assumption was made in the preceding discussion that the sample is reasonably homogeneous, i.e., that the atomic concentrations of species being analysed are the same throughout the active sample volume. If, however, there is a preferential concentration of an element on the sample surface, then the concentration of this element determined by the described procedure will be higher than the total (bulk) concentration (if lead to which the results are normalized is preferentially a bulk species). We will be returning to this question later in the text.

Let us illustrate the surface sensitivity of ESCA using a simple system consisting of a flat aluminum metal surface covered with a thin (≈10 Å) layer of aluminum oxide. The sample can be rotated about an axis changing thus the electron escape angle Θ (insert in Figure 1). A consideration of this arrangement shows that the effective sampling thickness, determined by the mean escape depth of electrons in the sample material, decreases with Θ. Therefore low escape angle measurements should yield much enhanced surface sensitivity, with perhaps only a few monolayers contributing essentially all of the photoelectron signal.[7,8]

In figure 1 we show the Al (2p) spectra from the aluminum specimen with a surface oxide layer measured at different angles Θ [8]. The chemically shifted oxide and metal peaks are clearly resolved. It is clear that the metal peak (representing the "bulk" species in this case) essentially disappears at low angles, with a corresponding increase in the relative intensity of the peak originating in the surface oxide.

Clearly, in the case of a non-homogeneous sample, the ESCA results will reflect the concentrations in the surface region. Advantage can be taken of this property by combining ESCA determinations with those made by bulk sensitive method such as XRF or a microchemcial analysis, so as to obtain information about surface species.

ANALYTICAL APPLICATIONS

Chemical States of Sulfur in Ambient Particulates.

The sulfur ESCA spectre of ambient samples, collected during the California ARB sponsored Aerosol Characterization Study (Summer 1972), were of varying degrees of complexity, sometimes covering the entire range of known sulfur binding energy shifts.[9] Individual binding energies, characteristic of definitive sulfur species were determined by a comparative study of large number of specimens. These binding energies are assigned to characteristic chemical states, with the help of ESCA results obtained with a number of simple sulfur compounds and with SO_2 and H_2S adsorbed on various solids. It is shown that the ambient particulate-sulfur spectra can be explained in terms of SO_4^{--}, SO_3^{--} and S^{--} ions, neutral sulfur, and surface bonded SO_2 and SO_3. Thanks to the measurable differences in binding energies between the bulk-type ions and the surface species, these can be distinguished in favorable cases.

In Figure 2 some representative sulfur 2p ESCA spectra are displayed in order of increasing complexity. Let us examine these spectra in more detail. Spectrum I consists of a single, well defined peak occurring at a binding energy of 169.2 eV. This peak is easily recognizable in all spectra (I through VII). This peak is denoted by A in Figure 2 and is assigned to the sulfate ion (see below). Spectrum II reveals an additional peak at a binding energy higher than that of the sulfate, which we denote as B. This peak must correspond to a sulfur species with a higher positive effective charge than that in SO_4^{--} ion. Peak B is also apparent in spectra VI and VII. An indication of the presence of a component (D) at a binding energy of about 167 eV is found in spectrum VII: namely, if the high binding energy group consisted only of peaks A, B and C, then one would expect a much more pronounced valley between this group and the low binding energy group, centered around 162 eV. We tentatively place component D at 167 eV.

Components A, B, C and D represent various oxidized forms of sulfur seen in ambient aerosol samples. In addition the occurrence of reduced sulfur species is seen in spectra V, VI and VII. By way of contrast we refer to spectrum IV, in which no trace of reduced sulfur is visible. Peak E with a binding energy of 164.2 eV is clearly seen in spectra V and VII. The same peak is only partially resolved in spectrum VII, where in addition two states, even more reduced, are evident. These are marked by F and G.

In order to assign these individual sulfur(2p) binding energies we have determined the sulfur binding energies of several common sulfur compounds, by using the same spectrometer and the same experimental procedure used for the ambient samples. These values are indicated in Figure 2 together with the value of elemental sulfur. Peaks A, D, E, and F of ambient spectra are therefore assigned to SO_4^{--}, SO_3^{--}, S° and S^{--} respectively. As a result of our measurements of SO_2 sorption on metal oxides[10], peaks C and B are assigned to SO_2^- and SO_3^- type surface species respectively.

In some ambient spectra, like the one in Figure 2-VII, an indication is found for the existences of a sulfur form (peak G) posessing a more negative charge than that of the common metal sulfides such as ZnS. Shifts between different sulfides can be expected because of the corresponding differences in electronegativities. We therefore assign peak G of the ambient spectra to an extremely negative sulfidic sulfur involving cations with lowest electronegativity.

Naturally not all of these seven species occur generally at all times and at all locations. In most instances, sulfates are found to be the dominant species, although concentrations of reduced forms of sulfur were at some locations comparable to the sulfate concentrations.

From these results stem implications of significance to aerosol chemistry. For example, finding that sulfur species other than sulfates are often present in significant quantities makes it necessary to reexamine certain analytical methods for sulfate determination. Some of these methods implicitly presume that the sulfates are the only sulfur compounds present in the aerosol samples. Methods based on the reduction of sulfates to H_2S are in this category.

It is therefore desirable to account separately for sulfates and non-sulfates. Our experiments indicate that ESCA provides a practical way to accomplish this objective. Care has to be taken, however, in interpreting the relative concentrations determined by ESCA because of the semisurface nature of the method, and therefore it is important to perform bulk analysis in addition to examining the surface, as a check on the method.

Recently Appel et al.[11] reported the results of an interlaboratory comparison of sulfate analysis by ESCA and by wet and microchemical methods performed on the same samples. Comparison of routinely obtained ESCA results with other (bulk) analytical techniques has demonstrated an agreement within a factor of two or better for sulfates. In selected individual samples the agreement was within about 20%.

As an illustration Figure 3 shows the diurnal variation of sulfates, obtained by ESCA and by a microchemical procedure (SRI), covering a 24 hour period at Pomona, CA. during a moderate smog episode. The ESCA spectra reveal practically only sulfates (and possibly small amounts of other oxidized forms of sulfur) with negligible amounts of reduced sulfur species. The intercomparison is therefore valid because of the nonselective nature of the microchemical procedure. The similarity between the two sets of data is obvious.

An additional comparison was made using 24 hour Hi-Vol filter sulfate data (barium chloride turbidimetric procedure) with the integrated ESCA results on 2 hour low-volume filters, collected through the same 24 hour period. Table I summarizes the results. The ratio of means of 1.0 supports the equivalency of ESCA to the barium chloride prodedure.

This agreement exists in spite of the surface nature of ESCA. It is therefore tempting to conclude that sulfates and lead, to which all ESCA measurement are normalized, are homogeneously distributed through the particle volume. To draw any definitive conclusions about the origins and causes of this apparent homogeneity would be, however, premature at this stage.

Chemical States of Nitrogen in Ambient Particulates.

A representative set of nitrogen(1s)ESCA spectra of ambient particulates is seen in Figure 4 I-III. Nitrate and ammonium peaks are recognizable in addition to a group appearing at lower binding energy, and therefore possessing a more negative net charge on nitrogen, than ammonium. This group appears to consist of at least two distinct peaks as evidenced by the spectrum in Figure 4-III. These negative nitrogen species are consistently observed in all ambient spectra and actually comprise the major fraction of the total nitrogen content.

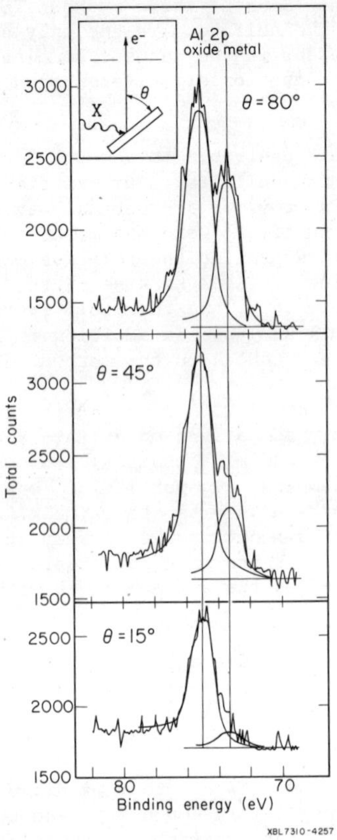

Figure 1. Al (2p) spectra from an Al speciment with a surface oxide layer at different angles Θ. The chemically shifted oxide and metal peaks are indicated. (From Ref. 8.)

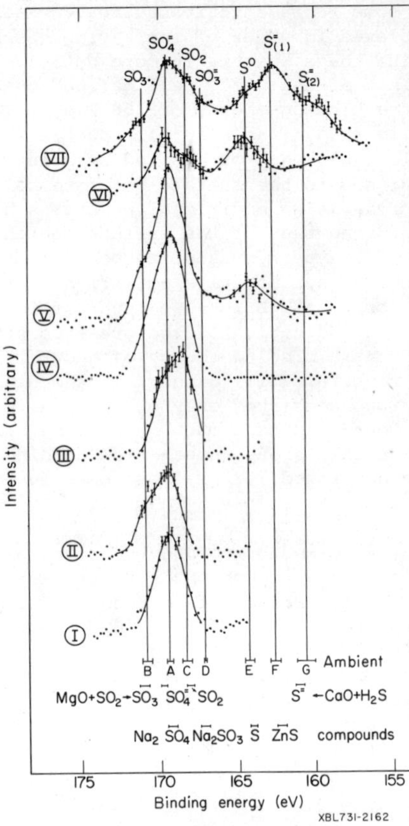

Figure 2. Sulfur (2p) photoelectron spectra of ambient pollution-aerosol samples. Sulfur 2p binding energies derived from ambient spectra are indicated together wtih sulfur 2p binding energies of some simple compounds and sulfur species produced by adsorption of SO_2 on MgO and H_2S on CaO. The binding energies of ambient samples are assigned to SO_3, SO_4^{--}, SO_2, SO_3^{--}, SO_3^{--}, S^0 and two kinds of S^{--} ions. (From Ref. 9.)

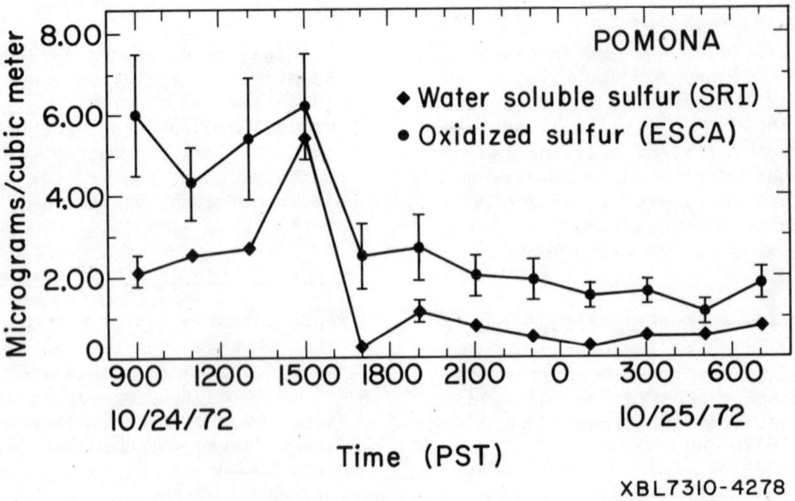

Figure 3. Sulfate diurnal variation as obtained by ESCA and by SRI microchemical procedure. (From Ref. 11.)

The nature and origin of these nitrogen species remains unfortunately unclear. In our early work[2] we have tentatively assigned them to two "organic" species such as amines nnd a heterocyclic compound similar to pyridine. This assignment was based on the correspondence between chemical shifts of the unknown species and those compounds. Our recent ESCA measurements made as function of the sample temperature show[12] that they are quite nonvolatile in vacuum up to about 350°C, and that therefore they are possibly not organic at all.

Our preliminary results on the exhaust particulates indicate that these nitrogen compounds might be produced in combustion. For example, in Figure 4-IV the nitrogen (1s) spectrum of particulates from an isooctane burning experimental internal combustion engine exhaust is shown.[13] The binding energy of this peak corresponds to the "strange" nitrogen in ambient samples.

In the previously mentioned interlabotatory comparison[11] total reduced nitrogen as determined by ESCA was compared with ammonium determined with wet chemistry. Lacking a suitable microchemical method for ammonium ion, integrated ESCA data on 2 hour filters were compared to the corresponding 24 hours Hi-Vol filter analysis by the indopehnol blue method. While the data are limited, the results indicated relative good agreement between the two sets of determinations. However, ESCA results show that the total reduced nitrogen cannot be equated with ammonium, we have therefore reasons to suspect the selectivity of the indophenol blue method.

The only serious disagreement between the results of ESCA and wet chemical analysis was found for nitrate. The ESCA results are low by about a factor of 5 with respect to the results obtained by xylenol procedure on the same samples. A likely source of the disagreement is the loss of volatile nitrate due to the spectrometer vacuum.

APPLICATION TO SO_2 CHEMISORPTION ON CARBON PARTICLES

In this section an outline is given of a proposed mechanism of chemisorptive formation of atmospheric sulfates.[14] This hypothesis is based on: a) ESCA measurements of ambient pollution aerosols; b) on ESCA sutdies of SO_2 chemisorption on graphite, activated charcoal and propane smoke particles, and c) on the measurement of the changes in light scattering induced by SO_2 chemisorption on propane smoke particles.

Ambient Measurements.

The measurements on ambient aerosols discussed here were made on samples collected on September 19 and 20, 1972, near the Harbor Freeway in downtown Los Angeles, CA. Our results and conclusions therefore apply to an urban atmosphere characterized by automotive and other anthropogenic pollutants, with reduced local visibility, but with low oxidant and low relative humidity levels. ESCA measurements were made on 2 hour total filters (TF) without particle size segregation, and on 2 hour after-filters of a Lundgren impactor (AF), containing mainly submicron particles. Pb concentrations, to which the ESCA measurements were normalized, were obtained by R. D. Giauque using XRF.[15] Carbon concentrations are the averages of determinations by ESCA and by a combustion technique.[16]

The sulfur spectra showed that sulfates were dominant in all cases, although in some samples reduced forms of sulfur -neutral and sulfides- were found in concentrations comparable to the sulfates.[9] In this discussion, however, we will be concerned only with sulfates.

The diurnal variations of sulfate concentrations in total and after filters obtained from a set of 2 x 12 filters covering a 24 hour period are shown in Figure 5a. Also shown are the variations of total carbon (TF) and lead (TF and AF) concentrations. The wind directions at sampling intervals are shown schematically in the same figure. An examination of Figure 5a shows that: a) there is a similarity between the patterns of sulfate and of carbon, while the lead pattern shows a different trend; b) changes in both sulfate and carbon concentrations are related to the wind direction, indicating sulfate and carbon sources that are both to some extent spatially localized; c) the difference between TF and AF sulfate trends can be explained by the influx of predominantly larger particles carried by the changed wind direction; and d) because of the changing wind direction the contribution of the Harbor Freeway is not the only significant factor in this episode. The early morning rise (0600 PST) in sulfate concentration, in addition to evidence just mentioned, and low oxidant levels, suggests that this sulfate is not produced by photochemical reacitons.

Two alternative mechanisms can be invoked to explain the similarity between the diurnal patterns of sulfates and carbon, and our data allow a choice to be made between them. Either there exists a "chemical link" between sulfates and carbon particles, or the carbon and sulfate particles were carried separately and noninteractively in the same air mass. In the latter case the sulfates, produced by photochemical and solution chemical processes, would be in the form of either H_2SO_4 or $(NH_4)_2SO_4$ or both.

ESCA measurements of ambient sulfates at elevated temperature suggest a chemical form much less volatile than liquid sulfuric acid. A detailed ESCA analysis show that the sulfates, in the samples studied, do not occur primarily as $(NH_4)_2SO_4$.

The ESCA analysis procedure is illustrated in Figures 5b, c and d. Figure 1b shows the nitrogen (1s) and sulfur (2p) ESCA spectrum of a sample of pure $(NH_4)_2SO_4$. The photoelectron peak positions (i.e. their binding energies) correspond to the ammonium and sulfate ions, respectively. The nitrogen/sulfur peak intensity ratio of about 1.4 reflects the atomic ratio of 2N/1S for this compound. Figure 5c shows the same spectral region for an ambient filter (TF 1200 PST). If all of the sulfate were present as ammonium sulfate, the corresponding nitrogen peak would appear as indicated by the dashed line in Figure 5c. In fact, the ammonium content in this, and other samples of the episode, is too low to account for the entire sulfate content. Rather, the dominant nitrogen peak corresponds to the mentioned species possessing a more negative

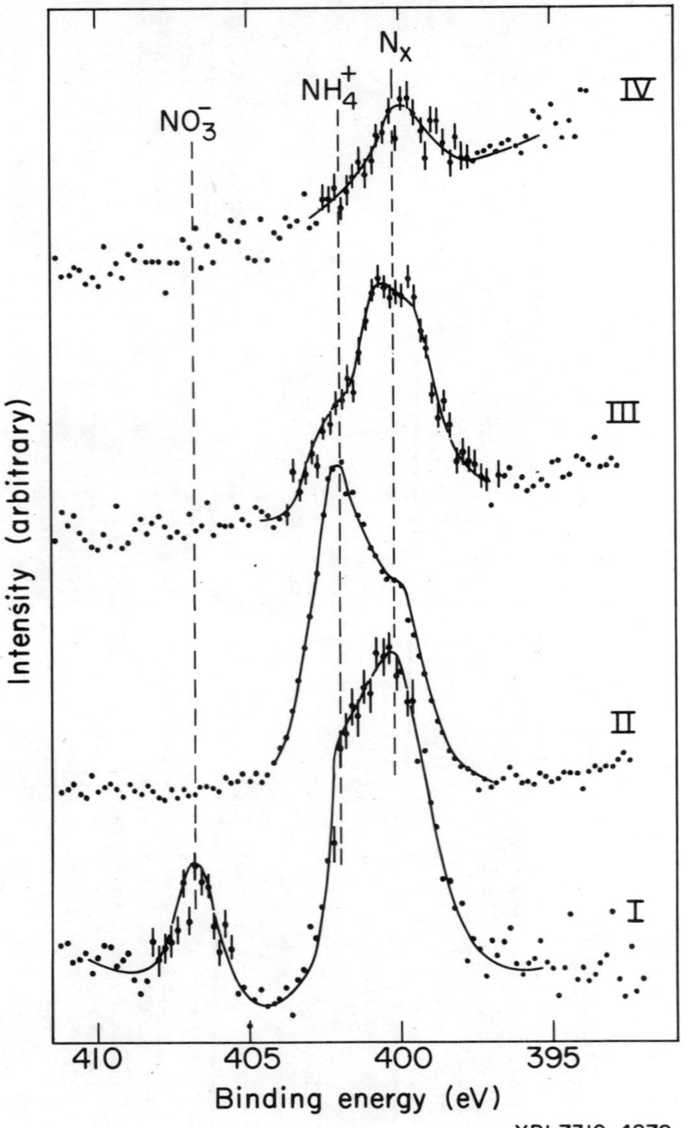

XBL7310-4279

Figure 4. Nitrogen (1s) photoelectron spectra of ambient pollution-aerosol sampler. Positions of NO_3^- and NH_4^+ peaks are indicated, together with the other negative nitrogen species (Nx).

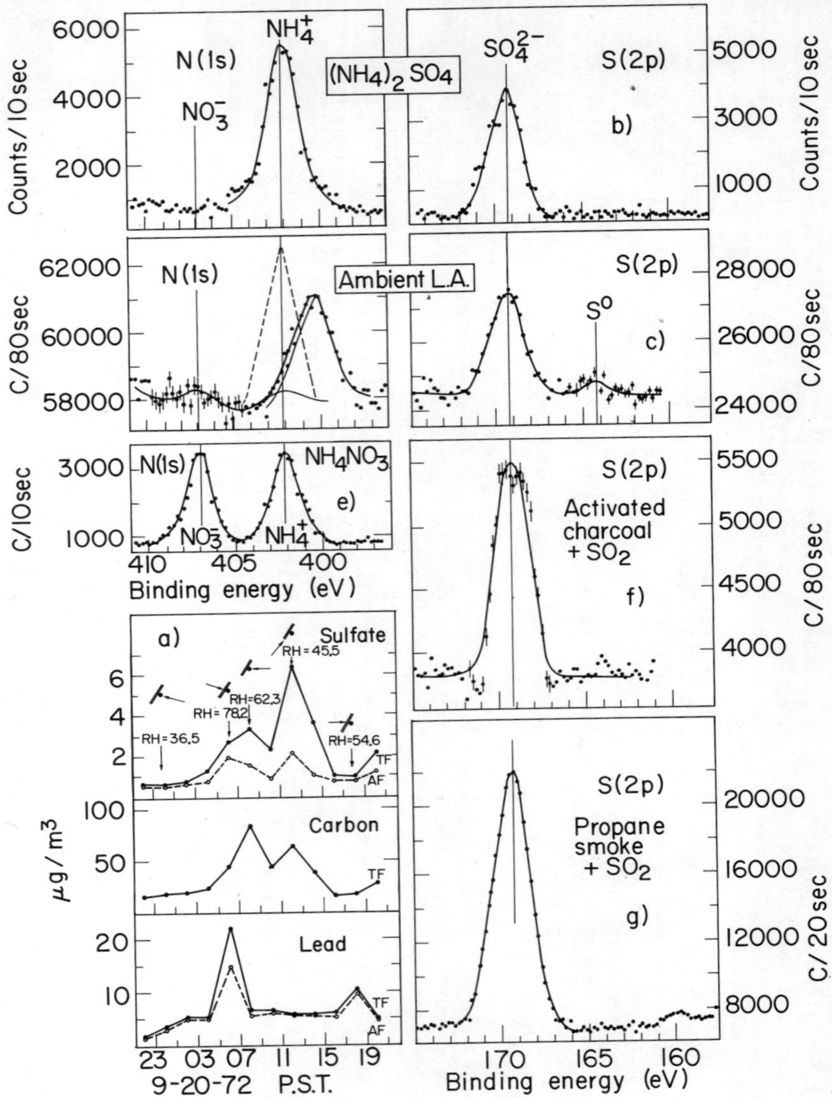

XBL736-3193

Figure 5. a) Diurnal variations of sulfate, carbon and lead. Sampling was done in Los Angeles, at the Harbor Fwy. on September 19-20, 1972. b) ESCA spectrum of nitrogen (1s) and sulfur (2p) of ammonium sulfate. c) Same region of an ambient sample (1200 PST). d) Nitrogen spectrum of ammonium nitrate. f) Sulfur spectrum of sulfate produced by SO_2 chemisorption on activated charcoal. g) Sulfur spectrum of sulfate produced by SO_2 chemisorption on propane smoke particles (generated by flow system, Fig. 6.)

charge than that of ammonium nitrogen.

Chemisorption Studies.

The results outlined thus far seem to indicate that there could be a genetic relation between carbon particles, which in the considered case are probably combustion produced, and sulfates. The question obviously is whether sulfate could be produced by chemisorption of SO_2 on carbon particles. Since our ESCA measurements on ambient samples at elevated temperatures up to 450°C) have shown that up to 80% of the ambient particulate carbon is non-volatile, probably in bulk elemental form, we have performed SO_2 adsorption measurements with anhydrous grade SO_2 on activated charcoal, propane smoke particles and graphite, representing different forms of elemental carbon.

In Figure 5f, a representative sulfur (2p) ESCA spectrum is shown of activated charcoal degassed in vacuum at 400°C and exposed to SO_2 (at STP). These data show that most of the sulfur remaining on the charcoal surface, after gaseous SO_2 has been pumped away, is in sulfate form. In some experiments, a small amount of reduced sulfur was also seen.

Similar experiments with smoke particles generated from propane were made with a flow system and SO_2 concentrations of 100 - 800 ppm. The apparatus, shown in Figure 6 was used both for sample preparation and for light scattering experiments.

Intense sulfate peaks were observed when SO_2 was introduced in the system upstream from the flame. A sulfur (2p) spectrum of a sample produced in this manner is shown in Figure 5g.

When SO_2 was introduced downstream from the flame, 6 to 8 times lower sulfate intensities were obtained for a given SO_2 concentration.

An insight into the mechansim of this surface-catalyzed conversion of SO_2 to SO_4^{--} was obtained by an ESCA study of SO_2 adsorption on single crystal graphite surfaces in UHV. Clean graphite (that is, showing no trace of oxygen) when exposed to about 10^{-6} torr-sec SO_2, revealed only sulfur, analogous to the reduced sulfur seen on some ambient and charcoal samples. On the other hand, if the graphite surface was initially oxygenated and subsequently exposed to SO_2, under conditions otherwise similar to the oxygen-free graphite, sulfate formation occurred. The oxidation (or oxygenation) was achieved by in situ exposure of a hot graphite surface to water vapor or by cutting the crystal in air to expose the fresh surface to oxygen and moisture.

Both propane-smoke and graphite experiments indicate that elemental carbon becomes efficient in converting SO_2 to sulfate when its surface is covered with oxygen, probably from surface OH^- groups. Evidence against consideration of liquid water as the cause of this conversion is provided by the flame experiments. Water condensation on the carbon particles is expected to occur rapidly downstream from the flame, where no sulfate conversion was obtained.

In addition to ESCA measurements on samples produced by the flow system, light scattering experiments were performed with an integrating nephelometer. Large increases in the scattering coefficient b_{scat} were observed between bare smoke particles and particles produced in the flow system when SO_2 entered the system upstream from the flame. Virtually no effect on light scattering was visible when SO_2 was introduced downstream from the flame. The interaction of SO_2 with particles before water condensation occurs, appears to be essential for both efficient oxidation and for an increase in light-scattering.

Scanning electron micrographs of the particulates formed in these experiments indicate that the increase in light scattering (visibility reduction) is caused by coagulation and agglomeration of primary submicron smoke particles (diameters $< 0.1\mu$) induced by surface sulfate formation.

Because of the availability of reactive carbon particles in combustion processes, it can be expected that sulfate-bearing particles are emitted into the atmosphere from primary sources, if the fuels contain trace amounts of sulfur. This was indeed verified by ESCA measurements of particulates produced by combustion of natural gas using a Bunsen burner. Intense sulfate peaks were detected for short periods of sampling (~10 min). The work on the determination of the ratio of sulfates to the total particulate mass is in progress.

Sulfate emission should occur in competition with the emission of gaseous SO_2. The SO_4^{--}/SO_2 ratio at a particular source would depend on the combustion regime, i.e. particle size, surface area, etc. Catalytic sulfate formation would also be expected to occur in the open atmosphere on reactive carbon particles. The observed strong effect of SO_2 chemisorption on light scattering indication that sulfate-coated carbon particles are, at least in part, responsible for visibility reduction in the polluted urban atmosphere. The proposed sulfate formation mechanism could occur in addition to other reactions, such as photochemical. We believe, however, that the chemisorptive mechanism outlined here may play a dominant role in urban situations that are characterized by large particulate carbon concentration. This implies that reductions of gaseous pollutants, such as NO_x and hydrocarbons in the case of automobiles, alone may not have the desired effect in improving visibility if fuels with sulfur content, as at present, are used and if no efficient primary submicron particulate reduction is achieved.

In the last paragraphs we have attempted to show, with a concrete example, how ESCA combined with other methods, may help in establishing the relation between chemical and physical properties (such as light scattering) of aerosols. It is recognized that only studies involving different physical and chemical methods may help in gaining a better understanding of the complex phenomena involved in air pollution. The purpose of this paper is to show that ESCA can be used to help achieve this goal.

ACKNOWLEDGMENTS

The results presented in this paper were obtained in

collaboration with Doctors S. G. Chang, N. L. Craig, A. B. Harker and W. Siekhaus. Reported experiments on surface sensitivity of ESCA were performed in cooperation with Professor C. S. Fadley. Discussions and close cooperation with Dr. B. R. Appel regarding the quantitative aspects of ESCA was most fruitful. All experiments described here were performed with skillful help of Mr. R. C. Schmidt. The author wants especially to thank Dr. J. M. Hollander for his encouragment and many useful comments.

REFERENCES

1. For a discussion of ESCA principles see: J. M. Hollander and W. L. Jolly, Accounts of Chem. Research, 3, 193 (1970). Specific discussion of chemical shifts of sulfur compounds is given in B. J. Lindberg, K. Hamrin, G. Johansson, U. Gelius, A. Fahlman, C. Nordling and K. Siegbahn, Physica Scripta, 1, 286 (1970)
2. T. Novakov, P. K. Mueller, A.E. Alcocer and J. W. Otvos, J. Colloid Interface Sci., 39, 225 (1972)
3. L. D. Hulett, T. A. Carlson, B. R. Fish and J. L. Durham, Proceedings of the Symposium on Air Quality, 161st. National ACS Meeting, Plenum Publ. Co., Washington, D.C. (1972).
4. Ambient aerosol samples were collected during the Aerosol Characterization Study (Summer 1972) sponsored by the Air Resources Board of the State of California
5. M. Klasson, J. Hedman, A. Berdtsson, R. Nilsson and C. Nordling, Physica Scripta 5, 93 (1972)
6. C. D. Wagner, Anal. Chem. 44, 1050 (1972)
7. C. S. Fadley and S. A. L. Bergström, Phys. Letters 35A, 375 (1971) and in Electron Spectroscopy, D. A. Shirley, Ed., North Holland, Amsterdam, 1972, p. 233
8. C. S. Fadley, R. Baird, W. Siekhaus, T. Novakov and S. Å. L. Bergström, J. of Electron Spectroscopy In press
9. N. L. Craig, A. B. Harker and T. Novakov, Lawrence Berkeley Laboratory Report, LBL-1584 (1973), Atm. Environ., In press
10. W. Siekhaus, A. B. Harker, N. L. Craig and T. Novakov, Lawrence Berkeley Laboratory Report, LBL-1583 (1973)
11. B. R. Appel, P. K. Mueller, J. J. Wesolowski, E. Hoffer, M. Fracchia, S. Twiss, S. Wall and T. Novakov, Presented at the 166th ACS National Meeting, Chicago, August 1973.
12. S. G. Chang and T. Novakov, unpublished data
13. A. B. Harker, T. Novakov and P. Pagni, unpublished data
14. T. Novakov, A. B. and W. Siekhaus, Lawrence Berkeley Laboratory Report LBL-2082 (1973) submitted to Science.
15. R. D. Giauque, unpublished data. R. D. Giauque L. Y. Goda and N. E. Brown, Lawrence Berkeley Laboratory Report LBL-1697 (1973).
16. Air and Industrial Hygiene Laboratory, unpublished data. Method described in P. K. Mueller, R. W. Mosley and L. B. Pierce, Air and Industrial Hygiene Laboratory Report AIHL-92 (1972).

TABLE I (From Ref. 11)

$\mu g/m^3$ SULFATE

ESCA Cellulose Ester Filters $\bar{\Sigma}$2 Hour Low Volume vs Wet Chemistry Whatman 41 24 Hour Hi Volume

Site	Date	Hi Vol. Wet Chemistry	$\bar{\Sigma}$2 Hr. Low Vol. ESCA	Hi Vol. / $\bar{\Sigma}$2 Hr. Low Vol.
San Jose	8-17-72	1.6±0.4	1.1±0.2	1.5±0.5
San Jose	8-21-72	1.0±0.2	1.4±0.3	0.7±0.2
Fresno	8-31-72	4.2±1.0	4.0±1.1	1.1±0.4
Riverside	9-19-72	5.9±1.5	6.4±4.6	0.9±0.7

Ratio of means = 1.0

Spearmans $\rho = 0.80$

Linear regression slope = 1.0 (intercept ≡ 0)

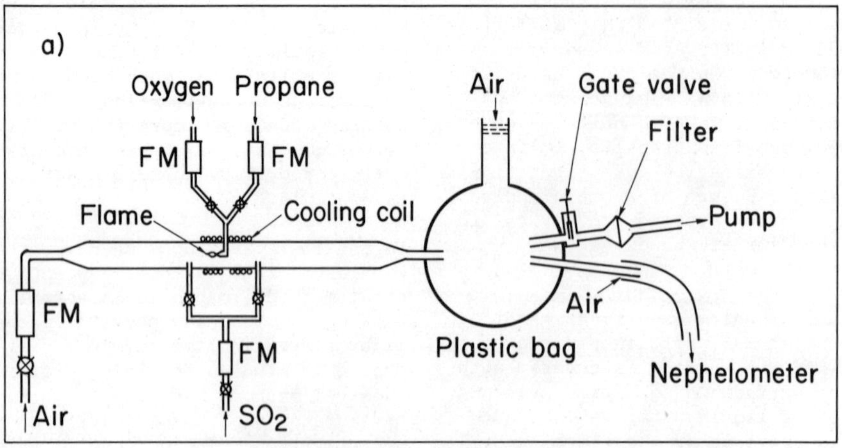

Figure 6. Flow system used for SO_2 chemisorption studies on propane smoke particles. The same apparatus was used for both sample preparation for ESCA measurements and for the study of the effects of SO_2 chemisorption on light scattering.

© 1973, ISA JSP 6694

ULTRAMICROANALYTICAL TOOLS FOR PARTICULATE POLLUTANTS

Walter C. McCrone
Walter C. McCrone Associates, Inc.
Chicago, Illinois

ABSTRACT

The successful identification of particulate pollutants requires specialized tools and techniques for sampling, fractionation, manipulation and characterization. The integration of these techniques enables the microscopist to identify almost any single inorganic particle larger than 10^{-15} g, organic particles must usually be considerably larger.

INTRODUCTION

The problem of studying particulate pollutants encompasses sampling, fractionation, characterization and identification. If we restrict our consideration to ultramicroanalytical tools we can assume that all of these operations are carried out by microscopists. This eliminates a larger number of so-called macro tools. The microscopist is not usually interested in particle-loading and, not too often, in percentage composition. He is interested in source and composition of major components and, sometimes, in specific trace components.

Sampling methods for microscopy can be more micro. Surfaces may be taped or wiped with cloth or tissue; tree leaves collect dust and may be removed for laboratory study; and finally, ambient air can be sampled with a micro vacuum cleaner. Manipulation of particles, one at a time or many together, is carried out by hand with specialized micro tools. Characterization and identification is nondestructive and involves polarized light microscopy, x-ray and electron diffraction, scanning and transmission electron microscopy, energy and wavelength dispersive x-ray analysis, electron and ion microprobe analysis and computer-interfacing, especially with the microprobes, for instrument operation and data reduction.

SAMPLING TOOLS

The microanalyst, because he is usually interested in composition, can sample by any means capable of delivering a composition-representative sample. He may also sample selectively such as picking out all colored, all magnetic or all high density particles. He may physically remove these special particles or he may select them visually in a microscopic field of view. Table I lists some of his sampling procedures.

TABLE I

Tabulation of particle collection and fractionation techniques

	Good for fractionating samples	
Good for gross sample	During initial collection	On collected gross sample
Tape	Particle-picking	Particle-picking
Snow surface	Electrostatic precipitator	Magnetic separation
Foliage	Thermal precipitator	Density gradient
Dust-fall jars	Impactor	Elutriator
Impinger	Centrifugal separation	Sedimentation
Cyclone	Cyclone	Low temperature ashing
Filter		Solvent extraction
		Sublimation
		Sieve

MANIPULATIVE TOOLS

Perhaps the least appreciated of the ultramicroanalyst's tools are his own fingers and the micro tools he uses to select individual particles and to prepare each for subsequent analysis. He must first remove the desired particle from all interfering substances, prepare it by cleaning, grinding or orienting and finally secure it with minimum adhesive to a microscope slide, glass fiber, TEM grid or polished metal surface for identification.

Tungsten Needles

Very few microscopists use mechanical micromanipulators. With practice, even submicrometer particles can be manipulated using steady hands and specially made very fine needles and pipettes. Fine tungsten needles are easily and quickly made as shown in Figure 1. Needles, ranging in tip diameters from 1-10 μm are standard in the art.

Polyethylene Needles

Occasionally, a very soft needle, Figure 2, is needed to pick up a contaminant particle without scratching the surface or picking up any of the substrate. Suitable soft needles are prepared from polyethylene as follows. A 50 mm length of 1/4 in. diameter polyethylene tubing is heated over an alcohol lamp whose heat is focused into a very narrow band by placing a metal plate with a narrow slit about 3 in. above the flame. When the tubing becomes transparent and soft, it is slowly pulled out to one third its original diameter. A tiny portion of the thinned-out tubing is then reheated and pulled out until you lose sight of the fiber. The fiber is cut, and the tip warmed for a fraction of a second well above the flame to round off the cut tip. A good needle should be about 2 cm long and 0.5 mm in diameter. To have strength, it should taper to 10 to 20 µm at the tip within the last 2 or 3 mm.

Microdroplets of Liquid

It is very important to be able to deliver tiny droplets of liquid onto any surface. There are two general methods for this. One, a glass fiber brush sealed into a capillary; and, the other, a polyethylene micropipette. Both are capable of delivering droplets as small as 50 µm in diameter.

Micropaintbrush

Figure 3 shows a bundle of 20 to 30 glass fibers, 5 to 8 mm long and about 20 µm in diameter, sealed with a microflame into one end of a glass capillary. A melting point capillary 1 to 2 mm in diameter and open at both ends is convenient. To use for small drop deposition, the capillary is partly filled (2 to 3 cm long column of the desired liquid) so that the sealed-in end of the glass fiber bundle is fully immersed. Droplets can then be "painted" onto any surface by touching the glass fiber "brush" to the desired spot. Drop size depends on the number and size of glass fibers, the liquid head in the capillary, the splaying of the fiber brush on the surface and any movement of the brush across the surface.

Normally, the glass fiber brush is used with pure liquids because evaporation concentrates a solution on the brush and may clog the fibers. This effect can be alleviated if the solution-filled capillary is stored with its brush tip in a bottle or test tube containing a little of the appropriate solvent soaked into a cotton ball at the bottom as shown in Figure 3.

Polyethylene Pipettes

These pipettes will deliver drops of organic solvent as small as 50 µm in diameter on any surface. They are unbreakable and so flexible and soft that they do not scratch even the softest surfaces. Polyethylene is inert and does not dissolve or swell in most organic solvents. They are made as shown in Figure 4.

Details of the operations used to make these tools and details of their use in small sample handling appears in Particle Atlas Two[1] from which most of the material in this paper was adapted. Those interested in these tools and techniques will find complete details in the 1127 pages of this four volume encyclopedia.

CHARACTERIZATION AND IDENTIFICATION

Small particles have an abundance of chemical and physical parameters that characterize and identify each one uniquely. A reasonably complete list of these parameters includes:

Chemical Properties

Elemental composition

Major elements	Chemical homogeneity
Trace elements	Microchemical tests
Molecular formula	

Physical Properties

General

Solubility	Magnetism
Vapor pressure	Piezoelectricity
Melting point	Ionization potentials
Boiling point	Oxidation potentials
Density	Mass/charge ratio
Hardness	Nuclear magnetic resonance
Molecular weight	Emission spectra
Fluorescence	Absorption spectra

Crystallographic (morphology)

Crystal form, habit and faces	Twinning
Crystal system	Cleavage

Crystallographic (optical)

Color (reflection, transmission)	Optic axial angle
Pleochroism	Dispersion (of indices, birefringence, optic axial angle)
Refractive indices	
Extinction	

Crystallographic (x-ray diffraction)

Powder data	Space group
Lattice spacings (a,b,c)	Line broadening
Lattice angles (α, β, γ)	Asterism

Not all of these properties can be measured on single small particles. Some require larger samples, great skill or expensive equipment. There are also other criteria on which to base selection of a method. It is difficult, in any case, to set down a series of rules or steps generally applicable to any analytical problem. The question is complex and should be placed in the hands of the supervisor of an instrumental analytical section. Obviously, he must know the capabilities and limitations of each tool and he must be free to apply any or all of the tools at any time on any problem.

"Superior numbers refer to similarly-numbered references at the end of this paper."

The choice of tools to use for a given analytical problem depends in the final analysis on which are available. If the laboratory is equipped with all or most of the tools as well as personnel trained in their use, the choice then is based on scope and limitations of each technique. The choice is not a simple one.

Most analyses begin (and often end) with the polarizing microscope (Figure 5). Many samples are small, e.g., contaminants in hydraulic fluids, photographic film emulsions, parenteral solutions, fuel nozzles, artificial fiber spinnerets etc. A cursory examination by an experienced microscopist will often identify the foreign material. If not, he can at least tell if the sample is homogeneous, crystalline, biological, organic, metallic etc. These observations help greatly in selecting the analytical tools to be used on that sample. Furthermore, if the sample is tiny, the microscopist will be able to isolate it and position it on a glass fiber for x-ray diffraction, on a polished beryllium plate for the electron or ion microprobe, on a rock salt plate for infrared absorption, or even on an electron microscope grid for microscopy or one of the microprobes.

Again, the choice of which tool to use depends on knowledge of what each tool can accomplish. Some major considerations are:

1. Destructive tests. It is usually desirable, if not necessary, to use only nondestructive tests. This is desirable:

   a. when you're not certain a given (destructive) test will give the needed answer;
   b. when the sample is unusually valuable or is one on which referee analyses may be necessary later. Particulate evidence in court cases might be one example; single radioactive fallout particles or single micrometeorites, two others;
   c. when the analysis must not destroy a functioning mechanical, electrical or optical part, e.g., an integrated circuit or an expensive inertial guidance system.

2. Compound vs ion identification. It may be sufficient to know the identity and percentage of one or more ions in a sample or one may wish to know what compounds are present and in what quantity. Ionic impurities in a semiconductor or a fluorescent material may be the problem, or identification and measurement of a sample in terms of the compounds present may be necessary.

3. Characterization of the solid state. It is not always sufficient to be able to say what compound is (or compounds are) present. The crystalline form may grossly affect the behavior of samples otherwise identical chemically. Graphite and diamond are both carbon but graphite is not a good abrasive nor is diamond a good lubricant. One crystal form of a drug may be much more efficacious than another form of the same drug, e.g., chloramphenicol palmitate, novobiocin, or methylprednisolone. The three crystal forms of titanium dioxide — rutile, brookite and anatase — vary greatly in their usefulness as paint pigments. In these and many other situations the analysis is incomplete unless the nature of the solid state is specified.

4. Organic or inorganic sample. Some analytical tools are best, or at least most useful, for organic samples, e.g., spectrophotometry and nuclear magnetic resonance (NMR). Others are best for inorganic samples, e.g., electron microprobe, emission spectroscopy, atomic absorption and neutron activation. A few are equally at home with either organic or inorganic samples, e.g., diffraction methods, optical crystallography and the ion microprobe.

5. Trace or ultramicroanalysis. The analysis to be performed may be a trace analysis (ppm or ppb concentrations, usually in a large sample) or ultramicro (a very small but reasonably pure sample). For example, we may wish to determine the trace amount of germanium in semiconductor silicon or, on the other hand, we may be trying to identify a single picogram particle of asbestos. Some tools are much less useful for trace analyses, e.g., microscopy and diffraction; on the other hand, the ion microprobe, emission spectroscopy, or neutron activation might be applicable. Other tools are ideal for ultramicro samples, e.g., diffraction, microscopy and microprobes.

6. Sample size. The minimum sample size that can be analyzed ranges from attograms (10^{-18} g) for the ion microprobe and selected area electron diffraction to microgram (10^{-6} g) for infrared absorption (and even larger for NMR).

7. Background data required. Some tools require that the literature have reference data for the compound to be identified, e.g., x-ray or electron diffraction, optical crystallography, and, to a somewhat lesser degree, absorption spectroscopy. Other absolute methods require no published tables of data for the compounds to be identified, e.g., mass spectroscopy, microprobe analyzers, emission spectroscopy and neutron activation. In general, the latter tools give an elemental analysis only; the stoichiometry may or may not then be apparent.

I have tried to outline some of the main considerations in selecting the proper tool for a given analytical problem. I realize that perhaps no laboratory except McCrone Associates has such a complete range of ultramicroanalytical instruments but a knowledge of what they can do shows, at least, that problems unsolvable 10 or less years ago can now be routinely solved.

ULTRAMICROANALYTICAL TOOLS

A brief discussion of scope and limitations for some of the major ultramicroanalytical tools is also helpful.

Polarizing Microscope

The only tool able to differentiate between different forms of the same chemical substance, e.g., silica as α, β and δ-quartz, α and β-cristobalite, α, β_1 and β_2-tridymite, coesite, keatite, amorphous silica, diatoms, sponge spicules, stishovite, melanophlogite, lechatelierite, chalcedony, pumice and opal. It is most generally useful for inorganic substances although biological organic substances can usually be identified. Particles in the pico- to nanogram range are the lower limit but identification often requires great skill in optical crystallography. The particle must usually have been described before so that the necessary background data (refractive indices, color, morphology etc.) are available for comparison.

Scanning Electron Microscope

With much better resolution and depth of field than the light microscope SEM can produce beautiful micrographs of particles 10 times smaller. It does, however, only "look at" the surface and requires elemental analysis capability to be generally useful. The latter, usually by energy dispersive x-ray analyzer, can be used down to about oxygen in the periodic table but good quantitative data are hard to obtain because a pattern of a single small area is not obtained. The electron beam bounces around the sample chamber and produces unwanted x-rays. Nearly every EDXRA pattern, for example, in Volume III of Particle Atlas Two, shows iron from the pole pieces in the SEM. Nearby particles in the sample will also contribute x-rays. Determining a reliable Si:Mg ratio of an asbestos fiber in an air pollution sample would not be possible.

Transmission Electron Microscope

The TEM shows a silhouette of any particle in the field. If thin enough, about 50 nm, surface and internal detail may be seen. If replicated, surface detail on larger particles is visible. Generally again, morphology is insufficient for the identification of small particles by TEM. Where the light microscope gives us optical properties and SEM gives us EDXRA, TEM gives us selected area electron diffraction (SAED). These patterns are obtained in seconds and can be measured for comparison with the ASTM diffraction data file(2). Particles as small as 20 nm in diameter will give excellent patterns.

X-Ray Diffraction

Not normally considered an ultramicroanalytical tool XRD will, with modifications described in Particle Atlas Two, give excellent diffraction patterns on samples measuring less than 5 µm in diameter and weighing less than 10^{-10} g. The limitation of the method is the possibility that the x-ray pattern for the particle analyzed may not be included in the ASTM data file.

Electron Microprobe Analyzer

EMA (Figure 6) has wavelength dispersive as well as energy dispersive x-ray analyzers and this makes possible better quantitative analyses, greater sensitivity and better sensitivity for elements, beryllium through sodium than the SEM. EMA then can analyze for all elements except H, He, Li (and sometimes Be) with sensitivities of at least 10^{-14} g with an accuracy of, at least, ±5%.

Ion Microprobe

This ion-sputtering mass spectrometer can detect a few thousand atoms of, and determine isotopic abundances for, any element in the periodic table. It is, in a sense, destructive but it consumes something like 1 μm^3/hr during which time dozens of complete quantitative analyses would have been completed. It is obviously useful for successive analyses as a function of depth. A good example of its sensitivity is the analysis of the vapor within 10 µm gas bubbles in glass or minerals. The analysis is made in the few seconds it takes the gas to leak out of the bubble when it is pierced by the ion beam. IMA is limited in its ability to make quantitative analyses since the ionization efficiency varies so much from element to element and varies with matrix composition.

Electron Microscope-Microprobe Analyzer

EMMA is a combination high resolution (1.0 nm) transmission electron microscope and a two-spectrometer electron microprobe. It gives both SAED and EMA data on any particle down to about 10^{-15} g. This very powerful tool therefore permits both diffraction data and elemental analysis on particles below the resolution limit of the light microscope. It is an ideal instrument for the identification of any single particle in a general sample, e.g., a chrysotile fiber in lung tissue.

Photoelectron Spectroscopy

Also called electron spectroscopy for chemical analysis (ESCA) this tool differs from all of the others so-far listed because it analyzes nondestructively organic or inorganic surface layers only 1.0-2.0 nm thick. Although it has some difficulty with the transition elements it not only covers the entire periodic table but measures shifts in binding energy for each given element thus, potentially at least, determining chemical composition. The sensitivity is of the order of 10^{-12} g but it must be spread in a thin, say 2.0 nm, layer over an area of about 1-5 mm depending on the depth of the layer of sample atoms generating photoelectrons.

Several general limitations of tools such as these are: (a) high cost (The aggregate cost of these tools as used at McCrone Associates was in excess of $1,000,000.), (b) training required by the specialized chemists and physicists (average salary in excess of $20,000 a year) and (3) the highly specialized sample handling problems. Only one of these ultramicroanalysts has the ability to pick out, clean and manipulate the samples analyzed by all of them. Without this added capability in sample manipulation much of the usefulness of these tools would be wasted.

REFERENCES

(1) McCrone, W.C. and J.G. Delly, THE PARTICLE ATLAS, EDITION TWO, Ann Arbor Science Publishers, Ann Arbor, MI (1973)

(2) X-Ray Powder Data File — ASTM

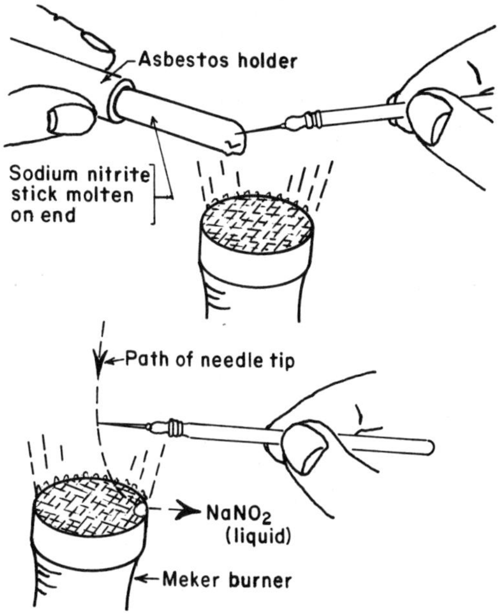

Figure 1. Preparation of tungsten needles (above); sodium nitrite stick (below) in a drop of sodium nitrite on a meker burner.

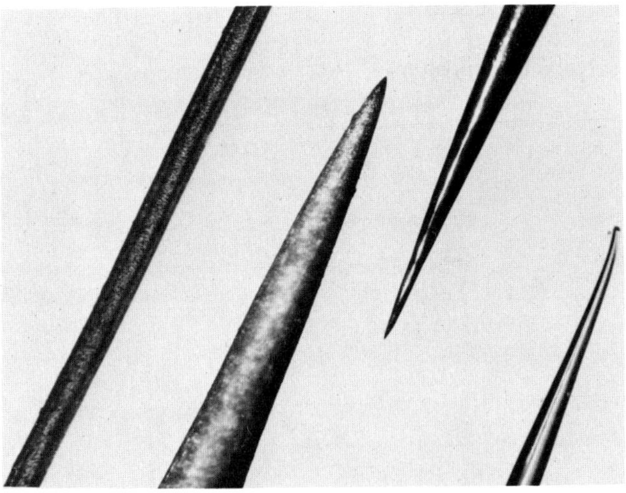

Figure 2. Human hair (left), two tungsten needles and polyethylene needle (right).

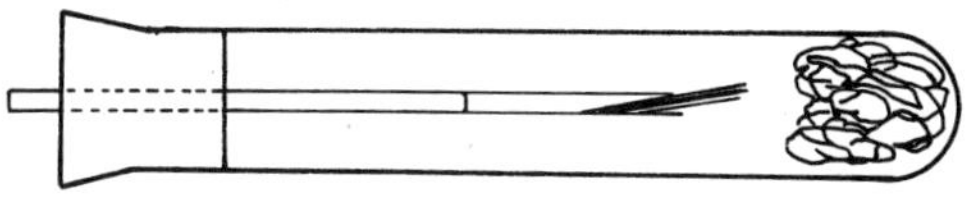

Figure 3. Protective container for glass fiber brush; a glass wool plug in the bottom of the test tube is soaked with solvent.

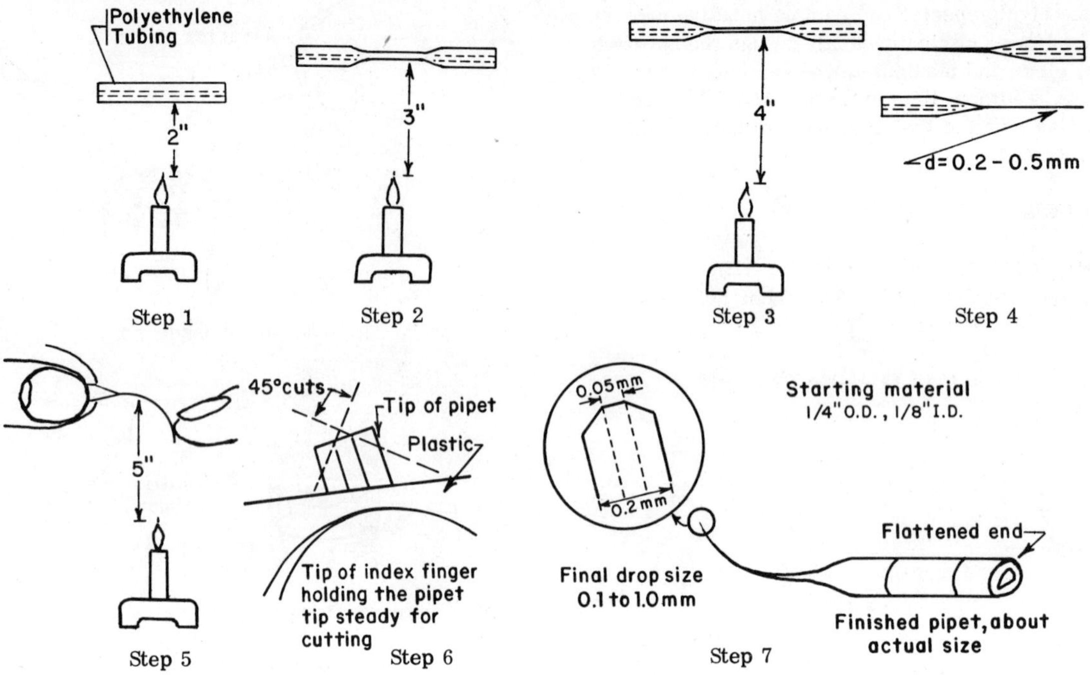

Figure 4. Steps in making polyethylene pipettes.

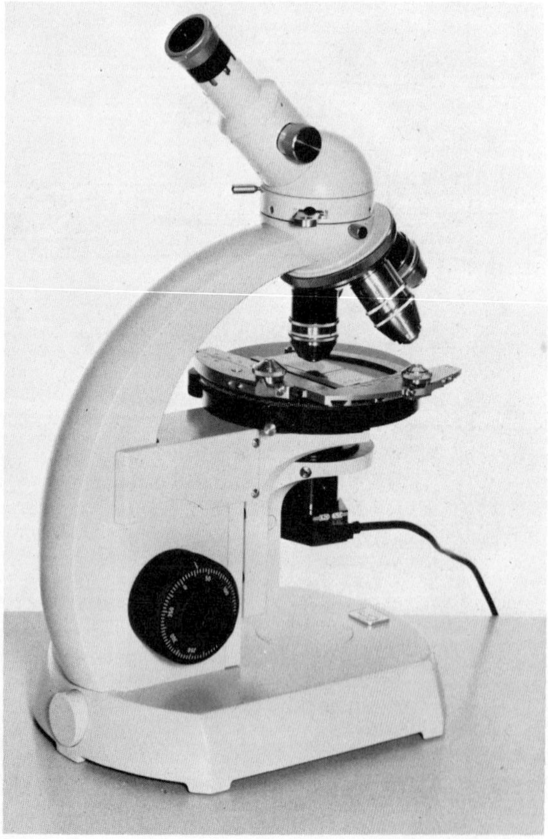

Figure 5. A typical low cost polarizing microscope.

Figure 6. The McCrone Associates electron microprobe equipped with computer interface and energy dispersive x-ray analyzer.

© 1973, ISA JSP 6695

APPLICATIONS OF X-RAY FLUORESCENCE TO PARTICULATE MEASUREMENTS

T. G. Dzubay and R. K. Stevens
Chemistry and Physics Laboratory
National Environmental Research Center
Environmental Protection Agency
Research Triangle Park, N. C. 27711

ABSTRACT

A system for determining the composition of atmospheric aerosol according to size as a function of time is described. Central to the system is an x-ray fluorescence spectrometer for determining the elemental composition of particulate matter collected on filters. In order to optimize the sensitivity for elements with atomic numbers above 12, each sample is analyzed by exciting it with characteristic x-rays from each of the secondary fluorescers Cu, Mo, and Tb. Samples have been collected in St. Louis using a device which manipulates up to 36 filters into an air stream for preadjustable collection periods. Particles were also collected in St. Louis using a new dichotomous sampler which uniformly deposits aerosol in the 0 to 2 μm and 2 to 10 μm size ranges on two separate filters suitable for x-ray analysis. The elements S, Mn, Zn, Br, and Pb were observed to occur mainly in the small particles, whereas most of the Si, Ca, Ti, and Fe occurred in the large particles.

INTRODUCTION

A major task in the field of air pollution monitoring is the development of devices for determining the mass and composition of airborne particutate matter as a function of size and time. In order to perform elemental analysis on an increasing number of samples in more than one size range, a rapid analytical procedure is required. The use of x-ray fluorescence spectroscopy is well suited to meet this need. A wide variety of elements can be determined in a few minutes without sample preparation. The composition of the sample is determined by irradiating with a beam of x-rays and observing the emission of characteristic K and L x-rays. For a number of years crystal grating spectrometers have been used to distinguish between the various wavelength components in the emission spectrum. Recently Birks and Gilfrich have evaluated multicrystal spectrometers which can simultaneously analyze up to 24 elements with excellent detection limits on air particulate samples[1].

Developments in high resolution - low background silicon detectors and electronics by Goulding et al[2] have greatly improved the alternative technique of analysis of samples by using solid state x-ray energy spectrometers. With this method the sample can be excited with beams of charged particles, x-rays from radioactive sources, or x-ray tubes. The relative merits of the various means of excitation have been investigated by Cooper[3]. From considerations of convenience, speed, and detectability, the use of a monochromatic photon source for excitation offers a number of advantages. Recently Jaklevic et al have developed an efficient low power x-ray tube which produces a nearly monochromatic x-ray beam by means of a secondary fluorescer[4].

EXPERIMENTAL

Description of X-Ray Spectrometer

Goulding and Jaklevic of the Lawrence Berkeley Laboratory have developed a new x-ray fluorescence system for the EPA, which incorporates three secondary fluorescers in order to optimize the sensitivity throughout the periodic table[5]. In this system, illustrated in Figure 1, an x-ray beam from the tungsten anode excites the secondary fluorescer which in turn excites the sample with its nearly monochromatic characteristic x-rays. The three fluorescers used are Cu, Mo, and Tb, for which the corresponding anode potentials of the x-ray tube are 35, 50 and 70 kV respectively. Each sample is analyzed for 5 minutes with each fluorescer with an electron current of 400 μA in the x-ray tube. Because of the very compact geometrical arrangement of the components in Figure 1, an adequate count rate is obtained with x-ray tube power dissipation of less than 30W. The use of a small computer to control the analysis station allows the analysis of up to 36 samples with any or all of the 3 fluorescers without operator attention.

The ability of the system to detect a wide variety of elements in the atmosphere is illustrated in Figure 2, which is a plot of the 3σ detection limits[5] superimposed upon the typical values of elemental concentration and their ranges in ambient air[6]. The detection limits (3σ above background) refer to samples collected for two hours at a flow rate of 7 liters/cm^2-min on a filter with 5 mg/cm^2 of mass per unit area and are analyzed under the conditions described above.

Sample Collection and Filters

The sample collection device must be designed with consideration given to the nature of the aerosol. Recent measurements by Whitby et al show that the volume and hence the mass of aerosol in ambient air has a bimodal distribution with a relative minimum at particle diameters of about 2 μm[7]. For this reason it is desirable to collect particles in at least two size ranges. Unless specific information on the shape of the size distribution is needed however, collection in only two size ranges is desirable in order to maximize the amount of collected particulates per unit weight of collection surface.

X-ray fluorescence analysis requires that the sample be uniformly deposited on the collection surface. Conventional impactors which produce an excessively thick deposit in a pattern of spots or slits are not well suited for this application. Also, collection may not be quantitative if particles bounce off of the intended impaction surface. These problems are avoided by use of a virtual impactor in which particles are impacted into a slowly pumped void[8,9]. Figure 3 shows a dichotomous sampler (two stages) containing a virtual impactor which uniformly deposits particles smaller than 2 μm on one filter and deposits those larger than 2 μm on a second filter.

For x-ray analysis it is necessary that particles be collected on a filter with very low levels of impurities. Figure 4 shows a spectrum of a Teflon membrane filter with a polypropylene backing which is excellent in this regard. Figure 4 also shows a spectrum of a clean glass fiber filter which has unacceptably large amounts of impurities. In addition to the filter being pure, other properties required for collection in ambient air are high flow rate, low mass per unit area, ability to accept a heavy load without clogging, and high efficiency for particles as small as 0.1 μm in diameter. High porosity filters are desirable for their high loading capability. The filter mass should not depend on relative humidity if gravimetric or β-gauge mass analysis is to be performed. Teflon membrane filters make an excellent choice, but for the work reported here teflon was not available, and membrane filters made of mixed esters of cellulose and having 0.8 μm pore size were used. Except for their Ca and Cl impurities and inferior gravimetric characteristics, these filters were quite satisfactory.

Calibration and Analysis

For aerosol samples it is assumed that an observed x-ray spectrum is a superposition of characteristic x-rays of each individual element in the sample. Individual spectra of up to 40 elements are accumulated and stored in the computer memory. These individual spectra as well as a stored clean filter background spectrum are compared with the unknown aerosol spectrum using a stripping procedure in order to determine the concentration of each element [5]. By making a judicious choice of the order in which the elements are analyzed, the problem of interfering K_a and K_β x-ray lines is largely eliminated.

The analyzer was calibrated using evaporated foils of Al, SiO CuS, KCl, CaF_2, Ti, V, Cr, Fe, Ni, Cu, Zn, Se, As, Mo, Cd, Sn, Sb, and Ba ranging in thickness from 50 to 150 $\mu g/cm^2$ and known to an accuracy of about ± 10%. A smooth curve was drawn through the calibration points and used to determine elements for which foils were not available or not accurately known.

Preliminary experience indicates that air filters can be analyzed to an accuracy of at least ± 15% for elements heavier than potassium (A=39). For lighter elements the energies of the characteristic x-rays are sufficiently low that significant self absorption can take place within the larger collected particles. The self absorption corrections needed for particles smaller than 2 μm are fairly small. Sulfur from combustion sources is expected to occur in the submicron size range. If particles are collected within the volume of the filter, errors due to attenuation by the filter medium can occur. However, this effect can be minimized by using highly efficient filters which have minimal penetration within their volume.

Evaluation With St. Louis Aerosol

Samples were collected for 24 hour periods in a St. Louis residential neighborhood in late August of 1973 using the dichotomous sampler of Figure 3. Also used was a device developed at the Lawrence Berkeley Laboratory for automatically changing filters every two hours. Figure 5 shows a photograph of two filters which were simultaneously used to collect particles in the dichotomous sampler, and Figure 6 is a plot of the x-ray spectra for these two filters. Table I shows the elemental analysis which is deduced from these samples. Figures 7 and 8 show plots of the total particulate samples collected every two hours and analyzed for S, Ti, Fe, Zn, Pb, and the Br to Pb ratio.

DISCUSSION

The filters shown in Figure 5 indicate the striking difference between small and large particles. The large particles produce a nearly white deposit, and yet they have almost as much total mass as the small ones, which make a black deposit. This is consistent with the notion that the large particles contain large amounts of earthen crustal material, whereas the small particles consist of combustion and secondary aerosols. From the analysis of the two filters in Table I it appears that the large particles contain at least 75% of the Si, Ca, Ti, and Fe in the atmosphere. At least 75% of the S, Zn, Br, and Pb is contained in the small particles.

The most abundant observed element is sulfur. Certain sulfate compounds such as zinc ammonium sulfate, sulfuric acid, and ammonium sulfate are known pulmonary irritants[10]. If one makes the reasonable assumption that all sulfur is in the form of sulfate, then one can deduce the maximum possible fraction of the sulfate that is associated with the various observed cations as shown in Table II. One notes that the biologically active zinc sulfate compounds are only minor components. Aluminum sulfate is omitted from Table II because verification of results for the light element aluminum has not yet been completed. Table II suggests, however, that significant amounts of sulfate may be bound to light element cations such as H^+ or $(NH_4)^+$. Additional work needs to be done to demonstrate the accuracy of the x-ray analysis for sulfur, to determine the percentage of sulfur in the form of sulfate, and to verify that the observed sulfur is not the result of SO_2 being converted to sulfate on the filter.

The diurnal patterns of the elements plotted in Figures 7 and 8 gives valuable clues about the source of these pollutants. For example, the enormous fluctuation in the titanium concentration is indicative of a local discrete source. The lack of major fluctuations in the sulfur concentration indicates a lack of a single local or discrete source. The Br to Pb ratio and the lead concentration seems to be influenced by rush hour traffic patterns. These results illustrate the utility of applying x-ray fluorescence spectroscopy to monitoring problems and to studies of the transport of pollutants.

ACKNOWLEDGEMENTS

The authors are grateful to F. S. Goulding, J. M. Jaklevic, B. V. Jarrett, and J. Meng for valuable advice and discussions, to J. A. Cooper for providing the plot of elemental abundances used in Figure 2, to W. E. Wilson for providing space in his mobile laboratory in St. Louis to operate the sampling equipment, and W. C. Peters and C. R. Sawicki and L. E. Hines for valuable assistance.

REFERENCES

(1) L. S. Birks and J. V. Gilfrich, "Development of X-Ray Fluorescence Spectroscopy in Elemental Analysis of Particulate Matter, Part II: Evaluation of Commercial Multiple Crystal Spectrometer Instruments", EPA Report No. EPA-650/2-73-006, June, 1973.

(2) F. S. Goulding, J. M. Jaklevic, B. V. Jarrett, and D. A. Landis, in Advances in X-Ray Analysis, Heinrich et al, ed., Vol. 15, 470, Plenum Press (1972).

(3) J. A. Cooper, Nucl. Instr. and Meth. 106, 525 (1973).

(4) J. M. Jaklevic, R. D. Giauque, D. F. Malone, and W. L. Searles, See Reference 2, P. 266.

(5) F. S. Goulding and J. M. Jaklevic, "X-Ray Fluorescence Spectrometer for Airborne Particulate Monitoring, EPA Report No. EPA-R2-73-182, April, 1973.

(6) J. A. Cooper, "Review of Workshop on X-Ray Fluorescence Analysis of Aerosols", Battelle Pacific Northwest Laboratory Report BNWL-SA-4690, June 1, 1973.

(7) K. T. Whitby, R. B. Husar, and B. Y. H. Liu, "The Aerosol Size Distribution of Los Angeles Smog," J. Colloid Interface Sci. 39, 177 (1972).

(8) W. D. Conner, "An Inertial-Type Particle Separator for Collecting Large Samples, APCA J. 16, 35 (1966).

(9) The device tested in this study was developed by Dr. Carl Peterson, ERC, 3725 N. Dunlap St., St. Paul, Minnesota.

(10) M. O. Amdur, "The Impact of Air Pollutants on Physiologic Responses of the Respiratory Tract", Proc, Am. Phil. Soc. 14, 3 (1970).

TABLE I. Analysis of St. Louis Aerosol on the filters shown in Figure 5. Unless indicated uncertainties are ± 15%. Also listed is the mass per unit volume determined gravimetrically.

Element	Below 2 μm ng/m³	2 to 10 μm ng/m³
Al	(561 ± 250)	(687 ± 250)
Si	403 ± 160	1254 ± 400
S	3081 ± 700	489 ± 200
Cl	54 ± 30	400 ± 100
K	150 ± 60	213 ± 80
Ca	103 ± 40	1560 ± 400
Ti	62	192
V	4 ± 2	<3
Cr	<3	<3
Mn	5 ± 2	6 ± 2
Fe	131	395
Co	<3	<3
Ni	<3	<3
Cu	<6	<4
Zn	45	15
As	20 ± 10	<4
Se	7	<2
Br	114	30
Rb	<3	<3
Sr	<4	4 ± 2
Cd	<21	<21
Sn	<21	<21
Sb	<21	<21
Ba	<45	<45
Pb	460	110
gravimetric	26,000	19,000

TABLE II: Maximum amount of various sulfate compounds deduced from observed elemental sulfur and metal concentrations in St. Louis for particles smaller than 2 μm on filter No. 5S (See Figure 6). It is assumed that all sulfur is in the form of sulfate.

Compound	Per cent of Sulfate*
K_2SO_4	<2.0%
$CaSO_4$	<2.7%
$Ti_2(SO_4)_3$	<2.1%
$Fe_2(SO_4)_3$	<4.0%
$ZnSO_4$	<0.7%
$PbSO_4$	<2.3%

*If only fraction F of the elemental sulfur were in the form of sulfate, then these percentages are to be multiplied by 1/F.

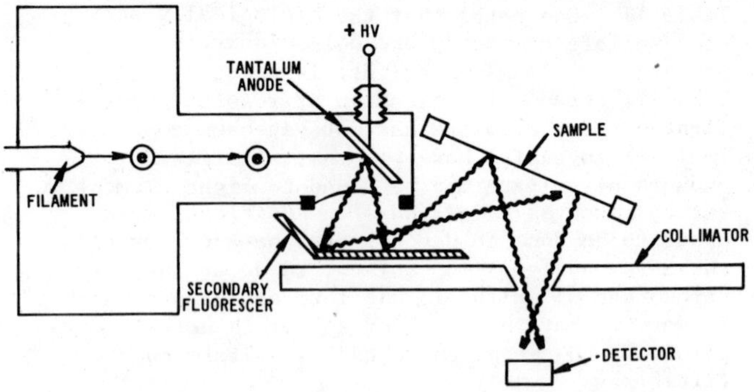

Figure 1. Schematic view of x-ray fluorescence spectrometer using a secondary fluorescer and a solid state detector.

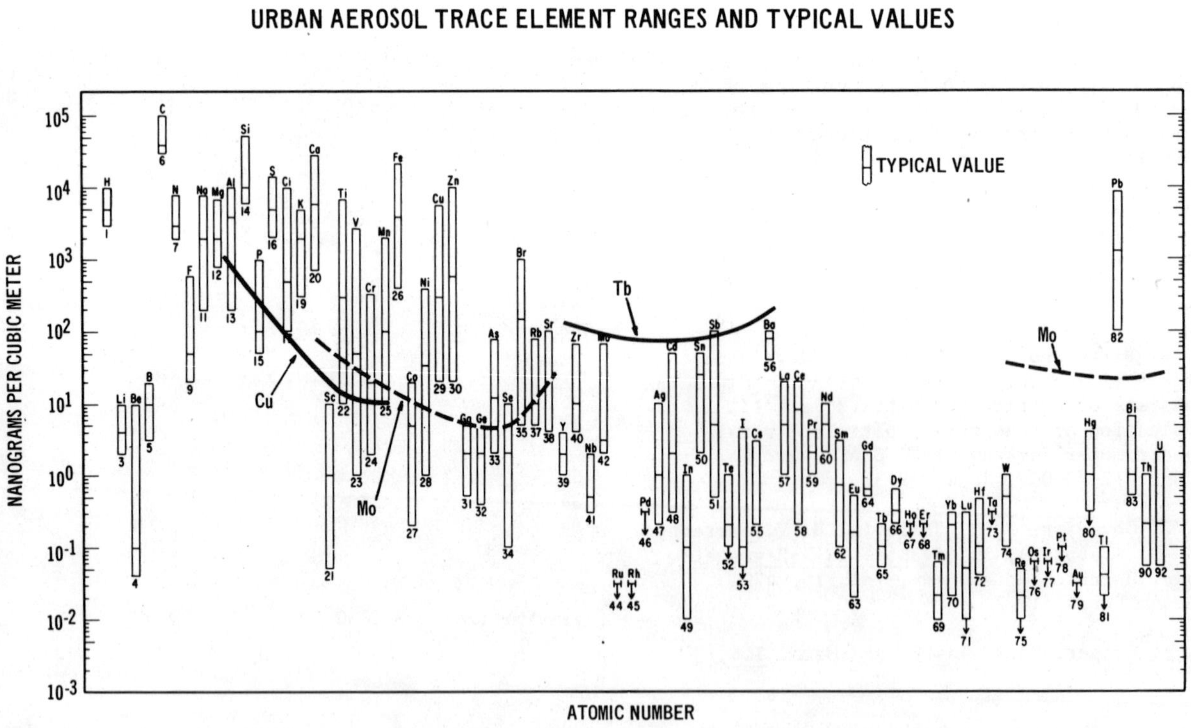

Figure 2. Detection limit curves using Cu, Mo, and Tb fluorescers plotted upon the ranges and typical values of urban trace element concentrations.

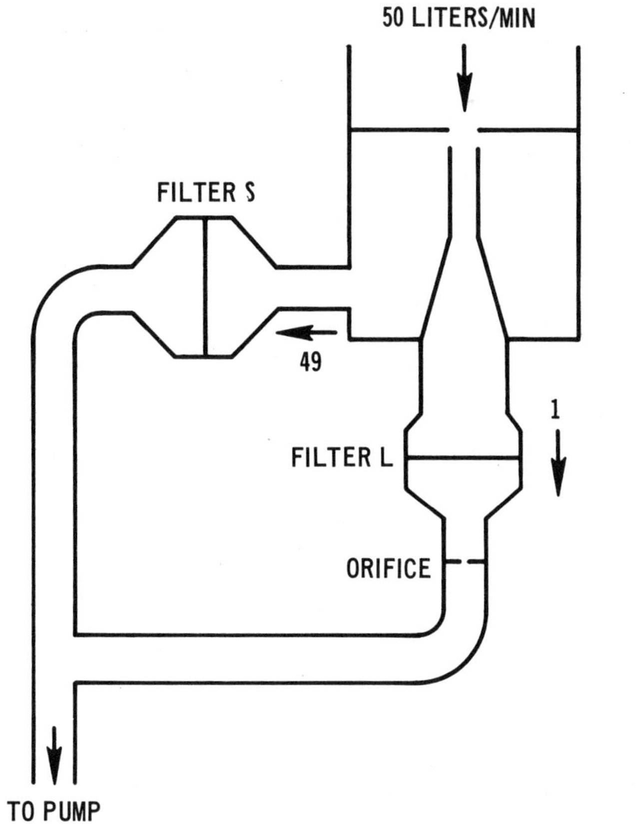

Figure 3. Schematic view of a dichotomous sampler which contains a virtual impactor. The flow rate at the inlet is 50 liters per minute, and the flow rates at the outlets are 49 and 1 liters per minute.

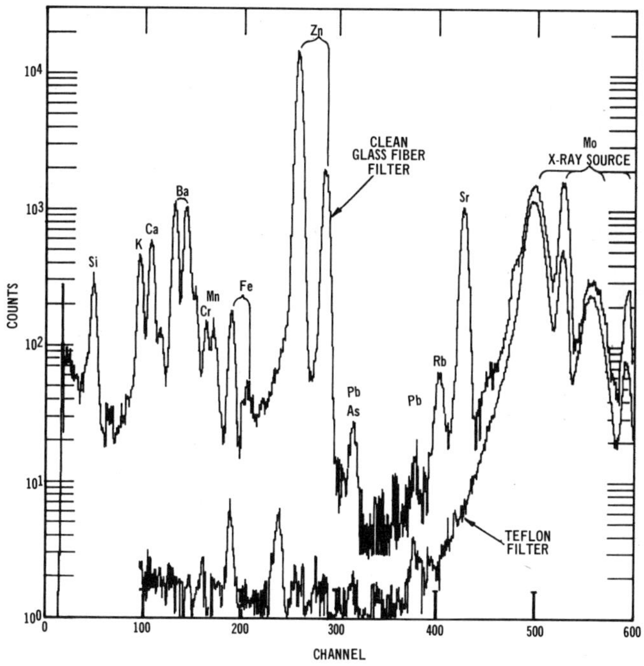

Figure 4. X-ray spectrum of a clean polypropylene backed Tefon filter and of a clean glass fiber filter using a molybdenum secondary fluorescer for excitation.

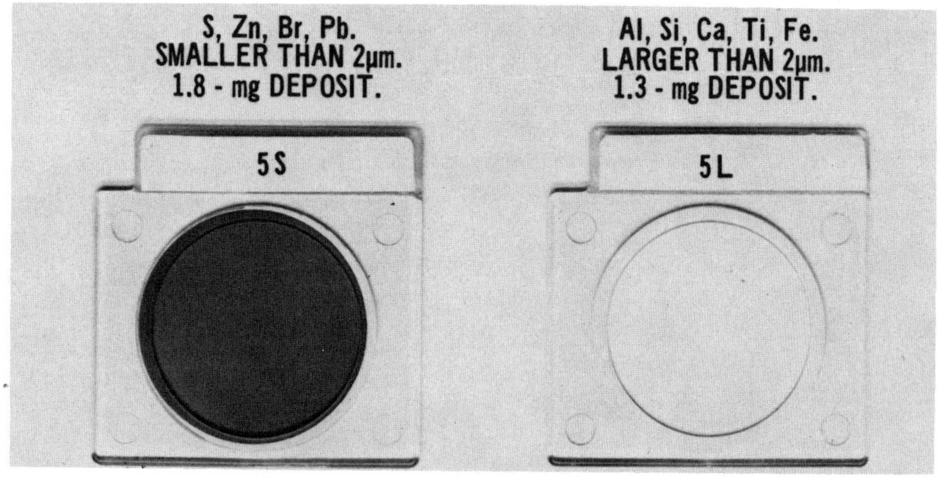

Figure 5. Photograph of filters used in dichotomous sampler for the 23 hour period beginning at 1015 hours, August 30, 1973 in a St. Louis residential neighborhood. The volume sampled is 68 m^3. The filters are mounted in 5.1 x 5.1-cm frames used for automatic manipulation within the x-ray analyzer.

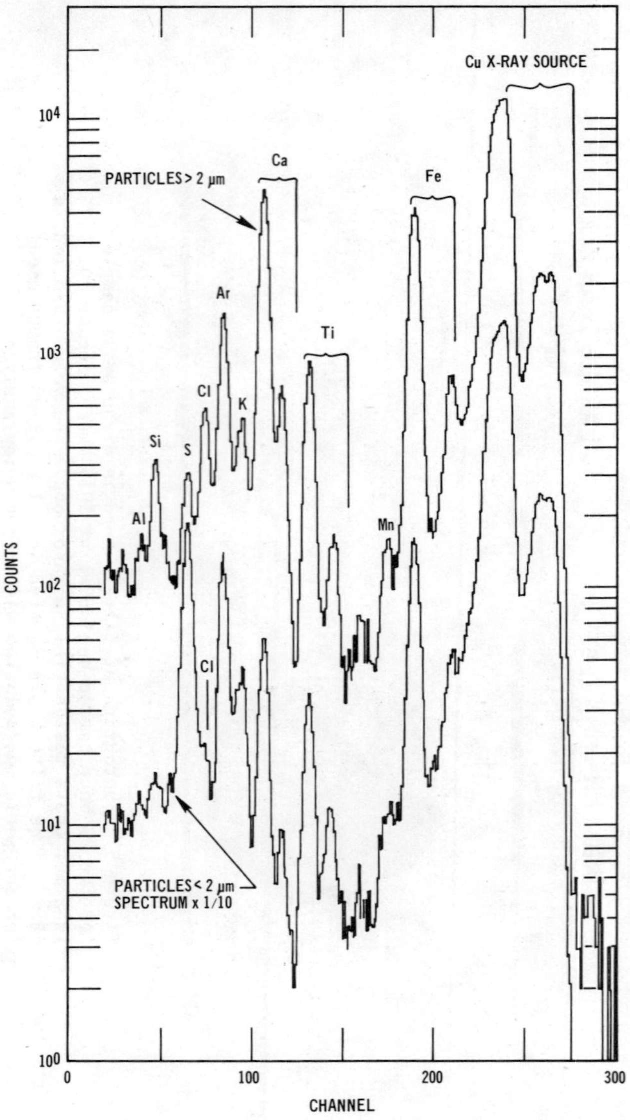

Figure 6. X-ray fluorescence spectrum with copper secondary fluorescer excitation for particles collected in two size ranges in dichotomous sampler (filters 5S and 5L of Figure 5).

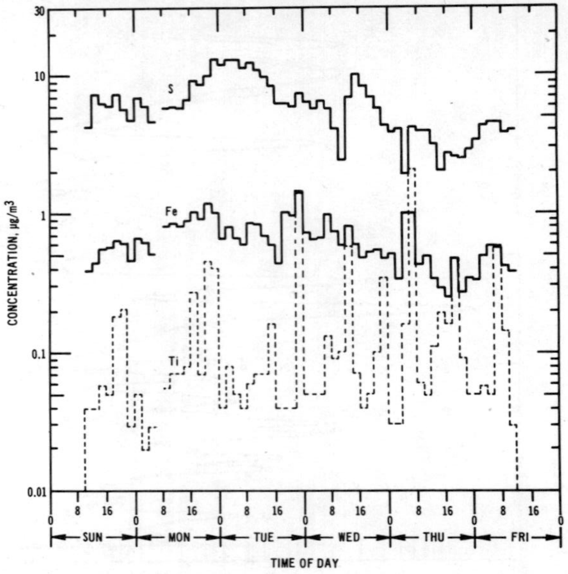

Figure 7. A plot of two hour average concentrations of S, Fe, and Ti for the five day period between Aug. 26 and Aug. 31, 1973 in St. Louis.

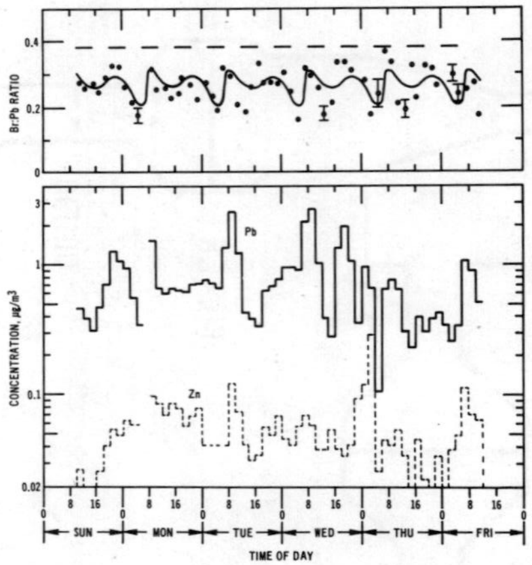

Figure 8. A plot of the two hour average concentrations of Zn, Pb, and the Br to Pb ratio for the period between Aug. 26 and Aug. 31, 1973 in St. Louis. The smooth curve is a repetitive 24 hour pattern and is a guide to the eye.

© 1973, ISA JSP 6698

THE WHAT SYSTEM: A NEW DIGITIZED RADIOSONDE AND DOUBLE THEODOLITE BALLOON TRACKING SYSTEM FOR ATMOSPHERIC BOUNDARY LAYER INVESTIGATIONS*

Paul Frenzen
Radiological & Environmental Research Division
Argonne National Laboratory
Argonne, Illinois 60439

Lad L. Prucha
Electronics Division
Argonne National Laboratory
Argonne, Illinois 60439

ABSTRACT

A semiautomatic system for obtaining measurements of wind and temperature through the lower several kilometers of the atmosphere with improved resolution and accuracy has been developed at this Laboratory. Two manually-operated theodolites have been equipped with high-resolution, pulse-type shaft encoders directly coupled to the elevation and azimuth axes of each instrument. Associated digital electronic and incremental tape equipment automatically records the four sighting angles in computer compatible form at preset intervals, as often as once each second. Balloons tracked for winds also carry transistorized radiosonde transmitters whose continuous temperature signals are recorded on the incremental tape along with the tracking data. A computer program subsequently averages several successive sets of sighting angles to smooth operator tracking errors, and calculates average winds, heights, and temperatures (hence the acronym), typically over periods of 10-20 seconds or through atmospheric layers 25-50 m thick. The system is fully portable and can be operated with storage batteries in the field.

INTRODUCTION

As the focus of atmospheric environmental quality assessment shifts from local to regional scales, increasing consideration is being given to the measurement and prediction of the long-term cumulative effects of atmospheric pollutants of low concentration. To investigate these problems the dispersive characteristics of very large atmospheric volumes (sometimes rather extravagantly termed "airsheds") must be examined. Potential pollutant receptor regions extend for hundreds of kilometers downwind of the multiple sources presented, for example, by large industrial/power complexes or major metropolitan areas. Obviously the weak concentrations associated with the increased time and space scales of these problems place special demands upon techniques of pollutant measurement. Similarly, the problem of pollution prediction (essentially one of atmospheric modeling) becomes more and more complicated because, in proportion to the increased times that contaminants remain airborne on long trajectories, progressively larger and larger circulation scales participate in the mechanisims of transport and dispersion. In particular, and unlike the essentially predictable cases of short range diffusion within a few kilometers of point sources, the dispersion of atmospheric pollutants on the so-called "mesoscale" typically involves circulations which occupy the entire depth of the planetary boundary layer.

Considerable theoretical work is presently directed toward the construction of computer models of this critically important, lowest 1 or 2 km of the atmospheric environment, a principal goal being the development of numerical methods for mesoscale dispersion prediction. But one difficulty that immediately presents itself is the lack of detailed observations of the density and flow distributions which characterize this special region. In a very real sense, the modelers are attempting to simulate atmospheric circulations we've scarcely seen.

Although adequate for ordinary meteorological operations, the methods employed for routine, twice-a-day observations by the world-wide upper-air network (single theodolite pilot-balloon tracking, rawinsondes, and standard radiosondes) do not provide either the accuracy or time/space resolution required for detailed planetary boundary layer study. Double-theodolite pilot-balloon tracking combined with frequent, serial radiosonde ascents are an expensive alternative sometimes adopted for mesoscale meteorological field experiments (e.g., Wangara, 1967;[1] METROMEX, 1971[2]). But the tedious conversion of the raw

*Work performed under the auspices of the U. S. Atomic Energy Commission.
"Superior numbers refer to similarly-numbered references at the end of this paper."

data to usable profiles of wind and temperature delays the final analyses excessively and adds disproportionately to the cost. To alleviate this aspect of the problem and more specifically, to make small, special purpose planetary boundary layer observation programs feasible, a semi-automatic, digitized balloon-tracking system combined with inexpensive, short-range radiosonde equipment has been developed at this Laboratory. The tracking equipment can also be used to evaluate transport and dispersion statistics in the free atmosphere.

EQUIPMENT DESCRIPTION

Figure 1 shows one of the two digitized theodolites constructed for this system. Except for the installation of optical shaft-position transducers (Optisyn, DRC Model 29), one of which is visible in the figure, these instruments are of standard U. S. Weather Bureau pattern (Warren-Knight Co., Model 85; shaft encoders installed by special arrangement with the manufacturer). The transducers are directly connected to the principal elevation and azimuth axes of each instrument through in-line, nonbacklash metal bellows couplings. The unit for sensing azimuth angles (not visible in the figure) is housed just below the leveling base plate, within the cylindrical extension of the tripod mount. Unlike other designs in which shaft-encoders are geared to the manual, vernier controls, the accuracy of the digital record of theodolite orientation produced with this direct-coupled arrangement is not disturbed either by backlash in the traversing gears or by disengaging and reengaging the manual controls during a run. Further, the presence of the shaft encorders installed as shown does not interfere with the use of the theodolites in the usual, nonrecording mode.

The Model 29 transducers used in these instruments produce ten thousand pulses per complete shaft revolution or, in this application, one pulse for each 0.036 deg change in theodolite orientation. Discrete pulse trains are produced in one or the other of two separate channels, according to whether the particular angle monitored is increasing or decreasing. Pulse trains corresponding to changes in the four tracking angles are continuously counted by the four up/down digital registers depicted in the first row of the control unit block diagram of Figure 2. At preset intervals determined by a crystal-controlled oscillator (set in multiples of either 0.01 s or 1.0 s by the four thumbwheel B.C.D. switches), the instantaneous numerical values of the up/down registers are duplicated (by parallel transfer) in the four associated memory registers shown just below. This transfer neither interrupts nor clears the up/down counters which continue to enumerate the + and - pulse trains representing the changing tracking data. The contents of the up/down registers are also continuously duplicated in 4-digit, Nixie tube displays that are shown at the bottom of each column of components in Figure 2 and which appear prominently on the control panel shown in Figure 3. This visible-readout feature is used when the registers are set at zero for the initial theodolite orientation and again when final angles are read at the end of a run as a check on the system performance.

The fifth register shown at the head of the last column of components in Figure 2 is a six-digit, timed scaler or frequency counter which, over the preset interval between instantaneous angle readouts, evaluates a temperature signal transmitted from a miniature radiosonde suspended below the balloon being tracked. At the end of each sampling interval, the contents of this scaler are also transferred to an associated memory register; but in this case, the scaler is reset to zero before resuming the next counting period.

The transistorized radiosonde used with this system is based upon a design developed by the Atmospheric Environment Service of Canada. Modifications made at this Laboratory increase the useful range of the transmitter and facilitate quantity production. As detailed in a recent ANL report,[3] the improved sonde offers unusually light weight (65 gm, including a 9 V battery of 216 format) and low cost (approximately $10 each for parts and assembly in quantities of 100). For measurements in the lower atmosphere, these lightweight sondes may be flown with ascent rates of 2 or 3 m s^{-1} using a standard 30 gm pilot balloon, a useful feature that simplifies deployment in the field and further reduces costs of operation by conserving helium.

The miniature radiosonde transmits a continuous audio tone (centered near 1 kHz) whose frequency gives a measure of the ambient temperature sensed by a small bead thermistor (Fenwall type GA45J1 or GA51J1; response time approximately 1 s). Thermistors are painted white to reduce radiation error, a rudimentary but sufficient precaution since this source of uncertainty on absolute measurements is not considered a serious limitation to single-point observations of boundary layer structure where the principal interest is the variation with height of the temperature gradient. Complications can of course occur on partly cloudy days.

Alkaline rather than the less expensive carbon-zinc batteries are used, these having been found to give excellent voltage vs. time performance in this application, after an initial 5 min period under load. Since the performance of both the battery and the unijunction transistor in the audio circuit are somewhat temperature dependent, the battery and transmitter circuit are thermally isolated from the changing environment in flight by being enclosed within an expanded polystyrene coffee cup. Tests indicate that these comparatively simple radiosondes can provide relative temperature data reproducible within a few hundredths deg C. Temperature gradients in the planetary boundary layer are therefore measured by this system with a high degree of accuracy.

After the tracking-angle and temperature data have been transferred to memory registers, the 32-point scanner-programmer activates the data gates sequentially, thus generating at the recorder drivers the serial input required by the incremental tape unit for each logical record. Blocks of records are counted and IR gaps are generated as required in the usual fashion, and an "end-of-file" is inserted by a manual push-button ("EOF", Figure 3) at the end of each run.

The final output of the system in the field is a computer-compatible magnetic tape record of the successive tracking angles generated during a programmed series of double-theodolite pilot balloon runs, each set of angles accompanied by a precision measure of the ambient temperature at the balloon location thus defined. From these data, successive measurements of wind, height and temperature (hence the acronym "WHAT system") are calculated by a digital computer, following the procedure outlined in the next section.

DATA PROCESSING

A method for computerized reduction of double-theodolite pilot-balloon data originally given by Thyer[4] has seen increasing use in recent years because of the adoption of this form of precision wind profile measurement for planetary boundary layer field studies. In the computer program written for the WHAT system, Thyer's method is used to calculate balloon positions from each logical record which, in this system, are obtained as often as once each second. These relatively uncertain positions are then averaged typically over five observation intervals to smooth the effects of operator tracking errors. Even without periodically being required to interrupt his telescopic view of a distant balloon in order to read elevation and azimuth scales visually, a theodolite operator is hard-pressed to maintain the cross-hairs of his instrument as close as the ±1 count digital accuracy of the WHAT system would seem to justify. The equivalent angular resolution of ±0.036 deg intercepts the nominal balloon diameter of 1 m at a range of 1600 m.

Due to small errors, the rays defined by the simultaneous orientations of two theodolites nominally tracking the same balloon need not exactly intersect. Recognizing this difficulty, Thyer's method computes the mutually orthogonal line of minimum distance between the two lines of sight and assigns a most probable balloon position to a point determined by dividing this line internally according to the relative slant-range distances from the balloon to the two theodolite locations. Further, by taking ±0.1 deg as a nominally tolerable error in each of the four angles, the method specifies a maximum permissible magnitude for this error distance between sight-lines. Expressed as a ratio with the actual error distance computed, the resultant factor gives an objective estimate of the accuracy of the given set of angles, error ratio values less than unity indicating results within the 0.1 deg angular sighting tolerance. With some care on the part of the theodolite operators, the WHAT system measurements consistently attain error ratios smaller than 1.0.

Average balloon positions thus determined are assigned the central times of observation; subtraction of successive average positions then gives average horizontal and vertical balloon displacements which may be identified as components of mean wind velocity and balloon ascent rates at the corresponding average heights of observation. Temperature data can either be collected over similar multiple sampling intervals to obtain averages through the same atmospheric layer as that used to determine the winds, or single values recorded nearest the central times of the wind observations can be extracted in order to approximate (within limitations imposed by a combination of balloon rise rate and sensor response) "instantaneous" measurements closely representing temperatures at the central levels assigned mean wind observations.

SOME APPLICATIONS

Because of the greatly increased facility with which the field data obtained can be reduced and the relatively small cost of the miniaturized radiosondes, the WHAT system makes feasible closely-spaced serial soundings capable of detecting developing phenomena in the planetary boundary layer. Preliminary and largely successful field tests of this new system were conducted in August from a site near St. Louis, in support of Argonne's acoustic sounder field program, the latter in its second year of participation in the Metropolitan Meteorological Experiment (METROMEX). At this writing, the detailed profiles of wind and temperature obtained with the WHAT system are being analyzed and closely compared with acoustic sounding charts recorded simultaneously. The St. Louis observations were conducted during the hours immediately following sunrise in order to examine the lifting of nocturnal, ground-based inversions, a daily atmospheric boundary layer occurrence of considerable importance to urban air pollution assessment.

In the St. Louis experiments, a theodolite base line of 600 m was used. During the program, a simple method was developed to cut the transmitter free from the balloon after 10 or 15 minutes of ascent. Descending parachutes were successfully tracked on a number of occasions, thus doubling the effective number of boundary layer observations in these runs by this "structure-sonde" technique.

Although primarily designed to measure mean transport aspects of pollutant dispersal (plus the density distributions that affect the transport), the precision tracking capability of the WHAT system could also be used to evaluate turbulent diffusion statistics in the free atmosphere. By tracking near-constant level balloons released at selected elevations, time-series measurements of three-dimensional, Lagrangian velocity fluctuations could be obtained. These data could then be used to compute Lagrangian autocorrelation functions and turbulence spectra which could be used in turn to examine the dispersive capabilities of the atmosphere in considerable detail. Although this procedure would obviously place great demands on the tracking abilities of the observers, some averaging could be introduced, provided due allowance is made for the loss of information in the higher frequencies. At the opposite end of the spectral band, the lower frequency limit is essentially defined by the length of time the balloon remains in view. Thus eddy periods between, say, 10 seconds and 10 minutes could be recorded in this way.

ACKNOWLEDGMENTS

The authors wish to express their sincere appreciation of the contributions made by several others to the development of this system. Richard L. Hart supervised much of the bench-testing while Bruce B. Hicks integrated the miniaturized radiosonde circuitry with the rest of the equipment. Alan W. Kittel and Marvin L. Wesely assisted in the field experiments, and Paul E. Hess contributed the computer program.

REFERENCES

(1) Clarke, R. H., with A. J. Dyer, R. R. Brook, D. G. Reid, and A. J. Troup, "The Wangara Experiment: Boundary Layer Data," Div. of Meteorol. Physics Tech. Paper No. 19, C.S.I.R.O., Australia, 1971, 316 pp.

(2) Ackerman, B., "A Field Program to Study the Urban Wind Field," Argonne National Laboratory Radiological Physics Division Annual Report, Jan.-Dec. 1971, ANL-7860, Part III, 162-182.

(3) Van Loon, L. S., R. L. Hart, and B. B. Hicks, "Some Improvements in Design of a 403-Megahertz Radiosonde," Argonne National Laboratory Radiological and Environmental Research Division Annual Report, Jan.-Dec. 1972, ANL-7960, Part IV, 88-93.

(4) Thyer, N., "Double Theodolite Pibal Evaluation by Computer," JOURNAL OF APPLIED METEOROLOGY, Vol. 1 (1962), 66-68.

Figure 1. Pilot-balloon theodolite equipped with shaft-encoders for digitized readout of tracking angles; transducer for elevation angles appears at the left, while that for azimuthal orientation is housed within the cylindrical extension below the base of the instrument.

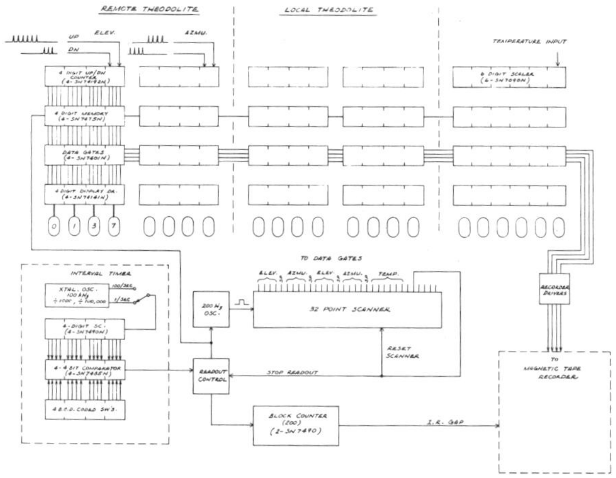

Figure 2. Block diagram of the WHAT system control unit.

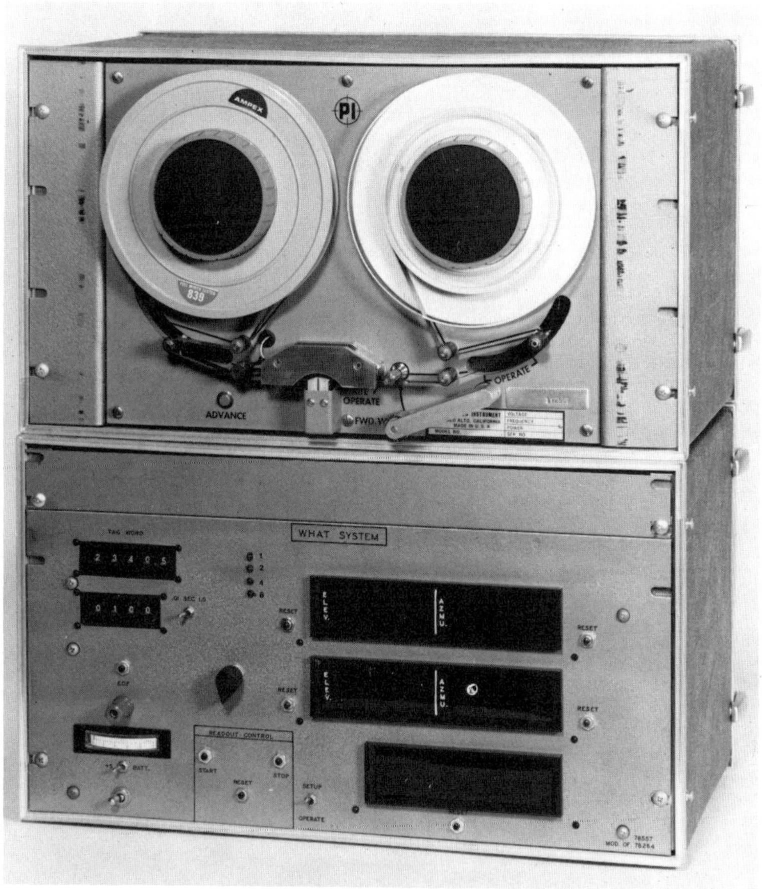

Figure 3. The WHAT system control unit and incremental tape deck; the units are fully portable and are powered by 12 V storage batteries in the field.

© 1973, ISA JSP 6699

A Comparison of Wind Speed and Turbulence Measurements Made by a Hot-Film Probe and a Bivane in the Atmospheric Surface Layer

S. SethuRaman and R. M. Brown
Meteorology Group
Department of Applied Science
Brookhaven National Laboratory
Upton, New York

ABSTRACT

Two independent systems for determining wind speed and turbulence levels are being used in an on-shore diffusion study on Long Island near Brookhaven National Laboratory. Results obtained from the two instrument systems are compared to illustrate the differences in the measured values of the vector wind, mean wind speed, variance, turbulence level and energy spectra.

Details of the physical characteristics and relative advantages of a commercial Vector Vane and a three-dimensional hot-film sensor are also presented. Measurements of the mean wind speed and the turbulence level compared well. The Vector Vane underestimated spectral densities for frequencies above 1 hertz.

INTRODUCTION

Wind speed and turbulence in the atmospheric boundary layer are important parameters related to environmental quality. A knowledge of their magnitude and character is important in the formulation of prediction models of environmental pollutants and for monitoring purposes. Frequently the results obtained from an experimental program are compared with those from another program or with observations at different locations in the same program. Some knowledge of the relative characteristics of the instruments involved is essential in order to understand the properties being observed. The relative comparison of the instruments in actual field conditions is helpful in interpreting the results obtained after giving due consideration to their characteristics and capabilities.

Over-water dispersion off the south shore of Long Island, New York, is being studied by the Meteorology Group of Brookhaven National Laboratory. This study will provide information for environmental impact analyses for possible siting of offshore power plants. Diffusion of oil fog smoke, released from an anchored boat off the coast, is measured at various distances downwind. Meteorological variables are measured with instruments mounted on a 16-m portable tower over the beach and from an aircraft. A bivane (Vector Vane, manufactured by Meteorology Research, Inc.) and a three-sensor hot-film probe with constant temperature anemometers (manufactured by Thermo Systems, Inc.) are two of the instruments used for measuring turbulence. The present study was undertaken to compare the characteristics and capabilities of the Vector Vane and the three-dimensional hot-film sensor exposed to the same flow conditions. Both instruments were mounted at the 16-m level of the tower, as close as possible, and carefully levelled. The hot-film sensor was aligned facing the direction of the wind. These two instruments formed part of an array on the tower consisting of cup anemometers, directional vane, and mean temperature measuring sensors. The arrangement of the two instruments being compared in this study is shown in Figure 1.

Research performed under auspices of the U. S. Atomic Energy Commission.

DESCRIPTION OF THE INSTRUMENTS

A brief description of the instruments, their operation, and calibration characteristics are given in this section. The mode of operation of the Vector Vane and the hot-film anemometer is so different that a brief comparison of each instrument's operation procedure and errors involved will be made. Photographs of the Vector Vane and the hot-film probe are shown in Figure 2.

Vector Vane

(a) Nature of Operation and Response Characteristics

The Vector Vane has a sensitive windmill-propeller with four light magnesium blades. The tail fins are made of plastic covered with a thin coating of aluminum. The vane is free to rotate 360° in the horizontal and ±60° in the vertical. Two potentiometers provide resistance changes proportional to the azimuth and elevation angles. A light beam chopper, attached to the propeller in combination with a miniature photocell and light source, provides a pulsed output proportional to the wind speed.

The dynamic response of the propeller can be represented by the differential equation for a first-order system

$$\tau\frac{dv}{dt} + v = f(t) \qquad (1)$$

FIGURE 1. A view of the hot-film probe and Vector Vane on the tower along with other instruments.

FIGURE 2. Hot-film probe with three constant temperature anemometers and Vector Vane.

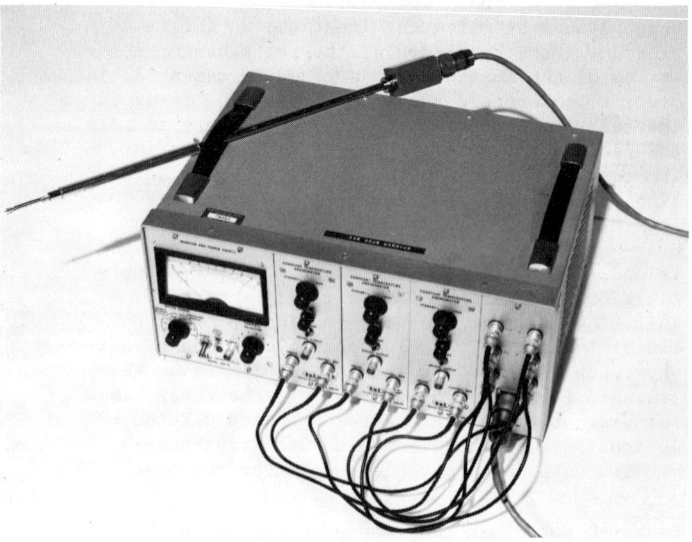

where τ is the time constant (time for the system to respond to 1 - 1/e or 63% of a step change),

v is the indicated wind speed,

t is the time, and

f(t) is a time dependent forcing function.

The dynamic response of the vane can be defined by the differential equation for a second-order system

$$\frac{d^2\theta}{dt^2} + 2\omega_n \zeta \frac{d\theta}{dt} + \omega_n^2\theta = f(t) \qquad (2)$$

where θ = angular displacement of the vane with respect to a fixed wind direction,

ω_n = natural or undamped angular frequency of the system, and

ζ = damping ratio, the ratio of the actual damping to the critical damping.

The Vector Vane has been studied extensively for its response characteristics[1,2] which are as follows:

Starting threshold	Speed	0.22 m/sec
	Direction	0.22 m/sec
Response distance	Speed	0.61 to 0.91 m
	Direction	0.61 to 0.91 m
Damping ratio	Direction	0.4 to 0.7

The response distance is defined as the distance over which the wind travels corresponding to 63% of a step function change and is independent of the wind speed. It can be shown that a system with a first-order response measures 81% of true energy for an input wavelength 16 times the response distance, L, for a sine input function in equation (1). For a wavelength of 3.7 L the indicated energy is 25% of the true value and for an input wavelength of L, the system indicates only 2.6% of the actual energy[1]. Although atmospheric turbulence does not necessarily follow a sinusoidal forcing function, the above figures give a rough indication of the importance of having response distances of small magnitudes in order to increase the frequency response of the instruments. The input wavelength λ may be defined in terms of the other pertinent variables as

$$\lambda = \frac{1}{k} = \frac{U}{f} = \frac{2\pi U}{n} \qquad (3)$$

where λ = wavelength (distance per cycle),

k = wavenumber (cycles per unit distance),

f = frequency (cycles per second),

U = wind speed (distance per second) and,

n = angular frequency (radians per second).

A knowledge of the response distance will be helpful to determine the frequencies above which the energies are under-estimated.

(b) Calibration

Speeds from the propeller of the Vector Vane are calibrated in a 0.61-m diameter, 6-m long, circular wind tunnel. Both the azimuth and the vertical angles are calibrated by moving the vane known angles in the horizontal and vertical directions. The calibrations are linear for the speed, azimuth, and vertical angles.

(c) Errors Involved and Practical Considerations

The propeller on the Vector Vane is calibrated in the wind tunnel over a range of steady, low turbulence, wind flows. In strong turbulence close to the ground, the ability of the instrument to measure the true wind speed will largely depend on its response characteristics. Due to the inability of the vane to continuously align itself with the vector wind, the propeller cannot always measure the true wind speed. When a propeller is present, the downwash from the propeller and the gradient of wind along the vane cause changes in the response characteristics of the vane. The response of the vane depends largely on the damping ratio. A damping ratio between 0.5 and 0.7 is considered reasonable with little overshoot and relatively fast response. The errors involved due to the above factors for the Vector Vane have been discussed by MacCready and Jex[1] and MacCready[2]. An error in mean wind speed of about 2% was computed for the Vector Vane by MacCready[2]. Errors involved in measuring the turbulent energy can be estimated from a knowledge of the distance constant or frequency response of the instrument. Based on the values of input wavelengths computed for a sine wave input function, a 75% energy under-estimation is possible for a wind fluctuation frequency of about 5 hertz. (Wind speed of 10 m/sec and distance constant of 0.6 m were assumed for this computation.)

From a practical standpoint, the Vector Vane is relatively rugged, easy to use, and holds a steady linear calibration for long periods of time. In addition, it is easily calibrated in the field before and after each experiment.

Three-Sensor Hot-Film Probe

(a) Nature of Operation and Response Characteristics

Hot-film sensors used for this study are quartz rods with platinum film on the surface. Gold plating on the ends of the rod isolates the sensitive area and provides a contact for fastening the sensor to the supports. The platinum film thickness is less than 1000 Angstroms and the diameter of the cylindrical film sensor is 0.025 mm. Figure 2 shows the probe consisting of three mutually perpendicular sensors operated by three constant temperature anemometers.

The detecting element of a hot-film anemometer is heated by an electric current. Ordinarily, the film is cooled by the wind which causes the temperature to drop, resulting in a decrease in

Superior numbers refer to similarly-numbered references at the end of this paper.

electrical resistance of the film. When a constant temperature anemometer is used, the electrical resistance of the film is kept as constant as possible. Any slight variation in temperature is immediately compensated for by an electronic feedback system. Voltage required to drive the necessary current through the sensor is obtained as output. For subsonic flow King's "potential flow" relation holds for heat loss. It is expressed as

$$\frac{E^2}{(t_s - t_e)} = A + B\ (\rho V)^{k/n} \qquad (4)$$

where E is the bridge voltage output,

V is the wind speed,

ρ is the density of the fluid,

t_s is the sensor operating temperature,

t_e is the environmental or fluid temperature,

A and B are constants that depend on fluid properties, and n is an exponent that varies with the Reynolds number of the flow. For air under subsonic flow conditions n takes a value of about two.

Although theoretical evaluations based on heat transfer properties are available for the response of the hot-film, direct calibration based on Equation (4) is commonly adopted since this eliminates the variability in the characteristics of the film material, supports, or other unknown factors. If the fluid temperature t_e happens to be the same during calibration and experimentation, no correction for the bridge voltage output is needed for a constant temperature anemometer. Most often, these temperatures are not the same, necessitating a correction for the voltage output. This correction factor can either be computed and applied to the observed values during the analysis or the sensor may be electronically compensated by using fast response temperature sensors mounted close to the speed sensors. For this study, temperature compensation was achieved by measuring the air temperature near the sensor and correcting the voltage during the analysis. Air temperature near the sensor remained constant throughout the experiment.

Frequency response of the hot-film anemometer was found to be near 1000 hertz and varied slowly with the mean wind speed. For a 10 m/sec mean wind this corresponds to a response distance of 1 cm as compared with 60 cm for the Vector Vane.

(b) Calibration

The three hot-film sensors of the probe were calibrated in the circular wind tunnel already mentioned with the flow at right angles to each of the sensors. As can be seen in Equation (4), the calibration of the hot-film sensors versus wind speed is not linear.

(c) Errors Involved and Practical Considerations

The hot-film sensor is directionally sensitive and errors are introduced in the measurements if the flow direction is not normal to the sensor. Champagne[3] found that the relationship between the actual mean velocity and the effective cooling velocity can be expressed as

$$V_c^2 = V^2\ (\mathrm{Sin}^2 \propto + K^2 \cos^2 \propto) \qquad (5)$$

where V_c = effective cooling velocity past the sensor,

V = mean velocity,

K = a constant that depends upon the fluid and the wind speed, and

$\propto$ is the angle the sensor makes with the mean wind direction.

K cos α is a measure of the effectiveness of the velocity parallel to the sensor.

Serious errors can be encountered using this system when large variations in wind direction occur. For the time periods involved in this experiment, the flow was fairly steady and the direction did not change appreciably. Other factors affecting the hot-film system such as conduction to supports, temperature gradient along the sensor, finite length of the sensor, and presence of water spray are to be taken inot account in interpreting the results. The relative importance of these errors depends greatly on the problem studied. For atmospheric studies most of the above errors turn out to be negligible. In locations where water spray is present, it is not advisable to use hot-film sensors.

RESULTS

The wind data from the bivane and hot-film was recorded simultaneously on magnetic tape in analog form. The analog record was digitized and then recorded at 0.1 sec intervals. The digitized data was analyzed using a CDC 6600 computer.

Comparison of the vector wind is made instead of individual wind components to keep the comparison as realistic as possible. This is due to the fact that the three-dimensional hot-film sensor was designed mainly to measure the vector wind rather than the components.

The following values are compared:

(1) Mean wind speed

(2) Standard deviation of the fluctuations of wind speed

(3) Turbulence level

(4) Energy spectra

(5) Energy dissipation rate

Mean and Standard Deviation of the Wind Speed and Turbulence Level.

Table 1 shows the mean wind speed, standard deviation (σ) of the wind speed and the turbulence level which is defined as the ratio of the standard deviation to the mean wind speed. These values were computed for three successive ten-minute periods when the wind speeds showed a tendency toward stationarity.

Table 1

Data	Hot-Film			Vector Vane			Turb.
Set	Mean m/sec	σ m/sec	Turb. Level	Mean m/sec	σ m/sec	Turb. Level	Level V.V./h.f.
1	8.31	0.66	0.079	8.13	0.55	0.068	0.861
2	7.64	0.72	0.094	7.74	0.64	0.083	0.833
3	7.63	0.79	0.103	7.91	0.71	0.090	0.874

The mean wind speeds as measured by the two systems are nearly the same. The vector wind speeds computed for the second and third ten-minute periods for the hot-film are somewhat less than the corresponding values for the Vector Vane. This may be due to a slight change in horizontal wind direction during these periods.

As expected, the standard deviations of the wind fluctuations as measured by the hot-film sensor were larger due to its higher frequency response. The Vector Vane measurements under-estimated the standard deviations by about 10 to 16% as compared with those measured by the hot-film. Because of the substantial difference in the frequency responses of the two instruments, this error is relatively small and probably not important for many practical purposes. Turbulence level is considered an important parameter in characterizing atmospheric conditions in meteorological measurements and analyses. The turbulence levels for the Vector Vane were about 13 to 17% less than those for the hot-film.

Energy Spectra

Energy distribution at various frequencies is generally used by meteorologists to provide information on eddy size distribution. The vector wind data for consecutive ten-minute periods were analyzed to determine the energy spectra. The spectral density S(n) is defined by

$$\bar{v}^2 = \int_0^\infty S(n)\, dn = \int_0^\infty S(k)\, dk \qquad (6)$$

where v is the fluctuation of the wind from the mean, and the wave number k is defined as $2\pi n/V$. Reciprocal of k represents actual length scales.

A comparison of the spectral densities near the high frequency end of the spectrum is shown in Figure 3 and in wave number domain in Figure 4. The relative under-estimation of the spectral densities at higher frequencies by the Vectpr Vame as compared with the hot-film anemometer can be seen. At frequencies above about 1 hertz, the under-estimated spectral density starts becoming significant.

It is often convenient to express the frequencies normalized with the height of the instrument and the mean wind speed as nz/V. Equation (6) can be rewritten as

$$\bar{v}^2 = \int_0^\infty n\, S(n)\, d \ln n \qquad (7)$$

A graphical representation of n S(n) versus $\log_{10} n$ has the advantage that the area under a segment of the curve represents the contribution to the energy in the corresponding log-frequency interval. Variation of normalized spectral density $nS(n)/\sigma^2$, is shown in Figure 5, where σ^2 is the variance obtained from Table 1. Normalized spectral densities obtained from the Vector Vane do not differ significantly for non-dimensional frequencies below one. An average mean wind speed was obtained from Table 1 for both instruments.

A graphical representation of the spectral densities estimated from the two instruments is shown in Figure 6. A maximum error curve has been drawn to show the maximum energy under-estimation by the Vector Vane as compared with the hot-film anemometer. Relative maximum error defined as the ratio of the difference in the spectral densities measured by the two instruments to that measured by hot-film is shown in Figure 7. The values expressed as a percentage were obtained from the maximum error envelope in Figure 6. The percentages have been computed with respect to the spectral density of hot-film and bivane and shown as separate curves. Thus, knowing the spectral density as estimated from bivane measurements, it will be possible to estimate the error involved as compared with the hot-film anemometer measurements. The percentage error has a tendency to increase with decrease in the spectral density which is associated with an increase in the cyclic frequency or radian wave number. The error becomes significant for cyclic frequencies of about one hertz.

Energy Dissipation Rate

An energy dissipation rate ε, obtained from Kolmogorov's hypothesis in the inertial subrange, is useful in determining the diffusive properties of turbulence. The approximate magnitude of error involved in the estimation of ε can be computed for the frequency range of interest. In the inertial subrange a relation of the form

$$S(k) = K^1 \varepsilon^{2/3} k^{-5/3} \qquad (8)$$

has been found to be applicable where K^1 is the Kolmogorov constant with a value of about 0.5. From Equation (8) it can be seen that the error involved in the computation of the energy

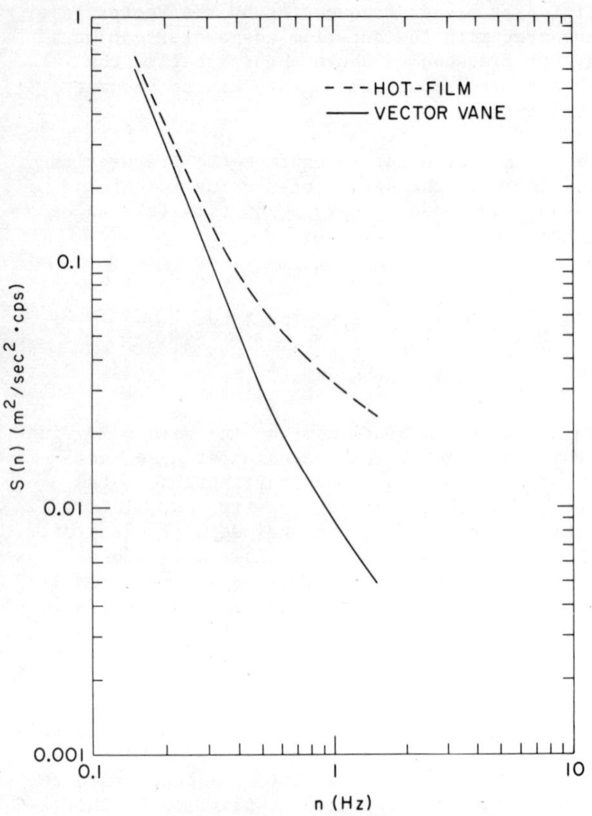

Fig. 3 COMPARISON OF SPECTRAL DENSITIES OF HOT-FILM AND VECTOR VANE IN THE FREQUENCY DOMAIN

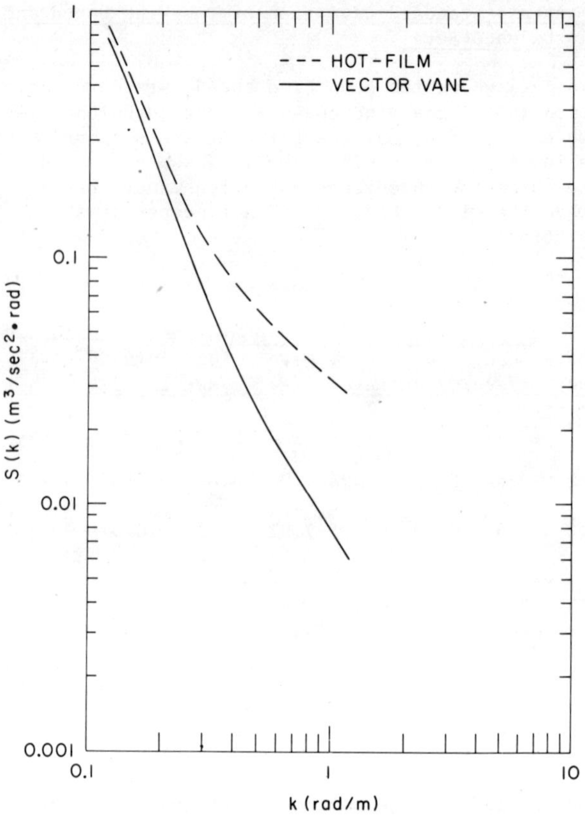

Fig. 4 COMPARISON OF SPECTRAL DENSITIES OF HOT-FILM AND VECTOR VANE IN THE WAVE NUMBER DOMAIN

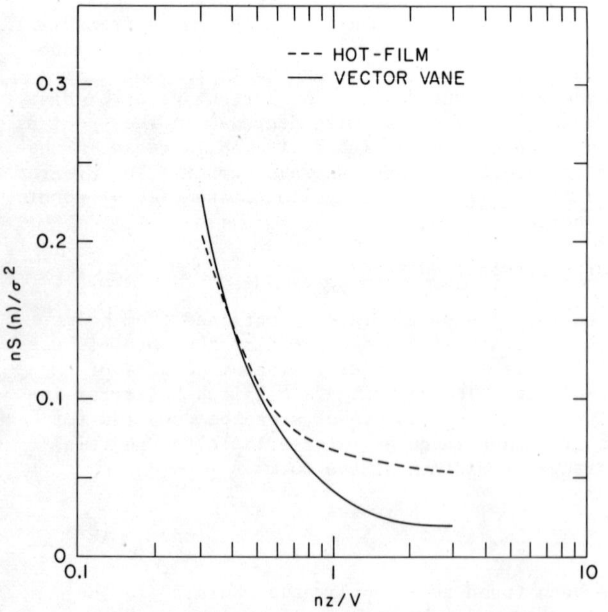

Fig. 5 NON-DIMENSIONAL ENERGY DISTRIBUTIONS FOR THE HOT-FILM AND THE VECTOR VANE

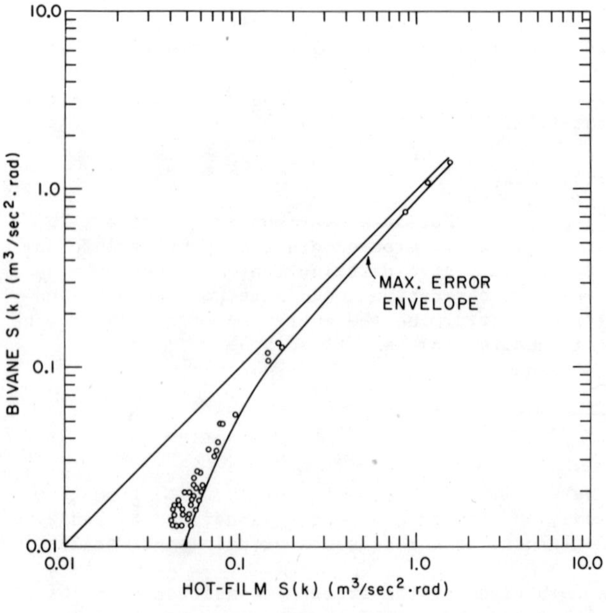

Fig. 6 SPECTRAL DENSITY OF THE VECTOR VANE VERSUS THAT OF HOT-FILM

dissipation rate ε would be $r^{3/2}$ where r is the relative error in the estimation of the spectral density. Since the beginning of the inertial subrange depends to a certain extent on the distance to the ground[5], a rough estimation of the error involved can be made for the instruments compared here. Assuming that the inertial subrange exists over the frequencies studied, a maximum error of about 18% in ε will occur if it is computed from frequencies below 0.5 hertz. The error tends to increase at higher frequencies.

CONCLUSIONS

Comparisons of the two instruments were made in the atmosphere using data measured by the systems exposed to relatively steady turbulence conditions. From comparisons of the vector winds made in near neutral strong wind conditions, the following conclusions can be made.

(1) The Vector Vane measurements of the turbulence levels are in reasonable agreement with those of the hot-film anemometer, with errors varying from 10 to 16%.

(2) Mean wind speeds obtained from both instruments were approximately the same.

(3) The Vector Vane was found to under-estimate the spectral densities above about 0.5 hertz and the errors involved were found to become significant above one hertz.

(4) Errors involved in computing other parameters viz. energy dissipation rate, friction velocity, etc. from the spectral densities depend on frequency range used for computation.

(5) The Vector Vane has the advantages of simple operation and ruggedness and will give reasonable results at low frequencies. For high frequency turbulence studies, the hot-film sensor provides more accurate information. But, a three-dimensional hot-film probe of the type used in this study should align itself continuously with the mean direction of the wind for best results.

ACKNOWLEDGEMENT

The entire Meteorology Group was involved in various aspects of this study. The material used in this paper is part of the data being collected in a continuing program to study the diffusive properties of the ocean-air, land-ocean environment, sponsored by the Division of Biomedical and Environmental Research of the U. S. Atomic Energy Commission.

REFERENCES

(1) MacCready, P. B., Jr., and H. R. Jex,"Response Characteristics and Meteorological Utilization of Propeller and Vane Wind Sensors," J. APPL. METEOR. Vol 3 (1964), 182-193.

(2) MacCready, P. B., Jr., "Mean Wind Speed Measurements in Turbulence," J. APPL. METEOR. Vol 5 (1966), 219-225.

(3) Champagne, F. H., C. A. Sleicher, and O. H. Wehrmann, "Turbulence Measurements with Inclined Hot Wires," J. FLUID MECHANICS, Vol 6 (1967).

(4) Lin, C. C., "Taylor's Hypothesis and the Acceleration Terms in the Navier-Stokes Equations," QUART. APPL. MATH., Vol 10. (1953), 295-306.

Fig.7 RELATIVE MAXIMUM ERROR AT DIFFERENT SPECTRAL DENSITIES

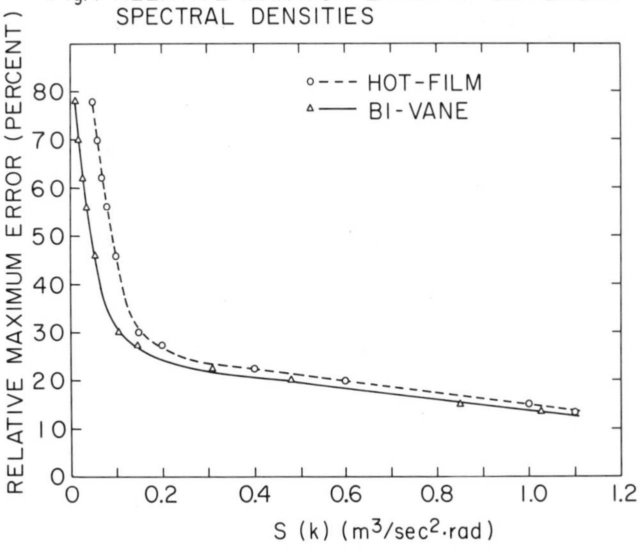

©1973, ISA JSP 6700

REMOTE SENSING APPLICATIONS IN AIR POLLUTION METEOROLOGY

D. W. Beran and F. F. Hall, Jr.

Wave Propagation Laboratory
NOAA Environmental Research Laboratories
Boulder, Colorado 80302

ABSTRACT

The application of ground based remote sensing systems in monitoring those meteorological parameters of importance in urban air pollution is considered. Typical system considerations include an analysis of general and specific site characteristics. Two examples of remote sensing systems centered on laser and acoustic methods are presented.

INTRODUCTION

This paper deals with the use of ground based remote sensing devices for monitoring urban meteorological parameters; however, we first consider the broader area of air pollution to define a framework within which the optimum remote sensing system must function.

For any urban region, we can equate the quality of the air at a given time to a large number of controlling factors, some constant or only slowly changing, as given in the list below. Following each item in the list are a number of questions that help in assessing its importance for a particular city. The list and the questions are not exhaustive and might be modified or extended for any given situation.

CONSTANT CONTROLS

1. Local and Surrounding Terrain: What is the orientation of valleys? What is the position of mountains, if any, relative to the city?

2. Type of Ground Cover: Will the local and surrounding ground cover contribute to the development of large turbulent eddies, enhancing diffusion. Is there a possibility of snow, which will reduce solar heating of the ground, thus prolonging temperature inversion lifetime?

3. Climatology: Is the region continental or maritime? What are prevailing wind directions; what are the important seasonal changes and influences? Are photochemical reactions important?

4. Location and Type of Pollution Sources: What is the relative position of major sources and population centers?

5. Political Boundaries and Regulations: Will the effluent from one region effect another which may have a different set of standards? Will local, State or Federal standards be used as a basis of control?

6. Demography: What is the size and distribution of the population and its relationship to pollution sources?

7. Total Effluent: What is the total quantity of foreign material that can be expected from the known sources?

8. Type of Effluent: Is the pollution primarily gaseous or particulate and what is the expected ratio of the two?

The items presented above must be checked periodically to determine significant changes that might alter the general pattern over a period of months or years. On the other hand, there is a second set of controls which vary from day-to-day and even from hour-to-hour. Intelligent air quality control decisions can only come about through effective real time monitoring and accurate short term prediction of these variables. A list of such variable factors and qualifying questions follows:

VARIABLE CONTROLS

1. Time Distribution of Emissions: How does the total effluent input rate vary between peak and normal traffic flow? What is the change from weekday to weekend effluent production?

2. Synoptic Weather Pattern: Does the synoptic pattern suggest enhanced or suppressed diffusion? Will a cleansing frontal passage occur before pollution levels reach unacceptable levels?

3. Mesoscale Flow Pattern: What are the local wind directions? Is there a convergence pattern over the city?

4. Inversion Topography and Depth: How does the inversion topography relate to the local terrain? Will pollutants be trapped by the juxtaposition of an inversion and irregular terrain? What is the

depth through which polluted air can mix?

5. Low Level Stability: How will the stability effect the rate of dispersion?

In many cases, farsighted planning with attention to the items in the above lists, especially the first list, would have eliminated the need for short term monitoring and prediction. It is unfortunate that so little attention has been given to proper urban planning and regulation, that the need for better monitoring and prediction has become urgent.

The use of sensors, both in situ and remote is largely confined to the measurement of those items in the second list. Two classes of sensors are required: those designed to observe and classify the actual pollutants, and those used to measure the geophysical parameters which control the distribution of pollutants regardless of the type of effluent. Both are important, and should complement each other. Knowing a type of pollutant is of little value unless some judgment about its future trajectory and concentration is made. Considerable effort is now being expended on the development of devices for monitoring pollution concentration, and many successful measurement systems are described in the literature. These devices will not be dealt with extensively here, although it should be noted that such instruments may find use as detectors of meteorological variables simply by using the detectable pollutants as tracers.

Following a section devoted to the controlling meteorological variables and the relevant spatial and time scales, the field of remote sensing is reviewed. These topics are then combined and conceptual remote sensing systems are proposed.

AIR POLLUTION METEOROLOGY

Pollution forecasting techniques are based on, and limited by, the available standard observations. These data usually consist of twice daily radiosonde profiles of the temperature, humidity and wind. In addition, hourly observations of such surface parameters as pressure, wind, temperature, humidity, visibility and cloud cover are available. Historically, these observations have been made in support of aviation and quite naturally are taken at or near an airport. As such, they may be totally unrepresentative of the urban region, perhaps several miles away. They have served well as input to forecasts of synoptic scale motions, the primary need of aviation, but are not adequate for mesoscale urban forecasts.

This inadequacy is documented by the schematic presentation of a variety of atmospheric phenomena in their time - and space - scale relationship, as described by the Committee on Atmospheric Sciences of the National Academy of Sciences(1) (see Fig. 1). The phenomenon of urban smog can be seen to have a typical time scale of between about two hours and two days and its spatial scale bridges the boundary between the meso- and synoptic scales. Synoptic scale observations are clearly inadequate for the needs of urban meteorology.

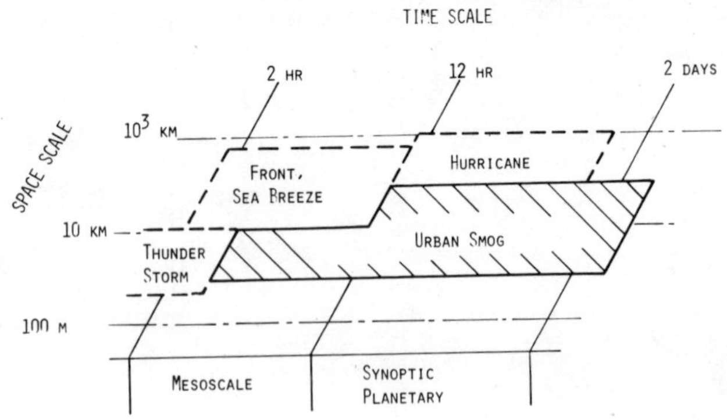

Fig. 1 Time and space scales of atmospheric phenomena(1). Note the region covered by urban smog.

If we now consider which meteorological parameters are important in controlling the level of pollution in a city, we find that only two can be classified as primary. For a given set of constant controls, the level of pollution will depend on the vertical variation of temperature and the direction and strength of the low level winds(2). Present observations provide this type of information at only a single location and usually only twice per day.

The optimum rate for obtaining the mesoscale meteorological data will be somewhat dependent on the particular parameter being measured and how it is to be used. In general, the data rate should not be so great as to clog the system with unnecessary detail, but should be rapid enough to depict significant changes taking place within the urban complex. For example, such changes as the shift from land to sea breeze, the onset or cessation of drainage winds, and the time of an inversion breakup are important for monitoring and predicting where and how severe pollution conditions will be. This time frame would require observations on the order of once every 10 to 20 min.

The optimum density of observing points again depends on the phenomena to be observed and the local terrain pecularities which might alter or modify the urban flow pattern. Grids of sensors spaced less than 10 km on a side have been proposed as optimum for standard in situ sensors.

In considering remote sensors, the criterion for both time and spatial scales must be viewed in a new light. An advantage of many remote sensing devices is their ability to rapidly scan large regions in three dimensions. Others can provide line integrals of certain parameters, giving spatially averaged values that may be nearer to the desired measurement than a time average from a single sensor. This

"Superior numbers refer to similarly-numbered references at the end of this paper."

difference in the output of remote sensors must be a primary concern in the design of a system for monitoring the urban environment and should result in a concept other than the commonly proposed grid of sensors.

Until now it has been assumed that the classical forecasting techniques employing only the temperature and wind field would be used to make the urban mesoscale forecasts. While remote sensing devices have demonstrated the ability to measure basic parameters, (temperature and wind) it should be noted that other, perhaps more valuable information is contained in the first order output of the sensor. For example, one of the more common products of a remote sensor is a range dependent returned signal intensity or frequency, which is used to derive temperature or wind speed. It is conceivable that the required information for providing a forecast of future pollution concentrations is contained in the more rudimentary sensor output and its conversion to classical meteorological variables is not always needed. Only careful planning and close cooperation between developer and user can avoid this potential waste of resources.

It is equally clear that remote sensor development should include research into new ways in which the remote sensor data can be used. This means going beyond mere speculation concerning its potential and actually developing and testing new forecasting techniques.

METHODS OF REMOTE SENSING

In the broadest sense remote measurement is the detection or observation of any parameter at a point other than the sensor. Now, let us narrow this view to sensors located on or near the ground and have as their primary function the detection or depiction of some meteorological variable. Aircraft mounted remote sensors have been omitted largely because, although they serve well as research tools, they are too expensive to use in an operational system. Satellite mounted remote sensors have demonstrated vast potential as an operational synoptic tool but they will not be discussed either because we are primarily interested in all-weather mesoscale observation networks which must function even when clouds may block the view from space.

Recent publications by C. G. Little and others have dealt effectively with the present status of ground based remote sensing and we shall borrow from these works to give a brief overview of the remote sensing field[3,4,5,6].

It is valid to ask why remote sensing should be used in place of, or in support of, conventional <u>in situ</u> sensors. Cost of a particular remote sensing instrument can be high, especially when compared with a single <u>in situ</u> sensor. A cost analysis is more realistic when the comparison is made with the total number of <u>in situ</u> sensors that would be required to perform a similar task. A better perspective can be gained by looking at some of the advantages of remote sensors; they are:

1. The sensor does not need to be carried into the medium to be measured.

2. Remote sensors often have the ability to scan the surrounding atmosphere in two or three dimensions, unlike the single point measurement capability of the <u>in situ</u> device.

3. High resolution time and space measurements are possible with many types of remote sensors.

4. More sophisticated parameters, such as the spectrum of turbulence and momentum flux are potentially available as a direct output of remote sensing instruments.

5. The measurement system does not modify the parameter being measured.

6. Integration of a given parameter over a line, an area, or a volume is oftentimes readily obtained as a direct output of the sensor.

7. A high level of automation can usually be achiev-achieved with only minimal effort, unlike radio-sonde operations where manpower is essential.

Most of the remote sensing devices considered in this paper rely on the interpretation of the interaction of some transmitted wave with the medium which is to be interrogated. The wave types used are both acoustic and electromagnetic. Various interactions of these waves with the atmosphere are exploited to make the wide range of measurements that are possible. For example, the strength of interaction of acoustic waves and the atmosphere is far greater than for electromagnetic waves[7]. This strong interaction is responsible for the rich detail found in most acoustic records. It is also this strong interaction which acts to limit the potential range of most acoustic devices through strong scatter and absorption of the energy in the wave. Radio waves, whose response to the atmosphere is much weaker, do not suffer this same rate of degradation and can be transmitted over longer ranges.

Assessment of the capability of various remote sensing devices is complicated by the wide range of potential measurements. The <u>in situ</u> sensor is capable of, at best, a one dimensional measurement (recognizing that the atmosphere advects past the sensor). Remote sensors have the powerful capability of multidimensional observations. This point is treated by Little[4] in his listing of the range of different types of measurements that are possible:

Line integral	-	the integrated value of the parameter along a line through the whole atmosphere.
Line average	-	the average value of the parameter along some line of known length.
Line profile	-	the distribution of the parameter along some line of known length.
Two dimensional coverage	-	the distribution of the parameter over a plane.

Three Dimensional coverage - the distribution of the parameter in space around the instrument.

Structure constant - a measure of the intensity of small scale fluctuations of the parameter in space and time.

Spectrum - the power spectrum of variability of the parameter in space - i.e. how strong are the variations of the parameter as a function of the spatial size of the irregularity.

Flux - the rate at which mass, momentum or heat is being transported.

Since our primary objective in this discussion is to relate ground based remote sensing instruments to problems of air pollution meteorology, it would be inefficient for us to go into a detailed discussion of all types of ground based remote sensing devices. Rather, we shall select instruments which appear to have the proper requirement based on our understanding of what is needed for the urban or mesoscale forecasting/monitoring problem.

It was previously developed that the primary meteorological parameters for urban forecasting were temperature and wind. Temperature profiles are important for determining low level stability. Hence, the first category of sensors to be discussed includesthose which have the potential of measuring either the exact temperature profile or providing an indication of the low level stability in a more direct fashion. The next category includes sensors which are capable of providing wind information. We then look at sensors which provide an indication of inversion height or mixing depth, without resorting to an absolute temperature measurement. Finally, as a separate category we consider those sensors which have the potential of areal probing.

Sensors of Temperature Profiles or Static Stability

a. Microwave Radiometry:

Passive, ground based radiometers can measure thermal emission of the atmosphere by operating in any frequency band where gaseous constituents are strongly absorbing. An example, is the operation of microwave radiometers in the 50-60 GHz oxygen band. The ability of such radiometers to measure quite accurately the temperature profile to heights of several kilometers[8] is shown in Fig. 2. By scanning in zenith angle while operating at one fixed frequency, or by operating in several fixed frequencies spaced throughout an absorption band, or by combinations of these two techniques, microwave radiometers have shown promise for monitoring the temperature profile in the clear atmosphere.

The sophisticated data processing, involving inversion of the radiative transfer equation, has made it possible to accurately determine temperature profiles up to and through ground based inversions. Elevated temperature inversions are, however, usually smoothed over and not clearly defined. In addition, the presence of clouds in the field of view complicates the data reduction leading to less accurate temperature profiles.

b. Laser Temperature Measurements:

A method for measuring the vertical profile of atmospheric temperature using Raman scattered laser radiation has been developed[9]. By measuring the ground level atmospheric pressure at a laser site, and, assuming that the hydrostatic equation can be used to predict pressure at greater heights, it has been shown that the atmospheric temperature profile can be directly obtained from the Raman backscatter from nitrogen. In one experiment, a pulsed nitrogen laser, operating at a wavelength 337 nm was used to measure temperature fluctuations at a height of 30.5 m. For comparison, the atmospheric temperature was also measured with a tower mounted thermistor.

The fluctuation in intensity of the Raman return showed good correlation with thermistor measurements, even when rapid fluctuations in temperature caused by wave motions in the stable nocturnal inversion were present. With the equipment used, the technique is capable of being extended to ranges of 5 or even 10 km, assuming that time resolution of several minutes is sufficient. This integration in time is necessary to obtain a sufficient number of returned scattered photons. When the good range resolution of pulsed lidars is coupled with the good spatial resolution provided by laser beams, a powerful research tool is available. A similar method employing energy level populations and Raman backscatter has also been demonstrated[10].

Another possibility of temperature probing with lasers has been demonstrated by Fiocco et al[11]. This technique requires the frequency distribution in the backscattered spectrum of laser light to be analyzed. Spectral width of the lidar return produced by molecular scattering can be interpreted in terms of the temperature of the scattering volume. A stable, single frequency laser and relatively high resolution spectrometric receivers are required. At this time, the technique has not found routine application because of limitations in range to several hundred meters with existing lasers and the requirement for complicated, high resolution and high quality spectrometric detectors.

c. Acoustic Echo Sounding:

An acoustic echo sounder relies on the scatter of sound from naturally occurring temperature and wind fluctuations in the atmosphere. In its simplest form (single axis, monostatic configuration), a short pulse of sound is transmitted, usually in the vertical direction, and the system is then switched to a receive mode. Sound scattered from temperature eddies (wind scattering does not contribute in the monostatic mode) is then received at the antenna. The record from a vertically pointed monostatic sounder is a time-height history of the regions of strong temperature fluctuations that are advected over the sounder. An example of this type of record is shown in Fig. 3. The enhanced acoustic scattering associated with temperature inversions is evident where the record is dark. During unstable periods

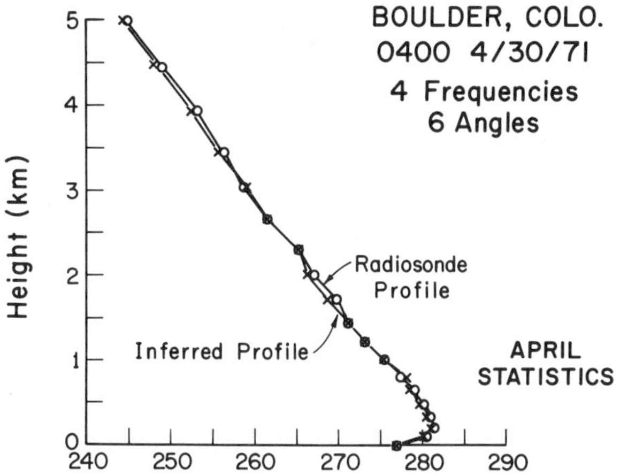

Fig. 2 Comparison of temperature profiles measured by a radiosonde and a microwave radiometer (Inferred Profile)[41].

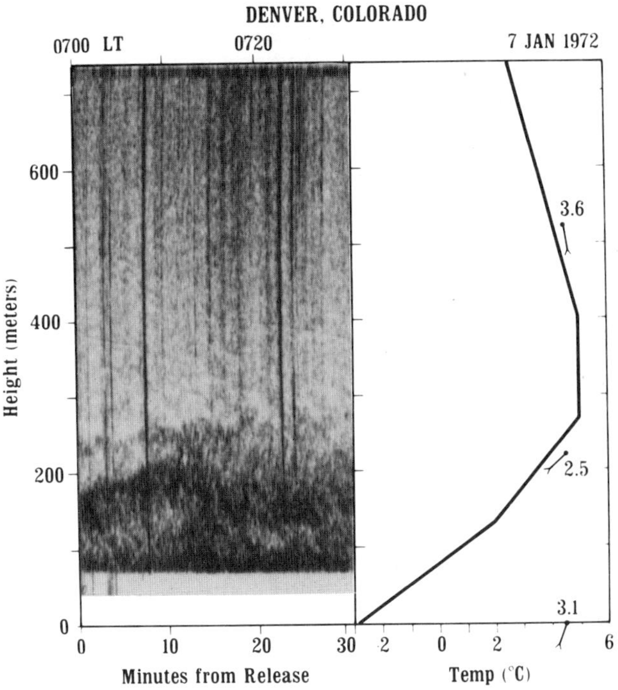

Fig. 4 Comparison of radiosonde temperature profile (right) with an inversion layer shown on an acoustic echo sounder record. The speed (ms^{-1}) and direction of the winds measured by the radiosonde are shown by the small arrows (north is up).

the record also shows dark regions, or regions of strong temperature fluctuation, which delineate the rising air in thermal plumes (see right side of Fig. 3). This type of information does not give an absolute measurement of the mean temperature. It can, however, be used to infer the stability of the boundary layer. Echoes from horizontally stratified layers, usually containing evidence of organized wave structure indicate a stable lapse rate. Neutral lapse conditions or regions where laminar flow exists will usually not produce a return and the record will be white. Vertically oriented dark regions, associated with convective thermal plume activity indicate unstable lapse conditions. A direct comparison between the acoustic sounder record of an inversion and a radiosonde temperature profile is shown in Fig. 4.

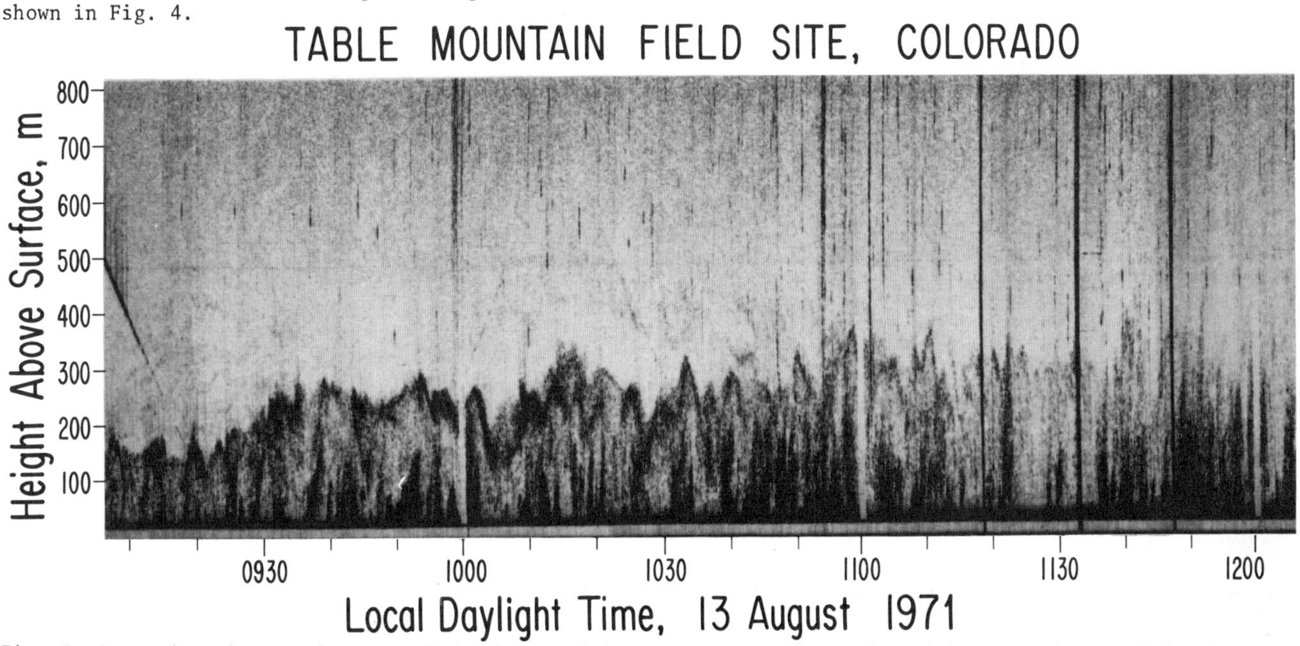

Fig. 3 Acoustic echo sounder record showing a rising temperature inversion with convective activity below. The inversion has dissipated after 11:15 local and the pattern is dominated by typical returns from thermal plumes

Attenuation of sound energy in the atmosphere is known to be a function of temperature, humidity, pressure, and scattering caused by turbulence. It is conceptually feasible to use the measured absorption of energy to derive both temperature and humidity profiles[12]. However, such techniques are still in the very early experimental stages and it will be several years if ever before they are ready for operational application.

d. RASS

The Radio Acoustic Sounding System (RASS)[13] is a hybrid device which uses an acoustic transducer to send a burst of sound upward into the atmosphere and a Doppler radar to track the sound wave and determine its propagation speed. Since the speed of sound is a function of the ambient temperature the Doppler information can be used to derive a temperature profile. Recent experiments have demonstrated the feasibility of this technique for measuring temperature profiles to heights of between 1 and 3 km. Errors caused by vertical velocities in the atmosphere can be eliminated by averaging the returned signal for periods of several minutes providing temperature information on a time scale commensurate with the needs of urban meteorology.

Wind Sensing

a. Wind Measurements with Lasers:

Aerosols are almost always present in the troposphere and especially in regions with air pollution problems. The backscattered light from a laser beam in such turbid atmospheres will always greatly exceed that from a pure molecular atmosphere; however, the aerosols are not uniformly distributed in the atmosphere. As the winds carry such nonuniform distributions along, it is possible to correlate the aerosol distribution seen at one point with that observed some time later some distance downwind. Using the so-called cross beam laser method (see Fig. 5), it has been shown that remote wind sensing is indeed feasible[5]. Two separated laser beams may be used to monitor the wind in a given plane, or a conically scanned laser (see Fig. 6) can measure the horizontal wind in any direction. By pulsing the laser, winds at different levels can be monitored. A range of greater than 10 km can be provided (in reasonable clear air) with nominal laser powers, but to date routine applications of this method have not been accomplished.

Another method for laser measurement of winds is to employ the Doppler shift of the backscattered light. This shift may be measured by beating the return signal against the propagated laser frequency in a heterodyne detection scheme[14,15]. Because of rapid degradation of coherence in the propagated laser beam through a turbulent atmosphere, it is advantageous to use as long a wavelength as possible, and most schemes now propose to use 10.6 μm CO_2 laser radiation. Demonstrated ranges of several hundred meters indicate the technique may have application in air pollution meteorology monitoring. A Doppler wind measurement which does not require a local oscillator reference has also been described[16].

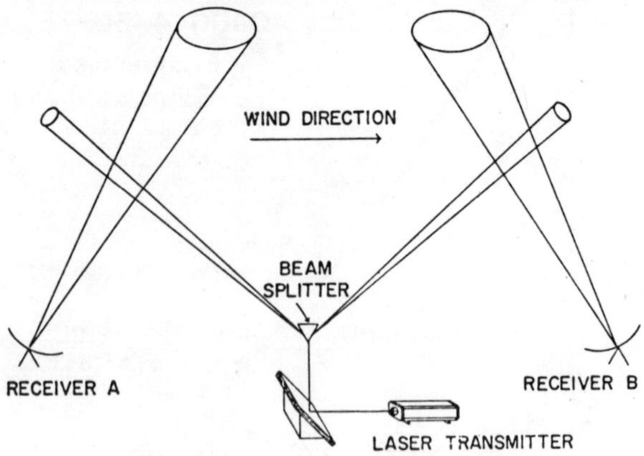

Fig. 5 Crossed beam wind sensing configuration employing a laser[5].

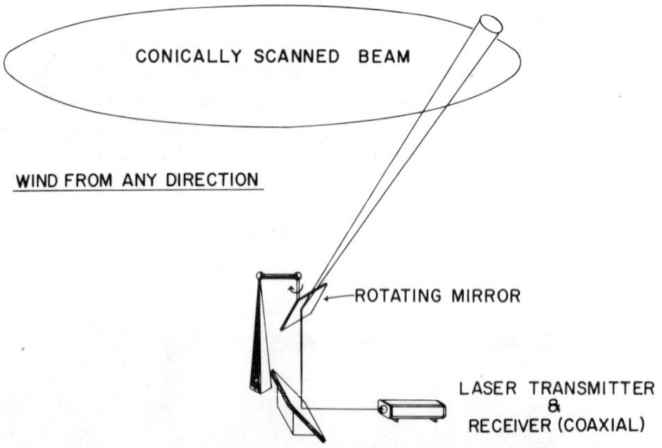

Fig. 6 Conically scanned wind sensing configuration employing a laser[5].

Instead, angled laser beams are used that intersect in a pattern of interference fringes where the beams cross. By monitoring the drift of discrete scatterers through these fringes, the transverse velocity may be sensed. Again, this method has been demonstrated in the real atmosphere but only over limited ranges of several hundred meters.

b. Radar

Pulsed Doppler radars have demonstrated their ability to measure wind fields whenever hydrometeor targets are present[17] (see Fig. 7). The pulsed Doppler method has also been used with artificial targets (chaff) to produce excellent wind measurements. The lack of natural targets greatly limits this method for operational use and it must be considered mainly as a cloud and precipitation research device.

More advanced radars, such as the FM/CW and the pseudorandom coded system may help to overcome this lack of natural targets. Both of these systems have been used to detect turbulent scatterers, and, at least with the latter, the Doppler information can be extracted[18]. The early stage of development of the coded system suggests that it will not be ready for operational application for several years. Even then, it must be demonstrated that the tracer availability is good enough to make it a viable system for operational air pollution use.

c. Acoustic

The vertical wind can be measured with the single axis monostatic system described earlier. The Doppler shift of the returned signal is related to the vertical velocity of the scattering volume, which is the vertical component of the wind[19]. By tilting the single monostatic antenna, a component of the horizontal wind can also be measured. A weakness in making wind measurements with a monostatic system is that the return signals are not always continuous; the lack of signal where neutral lapse conditions exist results in an intermittent sampling of the wind.

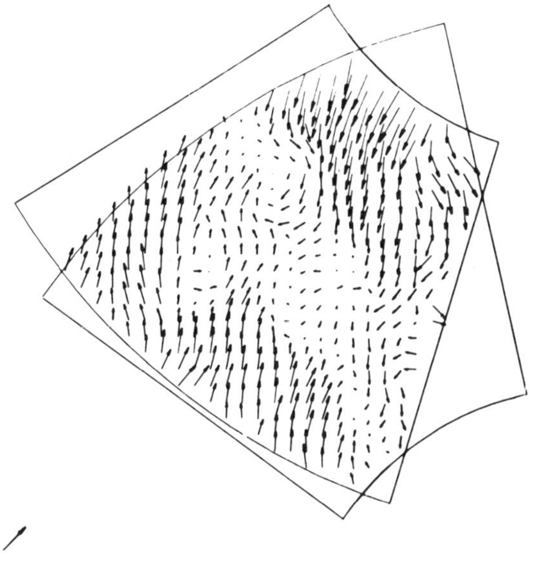

Fig. 7 Two dimensional eddy flow field derived by taking deviations from the mean wind field measured by coplaner Doppler radar. The vector shown in the lower left corner represents a wind speed of 0.2 m/sec[17].

The use of a bistatic system, where the transmitter and receiver are separated helps to solve this problem. Scatter from both wind and temperature fluctuations can be received with the bistatic system and the returns from wind fluctuations tend to fill in the gaps left by the neutral lapse areas which lack temperature eddies. Bistatic configuration like that shown in Fig. 8 have been used to measure the vertical profile of the horizontal wind up to heights of several hundred meters[20,21,22]. Antennas at A, A′, and A′′ in Fig. 8, transmit sound pulses which are scattered and received at points B and C. The Doppler shift is then related to the wind and profiles like those shown in Fig. 9 can be produced.

Another method of wind sensing with an acoustic sounder is to employ the angle of arrival of the returns from a vertically pointed monostatic system[23]. This is a relatively simple method of finding the line average of the wind to heights of several hundred meters; however, it also suffers from a lack of continuous signal under some types of thermal structures.

A passive acoustic method[24] that uses the infrasound generated by naturally occurring wave motion in the atmosphere has also been suggested. In principle, a spaced array of pressure sensors at the ground can be used to measure winds above the ground by sensing the hydrodynamic filtering effect of the atmosphere that causes high frequency pressure fluctuations to die out a short distance from the source, while larger scale features at greater distances are sensed. If the large scale frequency components are moving over the sensor array more rapidly than the small scale components, it can be concluded that the winds aloft are carrying the larger turbulent features past the sensors more rapidly than the winds near the surface

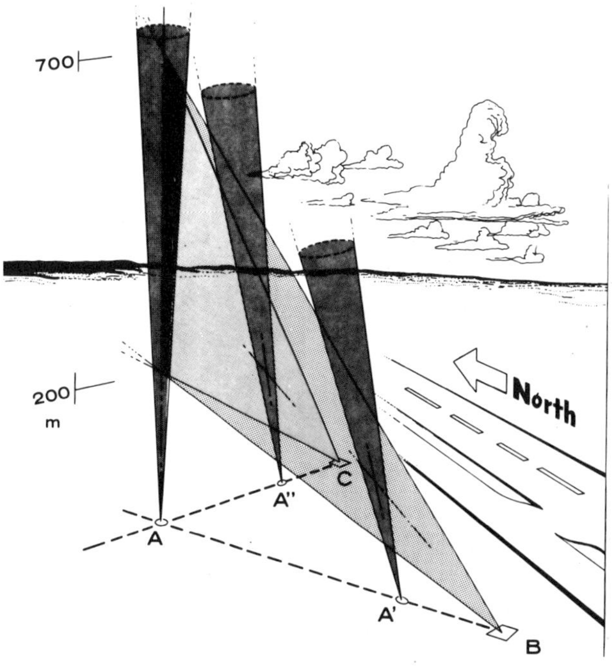

Fig. 8 A multiple transmitter acoustic antenna configuration used to measure a vertical profile of the horizontal wind. Transmitters at A, A′ and A′′ produce sound which is scattered to receivers at B and C.

and that the wind is increasing with height. A full cross spectrum analysis between the returns from multiple sensors can, in principle, provide the information necessary for deriving a vertical profile of the wind.

Measuring the Depth of Mixed Layers

a. Laser Measurements:

Most atmospheric aerosols originate from natural sources or human activities on the surface of the planet. Thus, turbidity, especially in polluted regions, is usually greater within the first kilometer or so of the atmosphere where mixing is strong. The ability of the laser to monitor the depth of these mixed layers through backscattered laser light measurements has already been noted and the actual utility of a pulsed laser system for studying the nature of polluted air has been documented[25,26]. One study in Chicago showed that it was possible to monitor the change in turbidity near the surface and aloft as the synoptic conditions produced air pollution situations. The backscatter returns obtained showed a strong correlation with wind direction, and the effects of the lake breeze from Lake Michigan could be determined. This study showed that the laser could be used in a semi-quantitative fashion by air pollution meteorologists to monitor particulate loading of the lower atmosphere and changes which occurred with varying wind regimes. The narrow beamwidths of lasers have been used to advantage in studying turbidity close to a mountain range and in studying orographic winds produced by the mountains, factors which influence pollution in the Los Angeles Basin[27].

Increase in the mixed depth above a laser station can be effectively studied by displaying side-by-side a number of laser backscatter returns, documenting the intensity of the return by brightness modulation of a cathode ray tube[28] (see Fig. 10). This technique was used during an air pollution study in St. Louis to provide continuous information on the depth of the mixed layer[29]. The mixing height has also been measured by a lidar in an experiment in Oregon[30]. Thus, the laser is already being used in mesoscale, coordinated experiments for studying the mixed depth, but as yet it has not been used in a routine manner to provide day-to-day information for local air pollution meteorologists.

b. Acoustic Echo Sounder:

The depth of the mixed layer is one of the most easily obtained outputs from a monostatic echo sounder (see Fig. 3). Experiments in the urban environment have shown that the inversion that caps the mixing layer can be continuously monitored on a real time basis and displayed on a simple facsimile recorder[31,32].

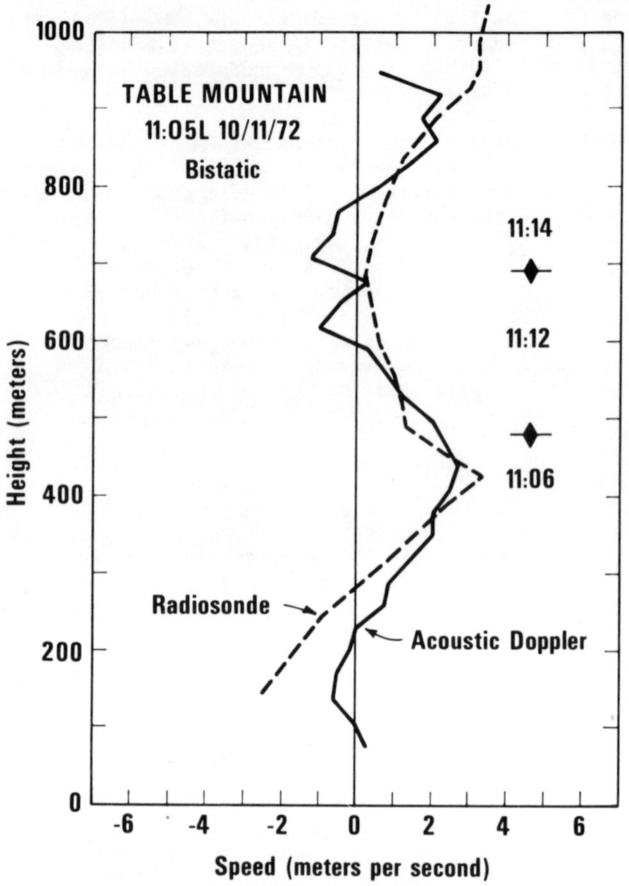

Fig. 9 Comparison of wind profiles measured by a radiosonde and acoustic Doppler system similar to that shown in Fig. 8[22].

An advantage of using the acoustic system for this type of measurement is that it does not rely on the presence of artificial tracers, but uses the naturally occurring temperature fluctuations associated with the inversion surface. This could provide a marked advantage over laser systems which require the buildup of aerosols before the mixing layer can be observed.

c. FM/CW Radar:

The marine inversion layer as well as layers associated with continental radiation inversions (see Fig. 11) have been detected with FM/CW radars[32,33]. The refractive index change that produces the returned signal to an FM/CW radar is probably more strongly a function of humidity than temperature. Nevertheless, this system shows a great deal of potential as a monitor of the capping inversion associated with air pollution.

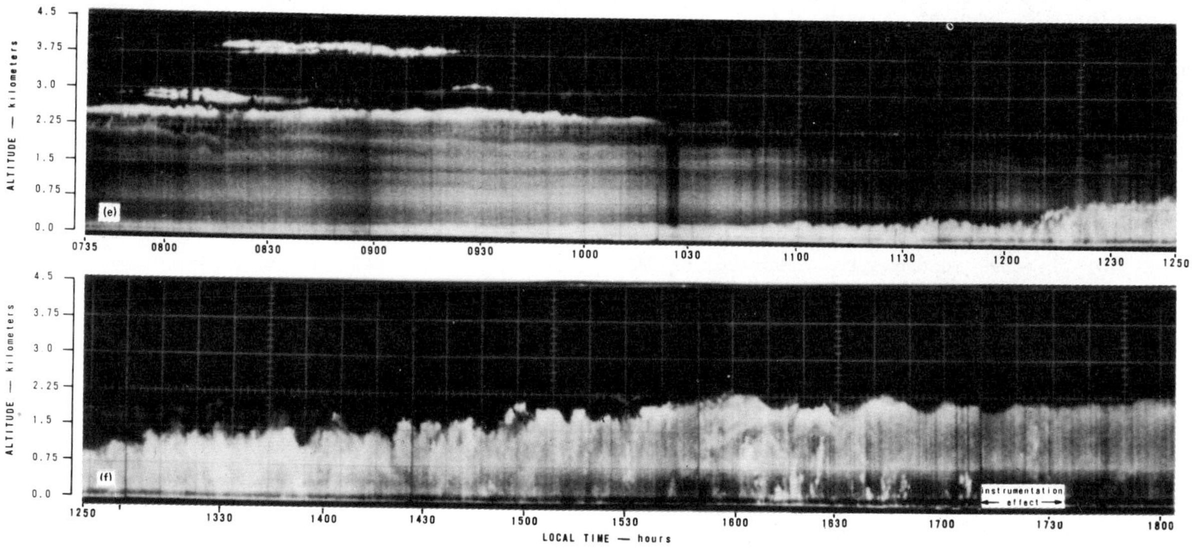

Fig. 10 Height/time section of the aerosol structure measured by a lidar system[29].

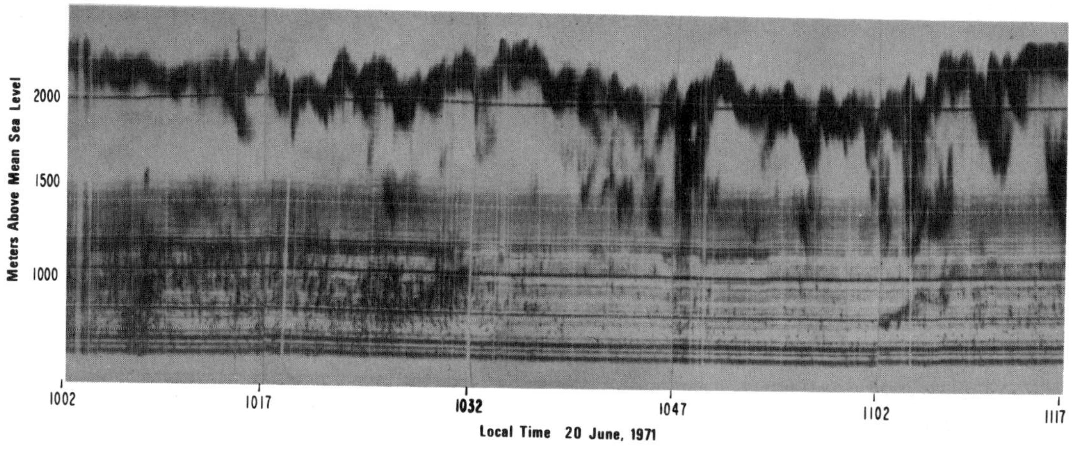

Fig. 11 FM/CW radar returns from an elevated inversion [33].

d. Radiometric:

Determination of the mixing layer depth has also been demonstrated with radiometers operating at 50-60 GHz[35]. This technique relies on the radiometric brightness temperature and equipment manufactured under the trade name of THERMASONDE.

Areal Probing

The ability of certain remote sensing methods to provide more than measurements along a given line-of-sight by scanning a considerable area around the station within defined horizontal ranges and altitudes adds greatly to the utility of the measurements. Operation in this manner is highly dependent upon angular resolution capabilities of the remote sensing device and its provision for a high rejection of unwanted returns in directions removed from the narrow probing main beam of waves.

a. Laser Methods:

Because of the high electromagnetic frequencies for laser probes and the inherently coherent wavefronts from the lasers, high angular resolution and good rejection in other directions is available. A laser beam can be used to illuminate a narrow column of aerosols which backscatter to the aligned receiver. With available lasers, returns from distances of 10 or 20 km may be obtained. Thus, the horizontal stratification in real time may be monitored(27). Until eye-safe lasers, or reliable safety precautions can be developed, rapid scanning of laser beams through large volumes of space over cities cannot be undertaken.

b. Radar Probing:

Powerful radars with large antennas (20 m apertures) are able to monitor the depth of the mixed layer and detect convection patterns over ranges of 10-20 km(36). When operated in the range-height indicator (RHI) mode, the spatial picture of the mixed layer and convective structure can be obtained along a vertical plane and when operated in plan position indicator (PPI) mode the regions where convective structures mix through elevated inversions is graphically shown. Because of the large size and cost of such radar installations, none have yet been used to monitor mixed layers in regions subject to air pollution problems.

c. Transverse Wind Sensing with Lasers:

Temperature variations in the atmosphere, which give rise to acoustic backscatter, also lead to variations in the optical index of refraction. These variations influence the propagation of laser beams, and are responsible for the familiar twinkling of stars. Because such temperature variations in the atmospheric structure have a finite lifetime, the irradiance measured from a transmitted laser beam with two closely spaced apertures will show a correlation in the irradiance when the signals are time lagged an appropriate amount. This time lag is a function of the wind, which advects the temperature structure across the two adjacent paths from the laser to the two spaced apertures. Such a technique has been used(37) to measure the average wind normal to a 15 km path. Real time readout of the transverse wind is possible through use of a signal correlation computer. Measurements along an instrumented 1 km path have shown excellent agreement between spaced, averaged propeller-anemometer readings and the laser derived wind (see Fig. 12). This averaged line integral of the transverse wind could provide information on the drainage across an entire valley, or with several laser paths, the wind flowing into and out of a city. The technique is now available to provide a highly valuable tool for air pollution meteorology studies.

In summary, it would appear that potentially all of the important meteorological variables associated with urban meteorology can be remotely sensed. It is also apparent that each variable could be measured by any of several different techniques. Table I is an attempt to give a rating to the various techniques for sensing temperature, wind and mixing depth. Each sensor or technique given in the second column is rated in various categories to draw some conclusion about its usefulness in an operational urban network of sensors.

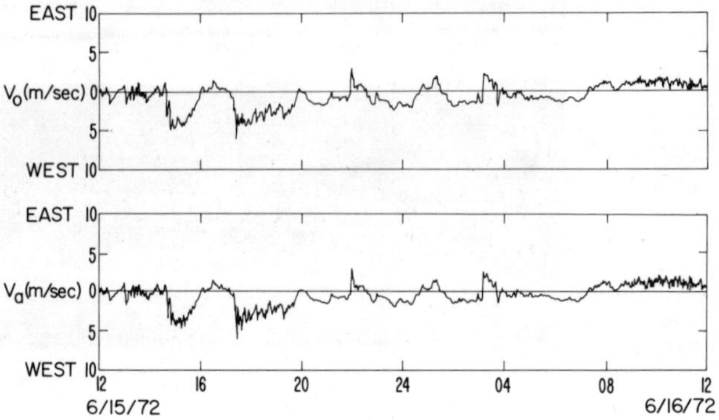

Fig. 12 Average wind speed (V_o) measured with a line-of-sight laser system compared with the average of 6 anemometers (V_a) spaced along the optical path(37).

The first category "stage of development" provides an indication of when the device might be ready for operational use. "Prototype" suggests that the system is ready for integration into a larger network; "near prototype" implies that the system would be ready within a year; "experimental" indicates that the system has been demonstrated to be feasible but significant engineering problems remain before it is ready for use, probably on a time scale greater then one year. Concepts other than those listed for remotely sensing various parametear have been identified; however, they are too far from being a viable competitor to warrant listing them at this time.

The second category "potential for all weather operation" considers such factors as the possible degradation of laser signal during heavy cloud or fog and the effect of rain noise on the acoustic system. In judging a system in this and in other categories, we must keep in mind the final objective of sensing the parameter for use in air pollution related meteorology. For example, the inability of the laser or microwave radiometer to penetrate thick clouds may not be a significant deterrent to its use for monitoring low level phenomena. In addition, the noise saturation from hydrometeors striking an acoustic antenna may be acceptable, since air pollution would probably not be a problem during or immediately after a heavy rain.

"Tracer availability" the third category, is a rating of how often a sensor will be able to operate It is a critical category if the sensor is being considered for continual operation and a low rating here indicates that the device is probably not acceptable for an urban air pollution application

Tracers are either natural or the result of inadvertent air pollution; the introduction of chaff, for example, is considered unacceptable for an operational system.

The categories of "spatial and time resolution" give an indication of how detailed the measurements by each sensor could be. Spatial resolution is in most cases, determined by antenna beam widths and pulse lengths. Time resolution refers to the averaging times required to produce acceptable measurements. Small scale turbulence measurement is not anticipated and readout of data is required on the order of every 15 to 20 min for an urban sensor network. All of the sensors considered have at least this capability.

"Accuracy," the next to last category, is difficult to assess because it is in many cases a function of the mode in which a system is operated. In addition, many of the instruments have undergone only limited testing and comparison with standard in situ measurements (a very difficult comparison to make) and accuracy estimates are based on the theoretical potential rather than actual calibration. As in the previous category, it is felt that all of the instruments listed can achieve accuracies suitable for the task considered here.

The final category, "Acceptability" refers to the potential nuisance or danger from a particular device. In other words, will it be acceptable for use in the urban environment? An example of a potential problem in this category would be lasers which might produce eye damage.

REMOTE SENSING SYSTEMS

Up to this point we have discussed the urban meteorological problem and the types of data that are required as input for short term forecasting. In addition, we have given a brief overview of the various remote sensing devices which could conceivably make some or all of the required measurements. It is obvious that some parameters can be measured by several techniques. What is not obvious is which of these sensors or collection of sensors would be optimum for the specific problem of improving our data gathering methods in the urban environment. This section is devoted to an assessment of the requirements for a measurement system and an attempt is made to provide guidance in selecting the optimum set of sensors for a city.

It is recognized that every urban region is a unique entity and the correct location and sensor distribution for one city may be totally unacceptable for another and should be dealt with on a case to case basis. Let us identify those features of a city or urban area that are common to all and use these features to generate a hypothetical city. We then suggest the types of sensors or sensor systems which are suited for this hypothetical city, keeping the analysis general enough so the results can be applied to specific cases with modification to fit unique local requirements.

First, all cities encompass a given area. This basic area will vary in size, shape and density of population, so for a general case, we will assume that the area is simply a circle of unit size. Fig. 13 shows the hypothetical model of a city made up of basic features.

Cities usually contain industrial regions which are a major source of pollution and as such need to be delineated in our model. Again, for the sake of simplicity we will designate the industrial region of an urban complex as a circle, contained within the larger "area" circle.

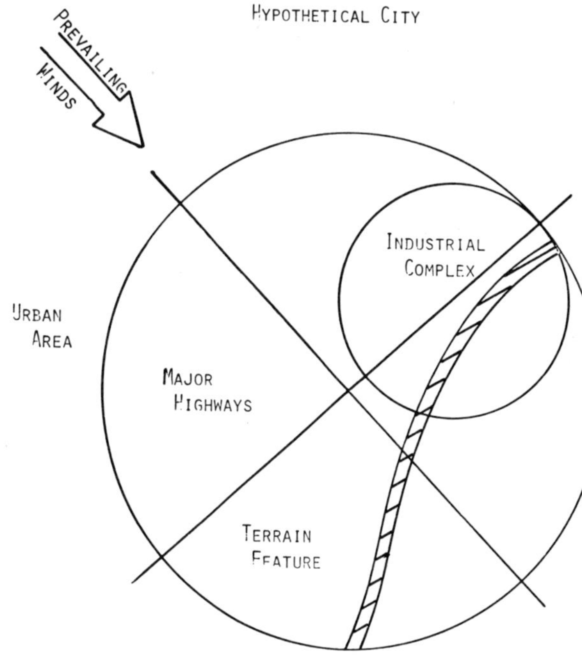

Fig. 13 A hypothetical city made up of features which are common to most urban regions.

In addition to a concentrated pollution source region, such as an industrial complex, all cities contain line sources, usually in the form of major arterial highways. For our hypothetical city we will show these simply as crossing diagonal lines.

Terrain features are another important influence on urban meteorology[38], many times controlling the local flow which advects or concentrates effluent in a given region. For example, a river valley is a natural place for cold air drainage, while a coast line produces local land and sea breeze circulation, alternately cleansing a region and concentrating pollution at the sea breeze front. Without specifying the type, we will include a terrain feature in our model as a cross-hatched band.

TABLE I

Parameter	Category / Sensor	Stage of Development	Potential for all Weather Operation	Tracer Availability	Spatial Resolution	Time Resolution	Accuracy	Acceptability
Temperature or Stability	Acoustic	Near Prototype	Medium	High	Medium	Medium	Low	High
	Laser	Experimental	Medium	Medium	High	Medium	Medium	Medium
	Microwave Radiometer	Prototype	High	-	Low to Medium	Medium	Medium	High
	RASS	Experimental	Medium	High	Medium	Medium	Medium	High
Wind	Acoustic	Near Prototype	Medium	High	Medium	Medium	Medium	High
	Laser	Experimental	Medium	Medium	High	High	High	Medium
	Radar	Prototype	Low	Low	High	High	High	High
Mixing Layer Depth	Acoustic	Near Prototype	Medium	High	High	High	High	High
	Laser	Near Prototype	Medium	Medium	High	High	High	Medium
	Radar FM/CW	Experimental	High	Medium	High	High	High	High

In the first section of this paper, climate was mentioned as an important influence on urban air pollution. It is difficult to build this feature into a totally general model, because of the wide range of climatic zones covered by the cities of the world. We can say, however, that all cities will have some prevailing wind direction and provide for this in our model.

Most urban regions can be reduced to a similar basic model by changing its shape and distribution of features. Recalling that the primary meteorological parameters for the urban region are the temperature profile (or some indication of the ambient stability), and the wind field within the boundary layer, we can now proceed with the instrumentation of our city. For temperature we can assume that conditions will be reasonably homogeneous over the urban area. The surrounding area will have different temperatures caused by urban heat island effect. The temperature profile should be monitored in both the urban area and outside of this area to have continuous information on the stability, including those changes induced by the urban heat island. Both temperature sensors should have a range capability which insures that information is collected throughout the depth of the boundary layer. Greater ranges would be desirable for synoptic scale input; however, a ground based temperature sensor with ranges greater than 3 km does not appear feasible at this time.

Measurement of the wind field over the urban area is a more difficult problem. In addition to a vertical profile of the wind, we would like to monitor the trajectories of polluted air parcels, which implies the need for either several sensors at strategic locations to measure the vertical profile, or the need for a single scanning sensor to map winds over a large area.

Based on experimental work to date, it is clear that the urban measurement system could be made up entirely of either acoustic or laser devices. As a starting point, we will look at how each of these families of remote sensors might be applied.

Looking first at acoustic devices, one can envision the core of a system being an acoustic sounder with full Doppler capabilities situated near the center of the city (see Fig. 14). This central system can also be operated in a monostatic mode providing the information on mixing depth and an indication of low level stability. The acoustic Doppler system can only provide a vertical profile of the horizontal wind. Single acoustic systems cannot monitor large regions. Therefore, in order to determine the trajectories of air parcels it would be necessary to establish several satellite stations around the city to sense the horizontal variations in the wind

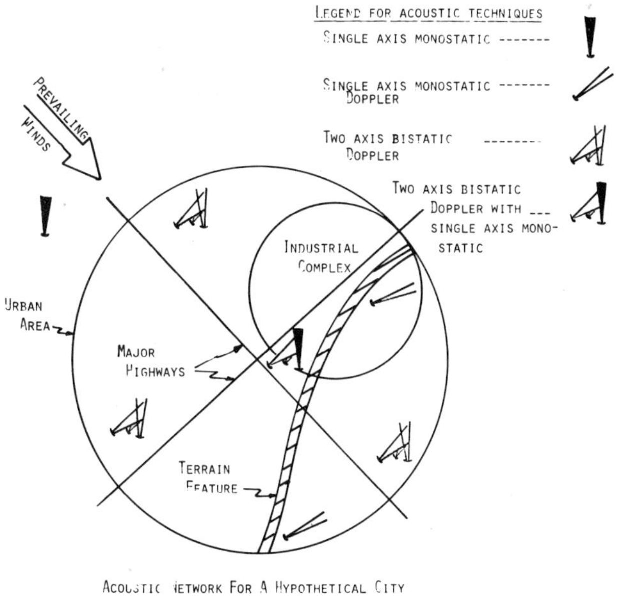

Fig. 14 Proposed network of acoustic sensors placed in the hypothetical city of Fig. 13

field. Here it is important to consider the orientation and type of terrain features and the city size before locating the satellite sensors. While the main core system should have a full wind profiling capability employing gating and measurement of at least two components of the horizontal wind, some of the satellite systems (for example, those along a river valley where only one component of the wind is needed) could conceivably be of a simpler design. An acoustic Doppler system which measures only the average wind from the ground to the capping inversion could provide information necessary for determining gross trajectories of parcels along or near a particular terrain feature. The depth of the mixing layer outside our urban area could be monitored by a simple monostatic system upwind of the city.

A similar acoustic system, based on the angle of arrival technique is also conceivable. Major drawbacks, however, would be less vertical resolution in the wind profile and a reliance on monostatic returns, which are dependent on strong signal from temperature structure alone.

Neither of these systems would provide a direct indication of the temperature profile. To do this, it would require the introduction of a RASS in place of the monostatic sounder located outside of the city and possibly a second RASS near the central acoustic Doppler system. The need for measuring the absolute temperature profile must remain as an open question for now. If this profile is used only to derive the stability of the lower atmosphere, one must ask if there are more direct methods which do not rely on obtaining the absolute temperature, but still provide the required stability information. For example, the variance of the wind is correlated with stability[39], and the monostatic echo sounder record provides a qualitative indication of the stability. Since both the vertical wind profile and a record of the lower atmospheric structure can be obtained from an acoustic Doppler system it is conceivable that a derived stability could be obtained which might eliminate the need for more elaborate temperature profiling techniques[40].

Turning now to an urban remote sensing system based entirely on the family of laser sensors we see some interesting differences. The ability of the laser to scan through a volume and to sense the line integral of a parameter can be applied in a variety of ways. For example, by locating a pulsed laser at some elevated point within the urban complex and scanning the beam (see Fig. 15) in a circular fashion, a complete plan position indication of the effluent over the city could be produced. It is conceivable that a time sequence of identifiable features in this type of information could then be used to produce a trajectory analysis of the air over the urban region. The addition of a second such laser at another location and the use of both lasers in a Doppler mode would make it possible to directly measure the wind field in the plane common to both laser beams. The temperature profile in the all-laser system could be obtained by either the Raman scatter technique or from the Doppler broadened widths of the backscattered spectrum of laser light. As in the acoustic system these devices should be located both within the urban area and in the surrounding area to measure the effect of the urban heat island. A possible weakness here would be the system located outside an urban complex where non-existence of tracers in the much cleaner air may hinder the operation of the longer wavelength Doppler laser system.

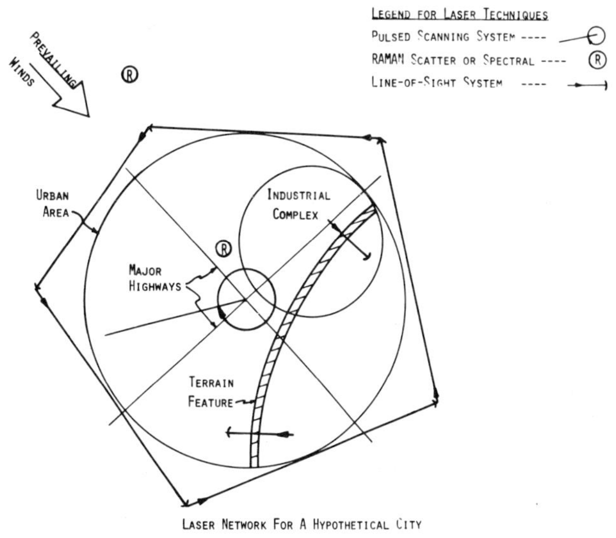

Fig. 15 Proposed network of laser type sensors placed in the hypothetical city of Fig. 13.

The depth of the mixing layer could also be observed by vertically pointed laser systems within the urban area; again the lack of tracers outside of the region could degrade the performance of a laser in that location.

The ability of line-of-sight laser systems to measure the average wind component normal to the beam paths could find many unique applications within the urban region. For example, if the major terrain feature in a city is a river valley, it is important to measure the very light drainage winds which transport pollutants along the valley. An ideal method for monitoring these winds is the line-of-sight laser. One or more of these systems placed so they sense the component of the wind along the valley floor would provide a continual monitor of the average transport of air into or out of a given region.

This same type of system could be installed in a ring around the city with perhaps four or five sets of transmitter and receivers to measure the low-level convergence into the urban area. The remote sensing system based on laser techniques is shown in Fig. 15

In developing the two "pure" systems described above, we have omitted one potentially very valuable technique for measuring temperature profiles, the microwave radiometer. The advanced stage of development of this technique and its potential range of 2 or 3 km makes it a highly competitive device. If the actual temperature profile, and not a more direct indication of stability is required, then the microwave radiometer should be considered as part of the final system.

To specify the optimum system for a given city, it would be necessary to consider such important factors as the cost of each system and the stage of development of each device. Costs are difficult to assess since the only basis for comparison must be made on the cost of present research oriented instruments. These are by their very nature more expensive than production models where the advantages of standardization can be had. In general, research oriented acoustic devices are less expensive to build and operate than lidar (optical radar) instruments. If these costs can be projected to production models, the conclusion must be that an acoustic system would be less expensive.

"Stage of development" is an equally difficult area in which to make predictions due to the uncertain nature of research projects. Some devices have been shown to be very close to the operational stage. It is safe to suggest that they should be considered first in planning for the immediate future. The microwave radiometer is one of the more advanced instruments, already having been tested in the urban region. Laser systems that are well advanced are those used to measure the depth of the mixing layer The line-of-sight laser has been demonstrated in several research projects and has been used to measure the wind component averaged over a 15 km path in the lee of a mountain range. Acoustic systems are also being tested in the urban environment under operational conditions. Single axis monostatic sounders have been used to monitor the inversion layers in cities and a full Doppler system has been placed in operation at an airport for the purpose of real time monitoring of the wind and wind shear.

In summary, the optimum system for any given city must be the product of many considerations. A variety of techniques are available for making the required measurements and selection of a system must be based on such factors as:

a) unique requirements imposed by terrain and demography of a particular city.

b) time available before the system must be operational

c) overall cost of the system

d) stage of development of the selected sensors.

It is concluded that the incorporation of remote sensing devices into urban monitoring systems has may advantages over presently used point sensors. In the future, remote sensing can play an important role in aiding the prediction, control, and monitoring of air pollution in cities.

REFERENCES

(1) Committee on Atmospheric Sciences, "The Atmospheric Sciences and Man's Needs; Priorities for the Future," NATIONAL ACADEMY OF SCIENCES, WASHINGTON, D. C. (1971), 88 p.

(2) Neiburger, M., "The Role of Meteorology in the Study and Control of Air Pollution," BULL. OF THE AMERICAN METEOROLOGICAL SOCIETY, Vol. 50, (1969), 957-965.

(3) Little, C. G., "Remote Sensing of the Atmosphere," ATMOSPHERIC TECHNOLOGY, NCAR, No. 2, June 1973, 51-56.

(4) Little, C. G., "Status of Remote Sensing of the Troposphere," BULLETIN OF THE AMERICAN METEOROLOGICAL SOCIETY, Vol. 53, (1972), 936-949.

(5) Derr, V. E., C. G. Little, "A Comparison of Remote Sensing of the Clear Atmosphere by Optical, Radio, and Acoustic Radar Techniques," APPLIED OPTICS, Vol. 9, (1970), 1976-1992.

(6) Derr, V. E. (editor) "Remote Sensing of the Troposphere," Wave Propagation Lab., NOAA Boulder, Colorado GPO (Catalog No. C55.602:T75) 809 p., August 1972.

(7) Little, C. G., "Acoustic Methods for the Remote Probing of the Lower Atmosphere," PROCEEDINGS OF THE IEEE, Vol. 57, (1969), 571-578.

(8) Westwater, E. R., "Ground-Based Determination of Low Altitude Temperature Profiles by Microwaves" MONTHLY WEATHER REVIEW, Vol. 100 (1972), 15-28.

(9) Strauch, R. G., V. E. Derr and R. E. Cupp, "Atmospheric Temperature Measurements Using Raman Backscatter," APPLIED OPTICS, Vol. 10, (1971), 2665-2669.

(10) Cooney, J., "Measurement of Atmospheric Temperature Profiles by Raman Backscatter," JOURNAL OF APPLIED METEOROLOGY, Vol. 11 (1972) 108-112.

(11) Fiocco, G., G. Beneditti-Michelangeli, K. Maischberger and E. Madonna, "Measurement of Temperature and Aerosol to Molecule Ratio in the Troposphere by Optical Radar," NATURE, Vol. 229 (1971), 78-79.

(12) Gething, J. T. and D. Jenssen, "Measurement of Temperature and Humidity by Acoustic Echo Sounding," NATURE, Vol. 231 (1971), 198-200.

(13) North, E. M., A. M. Peterson, and H. D. Parry, "RASS, A Remote Sensing System for Measuring Low-Level Temperature Profiles," BULLETIN OF THE AMERICAN METEOROLOGICAL SOCIETY, Vol 54 (1973), 912-919.

(14) Huffaker, R. M., A. V. Jelalian, J.A.L Thompson, "Laser Doppler System for Detection of Aircraft Trailing Vortices," PROCEEDINGS OF IEEE, Vol. 58 (1970), 322-326.

(15) Huffaker, R. M. A. V. Jelalian, W. Keene, C. Sonnenshein and J.A.L. Thompson, "Application of Laser Doppler Systems to Vortex Measurements and Detection," AIRCRAFT WAKE TURBULENCE AND ITS DETECTION (Ed. Olson, Goldberg and Rogers) Plenum Press, New York (1971), 113-124.

(16) Farmer, W. M. and D. E. Brayton, "Analysis of Atmospheric Laser Doppler Velocimeters," APPLIED OPTICS, Vol. 10 (1971), 2319-2324.

(17) Miller, L. J., "Dual-Doppler Radar Observations of Circulation in Snow Conditions" (1972) PROC. 15TH RADAR METEOR. CONF. AMER. METEOR. SOC.

(18) Reid, N., "A Millimeter Wave Pseudorandom Coded Meteorological Radar," IEEE TRANSACTIONS GEOSCIENCE ELEC. Vol. 7 (1969), 146-156.

(19) Beran, D. W., C. G. Little and B. C. Willmarth, "Acoustic Doppler Measurements of Vertical Velocities in the Atmosphere," NATURE, Vol. 230 (1971), 160-162.

(20) Beran, D. W. and S. F. Clifford, "Acoustic Doppler Measurement of the Total Wind Vector," AMS 2ND SYMPOSIUM ON METEOROLOGICAL OBSERVATIONS AND INSTRUMENTATION, March 1972, 100-109.

(21) Beran, D. W., "Acoustics: A New Approach for Monitoring the Environment Near Airports," JOURNAL OF AIRCRAFT, Vol. 8 (1971), 934-936.

(22) Beran, D. W., B. C. Willmarth, F. D. Carsey and F. F. Hall, Jr. "An Acoustic Doppler Wind Measuring System," THE JOURNAL OF THE ACOUSTIC SOCIETY OF AMERICA. (to be published Feb.1974).

(23) Mahoney, A. R., L. G. McAllister and J. R. J. R. Pollard, "The Remote Sensing of Wind Velocity in the Lower Troposphere Using an Acoustic Sounder," BOUNDARY LAYER METEOROLOGY, Vol. 4 (1973), 155-167.

(24) Priestley, J. T., "Correlation Studies of Pressure Fluctuations on the Ground Beneath a Turbulent Boundary Layer," NATIONAL BUREAU OF STANDARDS, Report No. 8942 (1966), 91 p.

(25) Barrett, E. W. and O. Ben-Dov, "Application of the Lidar to Air Pollution Measurements, JOURNAL OF APPLIED METEOROLOGY, Vol. 6 (1967) 500-515.

(26) Fernald, F. G., B. M. Herman and J. A. Reagan, "Determination of Aerosol Height Distribution by Lidar," JOURNAL OF APPLIED METEOROLOGY, Vol. 11 (1972), 482-489.

(27) Hall, F. F. Jr. and H. Y. Ageno, "Lidar Measurements of Turbidity in the Troposphere," LASER APPLICATIONS OF THE GEOSCIENCES, (ed. Gauger and Hall) Western Periodicals Co, North Hollywood, Calif. (1970), 17-32.

(28) Collis, R.T.H. and E. E. Uthe, "Mie Scattering Techniques for Air Pollution Measurement with Lasers," OPTO-ELECTRONICS, Vol. 4 (1972), 87-99.

(29) Uthe, E. E., "Lidar Observations of the Urban Aerosol Structure," BULLETIN AMERICAN METEOROLOGICAL SOCIETY, Vol. 53 (1970), 358-360.

(30) McCormick, M. P., S. H. Melfi, L. E. Olsson, W. L. Taft, W. P. Elliott and R. Egami, "Mixing-Height Measurements by Lidar Particle Counter, and Rawinsonde in the Willamette Valley, Oregon," NASA Tech Note D-7103, (1972), 78 p.

(31) Beran, D. W., F. F. Hall, Jr., J. W. Wescott, W. D. Neff, "Application of an Acoustic Sounder to Air Pollution Monitoring," PROCEEDINGS OF AIR POLLUTION TURBULENCE AND DIFFUSION SYMPOSIUM, New Mexico State Univ., Las Cruces, N. M. (1971), 7 p.

(32) Wyckoff, R. J., D. W. Beran and F. F. Hall, Jr., "A Comparison of the Low Level Radiosonde and the Acoustic Echo Sounder for Monitoring Atmospheric Stability," JOURNAL OF APPLIED METEOROLOGY, Vol. 12 (1973), 1196-1204.

(33) Gossard, E. E. and J. H. Richter, "The Shape of Internal Waves of Finite Amplitude from High Resolution Radar Sounding of the Lower Atmosphere," JOURNAL OF THE ATMOSPHERIC SCIENCES, Vol. 27 (1970), 971-973.

(34) Bean, B. R. R. E. McGavin and B. D. Warner, "A Note on the FM-CW Radar as a Remote Probe of the Pacific Trade-Wind Inversion," BOUNDARY LAYER METEOROLOGY, Vol. 4 (1973), 201-209.

(35) Anway, A. C., "Radiometrically Measuring the Mixing Layer Aids Air Pollution Forecasting," 2ND INTERNATIONAL CLEAN AIR CONGRESS (ed. England and Beery), (1970), 999-1003.

(36) Battan, L. J., "Radar Observations of the Atmosphere," Univ. of Chicago Press, Chicago, Ill. (1973)

(37) Lawrence, R. S., G. R. Ochs and S. F. Clifford, "Use of Scintillation to Measure Average Wind Across a Light Beam," APPLIED OPTICS, Vol. 11 (1972), 239-243.

(38) Turner, D. B. and J. L. Dicke, "Influence of Tropography on Transport and Diffusion," AIR POLLUTION METEOROLOGY, Institute for Air Pollution Training, HEW, Research Triangle Park, North Carolina (no date) 2-9 to 2-13.

(39) Singer, I. A. and M. E. Smith, "Relation of Gustiness to Other Meteorological Parameters," JOURNAL OF METEOROLOGY, Vol. 10, 121-126.

(40) Baynton, H. W., "Stability Inferences from Precision Rawins," MONTHLY WEATHER REVIEW, Vol. 96 (1968), 47-52.

(41) Snider, J. B., "Ground-Based Sensing of Temperature Profiles from Angular and Multi-Spectral Microwave Emission Measurements," JOURNAL OF APPLIED METEOROLOGY, Vol. 11 (1972), 958-967.

© 1973, ISA JSP 6701

VARIATIONS OF METEOROLOGY, POLLUTANT EMISSIONS, AND AIR QUALITY

George C. Holzworth*
Chief, Climatic Analysis Branch
Meteorology Laboratory
Research Triangle Park, North Carolina

ABSTRACT

This paper presents information describing various temporal and spatial variations in pollutant emissions, atmospheric transport/diffusion, and air quality that are broadly applicable to large cities in the United States. The overall impact on air quality of the interplay between diurnal variations in emissions and meteorology is described. It is concluded that complete explanations of air quality values measured continuously at specific locations require detailed emission and meteorological information.

INTRODUCTION

A general characteristic of the air quality records for any one of the common pollutants in urban areas is the variability of the concentrations. Concentrations are especially variable in time: from hour to hour, day to day, season to season, and year to year. In some cases the concentrations are highly variable over rather short distances. For those pollutants that react relatively slowly while they are airborne, their temporal and spatial variations in concentration are largely due to an interplay between similar variations in atmospheric transport/diffusion and pollutant emissions. For pollutants that more readily undergo reactions in the atmosphere (e.g., photochemical oxidant (Ox) formation or physical removal by precipitation) their concentrations also depend on the intensities of the phenomena involved. Temporal variations in transport/diffusion are tied to natural cycles of the weather and large spatial differences, to climatic effects. Variations in pollutant emissions are dictated by social/economic practices, and in some cases these are also influenced by the weather (e.g., space heating and cooling demands). Although emissions, weather, and atmospheric reactions may follow various diurnal, seasonal, etc., cycles, their phases, amplitudes, and periods are usually different. And this tends to confound explanations of observed pollutant concentrations. The objective of this paper is to describe some of the variations in pollutant emissions, atmospheric transport/diffusion, and observed concentrations.

* On assignment from the National Oceanic and Atmospheric Administration, U. S. Department of Commerce

EMISSIONS

Figure 1 shows hourly values of electrical generation, steam output, and natural gas sendout for plants in St. Louis in comparison to air temperature at the airport. This particular 2-day period of warming was selected to illustrate the impact of temperature on energy utilization. Aside from the effect of the overall warming trend, the diurnal variations of these curves are considered typical of winter. The gas sendout is mainly to households. Although natural gas is a clean-burning fuel, not everybody has it, but its consumption here is also indicative of the residential consumption of other fuels. On both days the gas sendout begins to increase around 0300 CST, reaches a maximum at 0800 CST, falls to a low value around 1600 CST, rises to a secondary peak at about 2100 CST, and then falls back to an early morning low. Although these cycles are clearly evident on both days, they are superimposed on a generally declining trend, in response to rising temperatures. The early morning and late afternoon increases in gas consumption are apparently due to space heating, cooking, and water heating that go along with starting and ending daily cycles of household activity.

The steam output shown here is for a downtown plant that mostly supplies business establishments. The steam curve is quite similar to that for gas, except the late afternoon rises and evening peaks in gas are absent. The steam curve falls rather rapidly from its morning peak around 0800 CST until late afternoon or early evening and then, apparently depending on the temperature, falls more slowly or remains nearly constant through the night until "start-up" the next day.

The electrical load shown here is for one of several generating plants that supply residential, commercial, and industrial customers. Although other generating plants in this system were operating at a rather constant (or base) load, the plant referred to here was carrying most of the variable load. The data for winter days other than shown here suggest that while there are some day to day similarities in the diurnal variation of electrical output at this plant, they are not as consistent as for gas or steam. Thus, in general the electrical load is relatively low in the early morning, rises rapidly to a plateau or peak by noon, often declines somewhat in the afternoon, returns by early evening to the level at about noon, roughly holds that level until about midnight, and then drops rapidly to the early morning low. The low values on Friday at 2200 and 2300 CST appear to be exceptional. Apparently, temperature has an effect on electrical demand. For example, the mid-afternoon electrical load on Friday was considerably more than on Saturday when the maximum temperature was 34°F greater, and similarly for the early morning low loads in comparison to minimum temperatures.

Figure 2 shows the average hourly traffic count for an expressway in Cincinnati. Although the counts for different types of roadways in different cities undoubtedly differ in detail, the diurnal variations shown here, especially for weekdays, are generally representative. The weekday curve only confirms the experience of most city motorists, morning and evening traffic peaks with a lull in between. Traffic counts for weekends are considerably different than for weekdays. Weekend curves for different cities and roadways are likely to differ considerably, and there may even be significant differences between Saturdays, Sundays, and holidays.

Thus, Figure 1 shows the impact of temperature and social practices on patterns of energy utilization by stationary consumers. The winter patterns shown are obviously modified in the summer. In some cities space heating requirements in winter are replaced in summer by space cooling with a peak demand for electrical energy in afternoons and evenings. Figure 2 illustrates the general diurnal patterns of emissions expected from autos. Since these patterns are mainly determined by driving practices, they are fairly constant throughout the year. It should be pointed out that the curves in Figures 1 and 2 only show patterns of energy utilization. Actual emissions depend on composition of the fuel, the operating mode in which it is utilized, and control devices that remove or alter pollutants before they are emitted.

AIR QUALITY

Figure 3 shows the yearly variation in diurnal patterns of SO_2 concentration by season in downtown Philadelphia. It illustrates the variations in seasonal-diurnal patterns that can occur from year to year. For example, for spring the SO_2 concentrations were considerably lower in 1965 than in the other 3 years, which all were remarkably similar. Also for spring, the 1965 diurnal variations were less evident than in the other years. On the other hand for autumn, the 1965 concentrations were consistently higher than in any of the other years and the usual diurnal variation was clearly evident. Similar comparisons can be seen for the other seasons.

The main causes of these variations in SO_2 concentration are variations in the patterns of emissions, which at least in part are related to the weather, and variations in atmospheric transport/diffusion. However, the dearth of available emission and meteorological information precludes the determination of quantitative relationships. Meteorologically, what is required is detailed information in the immediate vicinity and upwind of the air quality station, and in the immediate vicinity and downwind of each major source. In the case of SO_2, much of the emissions come from large specific sources. Consequently, slight changes in wind direction can have a significant impact on concentrations measured at a particular location. Since much of the SO_2 that is emitted comes from tall stacks, wind data above ground-level are required. Also, the vertical structure of atmospheric temperature, which ordinarily goes through a large diurnal variation near the ground, plays an important role in mixing elevated pollutant plumes down to ground-level. Obviously the usual weather observations made at an airport on the outskirts of a city are hardly adequate for such analyses.

Figure 4 shows the diurnal variations in concentrations of six major pollutants by weekdays, Saturdays, and Sundays in downtown Philadelphia. Notice that the variations are presented in terms of percentages of the average concentrations for all hours. As expected, most concentrations are lowest on Sundays and highest on weekdays. Most of the Sunday curves have only weak morning and evening peaks compared to weekdays, with Saturdays intermediate. The NO and CO curves suggest that traffic is heavier on Saturday nights than on other nights.

Although the concentrations of NO, NO_2, and total hydrocarbons, which are involved in the formation of Ox, are considerably less on Sundays and Saturdays than on weekdays, all three curves of total Ox are remarkably similar. It has been suggested that this may be because the amounts of these pollutants in the city atmosphere during the daytime are always adequate for undiminished photochemical generation of Ox, or at least during those months when the solar flux is sufficient to initiate the photochemical process.

The importance of solar energy in generating Ox in Philadelphia is clearly indicated in Figure 5 by monthly curves of its diurnal variation in concentration. During the 6 months October through March, when the potential solar flux is lowest, there is little generation of Ox. But in concert with rapidly increasing solar flux in April, the daytime generation of Ox becomes clearly established. The peak concentrations attain their highest values in June and July and then decline markedly through October as the solar flux decreases. The role of the weather in altering the potential flux of solar radiation plays an important role in assessing Ox air quality--in addition to the effects of transport/diffusion. Although as a general rule the dilution capacity of the atmosphere increases rapidly in the forenoon, it appears from Figure 5 that it is not sufficient to noticeably offset the rate of Ox formation.

The diurnal and seasonal variations in Ox concentrations contrast rather sharply with those for CO, as shown by comparing Figures 5 and 6. The basic source of both pollutants is the automobile, and both were measured at the same site in downtown Philadelphia. The reason for this difference is of course that Ox requires solar radiation for

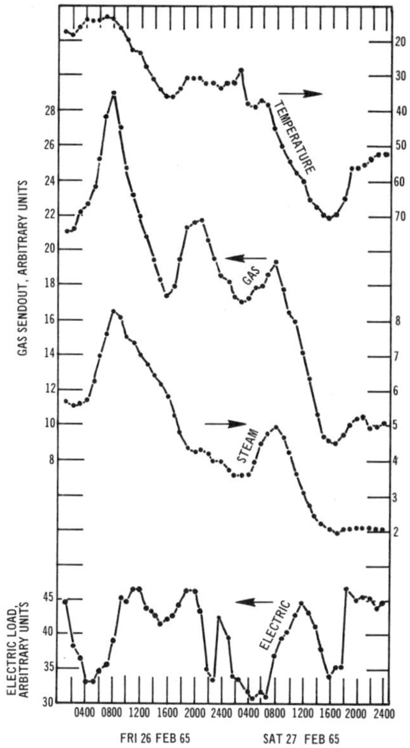

Fig. 1 Hourly values of electrical load, steam output, and gas sendout (all in arbitrary units) for plants in St. Louis, and hourly temperatures (F°) at the airport (after Turner [1]).

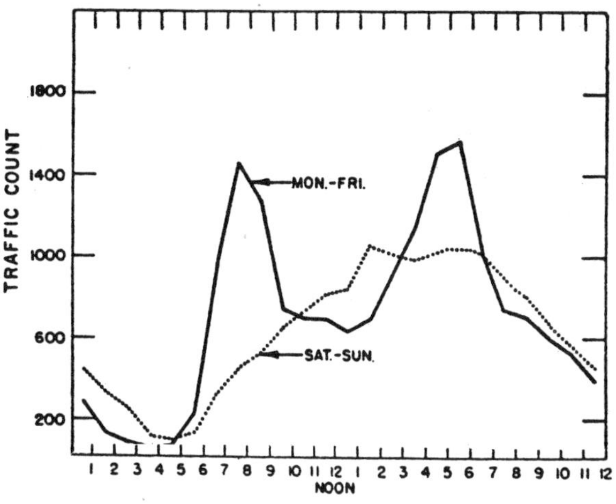

Fig. 2 Average hourly traffic count data for a main expressway in Cincinnati, 10-26 May 1960 (after McCormick and Xintaras [2]).

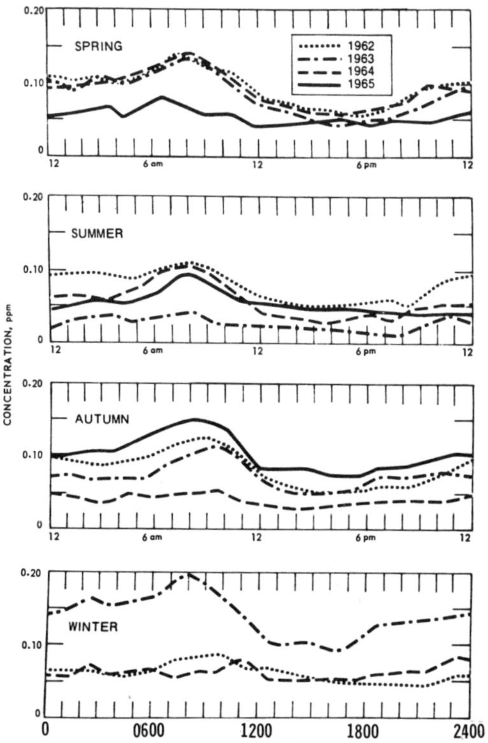

Fig. 3 Annual curves of diurnal variation of SO_2 concentration by season in Philadelphia (after U. S. Dept. HEW [3]).

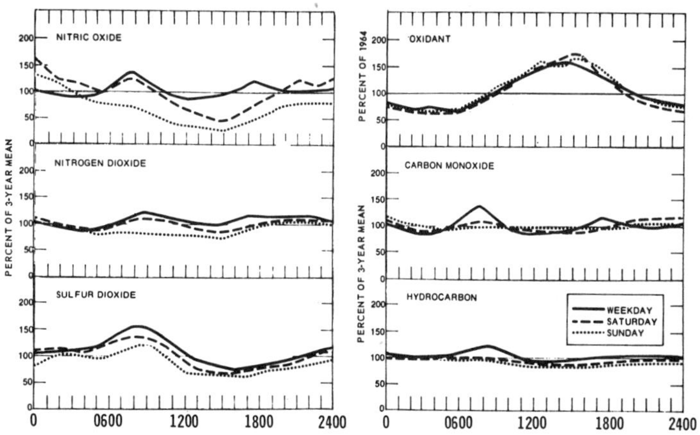

Fig. 4 Diurnal patterns of gaseous pollutant concentrations (in percentages of overall mean concentrations) for weekdays, Saturdays, and Sundays in Philadelphia, 1964 - 1965 (after U. S. Dept. HEW [3]).

its formation and some time in the atmosphere, whereas much of the measured CO comes directly from nearby autos. This CO, with a relatively short time from its emission until it is taken in by the sampler, has less opportunity to be diluted in the atmosphere than the CO that comes from more distant sources. Evidence for the general effects of dilution is shown in Figure 6 by the rather irregular month to month variations in the diurnal curves. For example, in March the CO concentrations are generally low compared to the values in February and April, and in September the concentrations, especially the morning peak values, are generally higher than in August or October. It seems highly unlikely that such concentration variations were caused mainly by variations in emissions. In spite of the implied monthly variations in atmospheric transport/diffusion, the direct effects of morning and evening peaks in CO emissions still show up clearly in the diurnal concentration curves. In fact, even the effect of switching between Daylight Saving and Standard Times can be seen in Figure 6. This is because all the concentrations are plotted against Standard Time, but the emissions followed Local Time and therefore were shifted 1 hour earlier during those months when Daylight Saving Time was observed. Thus, from November through March (Standard Time) the morning CO concentration peaks occurred at or slightly after 0800 EST but from April through October (Daylight Saving Time) they occurred around 0700 EST. Similarly, the evening peaks occurred at 1730 EST from November through March and nearer 1630 EST from May through October; April is only verv slightly earlier than March.

In addition to temporal variations, air quality may also vary significantly over rather short distances. Figure 7 shows the average distribution of CO concentrations (in ppm) at 3 meters (m) above sidewalk-level in the vicinity of a main street intersection in downtown San Jose, California for those hours when the wind just above the top of the tallest building was from the east. The building heights varied between 11 and 22 m. Among the five measuring stations the concentrations vary between 6.0 and 11.7 ppm, a factor of almost two. Just from one side of First Street to the other (at the bottom of Figure 7) the concentrations vary from 7.0 to 10.8 ppm. Notice in Figure 7 that the concentrations tend to be greater on the leeward (i.e., east) side of First Street than on the windward side. As found by Georgii [5] and illustrated in Figure 8, this distribution is due to the development of a vortex in the street canyon formed by tall buildings on either side. From Figures 7 and 8 it is clear that air quality and meteorological factors often vary significantly over very short distances.

METEOROLOGY

Small scale variations in meteorological factors of transport/diffusion, such as shown in Figure 8, are directly dependent upon meteorological factors that are representative of larger areas. For example, in Figure 8, the direction and intensity of rotation of the street vortex depend upon wind direction and speed over the tops of the buildings. Such meteorological factors representative of larger areas usually display rather systematic variations that depend on time and the general climate of the area.

Figure 9 shows the diurnal variation in hourly average wind speeds by season in Philadelphia. In all seasons the wind speeds are significantly greater around mid-day than at night, and this is generally true throughout the United States. It happens because during the day as the sun heats the ground the atmosphere is warmed from below, which causes instability and vertical mixing to extend to greater heights where the wind speeds ordinarily are faster. On the other hand, at night vertical exchange with faster moving air aloft is ordinarily cut off by the formation of temperature inversions at or near the ground. As shown in Figure 9 the rapid increase in wind speeds following sunrise begins earlier in summer than in winter; in the afternoons the rapid declines in speeds begin earlier in winter than in summer. The overall faster speeds in spring are typical of most places in the United States. The overall slower speeds in summer are more representative of the eastern United States; in much of the West the slowest speeds occur in winter or autumn.

The rapid increases in wind speeds during early mornings is indeed fortunate in view of the typical surges in pollutant emissions that occur at about the same times. On the other hand, the decline in afternoon speeds at times when some emissions are increasing is undesirable. Figure 10 shows a case where decreasing CO emissions (as inferred from the traffic count) were associated with increasing CO concentrations, apparently as a consequence of the offsetting effect of decreasing wind speeds.

As mentioned, at most places the vertical extent of atmospheric mixing tends to be greater in the daytime than at night. Figure 11 shows some estimated monthly mean mixing heights for afternoons and mornings for different places across the United States. These mixing heights are based on an analysis of atmospheric temperature soundings made daily by the National Weather Service. The morning estimates are roughly applicable to a few hours after sunrise. Earlier in the morning and at night the mixing heights would be somewhat lower. Figure 11 shows that at Denver in every month the mean afternoon mixing heights are higher and the mean morning heights are lower than at any of the other three locations. Each location is generally respresentative of its region of the United States. The Denver afternoon heights range from 1000 m in January to more than 3500 m in July; the morning heights range between only 100 and 400 m. To judge the significance of these values, one of the criteria that has been used to define episodes of high air pollution potential [7] is that the mixing heights should not exceed 1500 m.

The morning and afternoon curves for Washington and St. Cloud roughly follow the same seasonal variations as for Denver although the afternoon mixing heights for Washington and St. Cloud are

Superior numbers refer to similarly-numbered references at the end of this paper.

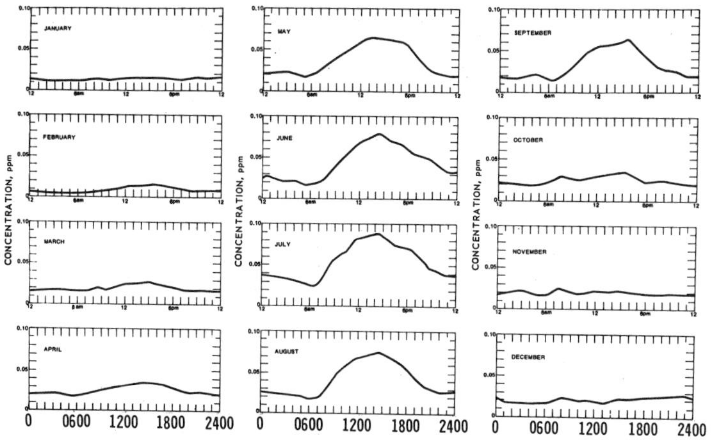

Fig. 5 Diurnal variation of oxidant concentration by month in Philadelphia 1964 - 1965 (after U. S. Dept. HEW (3)).

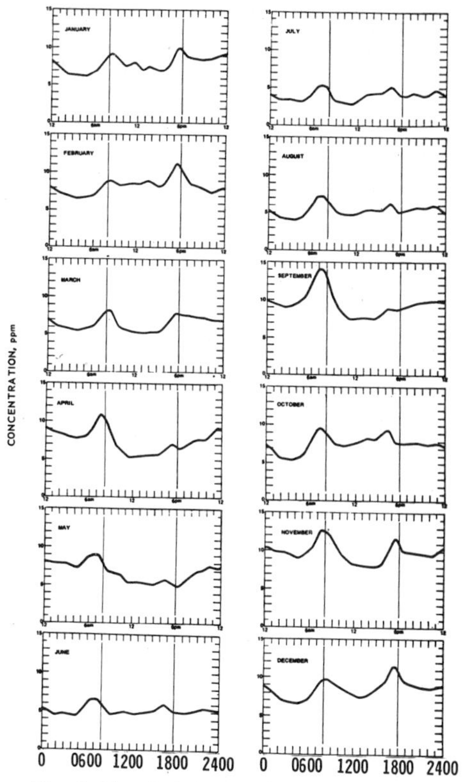

Fig. 6 Diurnal variation of CO concentration by month in Philadelphia, 1964 - 1965 (after U. S. Dept. HEW (3)).

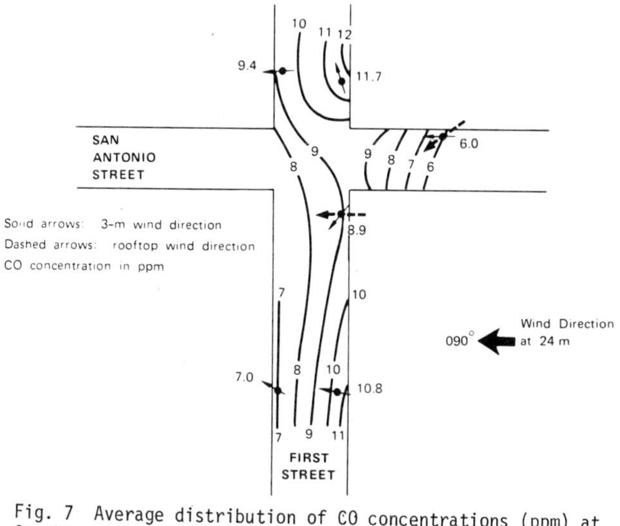

Fig. 7 Average distribution of CO concentrations (ppm) at 3m above sidewalk-level around a main street intersection in downtown San Jose, California for rooftop winds from the east (after Johnson et al. (4)).

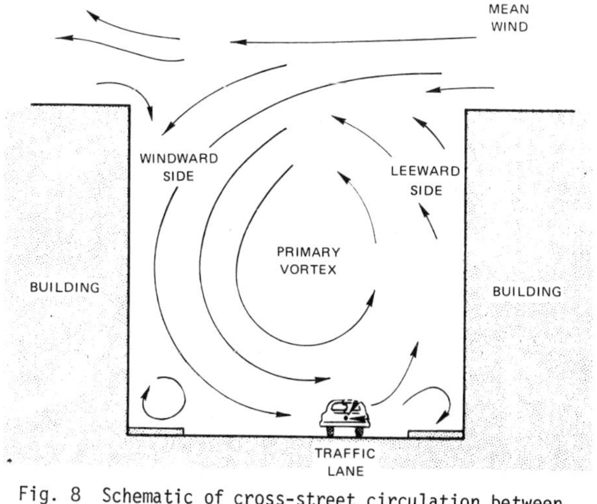

Fig. 8 Schematic of cross-street circulation between tall buildings (after Johnson et al. (5)).

considerably lower, especially in summer, and the morning heights are higher. The variations of the Oakland mixing heights are quite different from those for the other three locations. At Oakland in summer the unusually low afternoon mixing heights of about 650 m and the small morning-afternoon variation of about 150 m are typical of the California Coast; they are caused by a very intense temperature inversion that occurs at roughly 500 m above the surface both night and day during the warmer part of the year. These low afternoon mixing heights are an important factor in the West Coast Ox problem.

The previous figure indicates diurnal ranges in mixing heights averaged over several years. But it is of interest to consider the heights and air quality during an episode of air pollution. Figure 12 is for the 1966 Thanksgiving week air pollution episode in New York City. It is a time cross-section of estimated mixing heights, temperature inversion layers, and isotherms and also includes hourly average SO_2 concentrations. It is based on 6-hourly temperature soundings (soundings are usually made at 12-hour intervals) from JFK Airport on Long Island and continuous SO_2 measurements on Manhattan. First, notice that the SO_2 concentrations were indeed high. On the 24th (Thanksgiving day) and 25th the highest hourly average concentrations were 97 and 102 pphm, and the lowest were 18 and 17 pphm. On no day was the 24-hour average less than 16 pphm. Also notice that in general on each day the concentrations were low around midnight, peaked about 0600-0800 EST, were relatively low in the afternoon, and often rose again in the evening.

Now consider the mixing heights: they were relatively high in the daytime and low at night. The top of the relatively vigorous vertical mixing layer coincides with the base of a temperature inversion. In the morning as the sun warmed the ground, the ground heated the air from below. This caused instability, and eroded the stable inversion layer from below. After a surface-based inversion was thusly eliminated, the mixing layer rapidly increased to the next inversion above. Notice that on the 21st, 22nd, and 23rd a sinking inversion aloft caused a significant lowering of afternoon mixing heights. In the late afternoons and early evenings a surface-based inversion usually formed at JFK Airport to effectively cut off vigorous vertical mixing. Over Manhattan, however, the emission of man-made heat very likely maintained a nocturnal mixing layer (and correspondingly the inversion base height) at roughly a few hundred feet.

A comparison of the variation of mixing heights with the variation of SO_2 concentrations suggests an inverse relationship. However, at best such a relationship can only be rough because other factors are obviously important. Such factors include the diurnal and spatial variations in SO_2 emissions and in other factors of transport/diffusion (e.g., winds). In addition, it should be recognized that the estimated mixing heights are only generally representative of New York City, but the SO_2 measurements were made at a very specific location and therefore are likely to depend on detailed emission and meteorological data.

Although Figure 12 is for a particular air pollution episode, it is believed to be indicative of the variations in mixing height and thermal structure that occur during other episodes of atmospheric stagnation. However, the frequency of atmospheric stagnation (i.e., limited dispersion) days in the United States varies considerably from region to region. Figure 13, which is based on an objective analysis[9] of historical meteorological data shows the total number of stagnation episode-days in 5 years. A stagnation episode was defined as a period lasting at least 2 days during which:

(1) The morning and afternoon mixing heights did not exceed 1500 m.

(2) Average wind speeds in the mixing layers did not exceed 4.0 m/sec.

(3) Significant precipitation did not occur.

These criteria are similar to those that were used in a national program to forecast large areas with a high meteorological potential for air pollution[7]. The figure shows that in the East there is only a small area near West Virginia where the total number of episode-days in 5 years exceeds 100, about 1 day in 18. But in the West 100 days are exceeded at most stations and 200 days are exceeded over a large area. It is interesting that in the 5 years examined, the specified stagnation conditions did not occur once at Oklahoma City, Oklahoma or Dodge City, Kansas. In contrast, 563 episode-days occurred at San Diego. Although the climate of San Diego is similar to that of Los Angeles only 248 episode-days occurred at Los Angeles. This disparity is mainly because all criteria except wind speed were usually satisfied at both places but the speeds were slightly slower at San Diego. Figure 13 also shows that at most stations in the West the season with the greatest number of episode- days is winter, whereas in the East it is autumn.

To further illustrate the climatic variability of atmospheric transport/diffusion, mixing height and wind speed data have been used in a general way to quantitatively assess their overall impact on air quality[9]. This approach uses a very simple model that is much like a box. Assuming a uniform area emission rate over the city of 1 microgram (ug) /m^2-sec, the model gives the city-wide average concentration in ug/m^3 as a function of mixing height, wind speed, and city size (i.e., the distance across the city following the wind). Figure 14 shows that the city-wide average concentration that is exceeded on one-tenth of all mornings annually for a city size of 50 km

at New York City	is 63 ug/m^3
at Huntington, W. Va.	is 375 ug/m^3
at Chicago, Ill.	is 225 ug/m^3
at Oklahoma City, Okla.	is 77 ug/m^3
at Salt Lake City, Utah	is 95 ug/m^3
at Los Angeles, Cal.	is 188 ug/m^3
at San Francisco, Cal.	is 237 ug/m^3
at Portland, Ore.	is 300 ug/m^3

The isopleth pattern of Figure 14 for upper decile morning concentrations is significantly different from the pattern for median concentrations. Also, since mixing heights and wind speeds are generally greater in afternoons than mornings, the afternoon

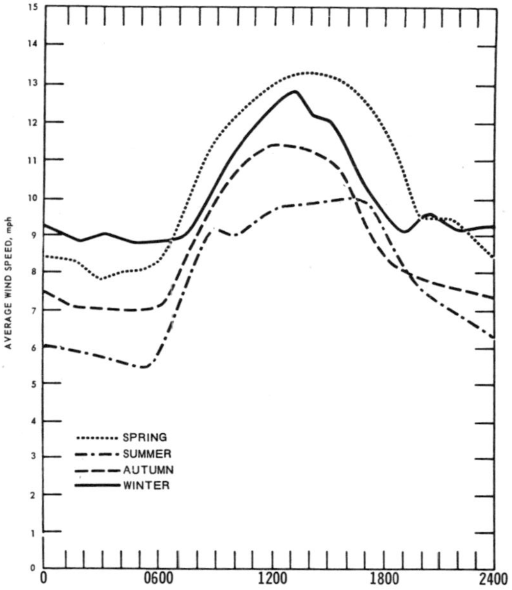

Fig. 9 Diurnal variation in horizontal wind speed, Philadelphia, 1962 - 1964 (after U. S. Dept. HEW (3)).

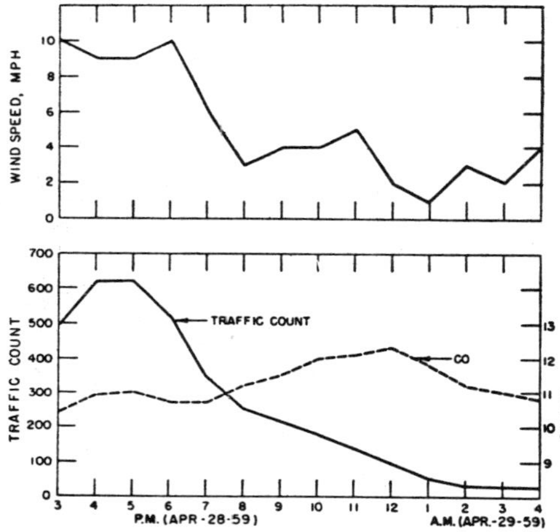

Fig. 10 Hourly values of CO (ppm), traffic count, and wind speed (mph) in Nashville, Tennessee (after McCormick and Xintaras (2)).

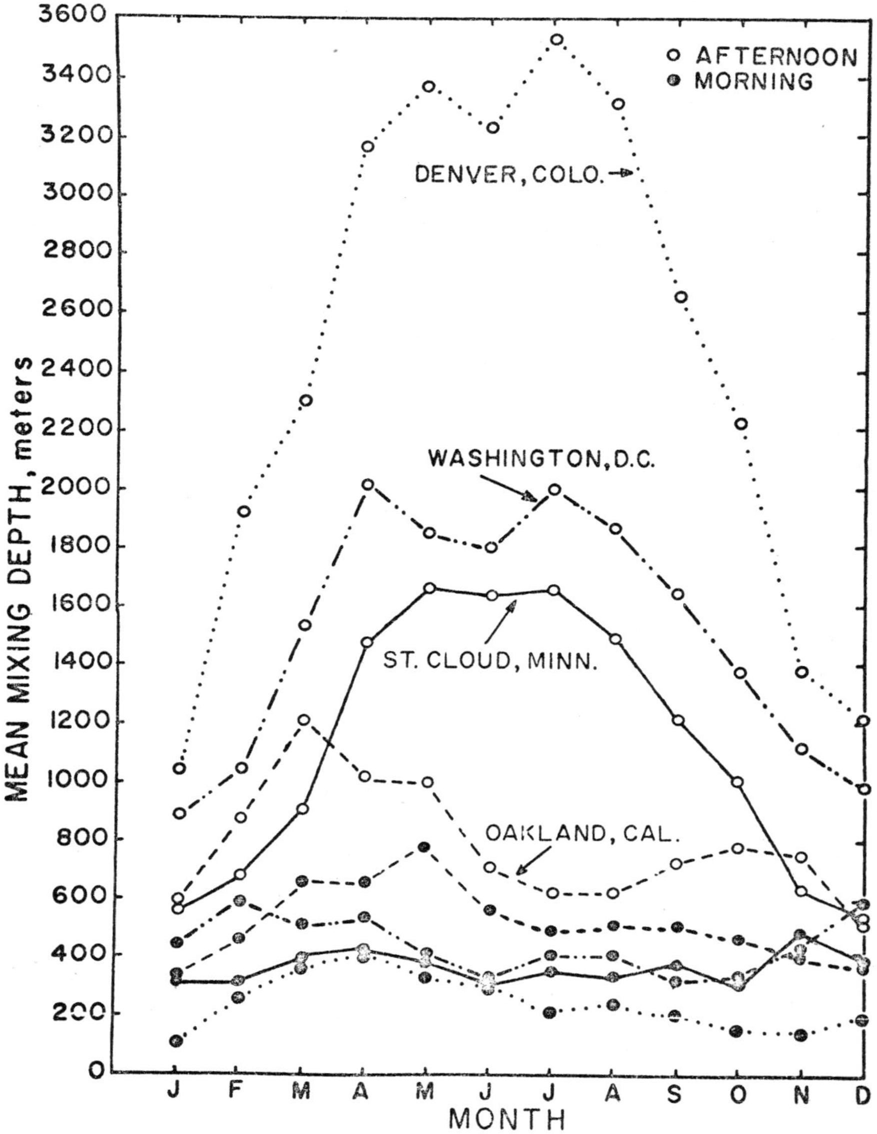

Fig. 11 Monthly mean afternoon (open circles) and morning (solid circles) mixing heights. All data for 1960 - 1964 except Washington for 1961 - 1964 (after Holzworth (6)).

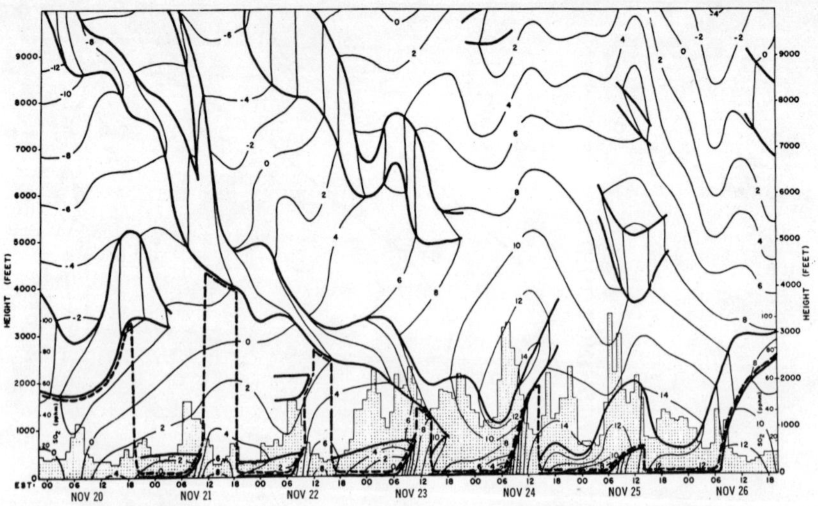

Fig. 12 Time cross-section of mixing height (dashed line), temperature inversion layers (between pairs of medium lines), and isotherms (°C, light lines) based on 6-hourly temperature soundings at JFK Airport, and hourly average SO_2 concentrations (pphm, stippled area) measured continuously in Manhattan (after Holzworth(8)).

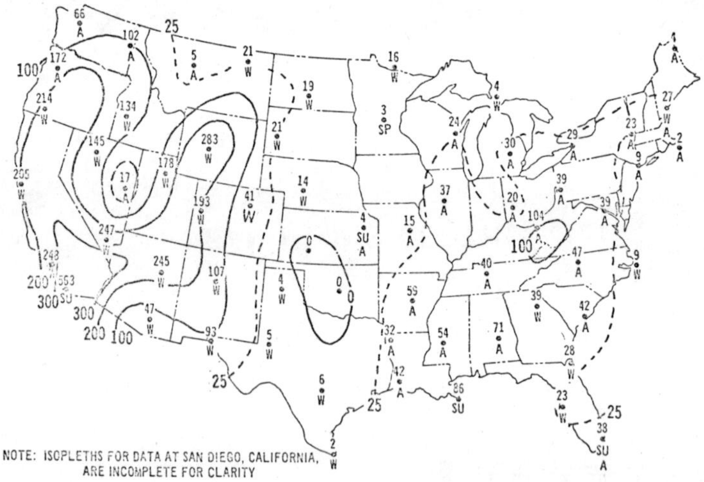

Fig. 13 Isopleths of total number of stagnation episode-days in 5 years with mixing heights ≤ 1500 m, wind speeds ≤ 4.0 m/sec, and no significant precipitation--for episodes lasting at least 2 days (after Holzworth (9)). Season with most episode-days indicated as Winter, SPring, SUmmer, or Autumn.

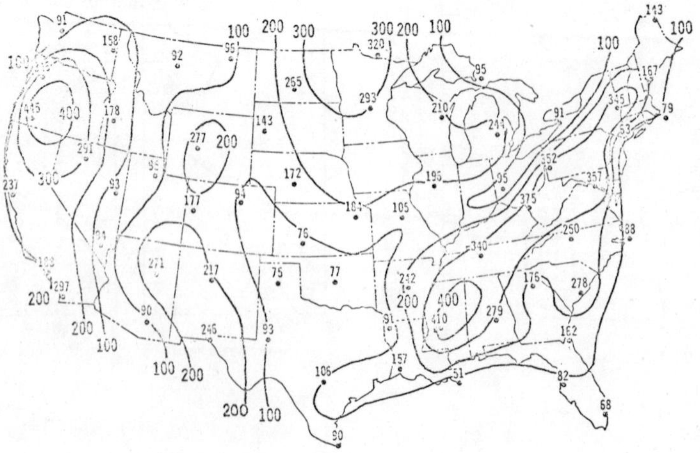

Fig. 14 Isopleths of theoretical city-wide average concentrations $\mu g/m^3$ that are exceeded on one-tenth of the mornings annually, assuming a city size of 50 km and a uniform area emission rate of 1 $\mu g/m^2$-sec (after an analysis by Holzworth (9)).

concentrations are usually much less than for mornings. And there are also seasonal variations. Thus, even in terms of a rather simple and general theoretical approach, the meteorological potential for air pollution is highly variable.

SUMMARY

This paper has provided some examples of temporal and spatial variations in pollutant emissions, atmospheric transport/diffusion, and air quality typical of specific locations in large urban areas in the United States. Very generally, pollutant emissions follow the cycles of human activity and energy utilization with rapid rises near the start of each day's activities, various degrees of decrease or increase around mid-day, increases in early evening, and declines late at night. Such energy utilization obviously depends on various characteristics of the weather. And the weather's transport/diffusion properties have a significant impact on air quality. At most places atmospheric dilution is greatest in the daytime and least at night but there are significant seasonal and regional differences. The overall effect on air quality of the usual diurnal patterns of emissions and dilution is that high concentrations at the beginning of a day's activities are mainly due to a morning surge in emissions; lower concentrations in the forenoon and afternoon are due to increased dilution and sometimes partly to decreased emissions; higher concentrations in the evening are caused by both decreasing dilution and increasing emissions; and low concentrations at night are mainly due to less emissions. Ox, which depends on solar radiation for its formation in the atmosphere, is of course an exception.

The foregoing descriptions are necessarily very broad because usually only very general information is available on pollutant emissions and transport/diffusion, especially over cities. The complete explanation of air quality measured at a specific point requires fine scale information on pollutant emissions and dilution factors in the vicinity of that point.

REFERENCES

(1) Turner, D. B., "The Diurnal Variation in Day-To-Day Variations of Fuel Usage for Space Heating in St. Louis, Missouri," ATMOS. ENV., Vol. 2 (1968), 339-351.

(2) McCormick, R. A. and C. Xintaras, "Variation of Carbon Monoxide Concentrations as Related to Sampling Interval, Traffic, and Meteorological Factors," J. APPL. METEOR., Vol. 1 (1962), 237-243.

(3) U. S. Dept. Health, Education, and Welfare, "Continuous Air Monitoring Projects in Philadelphia 1962-1965," National Air Pollution Control Admin. Pub. No. APTD 69-14 (1969), Cincinnati, Ohio, 364 pp.

(4) Johnson, W. B., F. L. Ludwig, W. F. Dabbert, and R. J. Allen, "An Urban Simulation Model for Carbon Monoxide," J. AIR POLL. CONT. ASSOC., Vol. 23 (1973), 490-498.

(5) Georgii, H. W., "The Effects of Air Pollution on Urban Climates," BULL. WLD. HLTH. ORG., Vol. 40 (1969), 624-635.

(6) Holzworth, G. C., "Large-Scale Weather Influences on Community Air Pollution Potential in the United States," J. AIR POLL. CONT. ASSOC., Vol. 19 (1969), 248-254.

(7) Gross, E. M., "The National Air Pollution Potential Forecast Program," ESSA Tech. Memo. WBTM NMC 47 (1970), Washington, D. C., 28 pp.

(8) Holzworth, G. C., "Vertical Temperature Structure during the 1966 Thanksgiving Week Air Pollution Episode in New York City," MONTHLY WEA. REV., Vol. 100 (1972), 445-450.

(9) Holzworth, G. C., "Mixing Heights, Wind Speeds, and Potential for Urban Air Pollution throughout the Contiguous United States," U. S. Env. Prot. Agcy, Office of Air Programs Publ. No. AP-101 (1972), Research Triangle Park, N. C. 118 pp.

© 1973, ISA JSP 6702

METEOROLOGICAL SENSORS IN AIR POLLUTION PROBLEMS

Daniel A. Mazzarella
Science Associates, Inc., Princeton, New Jersey

ABSTRACT

Meteorological sensors with applications in pollution problems encompass all the elements of weather. In the efforts related to diffusion and turbulence, the primary sensors are those that measure wind and temperature; the secondary sensors are those that measure dew point (humidity), solar radiation, pressure, precipitation, and such other variables of the atmosphere as are important, but do not usually provide inputs used in calculating the dispersion of an atmospheric contaminant. Of all the measurements made, or determined, the standard deviations of wind in the horizontal plane (σ_θ), and vertical plane (σ_ϕ), have been used most often as parameters in the equations for diffusion.

While most early investigations in this field were research-oriented, a major part of today's work is applications-oriented, and frequently includes measurements of air quality in urban and interstate networks. Another change is evolving. The analog strip chart recorder is still important but in a secondary role to the telemetered, automated, computer-compatible digital record.

The use of towers as platforms for sensors has continued and increased, with little change in the basic measurements, but most certainly, there is a greater recognition of the importance of sensor characteristics as part of the total system.

TERMINOLOGY

By convention and definition Gill and Hexter(1) have identified a sensor as the element that reacts to changes in the atmospheric variable, while the transducer is the device that converts energy from one form to another, most frequently into an electrical signal. Sometimes the terms sensor and transducer are used as if they were synonymous; always, they are part of the total instrument, which with the addition of a communications link, a data processor, a means of data storage, a readout device, or some similar combination of equipment evolve into a data acquisition system. In recent years it has become quite obvious that demands for more data, the use of many sensors, telemetering over long distances, and the expansion from independent to network installations make it essential to view the equipment used in air pollution problems as a total system. This approach has been discussed and recommended by Lansberg(2); Gant, Newman, Spiegler, and Tong(3); and MacCready and Lockhart(4).

THE METEOROLOGICAL ELEMENTS IN AIR POLLUTION - WIND

In writing about the gustiness of wind and the diurnal variation of atmospheric turbulence, Sutton(5) noted, "The most direct and easily recognized evidence of turbulence is the record of an anemometer which is capable of indicating instantaneous values of the wind speed, such as the Dines pressure-tube instrument ---. The outstanding features are that the velocity exhibits rapid fluctuations over most of the period, and that the amplitude of the oscillations changes considerably throughout the 24 hours, tending to a maximum in the midday period and falling to a low value during the night. This manifestation of turbulence is usually referred to as gustiness. Fluctuations of this type also occur in the direction of the wind, so that it is possible, with suitable instrumental aids, to evaluate not only the mean wind speed over a period but also the magnitude of the component oscillations referred to any convenient system of axes."

All investigators, whether concerned with research, operational, or climatological matters, are in agreement that the wind is the primary element in studying pollution problems. For climatological applications, it may only be necessary to obtain mean hourly values of direction and speed, with sensors that have slow response characteristics and to tabulate these as line bar graphs corresponding to the direction of the wind, in the form known as a wind rose.

For many operational requirements, and certainly for research programs, the dynamic characteristics of a wind sensor as reported by Gill(6), MacCready(7), and others influence the results.

Sensors for wind speed have included cup anemometers, propellers (windmills), pressure tubes and plates, heated wires and films, sonic devices, and innumerable research and experimental instruments. For measurements of wind passage as distinguished from wind speed, mechanical and electrical switches have been used most often to count each mile or fraction of a mile. Mechanical rollers with raised surfaces for marking on inkless paper have provided a record where the slope of the line within a given time period is easily interpreted as average wind, or wind flow. A similar record is now available through an ingenious electro-mechanical design, in which a cup anemometer is connected to the wiper arm of a potentiometer through a low-ratio gear system. In this construction, voltage is proportional to wind passage.

Wind direction has been measured with transducers that include a-c synchros; d-c synchros; capacitive-type devices; single, dual, and sine-cosine potentiometers; and other techniques for translating polar coordinates to Cartesian coordinates. Increasing use has been made in recent years of fixed arrays of propellers, hot wires, and sonic-type sensors for obtaining the components of wind in the X (downwind), Y (across), and Z (vertical) axis, to describe the three dimensional wind vector.

GUSTINESS

Sutton, as has been noted, observed that the analog record of speed and/or direction showed fluctuations in amplitude and frequency, usually in a pattern from day to night. Many investigators have studied this characteristic of turbulence; over the years there has been a persistent attempt to utilize wind records in seeking an engineering approach to describe the dilution of pollutants.

In January of 1950, at a Symposium on Atmospheric Pollution in St. Louis, Mo., Lowry and Smith, both of Brookhaven National Laboratory, presented papers making public what had previously been reported in AEC Technical Information Documents, and was later published. "A good approximation to the maximum ground concentration averaged over an hour," reported Lowry,(8) "may be found from the continuous record of a single anemometer and wind vane mounted at the level of the chimney top, if non-meteorological factors such as the nature of the terrain and the conditions of emission are neglected." In developing this idea Lowry used Brookhaven data, collected first from a flat plate wind vane and a cup anemometer at the 125m level of the meteorological tower, later from a propeller-type vane at the 108m level. Wind direction traces, manually reduced to 10-minute distributions from 10-second means, later to 1-hr. distributions from 10-second means, were classified into four types -- A,B,C,&D -- which by now are well known. By their characteristic appearance they were identified in terms of a turbulence index, with mean hourly wind speed used as a diffusion index. It is of interest to note this statement, "Sufficient accuracy will be obtained (in estimating the standard deviation of the wind) by taking one-fifth of the direction range, provided that isolated lateral gusts are ignored."

In a paper that dealt with the forecasting of micrometeorological variables, Smith(9) further defined the classification of gustiness types by adding limits to wind speed values; by specifying the vertical temperature gradient; and, by identifying turbulence as convective or mechanical. In a Microforecast Chart on the vertical temperature structure, Smith related lapse and inversion conditions to season, time of day, cloud cover, gradient wind, ground condition, type of advection, general air mass characteristics, sunshine, and state of the sky.

With the availability of additional data at the same site, but still from propeller-type anemometers, Singer and Smith(10) expanded and modified the earlier classification, dividing B into B_1 and B_2, and added a column for mean wind speed ($\bar{u}$). It should be noted that the revised classification, still in use today, was based on data covering two full years. The analog records, operated at a chart speed of 3"/hr. were manually reduced, using one hour averaging times.

THE STANDARD DEVIATION OF WIND DIRECTION

Numerous experiments to study the diffusion and transport of pollutants have been conducted in the years since the Brookhaven Gustiness Types were first presented and then revised. The results of many of the tests have been summarized by Islitzer and Slade(11). It was from data obtained at one of these, Prairie Grass in O'Neill, Nebraska, substantiated by earlier tests at Round Hill, Mass., that Cramer(12) showed in 1957 that plume dimensions and plume concentrations could be calculated for distances to 1.6km directly from the standard deviations of wind traces. Pasquill(13), who also used Prairie Grass data, found that it confirmed earlier work he had done in England; he suggested a plan for relating turbulence types to observed weather conditions, using only measurements of wind speed, wind direction, and the standard deviation of the direction. The A to F classification he proposed is in extensive use today, its appeal comes from the ongoing desire to describe the potential for pollution from standard observations. For the Atomic Energy Commission--which has adapted, expanded, and extended Pasquill's classification of stability types--the groupings are convenient, especially for tower measurements where site surveys are required. But, there still exists a lack of complete agreement about the temperature-difference spans in relation to the categories stable and unstable, as given in Safety Guide 23(14). It could very well be that the disparity results from the location of the reference temperature at 10m above the surface, as opposed to 2m for example, which has been used, and which is at a level that is much more thermally variable. While 10m meets an international standard for synoptic observations, and minimizes the statistical problems of dealing with roughness parameters, 2m for the air quality inspector is closer to the environment in which man breathes and works.

COMPUTING AND/OR ESTIMATING SIGMA

The first of the electronic Sigma Meters (sigma converters), designed to provide a real-time standard deviation from a wind vane transducer, was described in 1959 by Jones and Pasquill(15). Several models are now available commercially but data collected from these "black boxes" is frequently suspect. First, because the output is influenced by the dynamic characteristics of the vane. Second, because of the uncertainties related to the selection of electronic filtering circuits. And third, because of doubts about averaging and sampling times. Still, the search for acceptance continues. A non-commercial, experimental type electronic computing vane, made to supply standard deviation in support of atmospheric diffusion studies at Trombay, India was described in 1971, and considered "quite satisfactory" by Sachdev and Rajan[16].

Early in 1962, Markee[17] noted that few of the wind instruments used in urban air pollution surveys included equipment for computing the standard deviations of fluctuations, but there were many analog records on strip charts. To utilize these records with confidence required the testing of an idea which had been investigated rather briefly by several earlier researchers. It involved the relationship of range, or amplitude, to σ_θ, the standard deviation of the horizontal wind. Markee's study was conducted at two urban locations, Cincinnati, Ohio, and Nashville, Tenn. The commercially available equipment was installed at 10m, measurements were made during daylight hours, and the recorders operated at 6"/min. Averaging times of 1, 5, and 10 seconds were selected, and in addition instantaneous readings were made at 5, 10, and 15-second intervals. Computations of Sigma were made from sampling times of 15, 30, and 60 minutes. For the purpose of this report it is of interest to note this finding: "The longer the sampling interval (up to 1 hr.), the shorter the averaging time, and the higher the wind speed, the more reliable is the range as a predictor of the standard deviation." The range of wind direction surpassed the range of wind speed as a predictor of σ . While Markee's data for the test sites gave as an average relationship R/6.9, his comparisons with earlier pilot studies at Oak Ridge, Tenn.; Middletown, Conn., Shippingsport, Pa.; and Idaho Falls, Idaho, indicate ratios that varied from R/8.6 to R/5.0, undoubtedly representing variations in sensor and recorder characteristics, as well as in the height and exposure of the instrument. In 1950 Lowry had recommended dividing the range by 5; today, it is common practice, as indicated by Moses[18], to assume that the range is approximately 6 σ_θ, or after eliminating a few of the largest deviations on either side of the mean, about 4 σ_θ.

It is known that early attempts to describe the envelope of a plume by reference to the recorded oscillations of a poorly damped vane gave results which were in error by a factor of two, or more. Now that the literature is documented with the damping ratios of many direction vanes, as summarized by Mazzarella[19], and now that we have an increasing quantity of digital as well as analog data, it has become easier to show that the range of amplitude fluctuations in a wind direction trace is related to the damping ratio of the vane. That is, for a damping ratio of 0.6, range is approximately 4 σ_θ, at 0.4, R=4.8 σ_θ, at 0.2 R=6.5 σ_θ.

While devices that measure and compute have been suspect, some questions also have been raised about the Pasquill Stability Types as interpreted in applications related to tower measurements. After 13 months of wind and temperature observations made on a tower, and compared to observations at a nearby ground site, Luna and Church[20] concluded that the Pasquill categories are in the right sequence, A to F from instability to stability, but the neutral class D does not seem to be an indicator of zero lapse rate in temperature; moreover, the standard deviations are so large that almost any value of σ_θ and σ_ϕ could belong to any stability class. The large variability within any one category, concluded the authors, suggests the need for measuring σ_θ, σ_ϕ, and U (wind speed) directly --- with anemometers, wind vanes, bivanes, etc.

TEMPERATURE

The second most important element in air pollution problems has been identified as temperature, more specifically, the vertical temperature gradient. Departures from the adiabatic lapse rate of 1°C/100m (5.4°F/1000 ft.) which show an increase in temperature with altitude suppress convective and mechanical movements and are therefore identified as stable conditions. In terms of the sensor, it is convenient to identify turbulence as an index of diffusion measured by a wind instrument, and instability as an index of diffusion measured with temperature sensors in a vertical array.

ASPIRATED RADIATION SHIELDS

For maximum accuracy in measuring air temperatures outdoors, especially for remote measurements, a radiation shield with forced ventilation of a known rate is essential. There are, in the USA alone, at least six suppliers of off-the-shelf aspirated radiation shields. The designs vary, but most manufacturers claim radiation errors of less than 0.2°F; two make claims to a maximum error of 0.05°F, one of these is for a unique shield made from a double-walled, heavy-duty evacuated glass cylinder that is silver-plated on the inner surfaces. It is constructed to face open-end down; the other aspirated shield with similar limits of error faces horizontally, but is built with a baffle. Comparative temperature measurements in towers with sensors in different types of aspirated enclosures, with unequal rates of aspiration, have given disparate results, especially when the comparison is complicated by the use of sensors with different time constants. Until such a day as there is a design standard, these pitfalls will continue to produce questionable measurements.

In a recent paper, "Temperature Error of a Ventilated Thermometer in Dependence of Radiation and Wind," Slob[21] of the Royal Netherlands Meteorological Institute described a series of tests in which he studied the intensity and angle of incident radiation, the absorption characteristics of the material, the wind speed, the ventilation rate, and the effect of precipitation on the shield. These are, indeed, all important considerations, especially when examined in the context of the AEC's Safety Guide 23 for tower instrumentation which reads, "Temperature accuracy for time averaged values ±0.5°C. Temperature difference accuracy from either, difference between averaged temperatures, or average temperature difference, ±0.1°C."

TEMPERATURE SENSORS

R.T.D.'s, or resistance temperature devices, especially platinum wire sensors are being used extensively in meteorological research and operational installations. Their long-established characteristics of near linearity and stability have proven advantageous; while previously expensive, they are now available at prices almost comparable to nickel and copper elements. For maximum accuracy, especially where temperature-differences are to be measured, sensors with four leads are usually recommended. Three-wire sensors are essential for monitoring ambient temperatures on high towers where a temperature gradient of significant proportions can develop between the level of measurement and the length of cable. For convenience in processing the data, it is advantageous to work with a signal that is linear with temperature and there are now available numerous bridges with outputs in millivolts per degree. Some are designed to give temperature and temperature-difference simultaneously.

Thermistors, characteristically with an inverse resistance-to-temperature curve and a high resistance-to-temperature ratio, have always been of interest in meteorological measurements. The problems of long term stability, self heating, and extreme non-linearity have been limitations to usage, but manufacturers have reported progress in all three areas, the most significant being in the availability of a composite thermistor-resistor with a linear resistance-to-temperature relationship. Long term stability information is evolving and would seem to be dependent on the process of manufacturing and aging.

Thermocouples have been useful in meteorological research, mostly because they can be fabricated with a minimum of mass for fast response to changes of small amplitude and high frequency. Earlier problems of cold junction reference temperatures have been minimized by solid state devices, but the low level signal problem, especially for tower installations, compounded by R-F pickup interference, still exist and have discouraged usage. For some applications, thermopiles have provided a signal of sufficient magnitude, but in general the use of thermocouples has been limited to ground stations and short towers with elements of a small mass so that errors from radiation are minimized, without the need for elaborate radiation shields.

METEOROLOGICAL INSTALLATIONS

Several references have been made to short term and long term studies and programs involving meteorological sensors. A brief review of some of the facilities used in typical investigations will serve to identify equipment usage, to make comparisons and contrasts illustrating changes with time and project objectives, and to allow mention of the secondary sensors.

Hanford --- A description of the "Meteorological Equipment of the Hanford Engineer Works," the first of the AEC-related programs in meteorology, is given in a document by Church and Gosline[22], written about 1945, and declassified in 1947. In reviewing this report, it is interesting to note that current practice in tower instrumentation is basically the same as it was almost three decades ago; the changes and refinements since that time are in terms of accuracies, telemetering, data logging, and data processing.

The guyed tower at Hanford was described as an open framework, 12 ft. on a side, 408 ft. high, with instrument booms spaced 50 ft. apart, from 50' to 400'. Ground level measurements were made 3 ft. above the surface on a pedestal and boom arrangement 50 ft. NW of the tower. Air temperature at the tower levels was measured with nickel wire RTD's, installed in double-walled radiation shields; they were aspirated at 10 CFM by a single, centrally located pump installed at the base of the tower.

Wind speed at the Hanford site was measured with three-cup spherical-shaped anemometers and d-c generator-type transducers. Starting speeds were given as "--slightly below 2 mph and stopping speed - about 1 mph." In commenting on the wind equipment, the authors expressed the opinion that the anemometers used had a definite advantage for the program at Hanford in that gustiness as well as speed was recorded and that the anemometers embodied the main advantages of the bridled-cup turbulence integrator (designed by Gill; used at Trail, B.C.), and the standard weather service anemometer which measured wind flow. Wind direction was measured with a vane and an a-c synchro motor transducer. The recording of wind was in analog form at a chart speed of 3"/hr., with a d-c galvanometer for speed and a Selsyn (synchro) receiver for direction. Every thirty days, the inked traces for wind and temperature generated nearly 2,600 ft. of records. The Hanford tower, now part of Battelle NW is still in operation; it has been used as a platform for many sensors since it was constructed in the early 1940's.

Brookhaven --- Experience gained at Hanford was the basis for the Brookhaven towers which became operational in 1949. In this installation, three-lead 100-ohm copper wire RTD's, factory calibrated to ±0.1°F. were selected for measuring ambient temperatures sequentially at eight levels of a 420-ft. tower. Four-wire RTD's of the same type were used for the direct measurement and recording of temperature-differences from three levels on the same structure. The radiation shields used at Brookhaven were modifications of these at Hanford, but central aspiration was still employed. At a nominal aspiration rate of 21 ft./sec., the sensors were found to have a time constant (63% change) of 72 sec., as reported with other test data on shield design, circuitry, calibration procedure, etc. by Mazzarella and Kohl[23].

The smaller, 160-ft. tower at Brookhaven, subsequently used in limited applications, was equipped with the same wind sensors used at Hanford, but the heavy spherical anemometer cups were found to have a high threshold, to be non-linear, and to overestimate gusty winds. This led to the procurement of what was then a relatively new and commercially available windmill-type (propeller) anemometer and vane. While the threshold speed of this device was not as low as desirable, the output was linear, and the equipment was durable and dependable, and easier to install on a tall tower. Little was known about the dynamic characteristics of this instrument until wind tunnel tests were conducted to determine the effects of pitch and yaw on the propeller and the effects of angular displacement at different speeds on the vane. The results identified the instrument as reliable over a wide range of atmospheric conditions, responsive to eddies with wavelengths and frequencies of significance in some microscale, and all mesoscale and macroscale pollution problems, but inadequate at winds below 2 to 2½ mph. Studies related to light winds and/or high frequencies would have to be handled by other equipment.

In an early paper on the Brookhaven installation entitled "Stack Meteorology and Atmospheric Disposal of Radioactive Waste," Beers(24) identified the need for instrumentation to measure eddy velocities in atmospheric turbulence studies and indicated "---such equipment as has been used (both American and British bivanes essentially) is not available as a catalog item." From this interest there evolved at Brookhaven a primitive but functional remote transmitting bivane, with performance specifications comparable to those of the propeller type anemometers on the tower. The Brookhaven installation, a site of many experiments in turbulence and diffusion, and a location for the development and testing of many instruments, is still in operation. Originally, all tower data was recorded in analog form on strip charts; in 1959 Brown(25) described the development of an automatic system incorporating a punched tape output, in parallel with analog recorders.

ARGONNE NATIONAL LABORATORY

A third meteorological installation of an AEC contractor, that of Argonne National Laboratory, includes a 150 ft. meteorological tower. This facility, completed in 1950 and still in use, is well documented by Moses and Bogner(26). The wind sensors used on this self-standing and tapered structure at Argonne have included propeller-type anemometers and vanes, and conical cup-type anemometers which measure wind passage. Ambient temperature and temperature differences have been measured with copper-constantan thermopiles. A commercial lithium chloride dewcel has been used for dewpoint at four elevations on the tower, while a standard hygrothermograph has been located in a shelter near the base of the tower. Other measurements made routinely include direct and diffuse solar radiation with a 50 junction pyrheliometer; net radiation with a standard commercial adaptation of the Gier and Dunkle design; precipitation with a weighing type and a tipping bucket of standard design; pressure with a microbarograph, and, soil temperature with a 100-ohm copper element RTD.

In 1961 a data processing system for recording automatically on punch tape and teletypewriter was completed at this facility. By today's standards, the system is large and complex, but it represents some ingeneous approaches toward adapting existing analog outputs to a digital format. In this report on design characteristics of the data processing system, Moses and Kulhanck(27) have given suggestions and comments about the automation of meteorological data which are valid considerations in advanced, compact electronic systems. For example, one must weigh digital vs. analog, average vs. integrated, continuous vs. sampled.

NETWORK INSTALLATIONS

In contrast to the research and operational facilities of Hanford, Brookhaven, Argonne, and others, there are an increasing number of multi-station operationally-oriented urban networks in such locations as Chicago; New York City; Frankford/M, Germany; Los Angeles; Denver; Philadelphia; New Jersey; etc. In each of these meteorological measurements have been combined, on a routine basis, with air quality measurements. In these networks the data has been digitized, telemetered, and in general, adapted to computer techniques for processing.

The City of Chicago operates a twelve-station automatic real-time telemetry network for monitoring meteorological parameters and air contaminates. This system, initiated in 1962 with the assistance of the U.S. Public Health Service, has been described by Harrison(28) as the first real-time telemetry system on an urban scale. Telephone lines are used for transmitting measurements of wind direction and speed, using propeller type anemometers and vanes. Other measurements are made of temperature, humidity, and solar radiation.

In 1966 the Federal Government of West Germany agreed upon the installation of a monitoring system covering seven stations. Atmospheric pollutants, including exhaust emissions are measured along with the meteorological elements. Data processing and calibration is performed by a computer, as outlined by Georgii(29).

In 1968 the ten-station network in New York City, with its centrally located computer, was described in <u>Air Engineering</u>(30) as the most advanced system in air pollution monitoring. Wind speed, wind direction, and air temperature sensors were, and continue to be interrogated automatically over dedicated telephone lines once every five minutes for instantaneous values. The five minute scan cycle can be varied on demand--with the A to D conversion, storage of signal, averaging of data, tape punch, teleprinter readout, and direct computer entry accomplished at a central location.

AN INTERSTATE MONITORING NETWORK

An example of an extensive interstate network, primarily operational in purpose, and maintained by a coordinated group effort, can be found in the program of the Tennessee Valley Authority (TVA), functioning under the jurisdiction of the Federal Power Commission. In 1971 Montgomery, Carpenter, and Lindley[31] reported on a study in the vicinity of TVA's Paradise Power Plant, one of the world's largest coal-fired generators. It was a study relating peak and mean concentrations of SO_2. For this application measurements of temperature at 13 meters and 111 meters, made with platinum wire sensors, were taken as instantaneous values on the hour, telemetered, and recorded digitally on punch tape at the meteorological tower location, "---they were assumed to be representative of the preceding hour."

The results of that study are not pertinent to this report except that mean hourly temperatures might have been more representative, but it should be noted that the location of the investigation is only one of many in the TVA complex that includes eleven coal-fired and three nuclear power plants, all within seven states. As of July 1, 1973 the TVA's meteorological equipment in the Air Quality Monitoring Network[32] included thirteen locations with meteorological towers that vary in height from 150 ft. to 400 ft. The equipment list includes: 4 anemographs, 12 bivanes, 12 cup anemometers, 12 direction vanes, 43 platinum wire sensors in motor aspirated shields, 8 Li Cl dewpoint sensors, 9 hygrothermographs, 6 weighing type rain gages, 6 pyranometers of the 48-junction type for global sun and sky radiation, 5 total hemisphere radiometers of the Gier and Dunkle design, and 6 pressure sensors and transducers. The number of air quality sensors exceeds the meteorological sensors, but in each case the trend has been from analog to digital, paper tape and magnetic tape. The data logger used at the environmental data station for the Sequoyah Nuclear Plant, north of Chattanooga, Tenn., installed in 1971, is typical of the data gathering methods now in use. Each meteorological sensor input is scaled so that the values printed are in engineering units, at intervals considered to be representative of the selected parameter. The scan rates are: vertical wind, 2 sec.; horizontal wind, 5 sec.; wind speed, 15 sec.; solar radiation, 15 sec., temperature, 60 min.; pressure, 60 min.; precipitation, 60 min. The data logger computer program is written so that limit checks are made on each meteorological parameter input as it is measured. If the value does not fall within predetermined limits, this measurement is discarded. Since wind measurements are made most frequently, the standard deviation calculations of horizontal and vertical wind directions are performed for each 5 minute period and averaged at the end of each hour. A printout of averaged data is made on a teletype page printer each hour; information collected during the hour is put on punched tape.

A ZONAL MONITORING AND PREDICTION NETWORK

The parallel trends of network type installations; the use of meteorological towers and semi-portable trailers; the need for simultaneous measurements from meteorological and air quality instruments; the use of mini and central computers in supporting functions; and the effectiveness of real-time, monitoring-oriented applications in combination with historical, statistical, and prediction-oriented data is exemplified in the Air Quality Monitoring Analysis and Prediction (AIRMAP) system approach outlined by Gant, Newman, Spiegler, and Tong[3]. It has been described as a highly integrated hardware-software system which covers thousands of miles, linked by tone signals traveling over voice grade dedicated telephone lines. The air quality parameters measured are those that are required to meet Federal standards of SO_2, NO_x, and particulates. The meteorological parameters are those that are required to describe turbulence and diffusion: wind direction, wind speed, temperature and temperature-difference. For wind direction a vane with a wiper-type potentiometer has been used; for wind speed, a cup anemometer with a photo chopper transducer. Temperature and temperature-difference measurements are made with linear thermistors in aspirated radiation shields. In all cases, there is an A to D conversion; a specially designed transducer converts the digital signal to a tone for transmission over the telephone wires. There is a reconversion to voltage at the central receiving station, where two minicomputers operating in real-time gather data and handle programming and processing. All sensors are interrogated for instantaneous values at one minute intervals. Hourly averages, automatically processed, are printed on teletype and put on punch cards. Analog charts are used for back-up information at the field sites, and the minicomputer-processed data is fed to a large central computer which is programmed to compare real-time data with theoretical diffusion models.

A RESEARCH AND URBAN MODELING NETWORK

Among the most ambitious, most promising, and most advanced of the research programs now in progress is RAPS - the Regional Air Pollution Survey[33], in which a 30km radius around St. Louis, Mo. is the site selected for a meso-met. network for modeling an urban complex. The study will take five years for completion. Contract awards have been made, so this program in which a central computer is to interface with up to 25 stations and peripheral devices is now under way. Sensors will include a modified cup type anemometer with a threshold of 0.5 mph, a distance constant of less than 5 ft.; with a frequency to voltage conversion. The direction vane is also a modification of an existing device, updated to have a damping ratio of 0.4 at an initial displacement angle of 10 degrees, and a distance constant of 3.7 ft. Turbulence will be measured with an orthoganal array in which the lightweight propellers have a threshold of 0.3 mph and a distance constant of 2.7 ft. A signal conditioner will provide mean values in the form of one-minute averages for speed and direction. These will be programmed by the minicomputer at the station. Turbulence data from the orthoganal array will be sampled every ½ second. There will be a digital backup on teletype; at 5 stations the backup will be on magnetic tape, to insure 90% data recovery. Other meteorological sensors include linear thermistor probes in aspirated shields for temperature measurements at 5 meters at all sites; temperature differences will be measured between 5 and 30 meters at 12 locations. Temperature accuracies have been specified as ±0.5°C; temperature differences as ±0.1°C. Dewpoint is to be measured by a thermo-electric dewpoint hygrometer at all locations, to an accuracy of ±1°C. Barometric pressure, from 27 to 31.5 inches of mercury will be measured at seven stations with Ni-span C stacked diaphragms and a wiper type transducer. Solar radiation devices will be included at six stations. At four of these there will be equipment for measurements with 3 spectral pyranometers, 1 pyrheliometer, and 1 long wave pyrgeometer. At the other two sites there will be 2 spectral radiometers. In all, the list of measurements for RAPS includes 13 meteorological parameters and 10 air quality parameters. A nephelometer-type instrument for measuring the scattering coefficient has been included in the list of instruments for air quality. The RAPS study will involve sensors on 20 thirty-meter towers, and 5 ten-meter towers. All signals for meteorological data are to be conditioned for 0-5V d-c full scale.

METEOROLOGICAL TOWERS

Towers have become commonplace platforms for meteorological sensors, especially for pollution-related applications. Several investigations have shown that the geometry of the structure and the exposure of the sensor may influence the accuracy of measurements. This may be critical for research programs, but in most cases, the degree of inaccuracy is not as significant as the reliability or durability of an instrument selected to function in the difficult-to-reach and frequently hostile atmosphere of a tall tower. In Canada, where the Department of the Environment has a coordinated approach toward the selection, operation, and maintenance of a meteorological tower network, the selection of sensors has been standardized. In the Spring, 1972 issue of the Meteorological Tower Bulletin[34], there is a list of 17 Canadian towers. At least 4 more have been erected since early 1972, but among those then described, wind measuring equipment included 12 propeller type anemometers and vanes, 2 Canadian Met. Service Type U2A cup anemometers, 1 Canadian Met. Service Type 45 cup anemometer, and 2 R.W. Munro cup anemometers. All temperature and temperature-difference measurements are made with platinum wire RTD's located in aspirated radiation shields. While most of the data has been recorded in analog form, for ten minutes beginning on the hour, two commercial data loggers were put in use in 1971; two additional units in 1973.

In the United States the erection of meteorological towers has accelerated, in proportion to the requests for power plant licenses and in conformity with the mandatory requirements of the Environmental Impact Statement. While no precise figures on towers are available, the A.E.C. routinely publishes a map of nuclear power plant sites, and this is a reliable index of where there have been and/or are meteorological surveys in progress. The map dated June 30, 1973 identifies 172 reactors as operable, being built, or planned. In spite of the uncoordinated growth of towers and lack of standardization in the choice of sensors in the U.S., the AEC's specifications of accuracy, data recovery, and calibration schedules, summarized in Safety Guide 23, have established a basis for uniformity.

CONCLUSIONS

Sensors for wind and temperature have been of primary interest in measurements for air pollution problems, but the fluctuation of the wind as defined and classified from analog records, later analyzed from digital data, has received the primary attention of many investigators. Meteorological installations for studying air pollution problems are represented today by meso scale networks, designed with sensors for measuring both the elements of weather and the air quality. In most cases analog signals are converted to digital signals, tabulated, and analyzed by using computer-related technology.

Air pollution problems are varied and the complicated, frequently-frustrating task of designing and establishing an installation might be done more systematically by reference to the functional, geographical, technical, and logistical aspects of the measuring program as a total system.

REFERENCES

(1) Gill, G.C. and P.L. Hexter, "Some Instrumentation Definitions for Use by Meteorologists and Engineers," BULLETIN OF THE AMERICAN METEOROLOGICAL SOCIETY, Vol 53 (1972), 846-851.

(2) Lansberg, H.E., "Meteorological Observations in Urban Areas," METEOROLOGICAL MONOGRAPHS, Vol II (1970).

(3) Gant, N.E., E. Newman, D.B. Spiegler, and E.Y. Tong, "AIRMAP: A Cost-Efficient Air Quality Control System," Third Commonwealth Air Pollution Control Workshop, Virginia Beach, Va. (1973).

(4) MacCready, P.B. and T.J. Lockhart, "The Status and Future of Direct Meteorological Observations," ATMOSPHERIC TECHNOLOGY, NCAR, Vol 1 (1973), 70-75.

(5) Sutton, O.G., ATMOSPHERIC TURBULENCE, Methuen and Co., London (1949), 14.

(6) Gill, G.C., "Principles of Dynamic Response," METEOROLOGICAL INSTRUMENTATION IN AIR POLLUTION, Institute for Air Pollution Training, E.P.A., R.T.P., N.C. (1972)

(7) MacCready, P.B., "Theoretical Considerations in Instrumentation Design," SYMPOSIUM ON MET. OBS. & INSTR, AMS (1969), Washington, D.C.

(8) Lowry, P.H., "Micro-Climate Factors in Smoke Pollution from Tall Stacks," ON ATMOSPHERIC POLLUTION-METEOROLOGICAL MONOGRAPHS, Amer. Met. Soc., Vol I No. 4 (1951), 24-29.

(9) Smith, M.E., "The Forecasting of Micrometeorological Variables," ON ATMOSPHERIC POLLUTION-METEOROLOGICAL MONOGRAPHS, American Met. Soc., Vol I No. 4 (1951), 50-55.

(10) Singer, I.A. and M.E. Smith, "Relation of Gustiness to Other Meteorolgical Parameters," JOURNAL OF METEOROLOGY, Vol 10 (1953).

(11) Islitzer, N.F. and D.H. Slade, "Diffusion and Transport Experiments," METEOROLOGY AND ATOMIC ENERGY, USAEC (1968), 117-120.

(12) Cramer, H.E., "A Practical Method for Estimating the Dispersal of Atmospheric Contaminants," PROCEEDINGS OF THE FIRST NAT'L CONF. ON APPLIED METEOROLOGY, AMS, Hartford, Cn. (1957).

(13) Pasquill, F., "The Estimation of the Dispersion of Windborne Material," THE METEOROLOGICAL MAGAZINE, Vol 90 (1961), 33-49.

(14) Safety Guide 23, "Safety Guide 23," REGULATORY GUIDE SERIES, U.S. Atomic Energy Commision, Washington, D. C. (1972).

(15) Jones, J.I.P. and F. Pasquill, "An Experimental System for Directly Recording the Statistics of the Intensity of Atmospheric Turbulence," QUART. JOUR. ROYAL MET. SOC., Vol. 85 (1959), 225-236.

(16) Sachdev, R.N. and K.K. Rajan, "An Electronic Computing Wind Vane for Turbulence Studies," JOURNAL OF APPLIED METEOROLOGY, Vol 10 (1971), 1331-1338.

(17) Markee, E.H., "On the Relationships of Range to Standard Deviation of Wind Fluctuations," MONTHLY WEATHER REVIEW, Vol 91 (1963), 83-87.

(18) Moses, H., "Meteorological Instruments for Use in the Atomic Energy Industry," Chap. 6, METEOROLOGY AND ATOMIC ENERGY-1968. D.H. Slade, Ed. USAEC.

(19) Mazzarella, D.A., "An Inventory of Specifications for Wind Measuring Instruments," BULLETIN, AMS, Vol 53 (1972), 860-871.

(20) Luna, R.E. and H.W. Church, "A Comparison of Turbulence Intensity and Stability Ratio Measurements to Pasquill Turbulence Types," CONF. on AIR POLLUTION MET., RALEIGH, N.C. (1971).

(21) Slob, W.H., "Temperature Error of a Ventilated Thermometer in Dependence of Radiation and Wind," Royal Met. Inst., De Bilt, The Netherlands, CIMO VI SCIENTIFIC DISCUSSIONS, Helsinki, Finland, WMO, (1973), 47-50.

(22) Church, P.W. and E.A. Gosline, Jr., "Meteorological Equipment of the Hanford Engineer Works," USAEC, Oak Ridge, Tn., MDDC-841, (1947)

(23) Mazzarella, D.A. and D.K. Kohl, "Temperature Measurements on a 420-foot Tower," INSTRUMENTS & AUTOMATION, Vol 27 (1954), 1306-1309.

(24) Beers, N.R., "Stack Meteorology and Atmospheric Disposal of Radioactive Waste," NUCLEONICS, Vol 4 (1949), 28-38.

(25) Brown, R.M., "An Automatic Meteorological Data Collection System," JOURNAL OF GEOPHYSICAL RES., Vol 64 (1959), 2369-2372.

(26) Moses, H. and M.A. Bogner, ARGONNE NATIONAL LABORATORY - FIFTEEN-YEAR CLIMATOLOGICAL SUMMARY, Jan. 1, 1950-Dec. 31, 1964. ANL-7084 (1967).

(27) Moses, H. and F.C. Kulhanek, "Argonne Automatic Meteorological Data Processing System," JOURNAL OF APPLIED METEOROLOGY, Vol 1 (1962), 69-80.

(28) Harrison, P.R., "Considerations for Design and Operation of Monitoring Networks in Metropolitan Areas - Real Time and Manual," TECOMAP, WMO-WHO, Helsinki, Finland (1973).

(29) Georgii, H.W., "Design, Structure, and Performance of an Air Pollution Monitoring System in Area of Frankfort/M (Germany)," TECOMAP, WMO-WHO, Helsinki, Finland, (1973).

(30) Editorial , "Most Advanced System of Air Pollution Monitoring," AIR ENG., Dec. (1968), 22-24.

(31) Montgomery, T.L., S.B. Carpenter, and H.E. Lindley. Conf. on Air Pollution Met., Raleigh, N.C. (1971).

(32) T.V.A. Internal Report, "Data Logger Format for the Sequoyah Environmental Data Station," Report No. 72-16, Oct. (1971), "Aerometric Systems," Rev. July (1973); "T.V.A. Sampling Facilities," July 1 (1973).

(33) R.A.P.S. Bid Specifications, Exhibit B, "Central Computer Facility;" Exhibit C, Remote Monitoring Stations;" Exhibit D, "Telecommunications System."

(34) Meteorological Tower Bulletin, Dep't of The Environment, Atm. Env. Services, Downsview, Ontario, Canada, Nov. (1967) to Spring (1972).

TABLE I THE EVOLUTION OF BROOKHAVEN GUSTINESS TYPES

A. Lowry, P.H. (1950)

Type	Range of Trace	σ_θ (108m)
A	$> 90^o$	20°
B	15-90	12
C	> 15	5
D	< 15	< 1

Gustiness Types determined from manual reduction of Aerovane analog traces. Chart speed 3"/min., 10-sec. averages, 60 min. samples. Data for July 1949.

B. Smith, M.E. (1950)

Type	Range of Trace	σ_θ (108m)	ΔT^oC ($T_{108m}-T_{6m}$)	$\bar{U}$ (108m)
A	$> 90^o$	20°	$-1.5/-2.5^oC$	< 8 m/sec
B	15.90	12	-0.8/-2.0	> 5
C	> 15	5	-1.0/+1.0	5.25
D	< 15	< 1	$+1 >$	< 15

Gustiness Types determined from manual reduction of Aerovane analog traces. Chart speed 3"/min., 10-sec. averages, 60-min. samples. Data for July 1949, and other.

C. Singer, I.A. & M.E. Smith (1953)

Type	Range of Trace	% Occurrence	σ_θ (108m) *
A	$> 90^\circ$	1%	--
B_2	45-90	3	20°
B_1	15-45	42	13
C	> 15	14	7
D	< 15	40	4

Type	$\Delta T\,^\circ C$ ($T_{125m} - T_{6m}$)		$\bar{U}$ (108m)	
A	-1.25°	$\sigma = 0.7$	1.8m/sec	$\sigma = 1.1$
B_2	-1.6	0.5	3.8	1.8
B_1	-1.2	0.65	7.0	3.1
C	-0.64	0.52	10.4	3.1
D	+2.0	2.6	6.4	2.6

Gustiness Types determined by manual reduction of Aerovane traces, 1-hr. averages covering 2 years data, 4/50 - 3/52.
* σ_θ not shown in original manuscript, but determined from 6-sec. averages, 1-hr. samples as noted in "Meteorology and Atomic Energy-1968."

TABLE II PASQUILL STABILITY CATEGORIES

A Pasquill, F. (1958-unpublished), (1961), also P.J. Meade (1960) - determined from many experiments in England and U.S.A.

Wind	Daytime Insolation			Nightime Conditions	
(10m.)	(Strong)	(Moderate)	(Slight)	(Thin high ovc'st) (or > ½ low cloud)	(< 3/8 Cloud)
< 2m/sec	A	A-B	B	---	---
2-3	A-B	B	C	E	F
3-5	B	B-C	C	D	E
5-6	C	C-D	D	D	D
> 6	C	D	D	D	D

Above categories defined:

A Extremely unstable
B Moderately "
C Slightly "
D Neutral
E Slightly stable
F Moderately "

Stability Categories determined from data at Chemical Defense Establishment, Porton, England; National Reactor Test Station, Idaho Falls, Idaho; Project Prairie Grass, O'Neill, Neb. No reference to instrumentation, averaging times, sampling times, etc.

B. Pasquill, F. - As interpreted in Safety Guide 23 (1972)

Category	Classification	σ_θ (10m)	ΔT (°C/100m)
A	Extremely unstable	25.0°	< -1.9°C
B	Moderately "	20.0	-1.9/-1.7
C	Slightly "	15.0	-1.7/-1.5
D	Neutral	10.0	-1.5/-0.5
E	Slightly stable	5.0	-0.5/ 1.5
F	Moderately "	2.5	1.5/ 4.0
G	Extremely "	1.7	> 4.0

NOTE: The "Classification of Atmospheric Stability," Table 2 which appears in Safety Guide 23 is a hybrid, consisting of Categories A-F from Pasquill and Meade. G has been added; σ_θ values are from Chap. 3 of "Meteorology and Atomic Energy-1968;" ΔT values are A.E.C.-selected from experimental data at many locations.

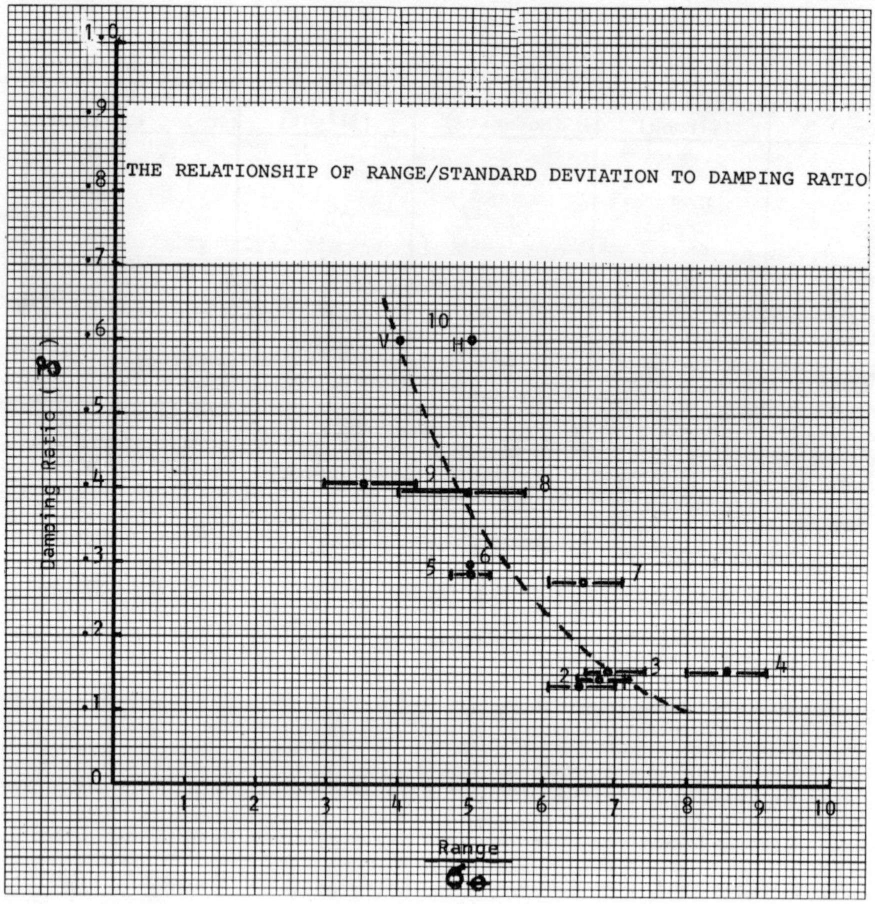

IDENTIFICATION OF DATA

Location	Elevation (m)	Data Reduction	Damping Ratio
Shippingsport	12, 16, 42	5 sec inst	.15
Oak Ridge	5, 16	10 sec inst	.17
Cincinatti	10	15 sec avg	.16
Nashville	10	15 sec avg	.16
Middletown	61	10 sec inst	.28
Brookhaven	108	10 sec avg - 1 hr sample	.28
Idaho Falls	6, 76	7½ sec inst	.28
Alabama	58	10 sec inst - 15 min avg	.4
Blue Hills Stn.	10	10 sec inst - 60 min avg	.4
California	76*	2 sec inst - 10 min avg	.6

* Bivane, V=vertical; H=horizontal

TABLE III

CLASSIFICATIONS FOR SELECTING SENSORS & ESTABLISHING OBJECTIVES FOR AIR POLLUTION PROBLEMS

I FUNCTIONAL	II GEOGRAPHICAL	III TECHNICAL	IV LOGISTICAL
A Atmospheric Scales 1 Micro 2 Meso 3 Macro B Time Scales 1 Research 2 Operational a Synoptic b Industrial 3 Climatological	A Location 1 Urban 2 Rural 3 Marine B Climate 1 Arctic 2 Temperate 3 Tropical C Area 1 Site 2 County 3 State 4 Nation 5 Globe	A Type 1 Electrical 2 Mechanical 3 Optical B Method 1 Direct 2 Indirect a Remote Sensing C Platform 1 Surface a Ground Station b Mobile Station c Tower 2 Aloft a Aircraft b Balloon c Satellite D Signal 1 Analog 2 Digital a Instantaneous b Integrated	A Availability 1 Commercial a Off-the-Shelf b Made-to-Order c R & D 2 Non-Commercial a Made-to-Order b R & D B Price 1 Competitive 2 Reasonable 3 Unimportant C Delivery 1 Acceptable 2 Non-acceptable 3 Compromise

© 1973, ISA JSP 6703

SOME INFLUENCES OF REGIONAL BOUNDARY LAYER FLOW ON ATMOSPHERIC TRANSPORT AND DISPERSION*

Larry L. Wendell†
National Oceanic and Atmospheric Administration
Idaho Falls, Idaho

ABSTRACT

The scales on which most atmospheric transport and diffusion work has been carried out have been either within less than a few kilometers or over long distances (i.e., >400-500 kilometers). Only recently has serious attention been turned to the intermediate scale of 10-100 kilometers. It is on this regional scale in the boundary layer of the atmosphere that air pollution problems are becoming more critical. Recent investigations in several areas of the western United States have shown that the atmospheric transport of pollutants is strongly affected by intermediate scale variations in the topography. This spatial variation in the flow can cause serious error in both short-term emergency forecasts of plume transport and long-term site evaluation studies based only on the source winds. Instances of large vertical shear in wind speed and direction may also be traced back to regional scale variation in terrain height. Thus, a knowledge of intermediate scale terrain configuration and associated wind patterns can provide valuable insight for the interpretation and analysis of pollution measurements.

INTRODUCTION

Until recently almost all experiments to determine atmospheric diffusion of pollutants were conducted on a scale of a few kilometers. In each experiment conditions were selected such that transport of the polluting material was accomplished by a mean wind of fairly constant direction and speed. The main purpose of the experiments was to study the nature of turbulent diffusion. However, the transport philosophy of these experiments seemed to carry over into diffusion models covering scales of tens of kilometers (mesoscale) as well as operational practices of determining the areas affected by an effluent from an accidental release or a long-term continuous stack release. The models assumed a steady transporting wind and the operational procedures were set up to use the source wind as if it applied for the entire distance of concern.

Results of some work by Wendell[5] have demonstrated both that the wind over a mesoscale region cannot be counted on to be uniform and that much of the spatial variability can be related to relatively mild terrain variation. Trajectory studies by Angell, et al.[1] have also revealed transport complexities directly bearing on pollution transport.

With the advent of sophisticated pollution models, advancing computer technology and the rapid development of pollution-sensing equipment, a system has been suggested in which an undesirably high concentration of some pollutant measured at a particular sensor could be quickly traced back to the offending industrial site. One of the weakest links in a system such as this would be the transport assumptions. The complexity of the flow in the boundary layer, especially under the influence of terrain variation, could make transport determinations from a single location so misleading that results could be completely erroneous.

The purpose of this paper is to review briefly some recent work showing mesoscale spatial variability in wind fields and the resulting effects on transport and also to present some terrain-induced mesoscale phenomena which appear strongly to influence transport.

MESOSCALE FLOW AND TOPOGRAPHIC FEATURES

In an attempt to determine the mesoscale horizontal variation of the wind in the lower portion of the planetary boundary layer, hourly averaged wind data from a randomly spaced collection of 21 wind stations within an area of 12,420 km^2 has been used to make a preliminary investigation of the problem.[5]

The scale of concern in this work is about an order of magnitude smaller than the synoptic scale. This is shown by an outline of the computational grid (rectangular box in southeastern Idaho) on a map of the western United States in Figure 1. The grid is approximately 86 km wide and 130 km long. The topography of the area is shown in Figure 2. The severe height variation is shown only qualitatively through a shadowed photograph of a relief map. The gentle

*Research carried out under primary sponsorship of the Atomic Energy Commission (Division of Reactor Development and Technology).

†Present affiliation: Senior Research Scientist, Battelle, Pacific Northwest Laboratories, Richland, Washington.

height variation over the relatively flat plain is depicted by the height contours. The area in which the terrain is less than 5,000 ft is stippled to emphasize the large fishhook-shaped depression in the plain. The solid black dots show the locations of the towers with the wind sensors. The sensor heights were 15 m or above. All calculations were restricted to the area to the right of the dashed line.

The basic calculation was an interpolation of the wind components measured at the nonregularly spaced towers to a regularly spaced grid of points. The interpolation technique involved a weighted averaging of the velocity components from several of the stations nearest to each of the grid points. The weighting factors used were the inverse squares of the distances from stations to grid points.

When the wind components are obtained at the grid points they may be used in several types of analyses. However, they may also be used to enhance the appearance of a plot of the raw data for visual observation. Computer produced plots of the wind fields at 6-hour intervals have been made by Wendell[6] for the entire year of 1969. An example of these plots is shown in Figure 3. The winds measured at the tower locations are plotted in the standard format of shaft and barbs. The winds interpolated to the grid points are plotted as vectors, with their length directly proportional to the speed. In these plots the 5,000 ft contour is included with the outline of the U.S. Atomic Energy Commission's National Reactor Testing Station.

One of the most interesting features of the 1969 wind field plots is the strong conformity of the flow patterns to the fishhook-shaped depression. Varying degrees of this conformity were observed during some portion of about 60 percent of the days of the year. The two series of plots shown in Figure 3 are examples of this phenomenon. During about 12 percent of the days of 1969 the cyclonic bending of the flow in the upper portion of the grid developed into a distinctive circular shaped eddy which persisted for several hours. Occasionally the formation of the eddy could be detected visually during daylight hours through the formation of a very striking and unusual cloud formation.

In Figure 3A a zone of convergence may be seen forming in the first two frames as northerly winds in the site's northern sector intercept the strong southwesterly flow. If the air from the southwest contains sufficient moisture, a cloud will form in it as it is lifted by the air moving down from the north. The cloud has some very striking characteristics. (An example is shown in Figure 4.) It seems to emanate from a stationary location and to grow as it moves rapidly downwind. This gives it the appearance of a plume from an invisible stack except that it does not decrease in density through diffusion because the convergence is apparently increasing downwind. The cloud serves to delineate this terrain-induced convergence zone (although it is several miles from any irregular terrain features) in the boundary layer in the same fashion that standing lenticular clouds indicate mountain waves at higher levels. It is dramatic evidence of mesoscale spatial variability in the boundary layer flow.

It is quite obvious that flow patterns of the type just discussed would be very poorly represented by the wind at a single location. Flow patterns that would have been well represented by a wind measurement at one location were observed to occur during some portion of only 21 percent of the days of 1969.

WIND FIELD-DERIVED TRAJECTORIES

To get a more direct observation of the effect of temporal and spatial variability of mesoscale wind patterns on transport, the hourly averaged wind data, interpolated to the regular grid, was used to construct trajectories of hypothetical particles introduced serially into the flow. Computer produced plots of trajectories of these particles, released once an hour, were also obtained for the entire year of 1969.[6] Figure 5 shows examples of these plots. These two series of trajectories indicate the path of 12 "particles" released one hour apart beginning at the times shown above each plot. The numbers at the ends of the trajectories indicate the order of release, and the letters along the trajectories indicate the locations of the particles at hourly intervals after release. Like letters may be connected to form a streak line and may be used to estimate the time change of the position of a plume center line.

A long series of trajectory plots of this type can provide an estimate of a climatology of the transport from a given release point for selected release periods.[5] From the data trajectory pattern types can be selected and ranked according to frequency of occurrence during the entire year, regardless of release period. However, an associated table[5] showing frequency of occurrence according to release period and season provides a fairly detailed picture of the variation of transport conditions for the given release location.

It may be noted from some of the trajectory pattern types that the spatial variability in the flow has played a large part in determining the pattern. To obtain a more direct comparison of the wind field and single station produced trajectories, several sets of trajectories were recomputed using only the source wind as if it applied over the whole grid. Figure 6 shows some of these results. The trajectories in the upper plots (WF) were obtained from the wind fields while those in the lower plots (SS) were obtained using the source wind only. The differences in the transport are quite apparent. Comparisons similar to these were also made using wind data from a nine-station network in Oklahoma.

The area of the study was an 80 x 112 km wind tower network centered around Oklahoma City, Oklahoma. The maximum terrain variation was somewhat less than for the Idaho location, but on the scale of one kilometer the terrain had more variation. A quantitative comparison was made of the difference between trajectories produced by wind fields and those produced by the source wind. Based on 250 comparisons the average difference was about 1.6 km at 16 km

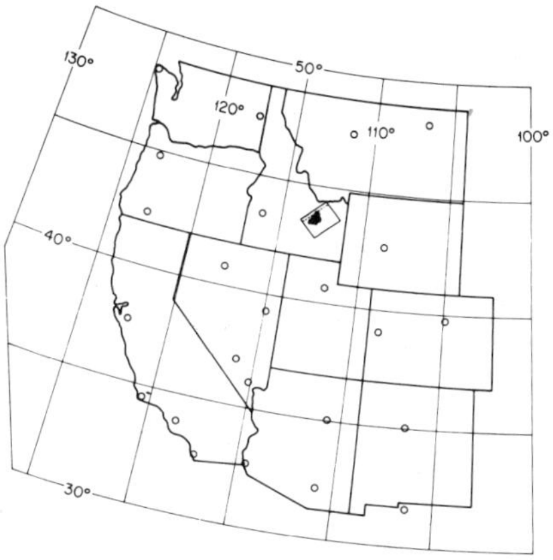

Fig. 1. Location and extent of computational grid for mesoscale wind observations. The solid border of the rectangular box in southeastern Idaho contains all the observation locations, but no calculations are carried out for the area northwest of the dashed line. The solid black area represents the geographical extent of the National Reactor Testing Station (Wendell).[5]

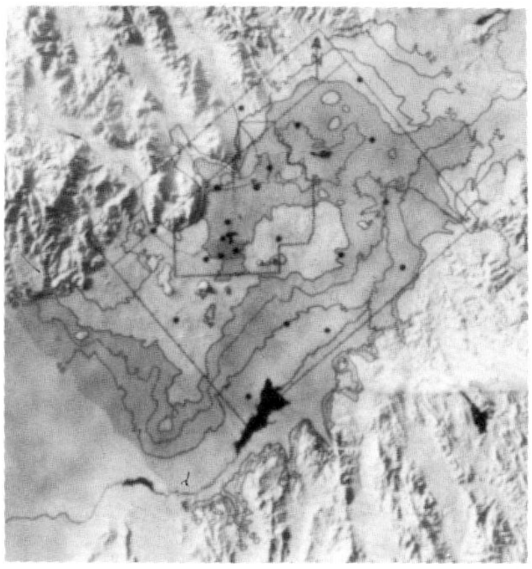

Fig. 2. Relief map of the Upper Snake River Plain in southeastern Idaho. The values shown on the few contour lines over the plain are in hundreds of feet. The stippled area within and adjacent to the grid indicates the terrain with an altitude less than 5,000 ft MSL. The tick marks along the border of the grid indicate the grid point separation, 8.53 km. The solid dots show the location of the wind towers.[5]

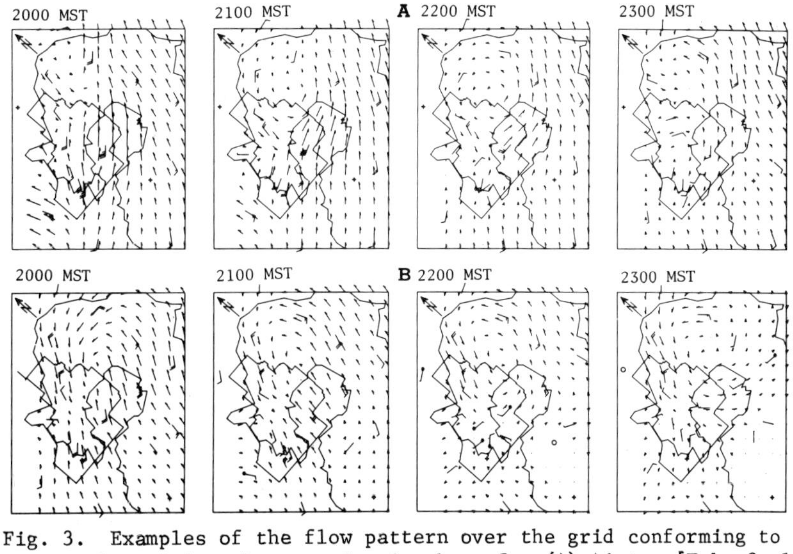

Fig. 3. Examples of the flow pattern over the grid conforming to the topography. A four-hour series is shown for (A) winter [Feb. 9, 1969], and (B) spring [May 13, 1969]. The 5,000 ft contour is included for reference purposes (see Fig. 1). The wind data from each station are plotted in standard form, and the interpolated winds are plotted as vectors at each grid point. The lengths of the wind vectors are proportional to the speed.[5]

from the origin, 8 at 32 km and 14.5 at 48 km. The separation is obviously not linear with trajectory length and indicates that even in mildly varying terrain the transport differences can become significant beyond a few km from the source.

It should be pointed out that the transport, calculated from wind measurements from these towers, is restricted to a layer in which these measurements are representative. During stable conditions most of the material released would remain in this layer and the tower derived trajectories should provide a reasonable estimate of the transport and generally a much better estimate than one from a single station.

WIND FIELD-DERIVED DISPERSION

To relate the wind field-derived transport of a contaminant to the measurements of an environmental pollution sensor one may incorporate these transporting winds into a K theory type of model. This has been done for short term cases by several investigators including Lamb and Neiburger[2] and MacCracken, et al.[3] An alternative approach is to superimpose the diffusion onto wind field-derived trajectories. This has been done by Start and Wendell[4] through a technique of dividing a continuous release into serially released discrete puffs which diffuse in a Gaussian manner as a function of distance travelled, mixing depth and stability class. This type of model is less complex and amenable to computer simulation of extended time periods without great expense. Figure 7 gives an example of puff releases which approximate a continuous plume. The dots correspond to puff center positions, and the circles represent contours of the minimum concentration of interest. The circles grow in size as the puffs diffuse. This release period is the same as the one shown for the March 4, 1969 case in Figure 6.

The plume, shown at two-hour intervals in Figure 7, undergoes a 90° bending about 10 miles south of the site boundary. As was demonstrated in Figure 6 this type of deformation cannot be produced with the source wind alone. A wind reversal at the source is indicated to have occurred just prior to 2300 MST 03/04/69. This new plume segment bends to the east as the oldest portion of the plume bends in the opposite direction toward it. By 0500 both new and old material are sweeping southward across the site as a line source.

To provide a quantitative presentation of the effect of the plume transport and dispersion, the total integrated concentration (TIC) is calculated on an array of grid points and is displayed in contour form at six-hour intervals in Figure 8, for the March 4 case. Eighteen hours after release the splitting and bending plume observed in Figure 7 leaves a doughnut-shaped ring of high TIC. It is difficult to imagine a pollution and wind measurements northeast of the site having any success at all in the location of the source.

TRANSPORT ESTIMATES FROM TETROONS

Using a network of wind towers to estimate transport is an indirect approach. A more direct method involves the tracking of a constant density balloon. A fairly complete list of references is provided by Angell, et al.[1] to the work done in recent years in which the radar-tetroon-transponder system has been developed and utilized to study various characteristics of air motion in the planetary boundary layer.

Tetroon trajectories have been compared with wind field-derived trajectories with widely varying results for individual comparisons. This is illustrated by the comparisons shown in Figure 9. In summary of observations thus far, if the tetroon flies low enough or there is little vertical wind shear the trajectories are generally quite similar. However, under the conditions of strong temperature inversions and local topographic anomalies, the trajectories can be very different. This does not mean that one type of trajectory is more correct than the other. It simply means that when the tetroon flight level is above the tower sensors, the difference in the trajectories is a reflection of the wind shear. This is dramatically illustrated in the next section by trajectory data from a pair of tetroons. These results suggest that the two methods of investigating mesoscale circulation of the planetary boundary layer should be used to complement each other.

TOPOGRAPHICALLY INDUCED VERTICAL WIND SHEAR

The increase in wind speed with height above the ground surface is familiar to everyone, and a certain amount of direction shear is generally expected. However, a drastic change in wind direction in the first 50 m above the surface is a curiosity, as evidenced by the notoriety received by the photograph of the smoke travelling in different directions from different levels on the Brookhaven tower. Photographs similar to this were taken by Air Resources Laboratories personnel during an early morning diffusion experiment in Idaho. One of these pictures taken from an aircraft is shown in Figure 10. Smoke was released from the surface as well as 30 and 60 m up on the tower. The smoke from the surface indicated a layer of wind from about 250° while the smoke at the other two levels indicated a wind direction of about 50°. This is a direction shear of about 160° between the two layers. Photographs and other observations from the surface indicated that the wind in the near surface layer was only about 30 ft thick. The transition between the layers was very abrupt and was primarily responsible for the dispersion of the lowest plume. The lower edge of this plume was about 3 m above the surface and very smooth.

The cause of this phenomenon seems to lie in the surrounding topography. Referring back to Figure 2, we can see that the overall slope of the terrain is downward from northeast to southwest, but that local slope in the vicinity of the tower (dot marked T) is primarily in the opposite direction. Thus when the surface layer in this area is cooled sufficiently, gravity will cause it to move toward lower ground even in opposition to the general drainage flow. This phenomenon was observed in data recorded several times since the smoke picture. This effect of fairly mild topographic variation on surface flow during

Fig. 4. Plume-shaped cloud in area of convergence during formation of mesoscale eddy. The visible moisture in the cloud was moving from left to right at 10 to 15 m S^{-1}, but the apparent source point and general shape remained essentially the same for over two hours.

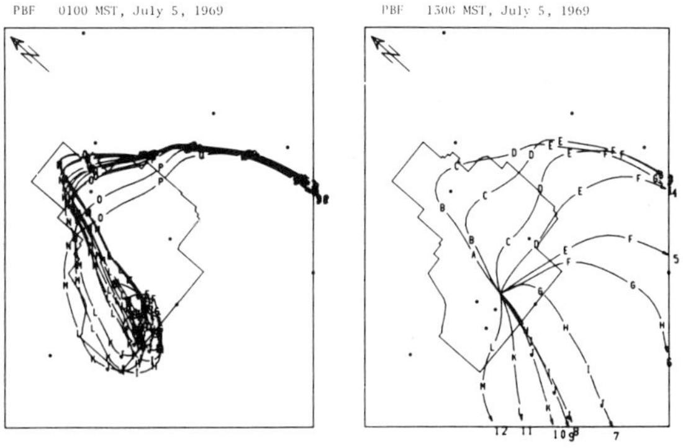

Fig. 5. Trajectories of hypothetical particles released hourly and transported by a time series of objectively interpolated wind fields. The numbers at the ends of the trajectories indicate the order of release, and the letters along the trajectories represent hourly positions.(5)

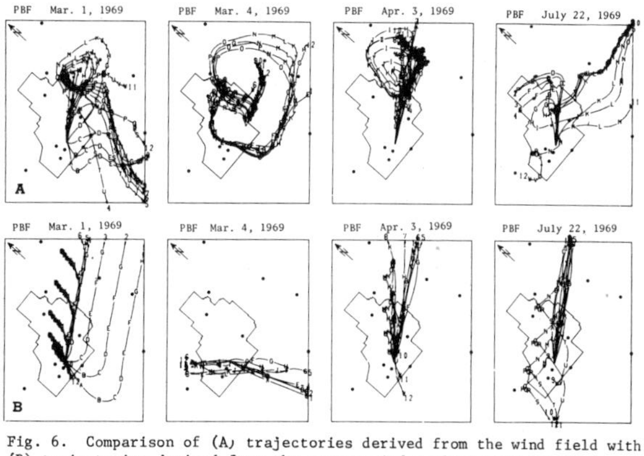

Fig. 6. Comparison of (A) trajectories derived from the wind field with (B) trajectories derived from the source wind applied over the whole grid for the corresponding time periods (1300 MST on days indicated).(5)

periods of low pressure gradient and high stability seems to be one that should be given serious consideration. It points up a need for routine wind measurements near the surface as well as at the height of the pollution source. Knowledge of source configuration is also shown to be important for transport determinations.

Another example of the effect of vertical shear on transport was discovered in the trajectories of a pair of tetroons released one minute apart about 10 km south of NOAA's Wave Propagation Laboratory tower at Haswell in southeastern Colorado. The tetroons were separately tracked by two ARL radars. Horizontal and vertical depictions of the two trajectories are shown in Figure 11. The horizontal trajectory plots begin with the tetroons quite close together but separating rapidly. Tetroon 30B proceeds in a north-northeasterly direction over the mild ridge in the topography, while 30A seems to follow the curve of the 4400 ft contour. The vertical depiction of the trajectories shows 30A dropping to near the surface about 30 minutes after the beginning of the track. The WPL tower is plotted on the vertical plot as a reference point. Flight 30B seemed to be reflecting the winds measured at the tower top for the entire flight, while 30A showed fair agreement with the winds measured by the mesoscale network of 15 m towers indicated by the crosses in Figure 11a. This case would also seem to show that a light prevailing wind can be overcome by a local drainage wind in the surface layer.

SUMMARY AND CONCLUSIONS

It has been demonstrated that regional boundary layer flow can be significantly more complex than the standard assumption of a mean wind speed and direction over the whole area. Under conditions of weak synoptic scale pressure gradients and high stability, even mild mesoscale terrain variations, can cause horizontal, as well as vertical, variations in the flow which are detectable only through a fine enough network of wind measurements. Trajectories constructed from these networks of wind measurements indicate the magnitude of the discrepancies which can exist in transport deduced from the source wind alone. Transport estimated from tetroon flights has also indicated a strong influence from local topographic variation as well as land-water proximity. The topographic variation has also been shown to cause a large enough vertical shear of wind direction to create large differences in atmospheric transport. These differences can greatly confuse any analysis of pollutant measurements, especially if the existence of the phenomenon is not suspected.

The regional scale transport and dispersion models now in existence have never been verified on more than a very short term, special case basis. With the development of new tracers and sensors a long-term verification program now looks like a definite possibility. This type of program should allow a quantitative assessment of the importance of some of the complexities in regional boundary layer flow on the transport and dispersion of pollutants. Since these complexities arise during the conditions of least dispersion, it would appear that an understanding of their effects on pollutant transport is essential.

REFERENCES

(1) Angell, J. K., D. H. Pack, C. R. Dickson, and W. H. Hoecker, "Three-Dimensional Air Trajectories Determined from Tetroon Flights in the Planetary Boundary Layer of the Los Angeles Basins," JOURNAL OF APPLIED METEOROLOGY, Vol. 11 (1972), No. 3, 451-471.

(2) Lamb, R. G. and M. Neiburger, "An Interim Version of a Generalized Urban Air Pollution Model," ATMOSPHERIC ENVIRONMENT, Vol. 5 (1971), 239-264.

(3) MacCracken, M. C., T. V. Crawford, K. R. Peterson and J. B. Knox, "Initial Application of a Multibox Air Pollution Model to the San Francisco Bay Area," LAWRENCE LIVERMORE LABORATORY REPORT, UCRL-73944 (1972).

(4) Start, G. E. and L. L. Wendell, "Regional Effluent Dispersion Calculations Considering Spatial and Temporal Meteorological Variations," NOAA TECHNICAL MEMORANDUM (in press).

(5) Wendell, L. L., "Mesoscale Wind Fields and Transport Estimates Determined from a Network of Wind Towers," MONTHLY WEATHER REVIEW, Vol. 100 (1972), 565-578.

(6) Wendell, L. L., "A Preliminary Examination of Mesoscale Wind Fields and Transport Determined from a Network of Wind Towers," NOAA Technical Memorandum. ERLTM-ARL 25 (1970).

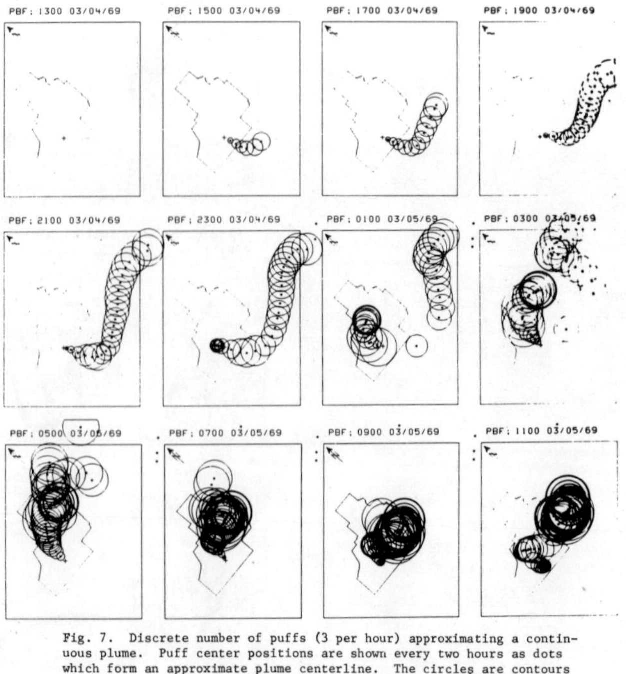

Fig. 7. Discrete number of puffs (3 per hour) approximating a continuous plume. Puff center positions are shown every two hours as dots which form an approximate plume centerline. The circles are contours of the minimum concentration of interest for each puff.[4]

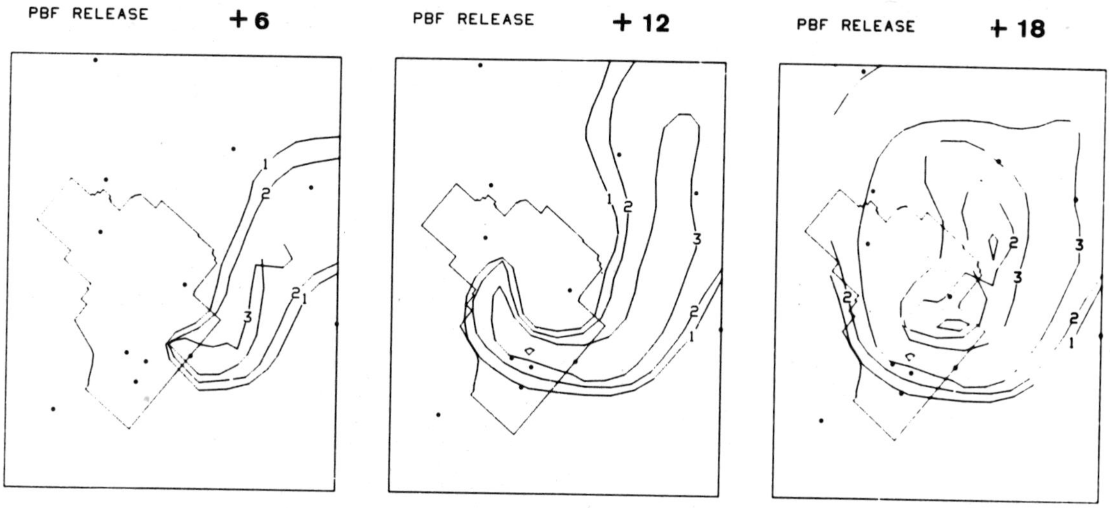

Fig. 8. Contours of total integrated concentration at six-hour intervals for the March 4, 1969 case shown in Figure 7. The numbers are indicators of relative concentration from a unit source (1 is the lowest TIC).

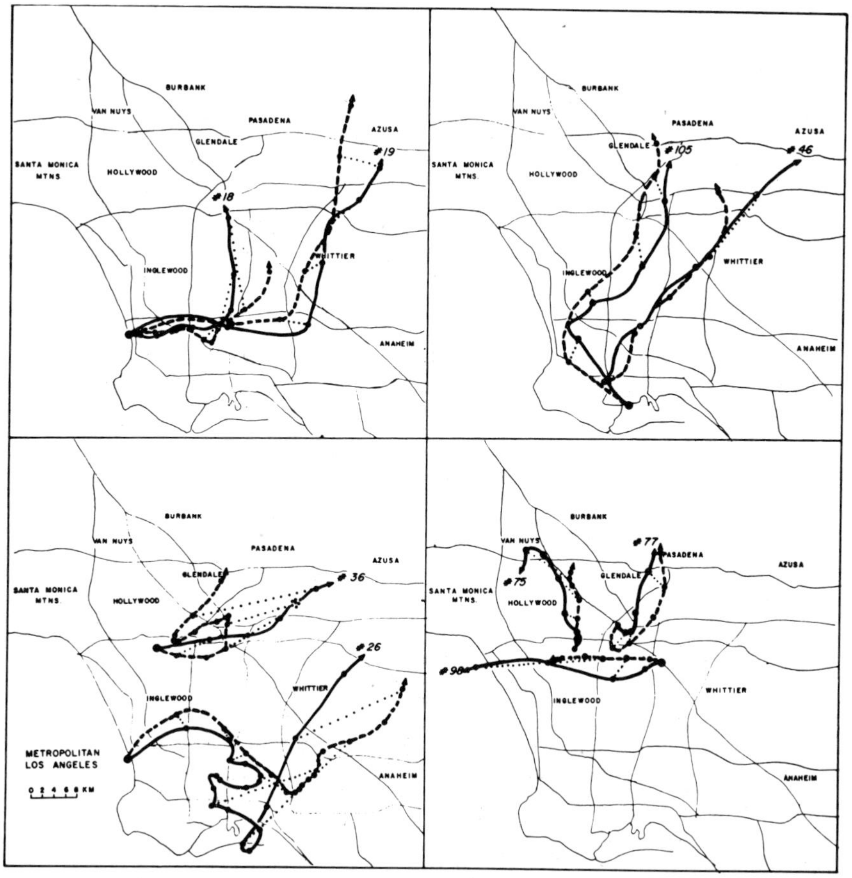

Fig. 9. Comparisons between tetroon (solid line) and wind field-derived surface (dashed lines) trajectories. The surface air trajectories were originated at the time and place of the tetroon trajectories. The dots along the trajectories are at one-hour intervals, while the dotted lines are isochrones connecting tetroon and surface trajectory positions for the same time. The tetroon flight numbers are given at the ends of the solid trajectories.[1]

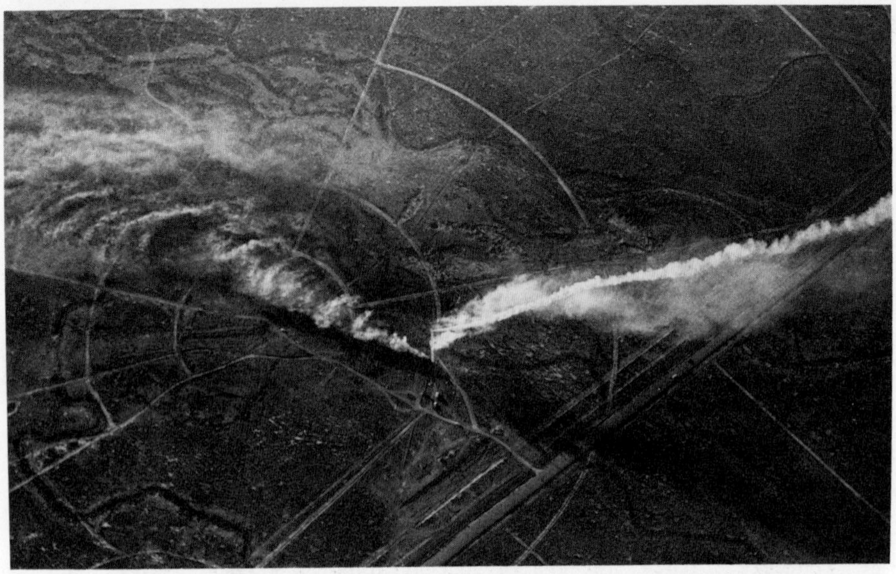

Fig. 10. Aerial photograph of smoke plumes generated at the bottom, middle and top of the 60 m ARL tower at the National Reactor Testing Station in Idaho.

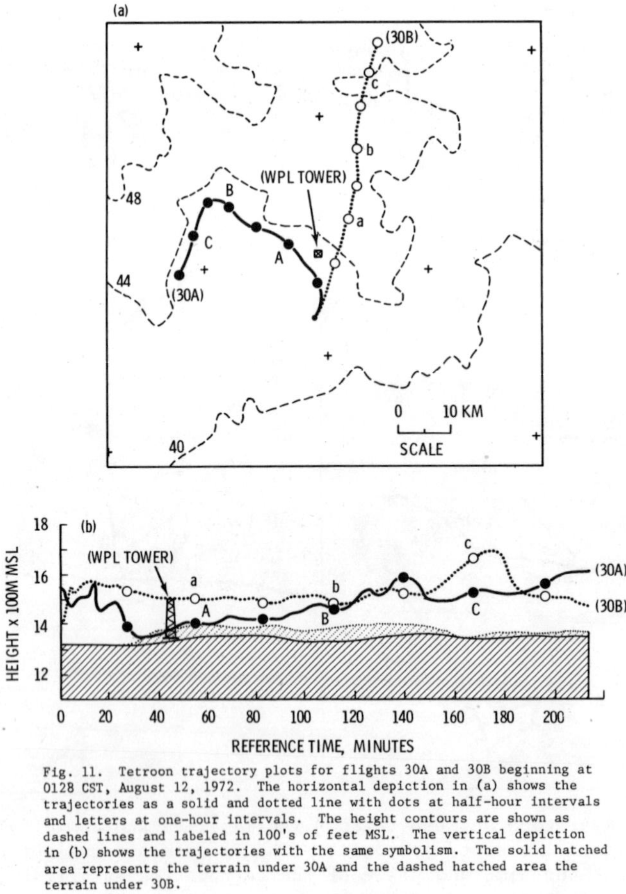

Fig. 11. Tetroon trajectory plots for flights 30A and 30B beginning at 0128 CST, August 12, 1972. The horizontal depiction in (a) shows the trajectories as a solid and dotted line with dots at half-hour intervals and letters at one-hour intervals. The height contours are shown as dashed lines and labeled in 100's of feet MSL. The vertical depiction in (b) shows the trajectories with the same symbolism. The solid hatched area represents the terrain under 30A and the dashed hatched area the terrain under 30B.

©1973, ISA JSP 6704

AIR POLLUTION METEOROLOGICAL OBSERVATIONS FOR SHORT-DURATION INVESTIGATIONS, STAGNATION EPISODES AND ACCIDENT EMERGENCIES

Paul A. Humphrey*
Chief, Special Projects Branch
Meteorology Laboratory
Research Triangle Park, North Carolina

ABSTRACT

Attention is directed to the requirements for instruments needed for air pollution meteorological observations during short duration investigations, stagnation episodes, and accident emergencies. The basic requirement is for wind direction and speed observations, with a secondary requirement for vertical temperature profiles and atmospheric stability determinations. Interest is focused in the first 2 - 3 thousand feet above the surface, or within the mixing layer during air pollution episodes. Aside from the fact that it is often necessary to make compromises because measurements cannot be made wherever and whenever desired, it can be said that meteorological instrumentation is available that can provide such information. However, with respect to the design of equipment that can be obtained commercially, there is a need for lighter weight, more compact instruments that are conveniently packaged for transport as baggage aboard passenger aircraft. Also, there is need for instruments more easily set up and operated under difficult field conditions, and that can be calibrated or oriented to satisfy more positively legal requirements that may be encountered during courtroom cross-examinations.

The comments included are based on actual experience obtained while responding to several actual or potential air pollution accidents and while participating in a variety of air pollution investigations.

The shortcomings or needless features of present instrumentation and performance characteristics required for air pollution emergencies are discussed, and new design features and devices are proposed.

Additionally, brief consideration is given to obtaining atmospheric sounding information under field conditions using miniature rockets, the radiometric thermasonde, acoustic radar, lidar, or other modern or futuristic devices.

In conclusion, it is suggested that there is a continuing need for quick response to air pollution emergencies and that meteorological instrumentation designed for emergency use is needed outside of the field of air pollution.

* On assignment from the National Oceanic and Atmospheric Administration, U. S. Department of Commerce

INTRODUCTION

Most of the papers being presented at the 2nd Joint Conference on Sensing of Environmental Pollutants are about some aspect of advanced instrument technology. This paper is somewhat different. In an age of instrument sophistication, when "computerization" and "remote sensing" are bywords, it calls for more attention to elementary requirements for basic observations, particularly for emergency situations when it is necessary to move and set up instruments quickly. My subject is air pollution instrumentation of the workhorse type, used for operational, not research, purposes. New, or futuristic, instruments will be mentioned only briefly.

The statements and suggestions made are based on more than 20 years of experience with many field operations, including firsthand contact with current EPA operational and research programs.

Meteorological instrumentation for air pollution purposes has already been described by Hewson[1], Moses[2], Gill[3], McCormick[4], and others[5]. This paper is of a more general nature. It is directed to the instrument community at large.

The design criteria of air pollution meteorology instruments are unique primarily with respect to the design of wind instruments. The requirement for precision is greatest at the low end of the wind speed scale. Exposure conditions may necessitate the use of a relatively heavy and rugged instrument such as an Aerovane, but sensitive, light-weight wind sensors are used whenever possible, with recorder speed scales that may end in the range of 25 to 50 miles per hour. Also, with respect to atmospheric soundings, or remote sensing, the dimensions of the phenomena to be observed in air pollution work are usually on the mesoscale. The air pollution meteorologist is most likely to need information from within the envelope of air covering a city, or within a few kilometers of an

Superior numbers refer to similarly numbered references at the end of this paper.

industrial plant.

SHORT-DURATION INVESTIGATIONS

Short-duration air pollution investigations require meteorological observations for many reasons. It may be necessary to describe the local air pollution climatology, correlate air monitoring results with meteorology, or to establish source-receptor-relationships for control purposes or for an effects study. Meteorological observation data are used to determine parameter values in dispersion equations, often part of complex mathematical models, for estimating air pollution concentrations. In most air pollution problems there is primarily a need for wind direction and speed observations. Secondarily, there is a need for observations to determine the dispersion parameters for the rates of spreading of a plume horizontally and vertically. Turbulence indicative of dispersion rates can be measured directly in various ways, for example, with bidirectional vanes, called bivanes, or ordinary wind vanes[(6)(7)], or dispersion parameter values can be selected from measurements of the vertical temperature lapse rate from a meteorological tower or temperature soundings[(8)(9)(10)]. As a last resort, it is possible to select dispersion parameter values from hourly, airport-type surface weather observations[(11)], taking into consideration only solar elevation, wind speed, and cloudiness. It is sometimes also desirable to have vertical temperature profile measurements as it may be necessary to take into account upper air temperature inversions that may act as a lid on a dispersing plume or trap air pollution over a city. If routine radiosonde observations are not available, special soundings may be taken by tethersonde or other means, for example, a helicopter.

Rossano[(12)] has described the community air pollution survey, and he has listed many such studies. These efforts were usually undertaken to assess the air resources of a locality by means of air quality monitoring, an emission inventory, and the air pollution climatology. Other examples of short-duration investigations, which determined the nature of particular air pollution problems, and papers describing the role of the meteorologist, are given in the reference list at the end of this paper[(13-21)]. Commercially available instrumentation advertised for air pollution work is generally adequate for short-duration investigations of an operational nature.

When it is possible to plan ahead, shipping, installation and operation of equipment present no special problems. There is likely to be a wide choice of masts or towers available; and vans, trailers[(22)], or portable shelters may be conveniently used. However, some short-duration investigations may have the same restrictions or needs encountered during air pollution emergencies.

Before leaving the subject of short-duration investigations, one comment will be made. Surface wind data are generally collected during studies of pollution from tall stacks. Wind recording instruments with sensors on masts only 5-10 meters high may be used to supplement soundings that can be made only certain times each day, or because such instruments may be the only feasible means of obtaining any wind data at all. Such wind data, if used alone can be misleading. Even meteorological towers are unlikely to reach high enough.

A better means of obtaining low-level wind soundings, say, in the first 3000 feet of the atmosphere, is the perennial need of the air pollution meteorologist. The efforts such as have been described at this Conference to develop remote sensing instruments for frequent, or continuous, low-level wind or temperature soundings that could be used in the vicinity of a stack are very welcome.

STAGNATION EPISODES

The stagnation episode occurs because of a prolonged period, usually lasting a few days, when there is a combination of low-wind speed and a shallow surface mixing layer. This results in a trapping action under an inversion lid. The effects of stagnation are primarily of concern in large urban areas where concentrations of particulates or sulfur oxides can build up. The observational need for stagnation conditions is for wind and temperature soundings through the mixing layer. Such soundings are needed centrally within the urban complex, or heat island, and not from a somewhat rural setting at a nearby airport, where meteorological observations are more likely to be available. The National Weather Service currently makes dedicated soundings for air pollution forecasting purposes in eight major cities. The nature of this program has been described by Kirschner[(23)]. Conventional radiosonde ground equipment and pilot balloon theodolites are used. However, only 100-gram balloons, which are smaller than standard size, are used to lift the radiosonde transmitter since a slow ascent is desired for more accurate data in the lower levels and it is unnecessary to ascend to stratospheric heights. The balloon is tracked visually because of the necessity for accurate wind data in the first few thousand feet. Usually, two soundings per day are made, and it is desirable to have these morning and afternoon at times of minimum and maximum temperature. Two men are normally required to make a sounding, so the program is relatively expensive. Actually, more soundings per day are desired, and it would be helpful to have more sounding locations in the same city, or in other cities where there are now none available.

A few years ago there was probably more emphasis on telemetering networks for episode response than there is today. Some cities, for example, Chicago[(24)], have such networks, which are basically air quality monitoring networks with meteorological instruments added. However, mathematical dispersion models that can make full use of such networks for forecasting purposes are not yet available. Also, forecasters have not used network observations as much as might be supposed. Air pollution forecasting for episodes, like forecasting tomorrow's weather, depends not so much on conditions on an urban scale as on the synoptic scale. Even so, detailed observations, which a network can provide, are needed when it comes to making dispersion estimates for the control of air pollution sources in some situations. It may be desirable to shut down industrial operations in only part of a city rather than act blindly and shut down industrial sources needlessly over the whole urban complex. Examples of an episode requiring control action that affected 21 industries occurred in Birmingham, Alabama, in November 1971, and another severe episode occurred in October of this year. The current concern about supplementary control systems, which includes meteorological control, for large courses of sulfur dioxide [25], and the occurrence of high oxidant episodes in various parts of the United States are expected to cause a resurgence of interest in the design of meteorological networks for air pollution purposes.

ACCIDENT EMERGENCIES

Accidents where a toxic cloud is released, or could potentially be released, occur more often than is generally realized. Several times each year there are news stories of industrial or transportation accidents causing the evacuation of people from urban or rural areas. The author of this paper was the meteorologist for four chlorine barge salvage operations and a train wreck with a burning boxcar of phosphorus pentasulfide. The chlorine barges were located near Natchez, Mississippi [26] [27]; Baton Rouge, Louisiana [28]; Louisville, Kentucky [29]; and Morgan City, Louisiana. The train wreck, which produced an enormous amount of acrid white smoke, was near West Lafayette, Ohio. In each case large-scale population evacuations occurred. For example, 4000 people were evacuated in Louisville, 2800 in Morgan City and nearby localities, and 1100 in the rural area downwind from the Ohio train wreck. Each chlorine operation required the setting up of a local observational program and a control center to provide hour-by-hour information on the most probable danger sector, and estimates of concentrations, assuming various amounts of gas might accidentally be released.

Coblenz and Roberts [30] have stated that the role of air pollution control agencies in accident emergencies is clearly limited to providing assistance and support to the agencies of government which are directly concerned with immediate health and welfare of the public. Generally, these are State and local Civil Defense, law enforcement, and fire fighting agencies. The EPA is the Federal Agency responsible for evaluating environmental hazards [31] and the author has been assigned the responsibility of heading the emergency response meteorological team from the National Environmental Research Center, North Carolina, where the Meteorology Laboratory is located.

At present the team would consist of one or two meteorologists and one or two meteorological technicians, depending upon the situation. The team would most likely respond to a request from one of the ten EPA Regional Offices; however, most EPA Regional Offices have meteorologists assigned, and some of them are acquiring meteorological and air monitoring instruments. Consequently, the probability that the Meteorology Laboratory emergency team will be called is now decreasing. It should be pointed out that the EPA, a Federal Agency, becomes involved when the emergency requires resources not available at the local or State level.

Two recording wind systems, a hand anemometer, and items for making pilot balloon observations are kept in reserve at the Meteorology Laboratory for the EPA emergency response team.

In the accident situation, the meteorologist is most likely to be concerned with dispersion over a distance of a few thousand feet, or a few miles. Vertically, he is primarily concerned with the atmospheric layer from the surface up to two or three thousand feet. Toxic clouds may rise because of the heat of a fire or an explosion, but the greatest hazard to a population is a spill of liquified gas, such as chlorine, that can remain in high concentrations at ground level even though it travels a relatively long distance.

Usually, the most important location for meteorological observations is close to the scene of the accident, to determine the initial conditions under which dispersion is occurring, or might occur. However, some sort of monitoring network may be needed because of terrain features, or so that changes in local wind conditions can be better anticipated.

It is most important to know the probable trajectory of the hazardous cloud. The most useful instrument is the one that helps to determine the three-dimensional air flow representative of an area: therefore, the choice of the pilot balloon system for the emergency kit. The surface wind recording system is included because at an emergency operational control center the wind recorder can be watched for changes in the wind between soundings, and it will provide a permanent record. A second wind system, if available, might be located nearer the scene of the accident or at some other location for better spatial coverage.

In the design of an instrument for emergency use the prime requirement is for dependability, but that means nothing if it is not possible to transport the instrument to where it is needed. Accident emergencies, if not close by, are likely to require that instruments be transported by commercial, or small charter aircraft. Consequently, the overriding factors to be considered in the design of instruments are those which must be considered for transportation.

The two wind systems currently being held in reserve for emergency use are battery-operated systems [32], with spring-driven recorders, and 100-foot lengths of electrical cable for connecting the anemometer cups and wind vane to the recorder. The usefulness of this type of system has been proven in emergencies and other field situations. However, its portability leaves much to be desired.

The recorder in its shipping case is relatively heavy, weighing 44 pounds. No carrying case is provided for cups, vane and cable. The translator box, containing circuit boards and batteries, contained in a carrying case, but one too small to sometimes ship separately. The wind mast was improvised from an ordinary 10-foot tripod roof tower available from suppliers of TV equipment. It is shipped unpackaged with padding at each end.

The orientation of a wind vane under emergency conditions presents problems not encountered at a National Weather Service station where there is plenty of time to use the sun's shadow at solar noon to determine true north. In an emergency, the orientation will be done most likely with a magnetic compass or from the bearings of landmarks; and it will be checked by sighting on Polaris if opportunity permits. A wind mast is needed that will facilitate the orientation procedure. Also, there are more locations than might be supposed where the reading of a magnetic compass is questionable. Therefore, alternative methods of determining true north positively and quickly are also desired.

Although a hand anemometer is likely to be taken to an emergency scene, it is used with caution. Wind observations made near eye-level above the ground or a rooftop are significantly less reliable than those made from the top of a short mast, even if the mast is no more than 3 or 5 meters high. The 8" plastic flowmeter-type of hand anemometer found in the Forest Service belt weather kit[33] is of questionable value. Its readings can fluctuate rapidly because of local eddies, and the tiny indicator ball may stick because of moisture or other foreign matter in the tapered flow tube.

With respect to pilot balloon observations, consider the weights of the items normally used:

	Pounds
theodolite with box	30
tripod	15
timer (9"x9"x3")	3
plotting board	15
graphing board	15
helium tank	125

The theodolite, now selling for about $1400, is overly designed as far as emergency use is concerned. The 21-power telescope, which enables an observer to follow a balloon to 20,000 feet, or more, under ideal seeing conditions, is much more powerful than is needed to follow a balloon to the altitude of 3000 to 5000 feet. A smaller, lighter weight version would be adequate for emergency situations. However, it should have a built-in compass and an easy method of leveling. Also, the theodolite might be designed to fit a lighter-weight tripod providing the tripod could be firmly anchored to the ground.

Smaller helium tanks are available, and the 40" size, which weighs 35 pounds, has been used. However, there is need for a suitcase-type carrying container for two, or perhaps one, of these tanks, plus valves and other accessories. Helium, in the form of a compressed gas is accepted for shipment aboard a commercial airline providing the tanks are properly tagged. It is possible to purchase or perhaps borrow helium locally in many areas. However, a small supply should be carried to the scene of the emergency for immediate use.

Ceiling balloons rather than 30-gram pilot balloons are used in emergencies. They rise slower, giving more detailed wind information, and use less helium. Ascension rate tables for 15- and 20-second intervals have been prepared by the EPA Meteorology Laboratory for use with these smaller balloons.

The plotting board is actually unnecessary for a short balloon run. The path of the balloon is easily plotted by using a simple chart and parallel rulers, as when navigating a ship. A piece of graph paper can also take the place of a graphing board.

The timer problem has been solved by electronic technicians of the EPA Meteorology Laboratory. They have constructed a pocket-size, battery-operated, solid state timer, which can be set to sound at 20- or 30-second intervals. It functions best when used in a shirt pocket where body heat can provide a somewhat constant temperature. The portability and convenience of this timer are very suitable for the emergency situation.

Instrumentation for making temperature soundings is not included in the emergency kit. However, a small battery-operated temperature meter, with connecting cable and sensor that could be easily attached to a light airplane or helicopter. available if needed.

At the risk of being branded a heretic, two statements will be made that will be perhaps surprising. One, the nearly indispensable item for an emergency situation is not an instrument in the usual sense; it is a topographic map, scale 1 to 24,000. Two, a meteorologist responding to an emergency should be prepared to work with no usual instruments at all. Baggage can be lost or delayed, and instruments broken. Also, the meteorologist may be forced to make a response before instruments can be set up. It is helpful to acquire a general knowledge of the synoptic situation before arriving on the scene, and to be able to use smoke behavior or other clues to estimate wind speed and direction and stability. Further, the meteorologist should be prepared to improvise devices to aid with such estimates. For example, slow-rising balloons are an excellent indicator of air flow.

Much more needs to be said about meteorological response to emergency situations than can be presented in this paper. A handbook of instructions based on previous experience is in preparation and copies will be distributed to EPA Regional Meteorologists.

NEW AND FUTURISTIC AIR POLLUTION METEOROLOGY INSTRUMENTATION

The National Weather Service is currently using the Loran-C balloon tracking system in its dedicated sounding program for the New York City area. It makes use of radio signals normally used as a navigation aid and it has automatic features that permit one-man operation. Also, the National Weather Service is experimenting with acoustic sounding equipment in Los Angeles to supplement its temperature soundings. Another acoustic system will soon go into operation in Philadelphia. Wyckoff, Beran and Hall[34] state "that the acoustic sounder is in its infancy." They say that an acoustic system with Doppler wind measuring capability could measure a profile of the vertical velocity. Also, they have found that it is feasible to operate the acoustic sounder in a city with little detrimental effect from interfering noise.

The Meteorology Laboratory EPA has supported the development of the Sperry Thermasonde (TM) Radiometric System[35]. It can make remote temperature soundings from the surface upward to about 1.5 kilometers by means of an elevation scanning technique, each sounding requiring about 20 minutes. Emission of microwave energy by oxygen molecules is directly proportional to the air temperature. The Thermasonde radiometer directly determines a brightness temperature profile that can be converted to a good approximation of the true temperature profile by means of an analog library. At the present time the analogs are developed from actual soundings made at, or near, the site where the thermasonde is to be used. Also, in its present state of development it can supplement radiosonde soundings, which are made once or twice a day. The device offers the most promise for routine use in a permanent installation where the analogs could be developed.

The nephelometer[36] and various other instruments for measuring visibility of atmospheric turbidity[37] have not been mentioned. However, such instruments are becoming more important as operational tools because of increasing national concern about the degradation of the atmosphere.

Lidar[38], which can show many details of the structure of the lower atmosphere, is a powerful tool for the air pollution meteorologist. However, it and balloon tracking techniques using radar[39,40] find application for the most part in research studies. Lidar is not readily available and is expensive and difficult to operate. Balloon tracking can be done only in certain locations where suitable radar is available.

A sounding system in a suitcase-type box is commercially available that offers the choice of minature rockets, or balloons, for lifting a tiny radiosonde-type transmitter[41]. The transportable balloon sounding system seems to offer more promise than the rocket system for the emergency situation. The rocket is desirable for some operations because it can be recovered and used again, and it does not require helium. Nevertheless, firing even a miniature rocket and hunting for it under emergency conditions might present some problems that would be eliminated by using a balloon. No more than one or two special soundings might be required, so the $50-$60 cost of the transmitters expended would be negligible; and helium is necessary anyway because of the requirement for wind soundings.

The need for upper wind information can occur when precipitation, low clouds, or fog prevent pilot balloon observations. Looking ahead, it is hoped that automatic balloon tracking equipment for low-level soundings under field conditions can become more generally available.

CONCLUSION

In general, available meteorological instrumentation is adequate for air pollution operations. However, the design of meteorological instruments seems largely dominated by electronic engineers, or others who seldom, if ever, are required to use the instruments under difficult field conditions. Too often fundamentals of instrument use are overlooked or forgotten. Much more attention needs to be given to design factors such as portability, rapid installation and start-up, positive orientation and calibration, and the other factors encountered in difficult operational situations.

There are various ways that a line of basic meteorological instruments in luggage-type containers could be developed. It would be most desirable if they could be developed by instrument companies in response to the demand of a sufficiently large market. The potential market should be explored, and it may be larger than is generally realized. The nation needs to be better equipped for air pollution accidents. EPA laboratories and Regional Offices, plus State and local air pollution control agencies with meteorological capability, have a need for more portable emergency instrumentation. Also, fire weather meteorologists, who are scattered throughout the United States, seem to have a similar need. At present a forecaster may travel to a forest fire in an instrumented van. There would be advantages to flying a meteorologist to the fire with a limited amount of instruments and other equipment and having the van arrive later. Both the fire weather and the air pollution meteorologist have communications and other field requirements that also require portability. It would be gratifying if the instrument community would respond to a basic need for emergency meteorological instruments.

REFERENCES

(1) Hewson, E.W., Meteorological Measurements, Chapter 24, Vol II, AIR POLLUTION, A.C. Stern, Academic Press, New York, London (1968), 329-391.

(2) Moses, H., Meteorological Instruments for Use in the Atomic Energy Industry, Chapter 6, METEOROLOGY AND ATOMIC ENERGY, D.H. Slade, Ed., TID-24190, U.S. Atomic Energy Commission, Oak Ridge, Tenn. (1968), 257-300.

(3) Gill, G.C., H. Moses, and M.E. Smith, Current Thinking on Meteorological Instrumentation for Use in Air Pollution Problems, J. APCA 11(2), (1961), 77-82,

(4) McCormick, R.A., Wind and Turbulence Instrumentation for Air Pollution Surveys, BULL. AM. MET. SOC., Vol 41, No. 3, April, 1960, 175-179.

(5) METEOROLOGICAL INSTRUMENTATION IN AIR POLLUTION, (EPA Training Manual), Manpower Development Staff Institute for Air Pollution Training, Research Triangle Park, N. C. 27711, (1973).

(6) Singer, I.A. and M.E. Smith, Relationship of Gustiness to Other Meteorological Parameters, J. METEOROLOGY, AM. MET. SOC., 10(2), (1953), 121-126.

(7) Jones, J.I.P. and F. Pasquill, An Experimental System for Directly Recording the Statistics of the Intensity of Atmospheric Turbulence, QUART. J. ROY. METEOROL. SOC., 85(365), (1959), 225-236.

(8) Safety Guide 23, ONSITE METEOROLOGICAL PROGRAMS U.S. ATOMIC ENERGY COMMISSION, Feburary, 1972, Table 2.

(9) Cleeves, G.A., T. J. Lemmons and C.A. Clemons, A Low-Level Air Sampling and Meteorological Sounding System, J. APCA, Vol 16, No. 4, April, 1966.

(10) Pack, D.H., et al, LOWER TROPHOSPHERIC SOUNDINGS (1966), Technical Note 77, World Meteorological Organization, WMO No. 192, TP 98.

(11) Turner, D.B., Relationships between 24-hour Mean Air Quality Measurements and Meteorological Factors in Nashville, Tenn., J. APCA, Vol 11, (1961), 483-489.

(12) Rossano, Jr., A.T., The Community Air Pollution Survey, AIR POLLUTION, A Comprehensive Treatise, Vol II, A.C. Stern, Ed., New York, Academic Press, (1968), 597-637.

(13) Hewson, E.W. and G.C. Gill, METEOROLOGICAL INVESTIGATIONS IN COLUMBIA RIVER VALLEY NEAR TRAIL, B.C. (1944), In report submitted to the Trail Smelter Arbitral Tribunal, U.S. Bureau of Mines Bulletin 453, 23-228.

(14) Frank, Sidney R. and Raymond E. Kerr (North American Weather Consultants), Role of the Meteorological Engineer in Community Air Pollution Surveys, J. APCA Vol 8, No. 4, February, 1959, 314-326.

(15) Hewson, E.W., E. W. Bierly and G.C. Gill, Measurement Programs Required for Evaluation of Man-Made and Natural Contaminants in Urban Areas, SYMPOSIUM, AIR OVER CITIES, SEC Technical Report A62-5, U.S. Dept. HEW, PHS, Cincinnati, Ohio, November, 1961, 251.

(16) Zeidberg, L.D., J.J. Schueneman, P.A. Humphrey, and R. A. Prindle, J. APCA, 11, June 1961, 289-297.

(17) MacCready, P.B., Design of Measurement Systems, SYMPOSIUM, ENVIRONMENTAL MEASUREMENTS, U.S. Dept. HEW, PHS, July 1964, 21-27.

(18) KANSAS CITY, KANSAS - KANSAS CITY, MISSOURI, AIR POLLUTION ABATEMENT ACTIVITY, Phase II. Pre-Conference Investigations, U.S. Dept. HEW, PHS, National Center for Air Pollution Control, Cincinnati, Ohio, March 1968, 180.

(19) KANAHWA VALLEY AIR POLLUTION STUDY, Prepared by Technical Staffs of National Air Pollution Control Administration and West Virginia Air Pollution Control Commission, U.S. Dept HEW, PHS, National Air Pollution Control Administration, Raleigh, N.C., March, 1970, 5 Chapters, 5 Appendices.

(20) Helms, G.T., et al, CHATTANOOGA, TENNESSEE-ROSSVILLE, GEORGIA INTERSTATE AIR QUALITY STUDY 1967-68, U.S. Dept. HEW, PHS National Air Pollution Control Administration, Durham, N.C., October, 1970, 120.

(21) Humphrey, P.A., Meteorology and Source-Receptor Relationships, Chapter 11, HELENA VALLEY MONTANA, AREA ENVIRONMENTAL POLLUTION STUDY, EPA, Office of Air Programs, Research Triangle Park, N.C., January 1972, 161-179.

(22) Foster, K.E., D.L. BROOMAN, AND P.A. Humphrey, Mobile Air Sampling Stations Integrate Air Monitoring Data, AIR ENGINEERING, 9(5), 12, May 1967, 14-15.

(23) Kirschner, B.H., Environmental Meteorological Support Units: A New Weather Bureau Program Supporting Urban Air Quality Control, THE 2ND INTERNATIONAL CLEAN AIR CONGRESS, PROCEEDINGS, December 6-11, 1970, Washington, D.C.

(24) Croke, E.J. and S. G. Booras, Design of Air Pollution Incident Control Plan, J. APCA, 20(3) (1970), 129-138.

(25) Environmental Protection Agency, (40 CFR, Part 51), Use of Supplementary Control Systems and Implementation of Secondary Standards, FEDERAL REGISTER, Vol 38, No. 178, Friday - September 14, 1973.

(26) Cottrell, H.B., et al, PUBLIC HEALTH SERVICE REPORT ON OPERATION CHLORINE, U.S. Dept HEW, PHS, (1963), 82

(27) AFTER-ACTION REPORT SALVAGE OF CHLORINE FROM BARGE WHICH SUNK IN THE MISSISSIPPI RIVER NEAR NATCHEZ, MISSISSIPPI, U.S. Army Engineer District, Vicksburg Corps of Engineers. Vicksburg, Mississippi, October, 1963, 31, exhibits, appendices.

(28) State of Louisiana, OPERATION SAFEGUARD, Operating Plan for Raising Sunken Chlorine Barge MTC 602.

(29) CRITIQUE OF THE CHLORINE BARGE INCIDENT, LOUISVILLE, KENTUCKY, MARCH - APRIL 1972. Office of Emergency Preparedness, Executive Office of the President, 73.

(30) Coblenz, Jack G. and John J. Roberts. Prevention and Control of Accidental Emissions of Hazardous Materials into the Atmosphere, Presented at the 66th Annual Meeting of the APCA, Chicago, Ill., June 1973.

(31) Letter from G.A. Lincoln, Director, Office of Emergency Preparedness, Washington, D.C., May 24, 1972, and reply from Sheldon Meyers, Director, Office of Federal Activities, EPA, Washington, D. C.

(32) Science Associates, Inc., Wind Survey System No. 441, Catalog No. 8, 55.

(33) GSA Supply Catalog, Federal Supply Service, October 1973, Meter, Air Velocity, Forest Service Spec. 5100-454, 170.

(34) Wyckoff, R.J., D.W. Beran and F.F. Hall, Jr., A Comparison of the Low-Level Radiosonde and the Acoustic Echo Sounder for Monitoring Atmospheric Stability, J. of APPLIED METEOROL. Vol 12, October, 1973, 1196-1203.

(35) Thermosonde (TM) Radiometer MK 11, Mod 0. Field Evaluation Test Report, Prepared for EPA, Sperry Microwave Systems, Microwave Electronics, Clearwater, Florida, June, 1973.

(36) Charlson, R.J., A New Instrument for Evaluating the Visual Quality of Air, J. APCA 17(7) (1967), 467-469.

(37) McCormick R.A. and D.M. Baulch, The Variation with Height of the Dust Loading over a City as Determined from the Atmospheric Turbidity, J. APCA, 12(10), (1962), 492-296.

(38) Johnson, W.B., Jr. and E.E. Uthe, LIDAR STUDY OF STACK PLUMES, Prepared for Dept HEW, NAPCA. Div. Met. Contract PH22-68-33. Stanford Research Institute, Menlo Park, Ca. 94025, (1969), 116.

(39) Pack, D.H. and J.K. Angell, A Preliminary Study of Air Trajectories in the Los Angeles Basin as Derived from Tetroon Flights, MONTHLY WEATHER REVIEW, 91(10-11), (1963), 583-604.

(40) Angell, J.K., D.H. Pack, C.R. Dickson and W.H. Hoecker, Urban Influence on Nighttime Air Flow Estimated from Tetroon Flights, J. APPL. METEOROL. 10, (1971), 194-204.

(41) Air Pollution Dispersal Prediction Using the Colspan Series 100 LARS Low Altitude Rocket Sonde and the Colspan Series 800 PBS Pilot Balloon Sonde, Colspan Environmental Systems, Inc., P. O. Box 3467, Boulder, Colorado 80203, (1973).

ADDENDUM TO "AIR POLLUTION METEOROLOGICAL OBSERVATIONS FOR SHORT-DURATION INVESTIGATIONS, STAGNATION EPISODES AND ACCIDENT EMERGENCIES"

INFORMATION PERTAINING TO SOLID STATE PILOT BALLOON TIMER

The solid state pilot balloon timer consists of a small aluminum box 4" x 2 1/2" x 1 5/8" containing a R-C (resistive-capacitive) controlled flip-flop timing circuit with an on-off switch and miniature speaker. The timer produces a brief audible tone at regularly spaced intervals. The duration of the interval between the sounding of the tone can be varied from 10 seconds to 1 minute by adjusting the R-component. Other components are an audio oscillator, a power amplifier, a volume control, an earphone jack, and a 9-volt radio battery which has a normal operating life of about one year. A circuit diagram for the timer is shown in Figure 1.

The advantages of this timer is its great size reduction over a commercially available AC-powered timer and its elimination of any need for AC-power, which is often unavailable in emergency field situations. The disadvantage is that cold weather affects the battery which in turn affects the timing interval. This effect can be reduced by placing the timer in a shirt pocket and using an earphone to monitor the timing tone. Also, the timing interval must be checked and adjusted periodically depending on the condition of the battery.

Other pilot balloon timers normally provide a warning sound to alert the observer two seconds prior to the sound that marks the instant the position of the balloon is observed and recorded. The solid-state timer could be improved by redesigning its circuit so as to prolong the tone from about a second to 2 or 3 seconds. The on-set of the tone could then alert the observer, and the observation could be made at the instant the tone stops.

SOLID STATE TIMER

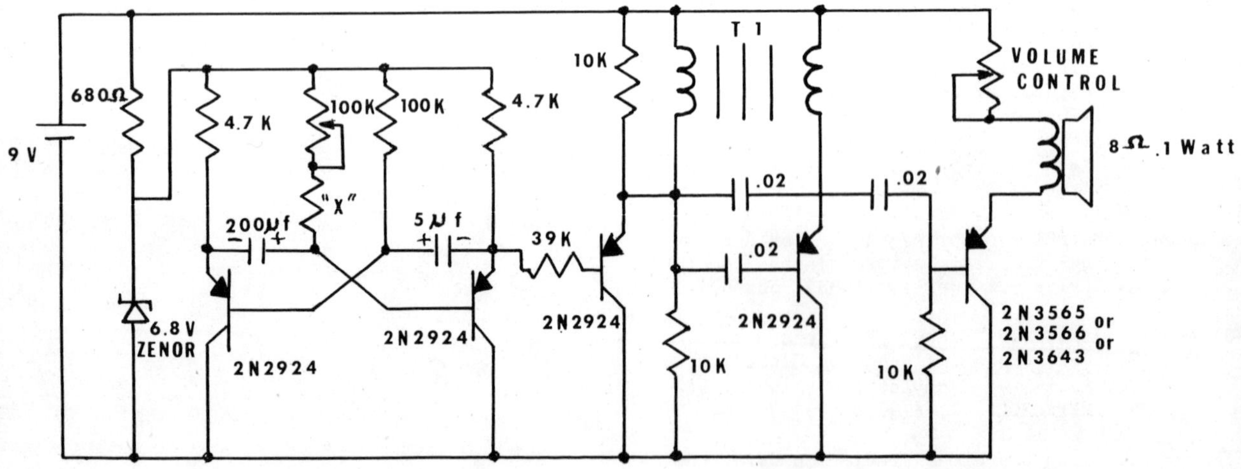

Figure 1.

"X" RESISTORS ARE SELECTED FOR SPECFIC TIME INTERVALS

© 1973, ISA JSP 6705

POLLUTANT DISTRIBUTIONS IN WEST COAST INVERSIONS

Albert Miller
Professor
Meteorology Department
San Jose State University
San Jose, California

ABSTRACT

Aircraft and surface measurements of the ozone concentration in the San Francisco Bay Area have been related to the atmospheric circulation and the dynamics of the elevated temperature inversion that prevails along the west coast. Case studies are available when the onshore flow of maritime air was very weak and also when there was a well-developed sea breeze. In the stagnant cases, the oxidant concentration increased at all levels from the surface to 2500 m or more immediately after the breakdown of the early morning temperature inversion and reached a peak in the late afternoon; in the evening, the concentration decreased much more rapidly below the radiation inversion than it did above the inversion. In the seabreeze cases, narrow, deep zones of high concentration move along the seabreeze front, inland of where the inversion disappears. Layers of high oxidant concentration have been found several hundred meters above the inversion base, under flow conditions that appear to preclude an explanation of horizontal transport, giving rise to speculation that there is a vertical transport into the inversion by wave action. Some observations of oxidant concentration delineate well-defined wave patterns. A newly instrumented tower in San Francisco may provide data on the flux of pollutants into the inversion by waves; some preliminary results will be presented. In order to measure the turbulent fluxes of pollutants by waves, faster response sensors will be required.

INTRODUCTION

At least three-fourths of California's sources of air pollution and population are located within 25 miles of the Pacific coastline. This narrow strip is also a zone of strong gradients of meteorological parameters as well as of pollutant concentrations. It is not at all unusual for the surface temperature to change by 35 F or more across this belt and for the oxidant concentration to change by 20 pphm. Typically, it is the interior edge of this belt that experiences the worst air pollution conditions.

Gradients in the vertical are even more dramatic(4,5). The very stable inversion layer that begins at 200-300 m above sea level and extends up to 900 m or more persists during more than half of the year, severely restricting vertical dispersion (6,7,8). The effect of this on a relatively-inert constituent, such as CO, is illustrated in Figure 1*: the concentration within the "marine" layer can be five times or more greater than that in the dry air aloft. However, the inversion layer is not simply a lid on the vertical dispersion. For one thing, its height and intensity vary considerably in time and space in the coastal belt which leads to a complex pattern of winds. Reactive contaminants, such as ozone, respond quite differently to the distribution of turbulent intensity in the vertical than do the more inert constituents: thus, the ozone concentration typically increases with height above the ground with a sharp peak often appearing within the inversion layer, as illustrated by Figure 2.

Meteorologists and chemists have been attempting to develop mathematical models of the atmospheric flow and photochemistry that affect the pollution levels within this coastal belt. The problem is extremely complex. Meteorologists are not even sure of the normal flow pattern on this scale, let alone the day-to-day and the strong diurnal variations. The photochemistry is also difficult, considering the long list of reactions that may be involved. We are going to have to learn

*The data contained in Figures 1 and 2 were provided by Hermillo Gloria, NASA/Ames.

"Superior numbers refer to similarly-numbered references at the end of this paper."

more about the time and space distributions of meteorological and pollutant parameters within this coastal zone and the relationships between them. We have done some observations, using airborne instrumentation[7], and I shall describe some of the patterns of ozone we have found (ozone is probably the most important contaminant of west coast smog) and relate them to meteorological conditions (particularly the inversion layer) within the coastal belt.

PATTERNS OF OZONE DISTRIBUTION IN THE BAY AREA

The elevated inversion of the west coast is a quasi-horizontal boundary over the Pacific Ocean; it is formed between weakly-descending, dry, potentially-warm air aloft and moist, cool air directly above the cold ocean water. At distances of 50 to 100 km offshore, the bottom of this transition layer is usually at 200-250 m and the top is near 900 m. The temperature increases from an average of about 13 C at the base to 22 C at the top, while the water vapor concentration goes from about 8°/oo to about 2°/oo. The average inland penetration of the marine air is about 40 km, but there is considerable variation. The inversion base slopes sharply upward toward the interior during the day, downward at night.

On occasion, the large-scale atmospheric circulation is such that the marine air does not penetrate the coastline at all. Under such circumstances, the warm dry air that is normally riding above the marine air reaches close to the surface in the coastal belt. Although there is a surface-based "radiation" inversion during the night, it is shallow and breaks down during the day. The afternoon maximum temperature under such conditions reaches into the mid-90's and sometimes gets over 100F (compared to the normal low 80's in this zone during the summer). Figures 3, 4, and 5 illustrate what happens under such conditions. As can be seen, the surface temperature reached abnormal values everywhere, except right over the Bay and Ocean waters and peak oxidant values were high throughout the heavily populated valley that encompasses the bays. The winds were extremely light and variable throughout the valley, so that horizontal transport was weak. A seventeen-hour series of vertical soundings (aircraft) of ozone and temperature were made at the southern extremity of the valley (San Jose); the vertical time section of Figure 4 is an analysis of these data. It can be seen that the ground-based inversion begins to lift shortly after sunrise and dissipates by midmorning; it reforms near sunset. The ozone concentration increased quickly beginning at around 10 a.m., when the inversion broke up at all levels up to 2700 m. The highest values throughout the day were invariably found at least 200 m above the ground.

The average ozone concentration and temperature between the surface and 2500 m, along with the total (hemispheric) radiation at San Jose are shown in Figure 5. Note that the peak total ozone in this 2500-m layer occurs less than an hour before sunset, when the radiation intensity is perhaps less than 10% of the noon value.

These observations, taken during a period when horizontal transport was extremely weak, appear to show that the ozone level depends on the turbulent intensity and the depth of vertical mixing (eddy sizes). Since the reaction rate producing ozone should diminish as the concentration of oxides of nitrogen and reactive hydrocarbons decreases, it appears that the high ozone level is due to a diminished destruction rate. The earth's surface and particulates in the air near the surface are sinks for ozone [2,3] and so ozone levels are most likely to increase when the extent of vertical dispersion is great and the intensity of turbulence low. These conditions are likely to occur, not when there is a strong inversion aloft, but when the atmosphere is near neutral and wind speeds are low. The isolated maxima that we so frequently observe aloft (Figure 2) are due to the sharp decrease in turbulent mixing that occurs at the base of the inversion; the inversion shields the ozone from surface destruction. These maxima within the inversion layer may occasionally descend to the surface when an unstable gravity-shear wave develops, which probably accounts for the anomolous nighttime peaks that have been observed at monitoring stations.

In the normal situation during the photochemical smog season, there is an onshore flow of marine air, which penetrates 50 km or more inland. The transition between marine air and the warm air over the interior occurs in a very small distance, as anyone who has traveled across it can confirm, and which is illustrated by maximum temperature lines of Figure 6. It is along the forward edge of this front where we observe the peak ozone concentrations. We have observed belts of high ozone from aircraft flown on a northwest-southeast course over the eastern side of the Bay (roughly, along the 90 F isotherm in Figure 6).

During the summer and early fall, a typical day's pattern of inversion location and ozone concentration are illustrated in Figures 7a, b, and c. In the early morning (Figure 7a), when the onshore flow is weakest, the inversion base slopes downward from 300-400 m at the coast to ground level at a distance of 100 km inland. Note the strata of ozone above the inversion base, although the concentration below the inversion is everywhere less than 2 pphm. Note also the undulation in the inversion over the San Francisco Peninsula (SFB-OAK); this wave is frequently present. On occasion it becomes a hydraulic jump having a width of about 25 km and with the ozone concentration within it 10 pphm or more higher than on either side of it.

During the morning the inland portion of the inversion breaks down but by midday the onshore flow intensifies. As it brings in fresh marine air, the inversion base progresses inland. By midday (Figure 7b) the forward edge of the inversion is about 20 km east of the Bay (40 km from the coast). A peak in the ozone concentration has developed ahead of the inversion edge at an altitude of about 600 m. By early evening (Figure 7c) the inversion has advanced some 35 km further inland and so has the peak oxidant concentration. On the following morning (figure not shown), the pattern looked very much like that of the previous morning (Figure 7a).

FLUXES INTO THE INVERSION LAYER

Strata of moist, polluted air have been found imbedded within the inversion layer (e.g., Figure 7) on almost every one of our observation periods. In some instances, they can be explained in terms of the circulation: air that has ascended in the region inland of the inversion edge has returned westward within the inversion layer. But there are many cases when the observed winds will not support a purely convective explanation. It now appears that there is mass and momentum exchange across the boundary.

Vertical fluxes within the inversion cannot be measured with aircraft so we are now attempting to do it on a television tower that was recently erected in the heart of San Francisco. The top of this tower extends to almost 500 m above sea level. Although the instrumentation has not yet begun to function smoothly, we have already observed some extraordinary oscillations within the inversion - e.g., temperature changes of 10C in periods of about 25 seconds.

There is no doubt that gravity-shear waves occur along the interface and that sometimes these waves become unstable[1,6]. They can be seen by observing the tops of the coastal stratus clouds at the base of the inversion. We have measured undulations with amplitudes as great as 200 m and with wave lengths as small as 100 m and as large as 30 km. An unusual example is illustrated in Figures 8a and 8b. One is of the ozone concentration and the other of moisture. At least two well-defined crests and troughs can be found. Within the crests the ozone concentration and the moisture reached 13 pphm and 10 g kg^{-1}, respectively, while in the troughs they dropped to less than 3 pphm and 4 g kg^{-1}, respectively. The maximum amplitude of these waves was almost 150 m and the wave length about 22 km.

SUMMARY AND CONCLUSIONS

Within the coastal transition zone in California, the temperature inversion is not merely a passive, quasi-permanent lid on the vertical fluxes of pollutants. There are pronounced diurnal and day-to-day variations in the intensity and inland extension of this boundary. In addition, it is constantly undulating and, on occasion, the waves break. There also appears to be some flux of pollutants through the inversion base to the warm dry air aloft.

Reactive contaminants such as ozone are strongly dependent on the vertical distribution of dispersion, by affecting not only the reaction rates but, more importantly, the destruction rates. Peak ozone concentrations are observed where the turbulent intensity is weak but the depth of vertical mixing is great. This occurs in advance of the inland edge of the inversion.

There are pronounced vertical and horizontal gradients of pollutants that are not merely a function of source strength distribution. Predictions of air pollution levels from mesoscale models will not achieve sufficient accuracy until we more fully understand the flow characteristics and their effects on both dispersion and atmospheric chemistry. Much of this understanding will come through carefully planned experiments within the atmosphere that involve simultaneous measurements of pollutants and meteorological parameters.

REFERENCES

(1) Gossard, E.E., J.H. Richter, and D. Atlas, 1970. Internal waves in the atmosphere from high-resolution radar measurements. J. Geophys. Res., 75, 3523-3536.

(2) Haagen-Smit, A.J. and L.G. Wayne, 1968. Atmospheric reactions and scavenging processes. Air Pollution, 1, ed. A.C. Stern, Academic Press, N.Y.

(3) Junge, C.E. and G. Czeplak, 1968. Some aspects of the seasonal variation of carbon dioxide and ozone. Tellus, 20, 422-434.

(4) Kruger, P. and A. Miller, 1966. Transport and radioactivity in rain and air across the trade wind inversion at Hawaii. J. Geophys.Res., 71, 4243-4355.

FIGURE 1.

VERTICAL PROFILES OF CARBON MONOXIDE IN BAY AREA

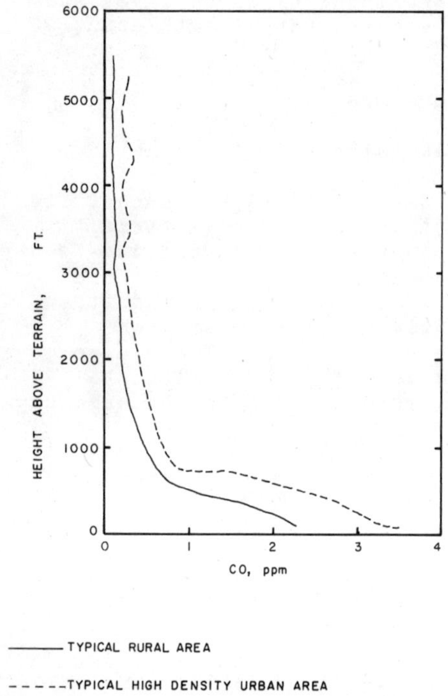

FIGURE 2.

VERTICAL SOUNDINGS OF OZONE AND TEMPERATURE (2000PST, JULY 26, 1973)

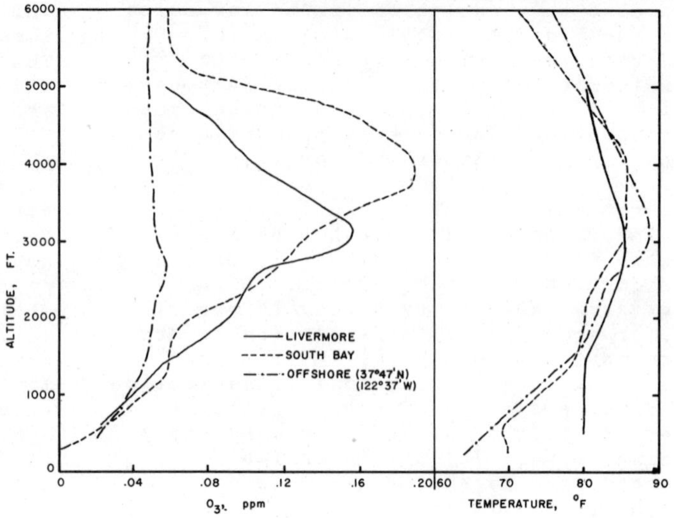

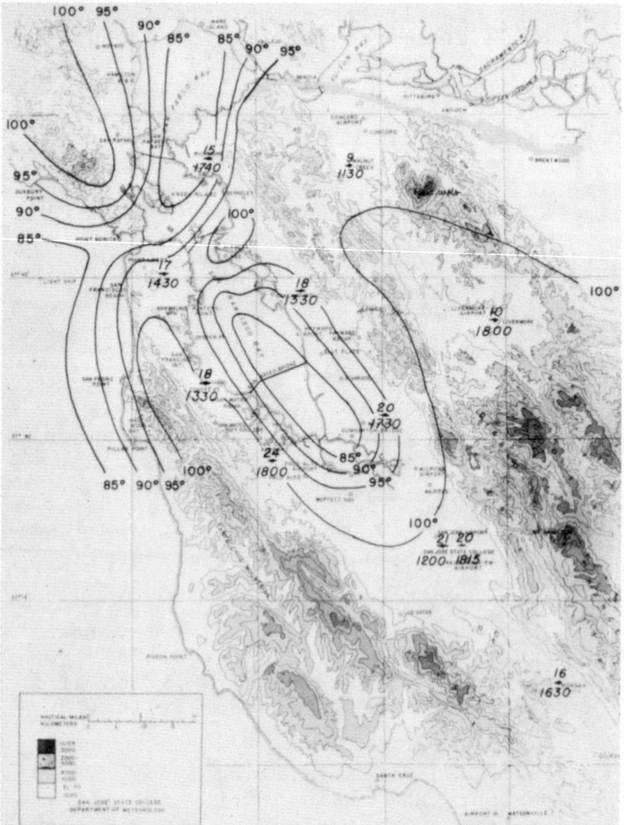

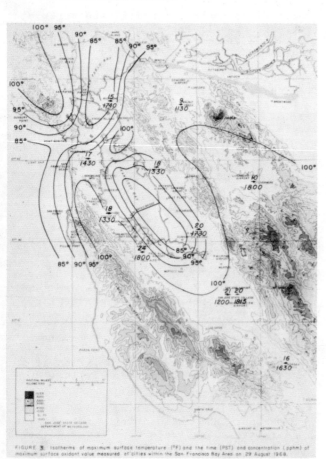

FIGURE 3. Isotherms of maximum surface temperature (°F) and the time (PST) and concentration (pphm) of maximum surface oxidant value measured of cities within the San Francisco Bay Area on 29 August 1968.

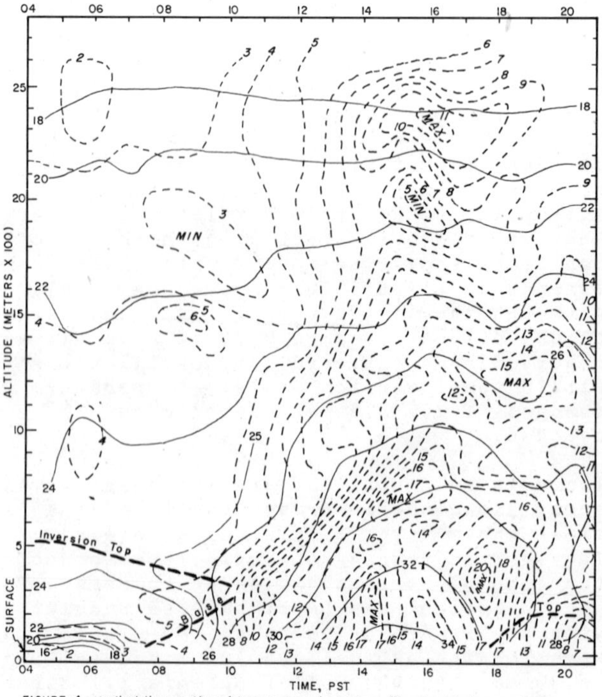

FIGURE 4. Vertical time section of temperature (solid lines, °C) and oxidant concentration (dashed lines, pphm) on 29 August 1968 at Reid-Hillview Airport.

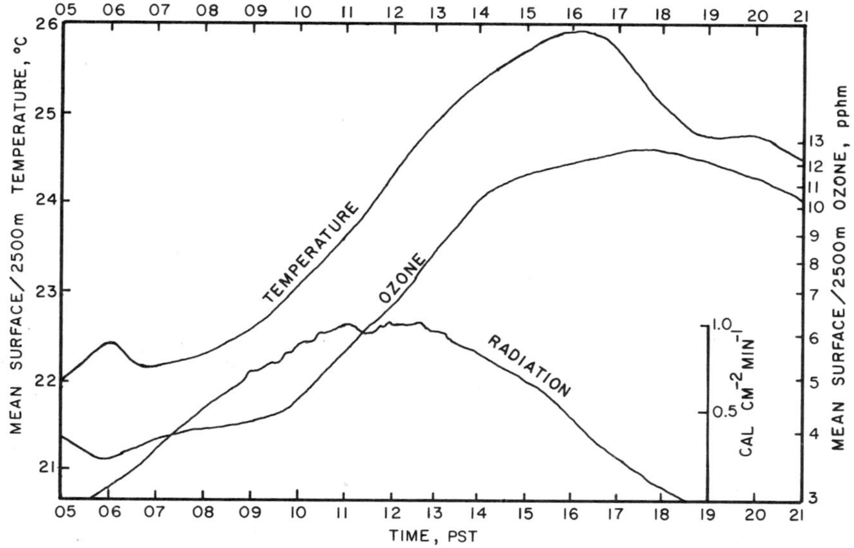

FIGURE 5. Mean ozone concentration and temperature from surface to 2500m, 29 August 1968, at San Jose; total (hemispheric) radiation at San Jose.

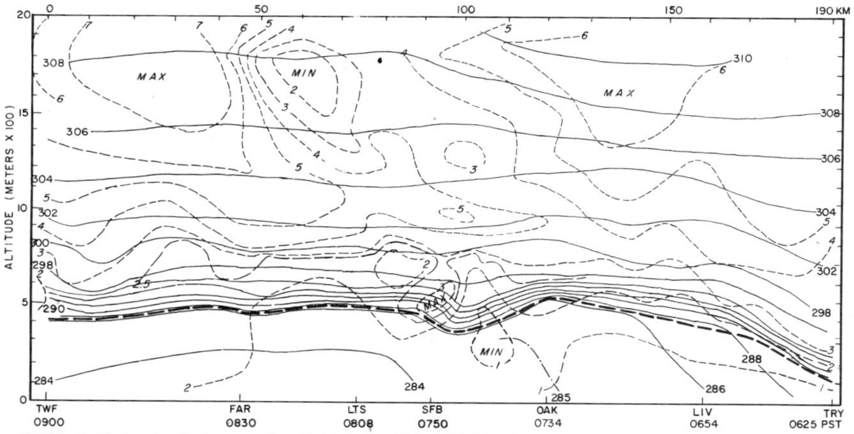

FIGURE 7a. East-west vertical cross section, 10 July 1968, 0625-0900 PST, (Solid lines: potential temperature, K; dashed lines: ozone concentration, pphm; heavy dashed line: inversion base. Sites and times of vertical soundings are indicated along lower border. Distance scale along upper border.)

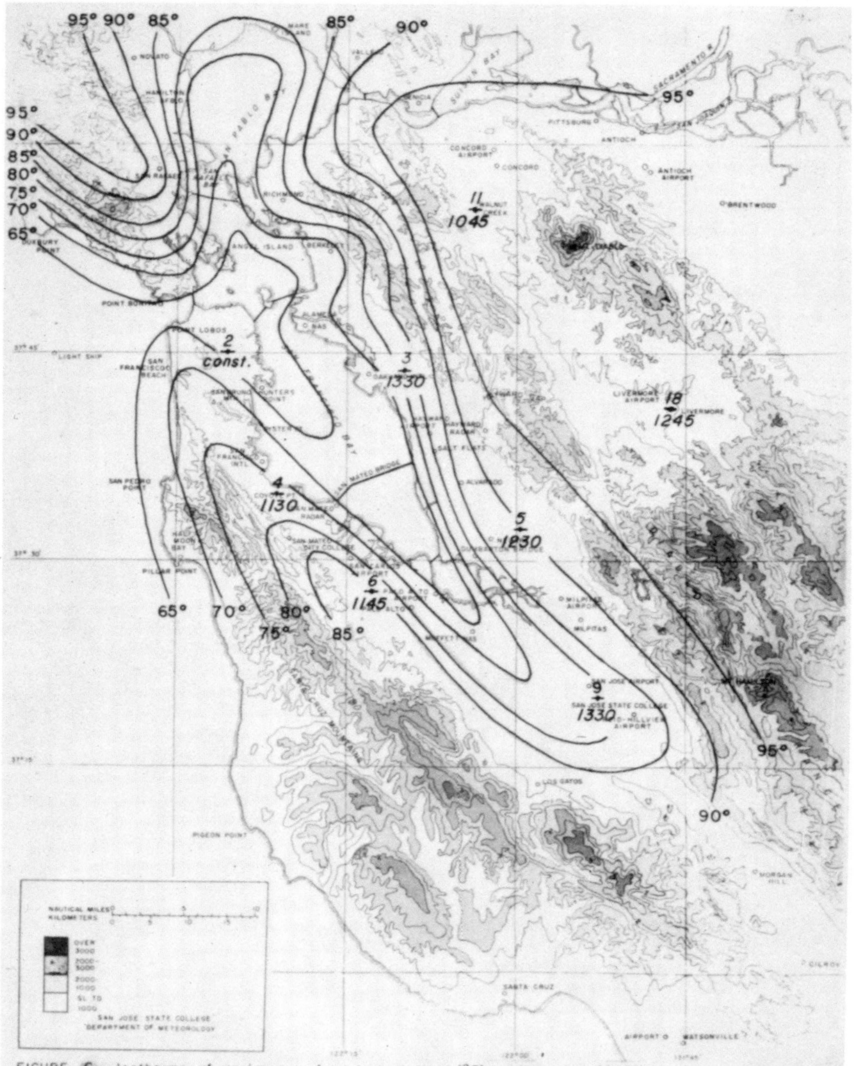

FIGURE 6. Isotherms of maximum surface temperature (°F) and the time (PST) and concentration (pphm) of maximum surface oxidant value measured at cities within the San Francisco Bay Area on 30 July 1968.

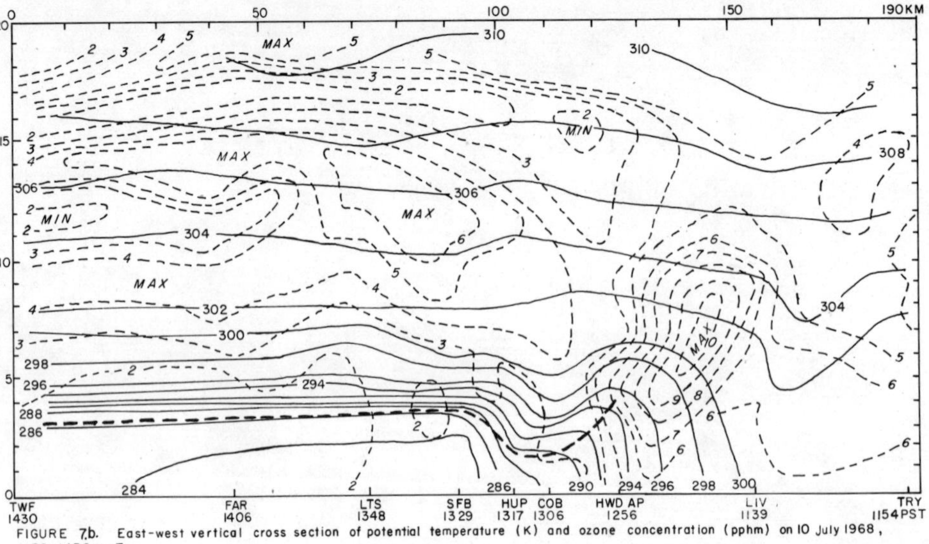

FIGURE 7b. East-west vertical cross section of potential temperature (K) and ozone concentration (pphm) on 10 July 1968, 1139-1430 PST.

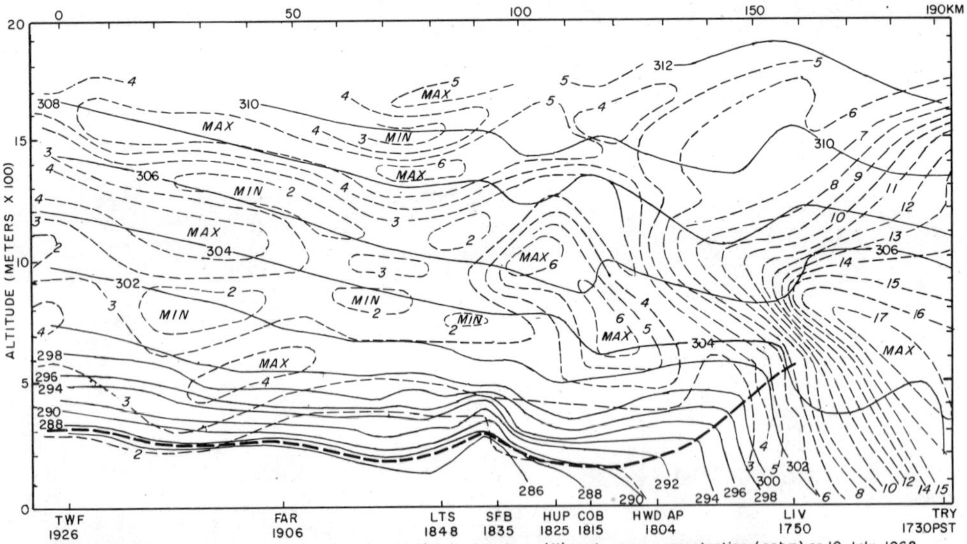

FIGURE 7c. East-west vertical cross section of potential temperature (K) and ozone concentration (pphm) on 10 July 1968, 1730-1926 PST.

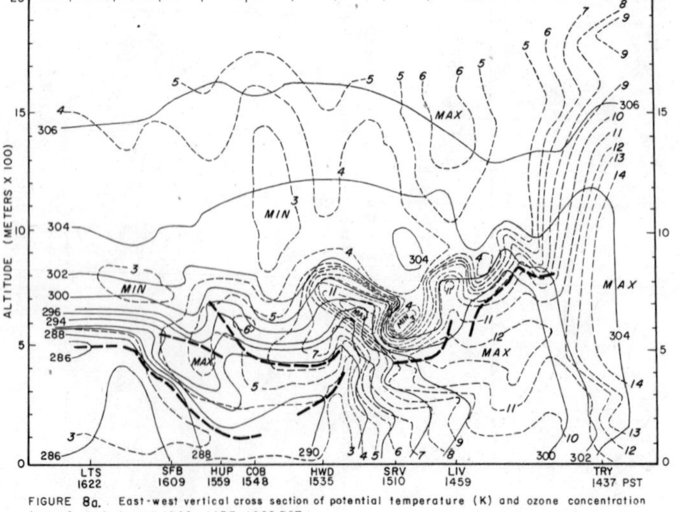

FIGURE 8a. East-west vertical cross section of potential temperature (K) and ozone concentration (pphm) on 8 August 1968, 1437-1622 PST.

(5) Lovill, J.E. and A. Miller, 1968. The vertical distribution of ozone over the San Francisco Bay Area. J. Geophys. Res., 73, 5073-5079.

(6) Miller, A., 1968. Wind profiles in west coast temperature inversions. San Jose State College, Meteorology Dept., Report No. 4.

(7) Miller, A., and D. Ahrens, 1970. Ozone within and below the west coast temperature inversion. Tellus, 22, 328-340.

(8) Neiburger, M., 1959. Meteorological aspects of oxidation type air pollution. The atmosphere and the sea in motion. The Rockfeller Institute Press, p. 158-169.

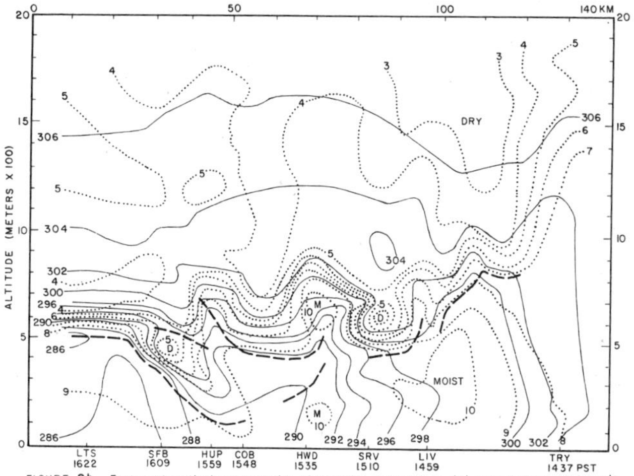

FIGURE 8b. East-west vertical cross section of potential temperature (K) and mixing ratio ($g\,Kg^{-1}$) on 8 August 1968, 1437-1622 PST.

© 1973, ISA JSP 6708

IN-SITU SAMPLING TECHNIQUES FOR TRACE ANALYSIS IN WATER

Harry B. Mark, Jr., Ronald J. Boczkowski and Kenneth E. Paulsen
Department of Chemistry
University of Cincinnati
Cincinnati, Ohio 45221

ABSTRACT

The study of the effect of changes in concentrations of metal ion pollutants on the natural environment is of major interest. However, most investigators agree that this problem cannot be resolved unless better analytical techniques are developed and their quantitative limits proved. Only when such quantitative numbers are provided can correlations with ecological and other parameters provide useful information and proof for clean up and control, etc. This research suggests that accurate quantitative data for concentration levels of metal ions in natural water systems can best be obtained by in situ techniques.

This research is concerned with a fresh approach to in situ measurement. It is believed that either controlled potential electrodeposition or chelating resins are possible approaches to simple techniques for in situ metal ion preconcentration which give a mechanically and chemically inert sample which can then be removed from the sample environment and analyzed by either neutron activation analysis, x-ray fluorescence or other means. The instrumentation, watertight packaging for in situ sampling, measurement parameters, and proposed study areas are discussed.

INTRODUCTION

A. EXAMINATION OF THE PROBLEM

One of the most important problems in oceanography and water resources science is the effect of the concentration and changes in concentration of the trace metal ions on the nature of the water systems.[1,2,5,6] Increasing concern has been expressed over the past few years about the abnormal increase, caused by man, of many metal ions, such as mercury, lead and iron, in the natural water systems. Such concern is, at this time, based partially on speculation, as we do not have in general the necessary trace analytical techniques that can give accurate values of the levels of increase in the concentration of such pollutants in natural systems. For example there has been tremendous publicity on the problem of mercury concentrated in edible fish such as Coho Salmon and Walleyed Pike in Lake St. Clair, Michigan.[7] But even in the extreme cases, there is considerable disagreement as to the true Hg level in fish analyzed. We had previously developed a Hg analysis method for blood analysis for the University of Michigan Hospital.[8] Thus, when the Lake St. Clair problem became known, we applied this technique to fish tissue. When it became apparent that our results were not in agreement with levels reported by many laboratories, we prepared, with the help of the Bureau of Commercial Fisheries, Ann Arbor, Michigan, a large Coho Salmon sample which was ground up and mixed many times to assure homogeneity (which was checked by numerous random samples using our method which showed less than a 10% standard deviation of all samples and by Dr. Philip LaFleur of the National Bureau of Standards).[9] This sample was sent to 15 different laboratories (industrial, government and academic). The results shown in Table I.[8] The spread of values obtained is certainly unacceptable from any point of view. Similarly, a more extensive comparative study of trace elements in sea water by Brewer[10] shows an even greater spread of analytical results using a carefully prepared sample that had no systematic built in sampling error. For example, the results obtained for Iron and Cobalt, two elements that analytical chemists feel are easily determined, are shown in Tables II and III. Thus, it is impossible to quantitatively define how such pollutants are affecting the environment.

Therefore, the initial problem encountered in studying the effect of changes of trace metal ion concentrations on the natural environment is one in analytical chemistry.[7,10] Once the nature of any natural water system is quantitatively defined, correlations between the changes in the ecology, etc., and pollutant levels in these systems are then possible to understand and explain. With this point in mind, we are now actively engaged in our study of the methods of quantitative trace analysis of metal ions. Our approach to improving analytical results is to develop reliable methods of in situ sampling which yield a sample representative of the exact environment from which it was taken and which is chemically and physically inert to change with time and handling. This approach is based on our belief that the sampling mode is the major or first problem of natural system trace analysis. Secondly, we wish to design this sampling matrix in such a way as to minimize (or eliminate if possible) any subsequent chemical manipulation[2] in the course of the actual analysis procedures. Clearly the two recent interlaboratory comparisons[7,10] cited above

Table I

Representative Results of Hg Analyses in Edible Fish Tissue

	Method	Results (ppm)	
Sample 1.			
	Atomic Absorption[a]	0.60	b
	Atomic Absorption[a]	1.0	b
	Atomic Absorption[a]	0.85	b
	X-ray Fluorescence	0.40	b
	Destructive NAA	0.87	b
	This Method	0.86 (10% standard deviation)	
Sample 2.			
	This Method	0.58 (10% standard deviation)	
	NBS (non-des. NAA)[c]	0.54	
	Average results of 15 Labs (various methods)[b]	0.51 (values ranged from 0.19 to 1.2)	

a. Atomic absorption results of three different laboratories.

b. Results supplied by Grieg Bureau of Comm. Fisheries, Ann Arbor, Michigan.

c. Results supplied by LaFleur NBS (1970).

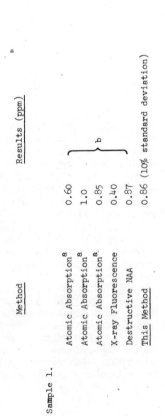

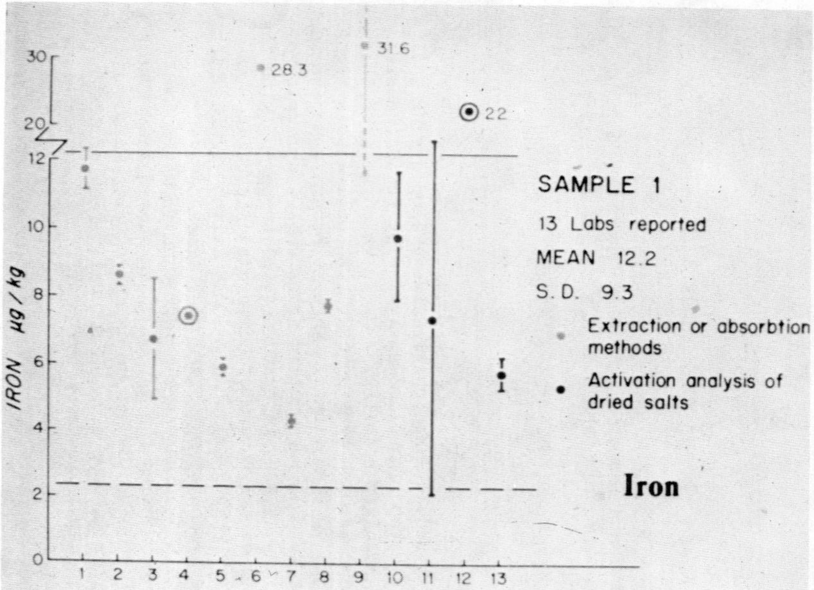

TABLE II

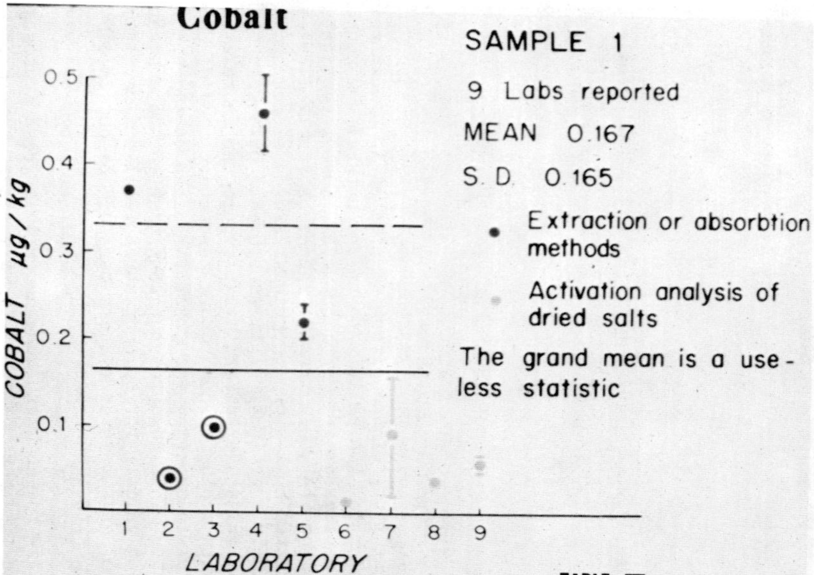

TABLE III

show that even if a common sample is provided, no two analysts get the same answer - usually not even close. These recent comparisons also indicated that things have not improved since the 1967 review by Hume.[1] This approach is opposite from the suggestion that the level of sensitivity and standardization of specific methods of analysis should be improved. We feel that presently there are a wide variety of analytical techniques that have suitable sensitivity and accuracy if one can provide a true and stable sample and minimize handling and manipulation.

Of the various trace techniques available in analytical chemistry, several of the well-defined techniques such as emission spectroscopy, anodic stripping voltammetry, neutron activation analysis, atomic adsorption spectrometry, electron microprobe spectroscopy, etc., and some mass spectrometry techniques are sufficiently sensitive to carry out trace analysis of metal ions in natural water systems. The question thus arises as to why these methods are not used routinely or reliably in such studies. These techniques seem to be ideally suited to the problem. Why has this type of problem not been solved in the past? A close examination of the analytical problem discloses why quantitative knowledge of natural systems is lacking.

The problem is not the quantitative analytical techniques but the quantitative sampling of the natural environment and sample storage.[1-4] In the past it has been the usual practice to take the sample (from any depth) in a metal or (better) a large plastic container and then carry out the analysis on the surface by classical methods. Obviously a normal metal bottle will contribute to the metal content of the sample, and even a plastic sample bottle will cause some problems.[1-4] It has actually been shown in trace analysis studies that plastic or glass sample containers both adsorb trace metal ions from the sample and/or contribute by surface dissolution other metal ions to the solutions.[1-4] Thus, the sample taken from the natural environment (unless extreme precautions are taken) cannot be analyzed because of the time dependent effects on concentration related simply to the nature of the container and conditions used to store the sample.

One other problem also arises; the nature of the sample with respect to metal ion valence and/or degree of complexation can be expected to vary with pressure,[11] and perhaps light intensity and energy, oxygen content, biological condition, etc. Thus, the sample would be best maintained at the same pressure, etc., but this would be virtually impossible if a container of water was taken and stored for subsequent analysis.

At first glance the only solution to this sampling problem, which arises from the simple need to maintain the sample in its initial environment or at least under the same conditions as it was when taken, is that the analysis in some way be made in situ directly and the system not removed or disturbed from its environment. In situ analysis has the further advantage of virtual instantaneous turn around of results. Thus, if anything unusual is found at a particular location, the investigator knows about it immediately and can investigate it thoroughly. If it was necessary to return to the laboratory before getting the analytical results of any test, the location of this unusual phenomena would be lost or it could have ceased or dissipated. There have been several attempts and suggestions for in situ methods.[12,13,72] However, complete in situ analysis presents many problems of trace element, as discussed below.

B. CONSIDERATIONS OF METHODS APPLICABLE TO IN SITU ANALYSIS OF METAL IONS

As mentioned above, a variety of analytical methods such as emission spectroscopy, neutron activation analysis, anodic stripping voltammetry, electron microprobe analysis, etc., are sufficiently sensitive for quantitative determination of metal ions at the levels to be expected in sea water and other natural systems. However, the inherent sampling problems discussed above require that metal ion analysis must be made in situ. In examining the available methods, spectrometric methods are obviously very difficult to adapt to in situ measurement, but electrochemical techniques, especially anodic stripping voltammetry is a very attractive prospect as it could be used directly in sea water (which has sufficiently high concentrations of electroinactive salts to ensure high conductivity). Simple anodic stripping voltammetry[14,15] has been shown to be applicable to the determination of concentrations of a specific metal ion in the 10^{-10}M range in highly specialized aqueous solution and/or systems.[16,17] By employing AC techniques on anodic dissolution,[18,19] direct pulse polarography,[20] or derivative techniques,[21] capacitance current contributions can be decreased and sensitivity levels somewhat improved.[20,21] On these sensitivity considerations as well as the natural simplicity of instrumentation[14,17] of anodic stripping voltammetry, this technique would be the most probable candidate for in situ analysis of metal ions in natural water systems. However, there are certain inherent disadvantages with respect to the dissolution peak resolution with mercury drop electrodes and multiple peaks[25] when using solid electrodes for identification and quantitative analysis which cannot be overcome on anodic stripping of the electrodeposited film of metals even when employing repetitive derivative techniques[21,22] or a small amplitude AC on "top" of the applied dc stripping voltage with phase selective current detection.[18,19] However, this resolution problem can be partially circumvented[16,23] using the mercury film electrode (plated on wax impreguated spectrographic graphite).[70,71] Because only about a one volt anodic dissolution "window" is available between cathodic deposition potentials and the anodic water salt breakdown potential limit, it is impossible in natural systems to obtain separate and distinct dissolution peaks for the large number of metal ions present with the magnitude concentration differences of natural water systems. Also, many metal ions have very similar dissolution peak potentials, which further makes identification and quantitative measurement impossible. This dissolution peak problem is further complicated if organic and inorganic species capable of forming complexes are present.[24,25] The net result is that for natural systems one would expect in many cases a broad undifferentiated anodic current-potential curve even with the use of AC or derivative techniques. Also any in situ analytical method involve either in situ data storage or data transmission to the surface. Obviously, such a system would be relatively expensive.

In view of the above argument on the difficulties of total in situ analysis, we must consider a compromise on this approach and specify a design that still removes the sampling and human error factors but is practical in operation and cost. These design specifications would be:

1. System sampled in manner to be rendered chemically and physically inert on removal from environment for analysis.

2. This sample must be in a form to be analyzed by a direct nondestructive method, one step preferably. (No chemical preparative steps.)

3. An in situ instrumentation must be very inexpensive.

4. High qualitative resolution for simultaneous identification of several elements.

5. Quantitative with respect to initial concentration of each element in original environment.

6. Samples must occupy a small space for convenient storage.

C. IN SITU SAMPLING APPROACH

Although stripping analysis has drawbacks with respect to dissolution peak "resolution" in complex systems, the use of the normal electrochemically controlled potential deposition step as an in situ preconcentration (sampling) step is perfectly applicable. Thus, a simple device for electrodeposition of trace metals was designed and employed in preliminary tests as an in situ preconcentration (sample preparation) unit in this research. The trace metals were deposited in the form of a thin metallic film on the electrode surface. Once the film is formed it has been found to be mechanically very stable and inert to chemical attack on removal from the environment of deposition. The film on the electrode was then returned to the laboratory for analysis by neutron activation analysis (NAA). Thus, the electrodeposition step, used only for the in situ sample acquisition, results in a sample having three desirable characteristics. First, it has been preconcentrated, making subsequent analysis by NAA or other more conventional methods fairly simple. Second, the sample is representative of the environment from which it was taken and finally, the sample can be handled and stored without causing a change in state or composition. With these objectives, we have continued the work with the electrodeposition sampling and also started work with other in situ sampling systems. In addition, we have carried out a preliminary examination of several ion exchange mattrices and surface adsorption techniques to examine the possibilities of these approaches to in situ sampling. A brief discussion of how these techniques satisfy our design specifications are also given here.

II. ELECTROCHEMICAL DEPOSITION FOR IN SITU SAMPLE AND RESULTS

A. PRELIMINARY APPARATUS

Preliminary studies of sensitivity parameters, effects of variation of electrolysis times and potentials, electrode pretreatment, film-electrode handling, etc., have been carried out using a watertight in situ electrochemical deposition unit designed and constructed in this laboratory. This section summarizes the design of this unit. For the in situ preconcentration technique we have built a very simple watertight electrodeposition device which was designed to be used for feasibility studies in relatively shallow water (surface to about 150 feet). The unit was built into a simple lucite watertight box similar to the types used to house underwater cameras as shown in Figures 1-3. With this simple design, only the three electrode surfaces [as indicated by AU, G, and REF (Ag/AgCl)] protrude through the lucite box through pressure "O" ring seals or epoxy resin. As the depositions were carried out at constant potential for these feasibility studies, no external control, power, etc., to the submersible lucite box was necessary. Thus, the box itself contained only an Hg battery power supply (PS) for the simple operational amplifier-based three-electrode potentiostat (A) and a simple mechanical timing circuit (T) for both starting and terminating the electrodeposition (this timing system was set on the surface before sealing the watertight lid of the box). The details of this controlled potential electrolysis device are shown in Figures 1 to 3 and the basic circuit diagram of the electronic potentiostat are shown in Figure 4.

The pyrolytic graphite working electrode holder (E) is a precision mechanism which allows for quantitative extrusion of the pyrolytic graphite cylinder from the end of the holder.[23,26,27] This design is similar to that employed previously for the NAA analysis of electrodeposited trace metal films.[23,27,30] After deposition, where the electrode surface is flush with the bottom surface of the unit, the cylinder is extruded a small distance (~0.2mm) and cleaved leaving a thin disk on which the metal film is deposited (after spraying with clear Krylon lacquer). This disk is easily stored for later analysis and the whole electrolysis assembly is again ready to be lowered for another run. This simple unit is of course hardly what one would want for routine data gathering in oceanographic research on trace metal content. However, since the mechanical (and expense) problem in the design and building of an automated system would be very severe at this time, we felt that it would be more logical to build this simple test system now to determine the basic influences of mass transport, organic surfactants, etc., on this approach to in situ preconcentration and to deter-

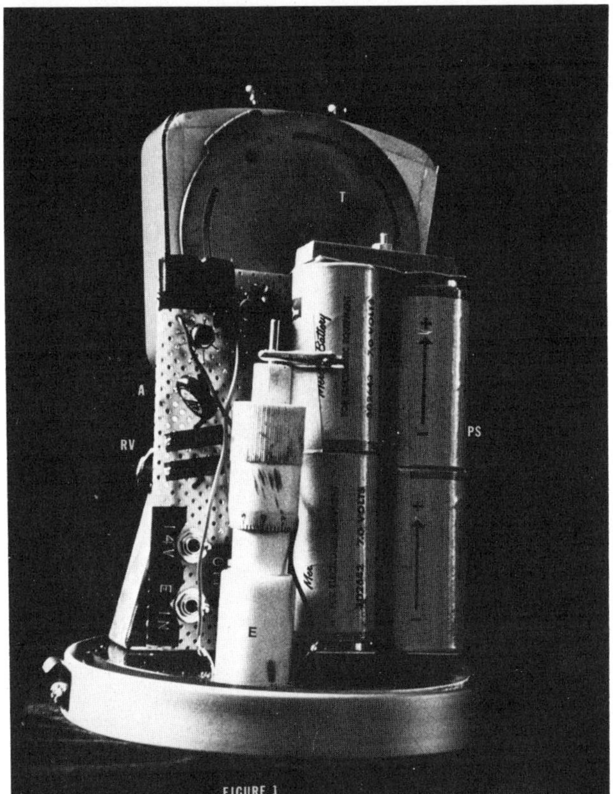

FIGURE 1

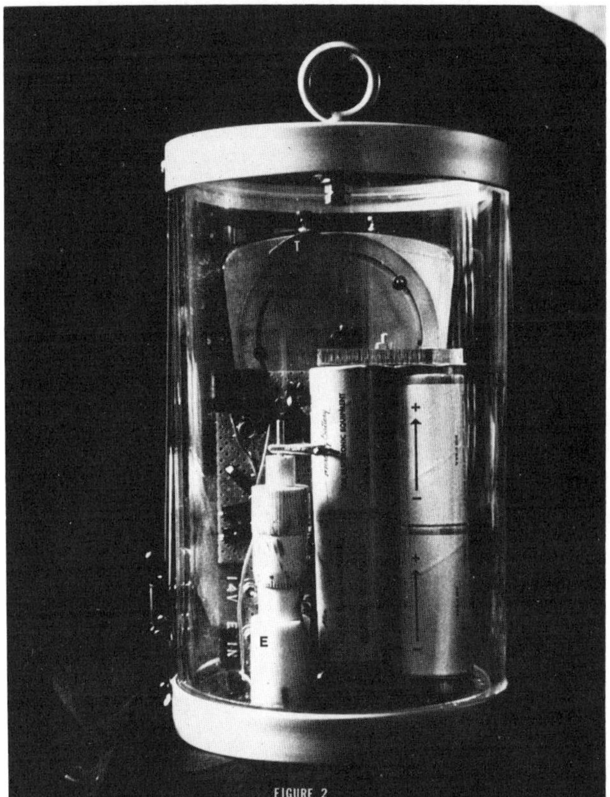

FIGURE 2

FIGURE 3

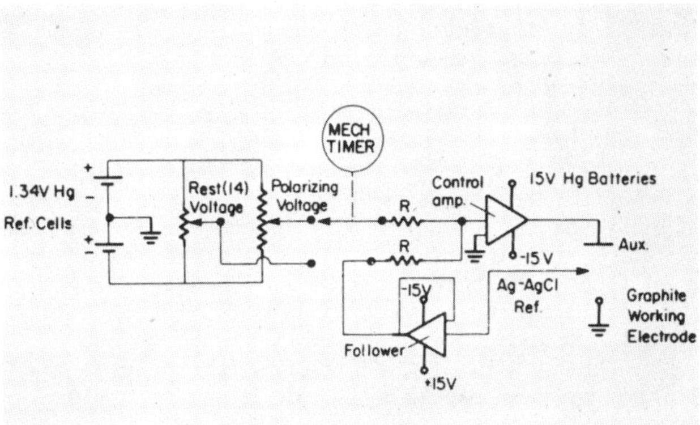

FIGURE 4: Controlled potential electrolysis circuit.

mine if it is a valid approach to the sample handling problems of trace metal analysis in natural systems.

Experiments[31,32,33] indicate that while the pyrolytic carbon as an electrode material appears to be an ideal deposition matrix (good electrochemical characteristics, cleavage properties, low background for analytical techniques such as electron microprobe, NAA, etc.), under static-no flow conditions plating efficiencies are very low. An example of low efficiency was an experiment of depositing Cu^{++} from a 1×10^{-3}M solution. No stirring was employed and the deposition time was 1 hour. The deposited amount of Cu was less than 0.1 mg (0.3% total efficiency for 500 mls of solv.). Since the concentrations of elements of interest are of the order of 10^{-7} - 10^{-8}M in sea water, electrolysis times of many hours would be required just to obtain a few mg or less of deposited metals. This small amount for some elements would be difficult to analyze even by NAA, although for those elements with exceptionally large cross-section to neutron capture the analysis should be satisfactory in reasonable length deposition times. However, for the analysis of the deposited films using other techniques such as x-ray fluorescence, atomic adsorption, etc. such long deposition times required to get sufficient sample material are not practical.

Another problem that must be considered is that of correlating amounts plated to actual in situ concentrations. The assumption here is that diffusion was the mode of mass transport and it should be constant in order to establish a correlation with in situ concentrations. In experiments raising and lowering the entire watertight potentiostat unit in a solution of known composition, we found less than a 10% variation in the amount plated over a range of raising and lowering rates. This suggested that the shape of the watertight case caused "buffering" or damping of the flow of solution adjacent to the electrodes so that effective convective mass transport did not occur to any significant extent. There is a good chance this would be true in actual conditions at sea also, even though small currents or ship movement might create transitory convective contributions. All in all, however, the uncertainty of the mode of mass transport (plus the fact that there is no way to monitor this parameter) would cause significant error in the back calculation of the in situ concentration such that you could never really determine how inaccurate or accurate the method was in real situations. Such uncertainty (and lack of confidence) would in itself lessen any value this method might offer for trace analysis. For this reason we have decided to abandon this simple static mode of deposition.

B. CONSIDERATIONS FOR ELECTRODEPOSITION

Controlled potential electrolysis is governed by the relationships:[34,35]

$$i_t = nFAD \frac{C_o}{\delta} \exp\left(\frac{DA}{V\delta} t\right)$$

or

$$i_t = i^\circ \exp\left(\frac{DA}{V\delta} t\right)$$

and in terms of fraction deposited

$$\frac{C_o - C_t}{C_o} = 1 - \exp\left(\frac{DA}{V\delta} t\right)$$

or in terms of concentration (fraction remaining)

$$\frac{C_t}{C_o} = \exp\left(\frac{DA}{V\delta} t\right)$$

where C_o and i° are initial concentration and current values (at t=0) respectively, C_t is the concentration of the ion of interest at any time t, D is the diffusion coefficient, A is the electrode area, V is the volume of the electrolysis solution, and δ is the Nernst diffusion layer thickness (inversely proportional to the stirring rate). Based on these relationships, the optimum conditions for electrolysis can be determined.

An efficient electrolysis cell will have a large electrode area to solution volume ratio (approaching "thin-layer" type electrolysis cell) and employ stirring in order to be able to plate as great a percentage as possible in the shortest period of time. These facts were observed experimentally during the preliminary work done prior to construction of a large volume cell. There are other factors which will affect electroplating which have not been previously considered.

The value of the diffusion constant, D, depends on temperature, T, and viscosity, η.

$$D = RT/6\pi \eta rK$$

With application to sea water electrolysis both these terms will not be constant. Temperature variations for sea water are in the range 0-25°C (273-298°K) depending on locale, depth, salinity, etc. For electrolysis this temperature range reflects a variation of about 82% in the amount plated as follows:

$$C_t/C_o = \exp(-\text{const} \times T)$$

$$\log C_t/C_o = \text{const} \times T$$

$$\frac{\log C_t/C_o}{\log C_t'/C_o} = \frac{T}{T'} = \log C_t/C_t' = 273°K/298°K = 0.915$$

$C_t/C_t' = 0.82$ or concentration of electrolysis solution after electrolysis for a period to under the same conditions will vary 82% as temperature varies from 0-25°C.

The amount plated (in conc. terms) equals C_o-C_t, so that this 82% variation also represents the relative difference between amount plated at 0°C as compared to amount plated at 25°C. The viscosity

$$\eta = \frac{\pi P r^4}{8lv} = \frac{\pi d g r^4 t}{8Q(1+\lambda)} \left(h - \frac{mv^2}{g}\right)$$

is dependent on a number of parameters; the main ones of interest here are pressure, P, and density,[36] d. In sea water pressure is a function of salinity and temperature. All these factors will affect an in situ electrolysis. We can, however, side step the influence of media, temperature, flow rate, stirring conditions, etc. and still obtain the desired concentration values. Returning to the basic equation:

$$\ln \frac{C_t}{C_o} = -\frac{DA}{V\delta}t$$

If one keeps the cell variables constant, the equation can again be expressed as

$$\log \frac{C_t}{C_o} = -kt$$

Since C_t is the concentration left in solution and what we actually measure is the amount plated, C_p, the expression can be changed using the relationship $C_o = C_t + C_p$

$$\log \frac{C_o - C_p}{C_o} = -kt$$

Since C_o, the original unknown concentration, and k are unknown, obtaining values for C_p at different t's will give us the results we desire. Therefore, all one need do is plate out at a constant flow rate for two different times and determine the amounts plated and C_o can be determined. Naturally, plating at more than two different times would improve the statistical basis of the analysis.

It is important to note that this approach is applicable only in the situation where we have a known, isolated volume of sample (for instance, 5 liters of sea water). It cannot, however, be used for direct in situ constant flow conditions where you presumably are sampling new solutions each second. The reason for this is that the expression is written such that a continual exponential decrease in concentration is expected in a finite solution volume. If you knew the value of k for sea water electrolysis - the efficiency of plating - you could determine the concentration directly. However, this value of k as mentioned above is dependent on temperature (through diffusion coefficient), salinity, viscosity; it may be difficult or impossible to determine for all the parameters for each sampling location.

The only other way to circumvent this problem is to have a specific volume encapsulating structure constructed with the in situ unit (for example, a 1 to 2 liter volume) and to electrolyze the known volume for a known length of time. For instance, 2 liters of sea water encapsulated and electrolyzed for 2 minutes, the electrodes are disconnected and 2 new liters of solution pumped in, the electrolysis begins again for 2 minutes and the process is repeated. The total amount plated is divided by the number of 2 liter volumes sampled and you obtain the number of mg.'s metal plated from 2 liters of solution in 2 minutes. By carrying out a second electrolysis using a new electrode where only the time of electrolysis is changed, $t_{(2)}$, will give you the necessary parameters to determine the in situ concentration from the two simultaneous equations.

$$\log \frac{C_o - C_{p(1)}}{C_o} = -kt_{(1)}$$

$$\log \frac{C_o - C_{p(2)}}{C_o} = -kt_{(2)}$$

C. THE CELL DESIGN

The cell design now under study (Figures 5-8) will electrolyze two samples (or more) from the same environment for different times. The solution in the cell will be static during each electrolysis period (repetitively filled and flushed in between deposition periods by the submersible pump) and the counter electrode (platinum sheet) will be connected physically to a high speed electromechanical vibrator. Such a system will have extremely high convection effects on the solution in the cell compartment and also will result in almost exhaustive deposition on the carbon disc working electrode of the trace metals in a few minutes.[37,38] As there is no directed flow of solution in this case for the parallel working-auxiliary electrode configuration, the deposited film will be quite uniformly distributed.

D. PREPARATION OF GRAPHITE WORKING ELECTRODES

The discs were boiled in concentrated HNO_3. They were then washed with a distilled H_2O and boiled with H_2O 3 times. The discs were then placed in a hot oven (>180°C) and the H_2O and HNO_3 driven off for about 12-24 hours. The discs were then placed in containers with molten parafin and a vacuum applied. After an hour or more the wax impregnation was finished. The cessation of gas bubbles from the discs under vacuum will signal that the wax impregnation is complete. The final stage was to place these discs in a large beaker (3 l) face down, cover them with molten wax and allow to cool. The wax was trimmed off each individual disc with a razor blade. The end faces are made conductive (free of wax) by polishing. It is also known that such graphite surfaces possess good electrochemical characteristics.[36] Polishing was accompanied by initial rough sanding (#350, 400 or 500 grade silica carbide paper), followed by finer sanding (#600 grade silica carbide), and final buffing (with filter paper).

E. ANALYSES METHODS FOR THE DEPOSITED FILMS

We are continuing to use neutron activation analysis (in cooperation with Mr. John Jones, Phoenix Memorial Laboratories, University of Michigan) and extended atomic absorption for the deposits on the carbon disc electrodes. We (Chemistry Department, University of Cincinnati) have recently purchased an

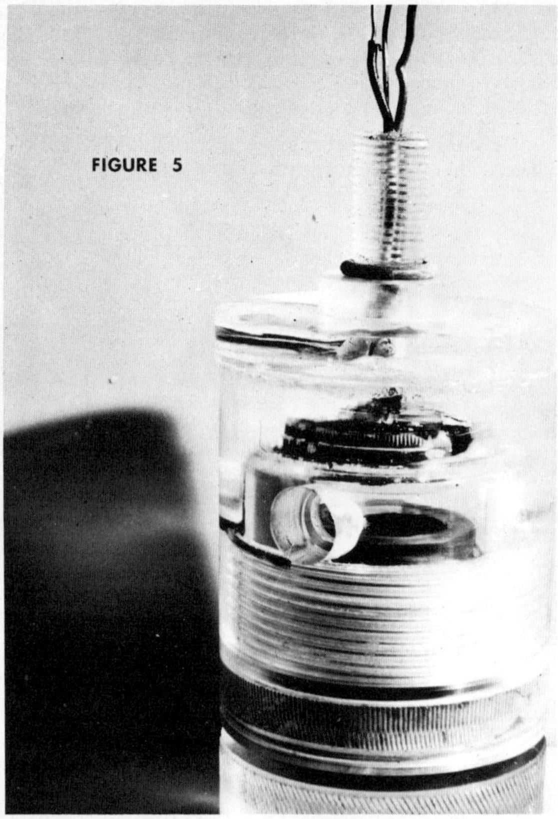

FIGURE 5

FIGURE 6

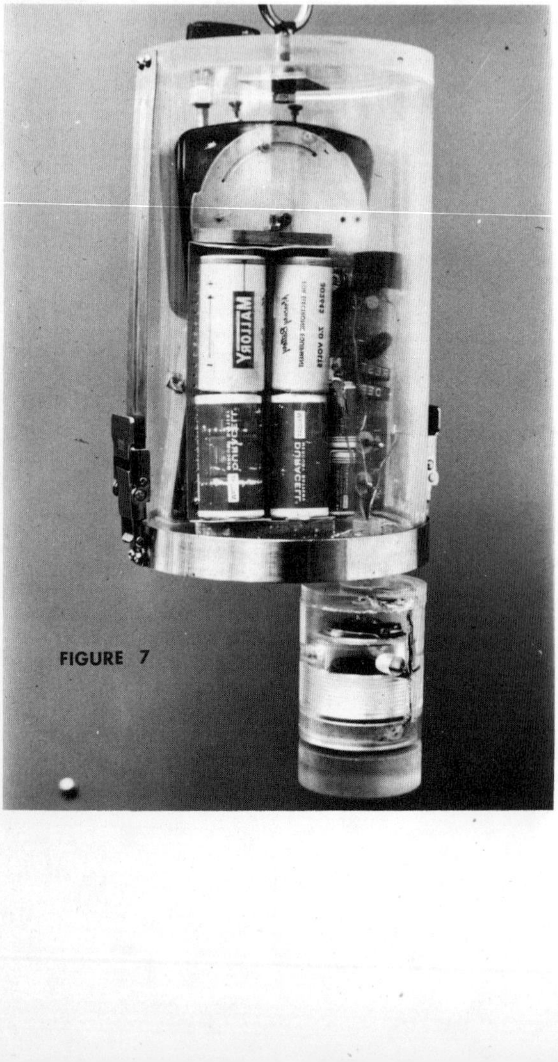

FIGURE 7

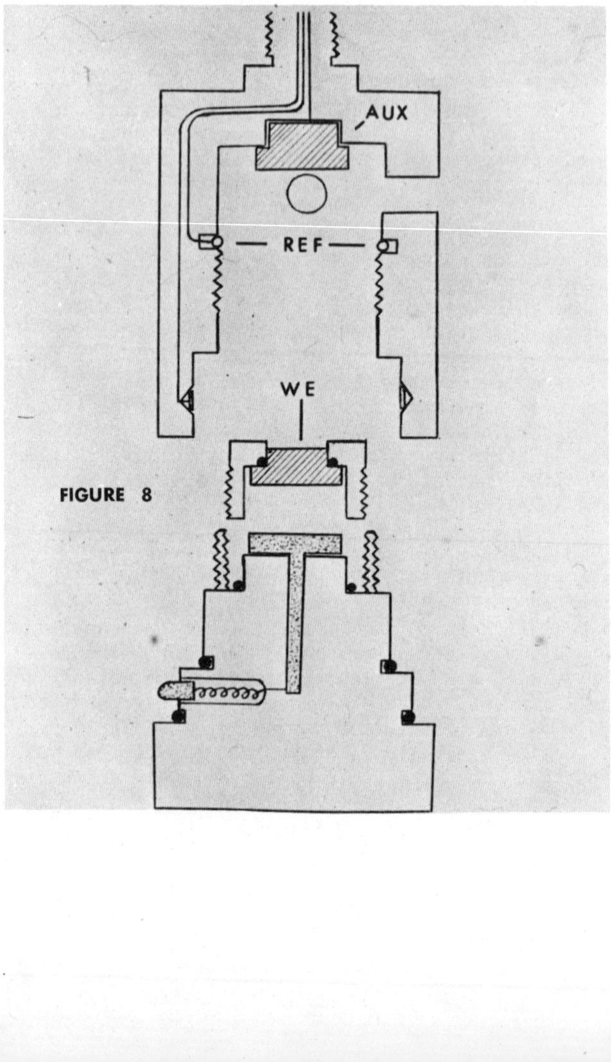

FIGURE 8

energy dispersive x-ray fluorescence spectrometer. This unit will be our major and general method of analyzing these thin deposited films on both the carbon disc and carbon gauze electrodes as it is fast and nondestructive. Now that we have probably solved the low plating efficiency problem, there will be no difficulty in obtaining sufficient material in convenient electrolysis times to fall within the sensitivity limits of this unit. It is especially convenient as all the elements deposited can be simultaneously determined[39,40,41] and as we are interfacing the system to our Raytheon Model 704 minicomputer for rapid multielement data reduction and analysis.

F. EXPANSION OF THE NUMBER OF METAL IONS THAT CAN BE ELECTROCHEMICALLY PRECONCENTRATED

The hydrogen overvoltage on pyrolytic carbon electrodes is relatively low, which means that the water decomposition potential (background) is only about -1.0 to -1.5 vs. SCE. While many of the metal ions of interest will electrodeposit at potentials more anodic to this limit, several important transition metals would be missed. It would be worthwhile to investigate the application of the mercury-plated film-carbon electrode developed by Roe, Carrit, and Matson[17] which has a much higher hydrogen overvoltage. Of course it would probably be impossible to do direct NAA, etc., analysis of the Hg film (in this case containing the dissolved trace metal atoms). It will be necessary to develop a method of separating the metals from the mercury film. We have carried out some preliminary studies of anodic dissolution of the metals from the mercury film-carbon electrode into a small solution volume, 100-250 μ liters, using a specially designed thin layer electrolysis cell.[31,42] Using radioactive zinc we have been able to determine that such dissolution is ~100% efficient. We are proposing to develop suitable analysis methods for such small volume samples for metal ions. We also propose to study electrode exchange by dissolution of the metal from the Hg film and deposition on graphite using the thin layer system with a nonaqueous solvent (which will eliminate the water breakdown problem).[31] This exchange of electrodes will then permit NAA, etc., analysis to be employed. (It should be pointed out that certain transition metal couples are totally irreversible at Hg electrodes and some other total chemical dissolution of the Hg-film must be developed to give a small volume solution containing the trace metals which can be analyzed. As the trace elements were preconcentrated, the actual concentrations of each dissolved into a small volume of solution is now high enough for ordinary methods of chemical analysis in many cases.)

III. APPLICATION OF AN ION EXCHANGE OR CHELATING RESIN, ETC. AS AN IN SITU PRECONCENTRATION MATRIX FOR TRACE ANALYSIS

Another approach to in situ sampling is the use of an ion exchange matrix. Again, the objectives are to preconcentrate the trace samples, have a sample which is representative of the environment from which it is taken, and to have a sample which is chemically and physically inert to allow storage until analysis.

We have done some experiments using ion exchange membranes as an in situ preconcentration matrix,[27-29] again looking for a simple direct method of obtaining a sample that is chemically and physically inert on removal from the site of the measurement and which can be directly (or simply) analyzed for trace element content. (We employed NAA of the membranes.) We found, however, that this membrane approach was not practical because too much time is required to reach distribution equilibrium (or even sufficient concentrations) of trace elements in the membrane matrix. Other problems were also found.[29]

Riley[43,44] and other groups[45-47,62] have employed ion exchange and/or chelating resins as preconcentration matrices for trace elements from sea water. We are adapting these chelate resin techniques of preconcentration to in situ sampling as it would be a simple matter to have a constant flow of water (using a submersible pump and motor) through a small column (made in the form of a replaceable cartridge) of such a resin assembled into a submersible sampling unit. The forced water flow through small mesh size will greatly increase the "sensitivity" of preconcentration when compared to the membrane experiments (Riley has carried out experiments on variation of mesh size, flow rates, etc.).[43] The resin containing the trace elements can be stored for subsequent analyses at a later date without fear of significant loss of the trace elements. We plan again to use NAA methods, but also plan to develop atomic absorption techniques (similar to those already developed),[43] x-ray fluorescence,[39] emission spectroscopy, etc. One possible advantage of the resin in situ concept is that trace elements present as particulates can be trapped out in the resin (or on a suitable filter) and, therefore, one will obtain a total concentration value for the trace elements. We are examining the "efficiency" of different column configurations and resins, to determine the distribution coefficients for the different trace elements, trying mixed bed techniques, etc., to evaluate and perfect this in situ sampling technique.

Recently Sugawara et al.[48,49] have developed a control-pore glass (CPG) immobilized chelate which has several attractive properties as an in situ preconcentration matrix for trace metals. An organosilane is strongly chemisorbed to the surface of porous (550 Å) glass particles (80-120 mesh). A selected chelating group is then reacted with the chemisorbed organosilane, coupling the chelating group to the glass surface and rendering it immobile. The most successful chelate so far has been 8-hydroxyquinoline. The resulting product has a working capacity in the range of 3 mg Cu^{++}/gm chelate at neutral pH solution. Because of the porous glass support, the CPG-immobilized chelate are resistant to compaction and swelling so are well suited to flow systems.

The selectivity of the control-pore glass product can be changed by changing the chelating group. It is only one step in its synthesis that requires changing so that a number of different products have been made[47] and others are possible and are being investigated by ourselves and others.[50] By synthesizing products specific for different groups of metal ions, it is possible that a mixed bed of such products would result in very high efficiency of chelation and accurate analysis of a large variety of trace metal ions.

We are presently analyzing the immobile chelate by x-ray fluorescence and plan cross checks by NAA and atomic absorption. X-ray fluorescence and NAA are useful in that they give simultaneous results for all metal ions present and without stripping the metal ions from the immobilized chelate. X-ray fluorescence is more accessible than is NAA. Even more accessible is atomic absorption, but conventional flame atomization would require an additional chemical step of quantitatively removing the metal ions from the immobilized chelate and subsequent analysis. Carbon rod atomization may be possible without this additional step and is being investigated.

IV. FUTURE AREAS OF STUDY FOR IN SITU SAMPLING

The approach taken in this proposed future research is to use separation and preconcentration methods in sequence to isolate selected constituents present in fresh waters and to adapt these methods to *in situ* use. Filtration, adsorption, and ion exchange media would be placed on a common flow line that is controlled and monitored by a specially designed pump system.

More specifically, the unit would be set up as a number of cartridges in series. The first would be a membrane filter holder assembly to trap particulate matter. Next would be a cartridge containing a macroreticular resin which would retain macromolecular and colloidal species with and without metal association. Following would be an activated charcoal container to isolate lower molecular weight organics and complexes. Finally, the last cartridge would contain a chelating ion exchange resin to complex trace inorganic ions.

The entire unit would be adapted for submersible use in conjunction with a portable battery-operated pump which may or may not be submersed. Since the bodies of fresh water of interest are relatively shallow, surface waters would primarily be studied and the unit would not be required to operate at depths greater than 50 feet. A magnetic flow meter would be used to monitor flow rate and its output conditioned for recording on magnetic tape. It may also be possible to use the flow meter output as a feedback signal for electromechanical control of the flow rate.

A. MEMBRANE FILTERS

The study of particulate matter in natural waters has most frequently been accomplished by the use of membrane filters.[51-53] These filters are composed of a partially cross-linked polymer of a cellulose substance whose porosity can be adjusted to a specified and uniform size. The membrane pore size most frequently used in these studies is the 0.45 μm filter since it represents a compromise between the achieving of a suitable flow rate and the trapping of the smallest particle possible.[51] Presumably a 0.45 μm filter will allow some colloidal matter to pass through, but since these particles are reasonably small, they can effectively be considered dissolved species. For the proposed work, the Millipore Company can provide membrane filters and holder assembly to satisfy all requirements.

Having obtained the particulate sample, the analysis can then proceed in a number of directions. At the simplest level the sample can be digested with acid and analyzed for the concentration of the various metals present. This could be accomplished with standard spectroscopic flame techniques (atomic absorption, flame emission, etc.), emission spectroscopic or even colorimetric methods. Such an analysis is essential for the understanding of the distribution of trace elements in natural waters.

Additional information on the nature of the suspended matter could possibly be obtained from an x-ray diffraction pattern of material isolated from the filter. Minerals could undoubtedly be identified in this manner. Chromatographic techniques might also be applied to isolate discrete metal-organic species.

B. MACRORETICULAR RESINS

Macroreticular resins represent a new generation of materials which can function both as normal ion exchange resins and as regeneratable adsorbents for organic matter in water.[54,55] These resins are now being produced with weak or no ion exchange capability which reduces their affinity for metal ions in solution.[55] In fact the scavenging of organics (decolorizing or polishing processes) from water for industrial use is now commonly and efficiently being accomplished with macroreticular or macroporous resins.[54,55] In comparison with gel type resins which also can be used to isolate organics from water, the macroporous resin has a larger pore diameter and smaller internal surface area.[54] These properties have been found to be more favorable for efficient removal of organics from water. Another advantage is that unlike the strong base anion exchange resins used in water treatment processes, macroreticular resins resist organic fouling, i.e., irreversible organic adsorption.[56] A more recent application of these resins was to quantitatively extract organic acids, bases, and neutral species at the part per billion to part per million level from portable water sources.[57] The removal of humic acid in this manner has also been reported.[58]

In the proposed application, macroreticular resins would trap a large variety of organics and organic complexes. A counter current extraction with alkali or sodium chloride solution would displace the concentrated organic matter from the resin. A total metal analysis can then be performed for complexed and/or organically associated metals. Using chromatographic techniques (thin layer, column chromatography) it may also be possible to isolate distinct metal-organic species for identification. Further work might include a study of the organic matter isolated.

C. ACTIVATED CARBON

The use of activated carbon for purification processes has been known for centuries and has more recently been applied in waste-water treatment.[59-61] For the purpose of this research activated carbon is intended as a secondary trap for organics that might pass through the macroreticular resin column. It is possible that these organics may be complexed to trace metals and so the carbon would be treated with acid for metal extraction and analysis. Perhaps metal complexes can also be extracted intact from the carbon and identification of these species made possible. Additionally, volatile organics adsorbed on the carbon might be analyzed by gas chromatographic and mass spectroscopic analysis.

D. CHELATING ION EXCHANGE RESINS

Since natural waters have rather large amounts of alkali salts, normal ion exchange resins could not be used to isolate metals present in trace quantities. Good success has been achieved using chelating ion exchange resins.[62-64] Biechler[62] has employed Dowex A-1 for isolating heavy metals in industrial waste-water. The resin Chelex 100 has successfully been applied to the analysis of trace metal ions in sea water.[63,64]

A similar application of chelating resins would be attempted in this work. Since this resin would be contained in the last cartridge no organic fouling should occur and the metals isolated should represent simpler ionic species. The analysis of the metals present would be accomplished as discussed above.

E. ANTICIPATED DIFFICULTIES

This work, naturally, may not be as uncomplicated or as successful as this proposal intimates. The basic approach, however, appears to be sound in view of the past success of each of the preconcentration methods. There are a few areas in which difficulties may arise and indeed are anticipated.

First is the metal analysis problem. This problem is precisely that a multielemental analysis of trace species is involved. It is essential that enough of each metal is present at each preconcentration stage for successful analysis. It is hoped that each preconcentration stage would isolate a sufficient quantity of material for the analysis. In terms of amounts, this would mean that at least 10 μg of each metal would have to be isolated if, for example, a flame spectroscopic method of analysis was employed.

Another problem is that of labile complexes. Since components are being separated out at each stage of the preconcentration process, equilibria are probably being disturbed. A change in the concentration of individual species governed by equilibrium processes would occur as separation took place. Using the methods described there would be no way of determining if this was actually occurring. As a check on this problem and also on the preconcentration sequence as a whole, it would be wise to take water samples for laboratory analysis and to adopt analysis procedures analogous to those described by Stiff.[51] Laboratory analyzed water samples would provide a standard to gauge the method proposed. The work of Stiff[51] suggests an indirect method whereby labile complexes might be detected.

F. INSTRUMENTATION

The main problem with using adsorption or exchange resins is that there is a dependence of separation efficiency on flow rate. In order to correlate the amount of material isolated with its *in situ* concentration, the separation efficiency has to be well characterized. Thus, if it is known that for a certain range of flow rates, separation efficiency is 100%, it would be wise to maintain the flow rate within that range in order to eliminate correction factors. In this work, therefore, it is essential that the flow rate be either controlled or monitored. The former is more desirable since it simplifies calculations. The latter can be used, but it would be difficult to apply and would involve a loss in accuracy. In addition to this efficiency problem it is necessary to know the total volume of water sampled in order to characterize the particulate distribution in the water system.

For these reasons, then, a monitored pumped system is required. The pump itself would be of high capacity and powered by a 6 or 12 V DC battery. Since the pump operation will occur at the surface, the pump need not be water-tight. In order to monitor the flow, an electromagnetic flow meter would be designed. Such flow meters have been employed in the past and have the virtue of producing an output signal whose magnitude is directly proportional to the flow.[65] A record of the flow meter output would be stored as a digital or conditioned analog signal on magnetic tape. Small portable multitrack magnetic tape cassette units are available and have actually been used for gathering physical and chemical data.[66-69] These units can then be interfaced with a computer for subsequent data treatment. Since these tape units are multitracked, other conditioned transducer signals can be simultaneously recorded. Such signals could come from temperature, pH, conductivity, and salinity monitors, since these parameters may have an effect on the preconcentration efficiency of the resins and also characterize the body of water studied. Thus, a permanent record could be obtained of conditions occurring during operation of the *in situ* preconcentration unit.

As stated above a constant or controlled flow rate would be preferred in the operation of the *in situ* unit. It is possible to use the flow meter output as a feedback signal for an electromechanical device that would control the flow. This device could simply be a selenoid physically blocking liquid flow in an on-line flow chamber. As the filter gets clogged with particulate matter the flow rate will decrease and the flowmeter would send a signal to the selenoid. The selenoid would back off, opening a wider passage in the chamber and compensate for the clogged filter thus maintaining a constant flow rate.

REFERENCES

1. D.N. Hume, "Analysis of Water for Trace Metals," Chapter in Advances in Chemistry, 67, 30 (1967); R.A. Horne, "Marine Chemistry," Intersciences Pub., N.Y., 1969, pp. 129-138.

2. "Trace Analysis," J.H. Yoe and H.J. Koch eds., J. Wiley and Sons., N.Y., 1957; D.E. Robertson, Anal. Chem., 40, 1067 (1968).

3. David E. Robertson, Anal. Chem., 40, 1067 (1968).

4. David E. Robertson, Anal. Chim. Acta, 42, 533 (1968).

5. E.D. Goldberg in "Chemical Oceanograph" Vol. I, J.P. Riley and G. Skirrow, eds., Acad. Press, London, 1965, Chapter 5.

6. D.F. Martin "Marine Chemistry," Vol. I and II, M. Dekker, N.Y., 1969 and 1970; T. Joyner et al., Environ. Sci. and Tech., 1, 417 (1967).

7. J.M. Rottschafer, J.D. Jones, and H.B. Mark, Jr., Environ. Sci. and Tech., 5, 336 (1971).

8. J.D. Jones, J.M. Rottschafer, H.B. Mark, Jr., K.E. Paulsen, and G.J. Partiarche, Mikrochim. Acta, 1971, 399-404 (1971).

9. P. Lafleur, Priv. Comm., Nat. Bur. of Standards, 1970.

10. P.G. Brewer and D.W. Spencer, "Trace Element Intercalibration Study," Ref. No. 70-62, Woods Hole Oceanographic Inst., Mass., 1970.

11. Werner Stumm and James J. Morgan, "Aquatic Chemistry," Wiley-Interscience, New York, 1970, 58-62.

12. J.P. Riley in "Chemical Oceanography," Vol. II, J.P. Riley and G. Skirrow, eds., Acad. Press, London, 1965, Chapter 21.

13. K.H. Mancy, D.A. Okun, and C.N. Reilley, J. Electroanal. Chem., 4, 65 (1962); A.G. Albin, Ph.D. Thesis (H. Freund), Oregon State Univ., 1969.

14. A.J. Bard, Anal. Chem., 39, 57R (1962); 36, 70R (1969).

15. I. Shain and S.P. Perone, Anal. Chem., 33, 325 (1961).

16. U. Eisner and H.B. Mark, Jr., J. Electroanal. Chem., 24, 345 (1970).

17. W.R. Mattson, D.K. Roe, and D.E. Carritt, Anal. Chem., 31, 1594 (1965); W. Mattson, Ph.D. Dissertation, Mass. Inst. of Tech., 1967; W.R. Mattson and D.K. Roe, Anal. Instrum., 4, 19 (1966); H.E. Allen, W.R. Mattson, K.H. Mancy, J. Water Poll. Cont. Fed., 42, 215 (1966).

18. N.L. Underkoflen and I. Shain, Anal. Chem., 37, 218 (1965).

19. A. Eisner, C. Yarnitzky, Y. Nemirosky, and M. Ariel, Israel J. of Chem., 4, 215 (1966).

20. E.P. Parry and R.A. Osteryoung, Anal. Chem., 37, 1634 (1965).

21. S.P. Perone and T.R. Meuller, Anal. Chem., 37, 2 (1965).

22. S.P. Perone, D.O. Jones, and V.F. Gutknect, Anal. Chem., 41, 1154 (1969).

23. H.B. Mark, Jr., J. Pharm. de Belgique, 25, 367 (1970).

24. B.H. Vassos, Ph.D. Dissertation, Univ. of Michigan (1965).

25. B.H. Vassos and H.B. Mark, Jr., J. Electroanal. Chem., 13, 1 (1967).

26. H.B. Mark, Jr. and F.J. Berlandi, Anal. Chem., 36, 2062 (1964).

27. H.B. Mark, Jr., F.J. Berlandi, B.H. Vassos, and T.E. Neal, "Proceedings of the 1965 Int. Conf. on Mod. Trends in Activation Anal., College Station, Texas (1966), p. 107.

28. U. Eisner, J.M. Rottschafer, F.J. Berlandi, and H.B. Mark, Jr., Anal. Chem., 39, 1466 (1967).

29. U. Eisner and H.B. Mark, Jr., Talanta, 16, 27 (1969).

30. H.B. Mark, Jr., U. Eisner, J.M. Rottschafer, F.J. Berlandi, and J.S. Mattson, Environ. Sci. and Tech., 3, 165 (1969).

31. J.M. Rottschafer, Ph.D. Thesis, Univ. of Michigan (1972).

32. J.M. Rottschafer, R.J. Boczkowski, and H.B. Mark, Jr. (In preparation for submission to Environ. Sci. and Tech.)

33. B.H. Vassos, F.J. Berlandi, T.E. Neal and H.B. Mark, Jr., Anal. Chem., 37, 1653 (1965).

34. P. Delahay, "New Instrumental Methods in Electrochemistry," Interscience Pub., Inc., N.Y., 1952, pp. 282-297.

35. C.N. Reilley and R.W. Murray, In "Treatise on Analytical Chemistry," Part I, Vol. 4, I.M. Kolthoff and P.J. Elving, eds., Interscience Pub., Inc., N.Y., 1963, p. 2219.

36. R.A. Horne, "Marine Chemistry," Interscience Pub., Inc., N.Y., 1969.

37. S. Bruckenstein, Private Communication, 1972.

38. A.J. Bard, Anal. Chem., 35, 1125 (1963).

39. H.A. Liebhofsky, H.G. Pheiffer, E.H. Winslow, P.D. Zeniang, "X-ray Adsorption and Emission in Analytical Chemistry," J. Wiley and Sons, N.Y., 1960.

40. B.H. Vassos, R.F. Hirsch, and H. LeHerman, 161st Meeting ACS, Los Angeles, Calif., April, 1971, Analytical Division paper No. 79.

41. R.O. Möller, "Spectrochemical Analysis by X-ray Fluorescence," Plenum Press, New York, 1972.

42. E.J. Berlandi and H.B. Mark, Jr., Nucl. Appl., 6, 409 (1969).

43. J.P. Riley and D. Taylor, Anal. Chim. Acta, 40, 479 (1968).

44. A.D. Matthews and J.P. Riley, Anal. Chim Acta, 51, 287 (1970).

45. C.M. Callahan, J.N. Rascual, and Ming G. Lai, U.S. Clearinghouse for Sci. and Tech. Inform., AD647661 (1966).

46. H. Nakagawa and F. Ward, P.H. Conf. Anal. Chem., Abs., p. 36 (1960).

47. S. Elemer, R.H. Zsuzsa, L. Alexandra, and K. Endre, Magyar Komiai Folyoirat, 75, 58 (1969).

48. K.F. Sugawara, H.H. Weetall and G.D. Schucker, 163rd Meeting Amer. Chem. Soc., Boston, Mass., April, 1972; Private Communication of Unpublished Results, 1972.

49. K.F. Sugawara, H.H. Weetall, and G.D. Schucker, Private communication of unpublished results, May, 1973.

50. D.F. Hercules, et al., Anal. Chem., 45, 1973 (1973).

51. M.J. Stiff, Water Res., 5, 585 (1971).

52. J.C. Laird, D.P. Jones, and C.S. Yentach, Deep-Sea Res., 14, 251 (1967).

53. D.W. Spencer and P.L. Sachs, Marine Geol., 9, 117 (1970).

54. R. Kunin and R. Hetherington, Ind. Water Eng., 6, (12), 34-39 (1969).

55. F. Martinola and A. Richter, ibid., 6, (12), 22-25 (1971).

56. J.J. Wolff and I.M. Abrams, ibid., 7, (1), 40-42 (1969).

57. A.K. Burnham, G.V. Calder, J.S. Fritz, G.A. Junk, H.J. Svec, and R. Willis, 162nd American Chemical Society Meeting, Div. Water, Air, and Waste, 3 (1971).

58. J. Seidl, "Ion Exchange Process Ind. Pap. Conf.," Soc. Chem. Ind., London, pp. 391-400 (1969); CA, 74, 67439w (1971).

59. J.B. Cendelman and S.C. Caruso in "Water and Water Pollution," Vol. 2, L.L. Ciaccio, ed., M. Dekker, Inc., N.Y., 1971, pp. 483-591.

60. J.W. Hassler, "Activated Carbon," Chemical Publishing, New York, 1963.

61. J.S. Mattson and H.B. Mark, Jr., "Activated Carbon," Marcel Dekker, Inc., New York, 1971.

62. D.G. Biechler, Anal. Chem., 37, 1055 (1955).

63. J.P. Riley and D. Taylor, Anal. Chim. Acta., 40, 479 (1968).

64. C.M. Callahan, J.M. Pascual, and M.G. Lai, U.S. Clearinghouse for Sci. and Tech. Inform., AD 647661 (1966).

65. G.P. Katys, "Continuous Measurement of Unsteady Flow," MacMillan Company, New York, 1964, pp. 135-151.

66. Data Sheet 53,54, Braincon Corp., Marion, Mass.

67. F.E. Snodgrass, Science, 162, 78 (1968).

68. M.D. Palmer and J.B. Izatt, Water Res., 4, 773 (1970).

69. J.S. Mattson and A.C. McBride III, Anal. Chem., 43, 1139 (1971).

70. W.R. Mattson and D.K. Roe, Anal. Instrum., 4, 19 (1967).

71. H.E. Allen, W.R. Mattson, and K.H Mancy, Jour. Water Pollution Control Fed. 42, 573 (1970).

72. D. Kester, K. Crocker, and G. Miller, Jr., Deep-Sea Research, 20, 409-411 (1973).

ACKNOWLEDGEMENT

This research was supported by the National Science Foundation, NSF GA 25563 and GP 35979.

© 1973, ISA JSP 6712

ATMOSPHERIC CONSTITUENT MEASUREMENTS USING COMMERCIAL 747 AIRLINERS

Porter J. Perkins and Gregory M. Reck
Lewis Research Center
Cleveland, Ohio

ABSTRACT

NASA is implementing a Global Atmospheric Monitoring Program to measure the temporal and spatial distribution of particulate and gaseous constituents related to aircraft engine emissions in the upper troposphere and lower stratosphere (6 to 12 Km). Several 747 aircraft operated by different airlines flying routes selected for maximum world coverage will be instrumented. The initial design of the system (location of equipment, aircraft interfaces, etc.) was conceptually defined in feasibility studies conducted by several airlines and an airframe company. An instrumentation system is now being assembled and tested and is scheduled for operation in airline service in late 1974. Specialized instrumentation and an electronic control unit are required for automatic unattended operation on commercial airliners. An ambient air sampling system was developed to provide undisturbed outside air to the instruments in the pressurized aircraft cabin. The data system has a flight data acquisition unit and tape recorder used in late model airliners. The inertial navigation system and air data system available on the newer jets will record the location and time of the atmospheric measurement as well as meteorological information such as wind speed and direction and air temperature.

INTRODUCTION

A NASA Program is underway to equip several commercial 747 airliners with special instrument systems to routinely obtain and record-in-situ measurements of several minor atmospheric constituents on a global basis. This paper describes the environment for the instruments, the data acquisition system for airliners, and the instrument improvements and modifications needed for automatic unattended operation on commercial airliners. Only existing atmospheric constituent measurement techniques are being used for this program.

The continuing concern regarding the effects of aircraft engine exhaust emissions on the natural troposphere and stratosphere indicates the necessity for obtaining relaible data on the background concentrations of a number of minor atmospheric constituents. Potentially harmful effects of several aircraft engine exhaust constituents including oxides of nitrogen, water vapor, and particulates have been suggested[(1)]. However, the actual effects of these constituents cannot be assessed without reliable atmospheric models and accurate knowledge of natural background levels.

Work has been underway for several years in the Climatic Impact Assessment Program of the Department of Transportation to collect data for verification of and input to atmospheric models related to the stratosphere. A number of universities and federal agencies have participated in this program by collecting data from rocketsondes, balloons, dedicated aircraft and ground and satellite stations using both in-situ and remote sensors. However, only a limited effort has been directed at the upper troposphere and the lower stratosphere. This region is of particular interest and importance because of the large quantities of exhaust emissions injected by the current commercial jet fleet, the possible long (weeks to months) residence times of emissions injected into the lower stratosphere, and the uncertainty over transport, mixing, and photochemical phenomena occurring in the vicinity of the tropopause.

The Lewis Research Center is implementing a Global Atmospheric Sampling Program (GASP) to examine the atmosphere between 6 and 12 kilometers[(2)]. This is a major effort in current NASA aircraft programs involving atmospheric monitoring. The GASP effort centers around the use of commercial airliners as an instrument platform to collect global air quality data on a routine basis[(3)]. The GASP objectives during the next five years are to (1) as rapidly as possible, determine the worldwide baseline concentrations of important minor atmospheric constituents in the troposphere and lower stratosphere, (2) identify any trends which could be related to the contribution of jet aircraft to possible atmospheric contamination, (3) establish an economical system for routine collection of air quality data. The data obtained with GASP systems will be used to develop more accurate atmospheric models, evaluate the potential effect of aircraft emissions on the environment, and to identify which engine exhaust constituents are most detrimental to the ambient environment. This information will then be used to guide future aircraft engine pollution reduction research. The GASP data may also be used to verify and complement data obtained from satellites equipped with remote sensors such as the Nimbus G scheduled for launch

Superior numbers refer to similarly-numbered references at the end of this paper.

later in this decade.

The approach which NASA Lewis Research Center is using to establish an operational system includes the following steps:

1. studies by airlines and airframe manufacturers to examine the technical, operational and economical feasibility of utilizing commercial airliners to collect air quality data,
2. evaluation of a number of commercially available candidate air quality instruments for the GASP application,
3. modification of selected instruments for automatic unattended operation and for approval by the airlines and the FAA,
4. system design, fabrication, and installation,
5. certification and validation,
6. data acquisition and analysis.

This paper describes the present status of the implementation of the GASP program and specifically addresses the constraints on the system, preliminary system design, instrument selection, and required instrument modifications. Step 1 is complete. Steps 2 and 3 are underway along with the system design of step 4. Data acquisition will start in late 1974.

ATMOSPHERIC CONSTITUENTS AND RELATED MEASUREMENTS

A number of minor atmospheric constituents have been identified as significant with respect to the contribution of jet aircraft. Table I gives a listing of important constituents which has been compiled with the assistance of the Air Resources Laboratory of the National Oceanic and Atmospheric Administration. The table lists the particulate and gas constituents related to pollution and of interest to meteorology between 6 and 12 kilometers altitude and other measurements of the atmospher that can be pinpointed at the time of the data taking. Table II lists the constituents selected for possible measurement in the initial airline installations, an estimate of the expected range of ambient concentrations, and the measurement principle chosen for the instrument. As noted in the Table, four constituents, ozone, water vapor, carbon monoxide, and particulates (count and size distribution) will be measured on the first two flight packages. These instruments were selected from among a group of candidate instruments on the basis of constituent priority, availability, minimum modifications, laboratory evaluation, and limited flight test results. Instruments to measure the other constituents in Table II are being evaluated in flight and laboratory tests. The system is being designed with sufficient space, power, and data handling capability to accommodate additional instruments as new or improved instruments become available to measure these or the other constituents.

In addition to the air quality data received from the instruments, a considerable quantity of additional data must be recorded to document instrument identities, system status, and supplemental flight data. The identity of individual instruments must be recorded so that the proper calibration information can be associated with each instrument channel during data reduction. The system status data includes measurements of sample pressure, temperature, and flow as well as other signals indicating valve positions, calibration cycle, or instrument temperatures. The supplemental flight data are obtained from the aircraft data computer system, and include aircraft position, speed, altitude, wind speed, wind direction and static air temperature.

AIRCRAFT SELECTION AND SYSTEM ENVIRONMENT

In order to establish the technical and operational requirements and the economics of using commercial airliners as instrument platforms, feasibility studies were conducted by an airframe manufacturer and several airline companies[4]. Initially, these studies served to identify the following constraints which would be imposed on the system by airlines operation:

1. No air crew duties would be imposed beyond operation of an on/off and emergency switch.
2. No revenue space would be used.
3. Limited servicing and maintenance would be performed on a noninterference basis.
4. An FAA Supplemental Type Certificate will be required.

These constraints dictate that the instruments be adapted for remote control (by an automatic on-board sequencer) and that they be reliable for periods of up to two weeks.

After examination of a number of aircraft in the commercial fleet, the Boeing 747 aircraft was selected to carry the GASP instrument system. The selection was based on (1) availability of nonrevenue space in an accessible location common to most airline configurations, (2) a route structure adequate for global coverage, (3) availability of an Inertial Navigation Systems (INS) to provide location information and a Central Air Data Computer (CADC) to provide flight data.

A specific area (shown in figure 1) adjacent to the nose wheel well below the passenger cabin was chosen for the GASP systems. The area is accessible through hatches from either the passenger compartment or outside the aircraft. Access may also be gained from the forward cargo compartment when the compartment is empty. The aircraft avionics are located in the same area just aft of the nose wheel well.

A study of possible inlet probe positions concluded that the probe should be located on the centerline just below the nose of the aircraft (also shown in figure 1). This location is sufficiently forward to avoid spray from the nose wheel well, sample contamination by the aircraft, and the likelihood of probe damage during ground operations. In this position, the probe will not interfere with the pressure, temperature, and angle of attack sensors located along the side of

TABLE I. - ATMOSPHERIC CONSTITUENTS AND RELATED MEASUREMENTS OF INTEREST IN NASA GLOBAL MONITORING PROGRAM

ATMOSPHERIC CONSTITUENTS		RELATED MEASUREMENTS
PARTICULATES	GASES	
RELATED PRIMARILY TO MAN MADE POLLUTION		
CARBON SULFATES NITRATES PROPERTIES OF INTEREST NUMBER DENSITY SIZE DISTRIBUTION MASS CONCENTRATION CHEMICAL COMPOSITION OPTICAL PROPERTIES	CARBON MONOXIDE HYDROCARBONS OXIDES OF NITROGEN (INCLUDING NITRIC ACID) OXIDES OF SULFUR (INCLUDING SULFURIC ACID) AMMONIA HYDROGEN SULFIDE	TEMPERATURE TURBULENCE PRESENCE OF CLOUDS WIND SPEED WIND DIRECTION TIME AND DATE POSITION ALTITUDE HEADING TRUE AIRSPEED RATE OF CLIMB
OF PRIMARY INTEREST TO METEOROLOGY		
AITKEN NUCLEI FREEZING NUCLEI CONDENSATION NUCLEI	OZONE WATER VAPOR CARBON DIOXIDE	

TABLE II. - ATMOSPHERIC CONSTITUENTS SELECTED FOR MEASUREMENT IN INITIAL AIRLINE INSTALLATIONS FOR GLOBAL MONITORING PROGRAM

ATMOSPHERIC CONSTITUENT	EXPECTED RANGE	MEASUREMENT PRINCIPLE
PARTICULATE PROPERTIES		
• NUMBER DENSITY	10^{-2} - 10/CC	{LIGHT SCATTERING
• SIZE DISTRIBUTION	0.5 - 10 MICRONS	
CHEMICAL COMPOSITION MASS CONCENTRATION		{FILTER/LAB ANALYSES
AITKEN NUCLEI	10 - 10^3/CC	CONDENSATION NUCLEI COUNTER (CNC)
• OZONE	0.01 - 2.0 PPM	UV ABSORPTION
• WATER VAPOR	3 - 10^3 PPM	AL OXIDE HYGROMETER
• CARBON MONOXIDE	0.05 - 0.2 PPM	FLUORESCENT NDIR
OXIDES OF NITROGEN	0.1 - 10 PPB	CHEMILUMINESCENT
SULFUR DIOXIDE	0.1 - 2 PPB	H_2SO_4 GENERATOR AND CNC
CARBON DIOXIDE	320± 5 PPM	NDIR (NEG. FILTERING)

• NOTE: MEASUREMENTS ON FIRST TWO INSTALLATIONS - OTHERS TO FOLLOW

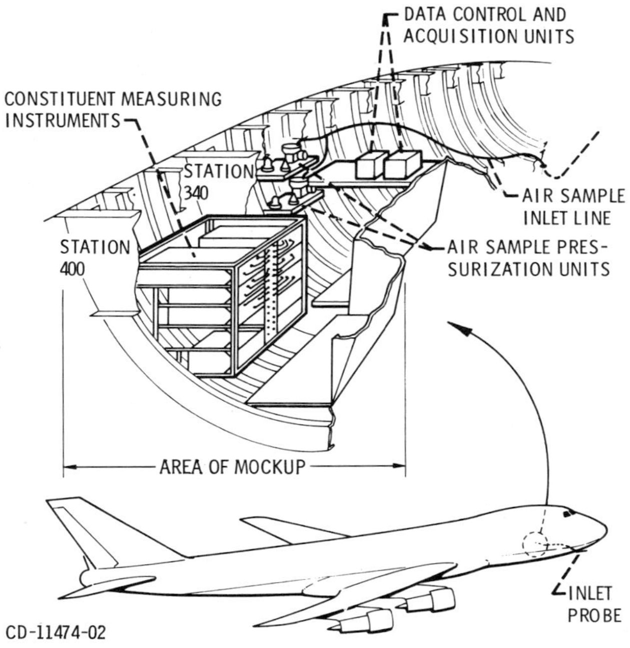

Figure 1. - Air inlet probe and equipment location on Airline 747 airplane.

the fuselage near the nose. Also, this probe location is a relatively short distance from the instrument location.

To assist in visualizing the area in the Boeing 747 where the system will be installed, a full-scale mockup of a portion of the airframe was constructed. Figure 2 is a photograph of the mockup including a conceptual model of the GASP instrument mounting system. The model shows a group of instrument cases mounted on a set of shelves installed in the aircraft. The mockup is being used for system design and layout.

The GASP system location in the Boeing 747 shares essentially the same environment as the passenger cabin during flight. The aircraft ventilation system circulates conditioned air through the cabin, forward and down into the nose wheel well area, then aft around the aircraft avionics racks into the forward cargo area. In-flight temperature will be about the same as that in the passenger cabin and the ventilation flow will be more than adequate for heat dissipation from the GASP instruments. However, the range of storage temperatures (during overnight and preflight periods) is likely to be more extreme and could extend from -10° to $+50^{o}C$.

The GASP location is maintained at cabin pressure during flight, however, this is less than one atmosphere and varies with altitude. Figure 3 shows the cabin pressure schedule and indicates that above 7.0 km (23,000 ft.) the cabin pressure is maintained at $6.1 N/cm^2$ (8.9 psid) above altitude pressure.

The Boeing Company vibration specifications for this location in the Model 747 are not severe and range from .025 g at 5 Hz up to 1.0 g at 1000 Hz. Mechanical shocks encountered during normal operations (landing, gusts, etc.) are considered in the structural design not to exceed 5 g's.

The feasibility studies also identified several constraints on the system and instrument operation. The system must meet safety standards established for the aircraft. No combustible or toxic gases can be carried. No flames or flammable materials are permitted. No cryogenics can be stored in the system. Although proper operation of the instruments is not required for aircraft safety, the system or instruments must not malfunction or fail in any way which could conceivably jeopardize the aircraft safety. In particular, the instruments must not generate electromagnetic interference. Instrument hold-downs, mechanical supports, and interconnects must be approved, and fire prevention techniques must be employed.

AIR SAMPLE FLOW SYSTEM

The air sample flow system for the GASP instrument package will provide both a pressurized and an unpressurized sample flow. A schematic diagram of the flow system is shown in figure 4. The unpressurized approach is necessary when pressurization would interfere with the constituent. However, special caution must be exercised since small leaks in the sample lines will contaminate the sample with the higher pressure cabin air. The pressurized approach is used for less reactive constituents, particularly when the instrument sensitivity is pressure-dependent and high sensitivity is required.

The air sample enters the system through a 2.5 cm diameter stainless steel external probe. The probe is closed with a motor-driven cap during ground operations and at lower altitudes to prevent contamination (as shown in figure 4). As the aircraft ascends through 6 kilometers, the cap is withdrawn and flow is initiated. Inside the aircraft, the sample velocity is slowed as it passes through an expansion section and the flow is divided into three ducts.

Approximately 1000 std liters per minute at 12 Km or 3000 std liters per minute at 6 Km flows through a unit of several individual filters which can be sequently exposed to collect particulate samples for subsequent laboratory compoisiton analysis. A prototype filter assembly being developed for airliner operations is shown in Figure 5. The assembly using only 1 filter unit is being tested in a wing pod mounted on a NASA-Lewis F106 aircraft. Each filter element is enclosed within a cartridge to prevent contamination. A special actuator has been developed to insert a filter element, expose it for a period of time, retract it into its cartridge, and index to the next cartridge. The filter tube is purged with ambient air between filters. Between exposure periods and purge cycles when the unit is not in use, it is sealed with isolation valves at each end. Flowrate through the filter unit is measured with a venturi downstream of the filter. Downstream of the venturi unit the flow is discharged overboard through a flush static port in the fuselage. Laboratory techniques such as neutron activation and wet chemistry will be used for analysis of deposits on the filters. The flight investigation on the F-106 filter unit is to define the exposure cycle and to develop filter handling and analysis techniques.

A smaller portion of the inlet sample is ducted to the particle and nuclei counters. A pump downstream of the particle counter sensor maintains an unpressurized sample flow through the instrument and a mass flow meter is used to measure the flowrate. Sample pressurization would result in particle losses in the pump. A special pressure controller is being developed to supply the nuclei counter. This controller uses filtered cabin air to raise the sample pressure to an acceptable level for the operation of the condensation type nuclei counter. Both of these instruments are located as close as possible to the inlet probe to avoid impaction and electrostatic particle losses in the sample line.

The remainder of the sample flow is ducted through a separate line to the gas analyzers. The oxides of nitrogen instrument draws an unpressurized sample flow since the chemiluminescent detector is more sensitive at low pressures. Sample flow to the balance of the instruments is pressurized with

Figure 2. - Full scale mockup of area selected in 747 for GASP measuring equipment.

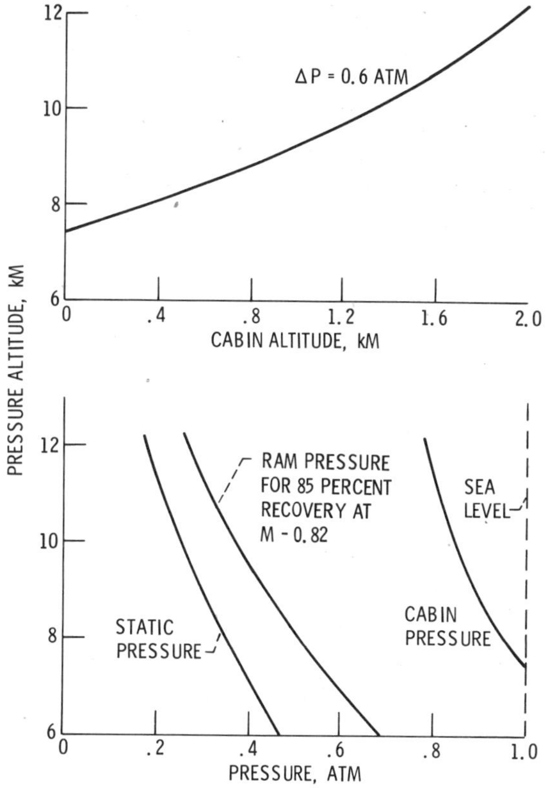

Figure 3. - Pressure environment for air sampling system installed in 747 aircraft.

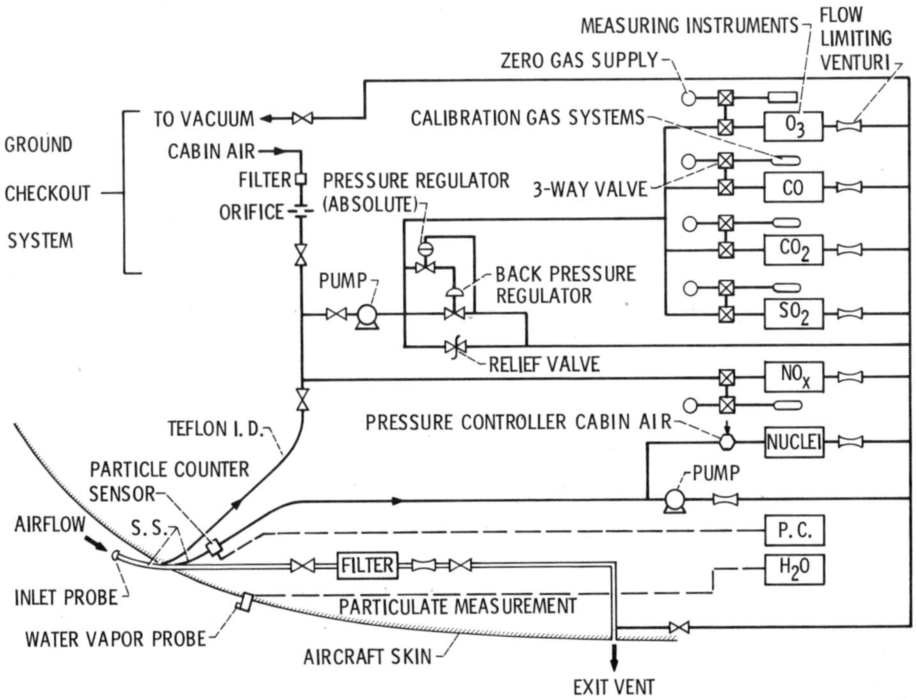

Figure 4. - Basic 747 air sample flow system for GASP

a single-stage diaphragm pump. A pressure regulation system described in ref. 5 is used to maintain a pressure of one atmosphere at the inlet manifold which supplies the various instruments. A backpressure regulator controls the inlet manifold pressure. An absolute pressure regulator which contains a sealed, evacuated bellows reference cell supplies a constant reference pressure to the dome of the backpressure regulator. Downstream of each instrument, a choked venturi establishes the instrument flowrate. Thus, the inlet manifold sample pressure is independent of pressure altitude and cabin altitude but the instrument flows are not required to pass through a regulator upstream of the instruments. The discharge flows from all the instruments are collected in a common line and exhausted overboard through the vent port. A prototype version of this pressurization system was successfully flight tested on the NASA CV-990 during three flight periods in 1972[5].

An important consideration in the design of the flow system is the component material selection. Some reactive constituents such as ozone are compatible with only a few materials and an improper choice can result in an excessive reduction in the constituent concentration[6]. For this reason, the sample lines leading from the inlet probe to the pressurization pump and all the components in the pressurized portion of the system upstream of the instruments are fabricated from Teflon or are Teflon-coated on the inside. Ozone destruction can be minimized by minimizing sample residence time, keeping the system clean, and conditioning the system with high ozone concentrations.

The schematic diagram in figure 4 indicates that the sensor for the water vapor measurement is not located in a sample flow line. Experimental work has shown that surface effects inherent with in-line installations cause excessive interference with the measurement. As a result, the sensor is located in a modified total air temperature probe mounted externally near the inlet probe. The probe is specially designed to prevent water droplets and particles from impacting the sensing elements.

Wherever possible and necessary to assure reliable data, a means of in-flight calibration is planned as shown in figure 4. The ozone instrument inlet has a bypass line containing an ozone scrubber for producing a zero gas and an ozone generator for a span check. The carbon monoxide analyzer generates a zero gas internally. Three-way valves are located in the inlet lines of most of these gas analyzers to input calibration and zero gases. A gas scrubber will be used to provide a zero gas and gas bottles or permeation tubes may be carried as span gas sources, depending on safety considerations.

A provision is also included for ground checks of the sample flow system and instruments. A vacuum pump is carried out to the aircraft and connected to the exhaust port. The ground check inlet valve is opened in lieu of the inlet probe. A limiting orifice in the ground check line restricts the inlet flow, simulating an altitude condition, and the filter prevents system contamination.

Some instrumentation is included to verify that the flow system is functioning properly. An absolute pressure transducer (strain-gage type) will measure the inlet manifold pressure. These pressure data will be required for several instruments whose calibrations are pressure-dependent. A thermistor will be used to measure the inlet manifold sample temperature. Individual instrument flows will be determined by measuring the pressure between a fixed orifice and the choked venturi downstream of each instrument.

An additional instrument which will not be included in the early systems but may be added as it becomes available is a cloud detector. Data on the presence of clouds is helpful in identifying local meteorology and also in interpreting data from the particle counter. Entrained water droplets can interfere with the particle count data. The water vapor sensor should provide some information; but an infrared radiometer cloud detector will eventually be included to give a positive indication.

DATA MANAGEMENT AND CONTROL SYSTEM

One of the constraints on the GASP system imposed by the airlines is that the system must not interfere with or place any additional duties on the flight crew, which leads to the requirement for automatic unattended operation. A schematic diagram of the data acquisition, management, and control system which provides this capability is shown in figure 6.

The heart of the system is the data management and control unit (DMCU) which is specifically designed for the GASP system. It contains the programming necessary to automatically direct the activity of the system. However, a manual override is provided for operation during ground checkout and flight validation tests. During manual operation, the system is directed and monitored from a control and display panel located in the cockpit during the flight validation tests or in later installations located in a ground checkout console.

There are essentially three modes of operation: standby; calibration; and data. During standby operation, power is available and portions of the system are warming up; however, the sample flow system is not operating and data are not recorded. The system is in the standby mode at least one hour before initial calibration begins. The calibration mode of operation is cyclical. It consists of 5 minutes each of (1) zero gas and outside air purge, and (2) span gas flows. The system performs a calibration cycle as frequently as determined necessary which could be at the beginning and end of the data portion of each flight as well as periodically during the data portion. The calibration data are recorded and used in data reduction. The system begins the data mode of operation when the aircraft climbs through 6 kilometers and after the initial calibration cycle is completed. Data are recorded at 10 minute intervals (except when necessary during calibration cycles) until the aircraft descends through 6 kilometers.

A number of programmable functions are available for automatic control. These include the timing of the calibration and data modes and sequences and of data acquisition during these modes. Altitudes for inlet probe opening and closing and initiation of filter exposure are programmable as well as the filter exposure interval.

In addition to its other control functions, the DMCU also manages the data flow between the Flight Data Acquisition Unit (FDAU) and the data recorder. The FDAU is essentially a standard data handling and conditioning unit used extensively on late model airliners. It accepts the output signals from the air sampling instruments, the flow system instruments, and the Central Air Data Computer (CADC). On command from the DMCU, it supplies these data to the recorder. The FDAU cannot couple directly to the Inertial Navigation System (INS) input but requires an INS interface unit and conditioning equipment in the DMCU. The INS data contains latitude, longitude, windspeed, and wind angle. The FDAU can couple directly to the CADC to receive altitude, airspeed, and static air temperature data. The FDAU is capable of processing d.c. analog, digital, and synchro signals and converting the analog and synchro signals to digital signals.

The data recorder is also a standard cassette-loaded serial-digital tape recorder used on today's airliners. The recorder capacity is established on the basis of a 14-day mission with system expansion to include up to 10 air sampling instruments. The use of this recorder requires a ground-based transcriber to translate the serial digital data tape into a computer-compatible digital tape.

INSTRUMENT DESCRIPTION

The instruments used for the measurements of the atmospheric constituents given in Table II are either being modified or are under serious consideration for eventual inclusion in airline 747 GASP systems. The first four instruments have been selected to be included in the first two GASP systems. A brief description of the operating principle and unique features of each of these four instruments is given below.

Ultraviolet Absorption Ozone Monitor

This instrument alternately passes the sample gas and then an ozone-free zero gas (obtained by passing sample gas through an ozone destruction filter) through a 71 cm long tube and measures the difference in intensity of an ultraviolet beam traversing the same path length. The difference is converted into ozone mixing ratio. The instrument employs a reference system which compensates for variations in the optical components, interfering gases, and variations in the ultraviolet source[7].

Aluminum Oxide Hygrometer

This instrument consists of two parts: a sensing element and an electronics package. The sensing element is a small strip of aluminum which is anodized by a special process to provide a porous oxide layer. A very thin coating of gold is evaporated over this structure. The aluminum base and the gold layer form the two electrodes of a capacitor. The amount of water adsorbed on the oxide layer determines the conductivity of the pore wall which becomes a measure of the water vapor[8].

Fluorescent Carbon Monoxide Monitor

This instrument is a single-beam non-dispersive infrared absorption analyzer[9]. A source alternately irradiates two gas cells mounted on a rotating wheel. Each gas chamber contains a carbon monoxide gas mixture which then fluoresces into the sample tube and the transmitted beam intensity is measured at the opposite end of the tube. One of the gas cells contains a common isotope of CO while the other contains a rare isotope. The ratio of transmitted intensities can be converted to CO mixing ratio. This instrument has a particularly stable span concentration although it is sensitive to changes in the ambient temperature. A heated Hopkalite filter is included in the instrument to produce a zero gas for in-flight calibration.

Light-scattering Particle Counter

This instrument consists of two parts: (1) a sensing package which handles the sample flow and contains the light source, optics, and a photomultiplier tube, and (2) an electronics package which includes pulse height discriminating and counting circuits(10). Each particle larger than 0.5 microns scatters light from the source as it passes through the sensor which is focused to pick up the forward scattered light on the photomultiplier tube. The height of the resulting pulse is porportional to the particle diameter and the pulse train from the tube is discriminated and counted to yield a size distribution.

Instrument Modifications

Since most of the instruments selected for the GASP system have not been designed for operation in aircraft, or remote or automatic control, a number of modifications are required to adapt the instruments to the system and to the environment.

The first requirement is that the instruments must be repackaged for mounting in the aircraft. A number of considerations were involved in selecting a package design. The case must be capable of supporting instruments weighing up to 50 lbs. at the specified shock and vibration levels. The cases should be somewhat uniform and easily replaceable. In particular, it is desirable that the case design be familiar to the airlines. The standardized "ATR" case and the support structure which are used for much of the avionics on commercial jetliners were selected. The ATR case and the supporting rack have been standardized by the Aeronautical Radio Incorporation and they are available in a variety of dimensions. Photographs of a typical ATR case and its insertion in a hold-down rack are shown in figure 7. As is shown the ATR case is easily handled and can be quickly removed and replaced. Electronics for input and output are provided by a connector on the rear of the case. Plumbing connectors for sample flow, a circuit breaker, and a rotameter to indicate instrument flow will be installed on the front face of the ATR case.

The electrical power available on the Boeing 747 resulted in modifications of most of the instruments. A 400 Hertz, 115 volt system is the primary power supply. For most instruments, only minor modifications were necessary although most motors had to be replaced.

The requirement for remote control resulted in only a few modifications to several instruments. Several solenoid valves and relays were installed for flow and instrument control. A thermistor was installed in each instrument case to protect against overheating.

Two resistors were installed in each instrument and their values are recorded along with sampling data on the data tape for identification purposes. This permits individual calibration curves for various models of a given instrument to be stored in the data reduction program. When an instrument consists of two or more separate packages, a corresponding number of identification channels will be established and the performance record of each unit will be traceable.

TESTING AND CERTIFICATION REQUIREMENTS

The feasibility studies determined that addition of the GASP system to 747 airline aircraft presented a substantial modification to the aircraft and that a Supplemental Type Certificate (STC) from the FAA would be required. This certification of the GASP system will involve several elements. The system design and operation must be thoroughly documented, including a detailed failure analysis. Then ground and flight test programs must be developed and performed. The STC may then be granted on the basis of the documentation, an inspection of the system installation, and successful flight test results.

The ground tests will consist of a series of environmental tests followed by a series of system tests. Detailed environmental test procedures for electrical and electronic equipment intended for commercial jet aircraft have been established by the Radio Technical Commission for Aeronautics[11]. The Boeing Company has also developed environmental test specifications for the Model 747 airplane. The following group of environmental tests are pertinent to the GASP system:

1. Temperature and Altitude. This series of tests includes hot and cold soaks, operating temperature extremes, temperature variations, pressure variations, and decompression.

2. Vibration. This test procedure involves three types of tests; a preliminary sinusoidal scan of the vibration envelope, a period of dwell on the four worst resonances located during the scan, and sinusoidal sweeps through the envelope.

3. Shock. Two tests are required; an operating shock test after which the equipment should perform within specifications, and then a crash shock test with dummy components to verify the mechanical integrity of the case.

4. Humidity. High relative humidities are maintained with the highest operating temperature for several days.

5. Power Input. The power supply frequency and voltage is varied over specified ranges, then the instrument response to voltage transients is examined.

6. Electromagnetic Compatibility. The equipment is examined for both radiated and conducted susceptibility and interference.

The GASP system equipment is in an unusual aircraft operational situation insofar as no aircraft systems rely on its operation. A malfunction or failure of a GASP instrument will have no effect on aircraft operation or safety, and can only result in erroneous sample data. Viewed in this light, the only significant environmental tests are the shock and vibration tests which show that the equipment will not come loose, and the electromagnetic interference tests which verify that the system will not affect the aircraft instruments. However, since NASA is interested in retrieving valid data and minimizing loss of data due to instrument failure between check periods, one of each of the instruments will be subjected to the extent possible the full complement of environmental tests.

Following the environmental tests which will primarily involve components, the system will be assembled and checked through its various modes of operation. Various failures will be simulated to ensure appropriate system response.

Three sets of flight tests will be performed. A series of flights on the NASA CV-990 based at Ames Research Center will demonstrate system operation in a flight environment and system compatibility with other aircraft systems. Also, an attempt will be made to obtain comparison sampling data from other sensors operating from other platforms. Next, the system will be installed, inspected by FAA, and flight tested on a Boeing 747 airplane. No more than one flight test should be required. Finally, the system will be closely observed for a three-month validation period during which the aircraft will be in routine commercial airline service. NASA will follow the first GASP system closely for this initial three-month validation period. During this period, an observer will accompany the system on most routine commercial flights and monitor its operation with the control and display panel in the cockpit. The output data will also be scrutinized on a daily basis.

CONCLUDING REMARKS

A program to determine the temporal and spatial distribution of minor atmospheric constituents associated with gaseous and particulate pollutants in the upper troposphere and lower stratosphere (6 to 12 Km) is currently being implemented by NASA. Continuous measurements will be obtained from several instrumented 747 airliners in commercial service. Those constituents affecting air quality that are related to aircraft engine exhaust emissions will be measured.

Feasibility studies established certain constraints imposed by airline operations on a non-interference basis. The measurement system requires no duties of the air crew, no use of revenue space, and no servicing or maintenance that would interfere with routine operations. Because a significant modification to the aircraft is required, an FAA Supplemental Type Certificate (STC) of airworthiness must be obtained. A completely automatic system to meet these requirements is being designed and fabricated. Instruments were selected on the basis of suitability for this application and are being modified and tested to comply with FAA and airline requirements.

Considerable effort is required to obtain data automatically on commercial airliners. Instruments designed for use on the ground must be modified and systems must be built to collect and control outside ambient air. Nevertheless the concept will provide large quantities of data on a world-wide basis much less expensively than could be obtained from dedicated aircraft.

REFERENCES

(1) Johnston, H., "Reduction of Stratospheric Ozone by Nitrogen Oxide Catalysts from Supersonic Transport Exhaust," Science, Vol. 173, No. 3996, Aug. 6, 1971, pp. 517-522.

(2) Rudey, Richard A., and Perkins, Porter J.: "Measurement of High-Altitude Air Quality Using Aircraft," Paper 73-517, June 1973, AIAA New York, New York.

(3) Steinberg, Robert: "Role of Commercial Aircraft in Global Monitoring Systems," Science Vol. 180, No. 4084, April 27, 1973. pp. 375-380.

(4) Stepka, Francis S., and Perkins, Porter J.: "Use of Commercial Air Transports for Atmospheric Constituent Sampling," Proposed NASA TN.

(5) Reck, Gregory M., Briehl, Daniel, Perkins, Porter J.: "Flight Tests of a System to Measure Minor Atmospheric Constituents from an Aircraft," Proposed NASA TN.

(6) Altshuller, A. P., and Wartburg, A. F., "The Interaction of Ozone with Plastic and Metallic Materials in a Dynamic Flow System," International Journal of Air and Water Pollution, Vol. 4, Nos. 1/2, 1961, pp. 70-78.

(7) Bowman, L. D., and Horak, R. F., "A Continuous Ultraviolet Absorption Ozone Photometer, "Instrument Society of America, AID 72430, 1972, pp. 103-108.

(8) Goodman, P. and Chleck, D.: "Calibration of the Panametrics Aluminum Oxide Hygrometer." Analysis Instrumentation. (7) 1969, p. 233.

(9) Link, W. T., McClatchie, E. A. Watson, D. A., and Compher, A. B.: "A Fluorescent Source NDIR Carbon Monoxide Analyzer," Paper 71-1047, Nov. 1971, AIAA, New York, New York.

(10) Joenicke, R., "The Optical Particle Counter: Cross-Sensitivity and Coincidence," Max Planck - Institut Fur Chemie, Mainz, Germany and National Center for Atmospheric Research, Boulder, Colorado, 1973.

(11) RTCA SC-112, "Environmental Conditions and Test Procedures for Airborne Electronic/Electrical Equipment and Instruments," Document No. DO-138, Radio Technical Commission for Aeronautics, Washington, D. C. June 1968.

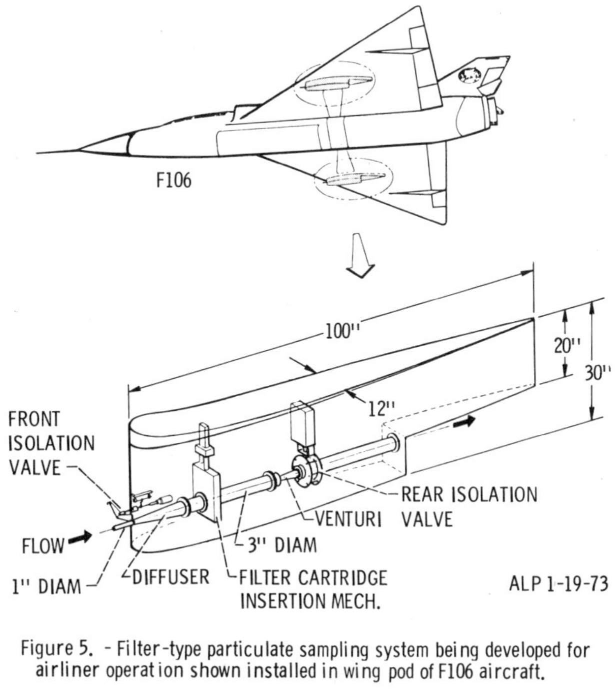

Figure 5. - Filter-type particulate sampling system being developed for airliner operation shown installed in wing pod of F106 aircraft.

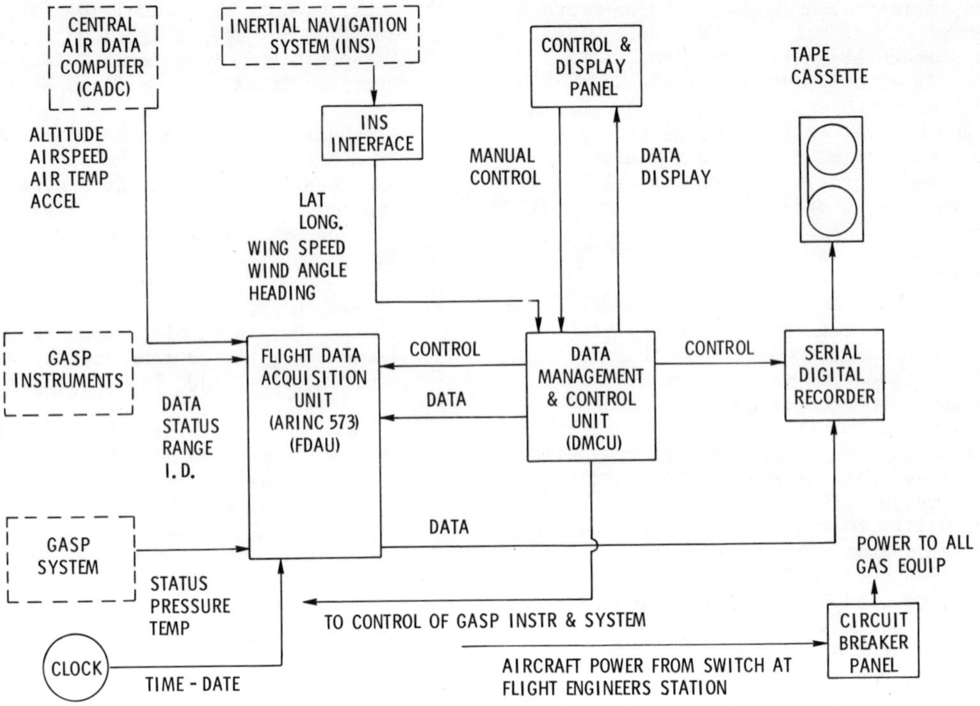

Figure 6. - Basic 747 air sample data acquisition, management, and control systems for GASP.

(a) CASE PARTIALLY INSERTED IN RACK.

(b) CASE SECURED IN RACK.

Figure 7. - Typical airline instrument case and hold-down rack used in repackaging atmospheric constituent monitoring instruments.

© 1973, ISA JSP 6713

GLOBAL MONITORING OF POLLUTION ON THE SURFACE OF THE EARTH

G. B. Morgan
E. W. Bretthauer
S. H. Melfi
U. S. Environmental Protection Agency
National Environmental Research Center
Las Vegas, Nevada

ABSTRACT

The need to develop an adequate data base for the scientific management and improvement of our natural resources on a global basis has become apparent over the past two decades. During the middle 1940s, air pollution was regarded as a local nuisance. Public indignation was focused on dirty shirts rather than dirty lungs. Global monitoring of the environment of the planet has become more important as man increasingly exerts his influence on our biosphere. There is an urgent need to assess ecological problems which now exist and to be able to predict potential dangers to the ecosystem.

Global monitoring must produce detailed knowledge of the composition, change, and trends of pollutants in the atmosphere, water, and soil. It is also necessary to understand the kinetics of pollutants in the global environment, the chemical and physical changes that they undergo, and their effects on all living things. Man now suspects that a minute change of pollutants in the oceans may greatly affect his very existence.

The objective of a global monitoring system is to provide information for effective management of environmental quality and to assist in pollution control. In the effective management program, data are needed to determine the current state of pollution in air, water, land and biological materials; to establish standards for pollutant concentrations, to determine long- and short-term trends in environmental quality, to determine the extent to which man, animals, vegetation and all other elements of the environment are affected by present and potential pollutants, and to develop strategies for the control of pollution. The existence of global pollution does not imply the necessity of a global solution. The main sources of pollution are the activities of man and can be quite effectively controlled at the source. The control will have to be implemented at national or regional levels and through international cooperation.

Global monitoring from the surface of the earth is accomplished by measuring pollutants with contact sensors, remote sensors, and by collecting environmental samples which are returned to a laboratory for analysis. Environmental samples as used here include gases; atmospheric particulates; precipitation; and soil, water, and biological samples.

This paper describes some of the systems most commonly used, the analytical procedures, and representative data.

INTRODUCTION

Man's technological advances and increasing population have caused a significant impact on our environment. We are now faced with the arduous task of describing on a global basis those factors that impinge on environmental quality. This information is needed by both the researcher and the official who must not only assess environmental quality but also control emissions into all aspects of the environment.

Environmental monitoring may be defined as the systematic collection and evaluation of physical, chemical, biological, and related data pertaining to both the environmental quality and pollutant discharge into all media. The main elements of such a program are:

(a) Design of suitable monitoring systems;
(b) Instrumentation and methodology standardization and quality control;
(c) Field sampling and measurements of environmental quality, emissions and effluents;
(d) Laboratory analysis and evaluation of field samples;
(e) Information synthesis including operation of technical information systems, data analysis and preparation of reports.

Global monitoring is needed for the following reasons:

(a) To observe long-term effects of pollutants on climate;
(b) To determine background concentrations and trends of pollutants for which standards are being considered;
(c) To quantify global transport of pollutants;
(d) To understand the chemical and physical processes of the global environment.

Satisfying these needs will provide the comprehensive data and information required to assess the state of the atmosphere, oceans, marine resources, inland surface waters, and land. A pollutant-oriented integrated global monitoring system is also necessary to determine if sensitive human populations, environmental receptors or environmental conditions are exposed to concentrations of a pollutant that are sufficiently high or of sufficient duration to induce adverse effects. In addition to the system design there are three areas upon which the success of a global monitoring effort is directly dependent, namely, comparable measurement methodology, comparable data synthesis systems, and a strong quality control program.

Globally the time scale for which changes must be monitored is extremely long and the variables are numerous, but the data are of great importance in determining the impact of global changes on human lives and the adjustments that man must make to survive in his environment.

In order to establish and operate a successful global monitoring system it is necessary to agree upon certain institutional principles such as:

(a) Establishment of specific objectives for the system;
(b) Coordination through an international agency;
(c) Maximum utilization of national monitoring programs;
(d) Establishment of a mechanism for the routine exchange of information.
(e) The design of monitoring systems to meet clearly defined goals and objectives;
(f) Agreement on national participation and the extent of such participation;
(g) Establishment of pollutant priorities.

GLOBAL ENVIRONMENTAL QUALITY

The chemical composition of the environment is always changing. This change involves the introduction of a number of new substances from industrial and other man-made sources and a redistribution of naturally occurring substances. A summary of the global concentrations and sources of some selected atmospheric pollutants is shown in Table 1. Specific environmental problem areas are outlined as follows.

Atmosphere

The carbon dioxide content of the lower atmosphere is increasing and there is concern that if this trend continues the heat balance of the earth could be affected. Results from the Mauna Loa laboratory and other sources indicate the level has risen from a concentration of about 290 ppm in the pre-1930's to a level of approximately 320 ppm in 1968 and is expected to reach a level of 375-390 ppm by the year 2000.

The concentration of carbon monoxide varies with season, altitude and latitude. Normally the concentration at remote "ground Level" ranges from 0.05 to 0.20 ppm--much below the level at which health effects have been observed. In addition to incomplete combustion the oceans are a source of carbon monoxide. Soil bacteria and oxidation in the stratosphere are two of the major sinks for this pollutant.

The more important sulfur compounds on a global basis are H_2S, SO_2 and particulate sulfate.(1)

Superior numbers refer to similarly-numbered references at the end of this paper.

The H_2S is primarily derived from the decay of organic matter, whereas SO_2 results from the combustion of sulfur containing fuels. Both compounds are oxidized to SO_3 which eventually becomes sulfuric acid or its salts. Both SO_2 and its oxidation products have been found to have effects on plants and animals. The oxidation products can also form aerosols which can affect the global heat balance.

Nitrogen oxides occur on a global basis at levels on the order of 0.001-0.003 ppm.(2) In the presence of ozone, hydrocarbons, and water vapor they may form nitric acid or its salts which occur in the aerosol form.

The fine particulate portion of the background atmosphere constitutes a very important group of air pollutants. In the remote areas these particles range in size from 0.001 to 100 μm with the majority by both mass and numbers falling within the range of 0.05 to 8 μm. Particles in the range of 0.05-1 μm result from condensation of combustion products, oxidation of gases such as SO_2 and NO_2, and from photochemical reactions involving hydrocarbons.

The particulate portion of the atmosphere also contains a number of specific pollutants such as DDT, PCB, asbestos, Pb, Hg, and other heavy metals.

Hydrosphere

The hydrosphere receives pollution from both air and land masses. Many pollutants such as organochlorines, heavy metals, and other toxic compounds are emitted into the atmosphere and are present in estuaries. These pollutants eventually find their way to man's food supply in bays and oceans. This stresses the need for a pollutant oriented integrated approach to monitoring the global environment so that one can determine whether changes in background concentrations are significant or merely represent a change in distribution within the biosphere.

Precipitation

Over the past 15 years precipitation has shown a trend toward increasing acidity which primarily results from SO_2 and its oxidation products. Europe has for almost 20 years reported rain in the range of pH 4-5. The United States is also experiencing acid precipitation.(3) Figure I shows that the northeast section of the United States is quite acid. Precipitation of this acidity can do considerable damage to structures. Even more important, it can solubilize calcium and magnesium salts in the weakly buffered glaciated soils of the Northeast. Surprisingly, however, the lowest known pH readings were recorded in the West. In 1960 at Pocatello, Idaho, rain with a pH 3.2 was recorded, and in 1964 the lowest reading was 2.5 at Glasgow, Montana. In Philadelphia the overall pH range for 1960 was from 3.7-4.4.

Heavy metals have also been found in precipitation. Lazrus et al. reported concentrations of Pb, Mn,

Table 1. Global Background Concentrations and Sources of Selected Atmospheric Pollutants

Pollutant	Major Sources: Man Made	Major Sources: Natural	Concentration
SO_2	Fossil fuel combustion	Volcanoes	0.2-0.4 ppm
NO_2	Combustion	Bacterial action	0.001-0.003 ppm
Fine suspended particulates	Combustion	Volcanoes	1-10 $\mu g/m^3$
Pb	Fuel additives, Smelters, Combustion of fossil fuel	Negligible	0.1 $\mu g/m^3$
CO	Combustion, Auto exhaust	Forest fires	0.1 ppm
CO_2	Combustion	Biological decay	320 ppm
SO_4	Oxidation of SO_2	Volcanoes, Ocean spray	0.1-0.3 $\mu g/m^3$
NO_3	Oxidation of NO_2	Negligible	0.02-0.1 $\mu g/m^3$
Hydrocarbons:	Combustion, Exhaust, Chemical processes	Biological processes	
non-CH_4			0.001-0.001 ppb
CH_4			1-1.5 ppm

Figure I

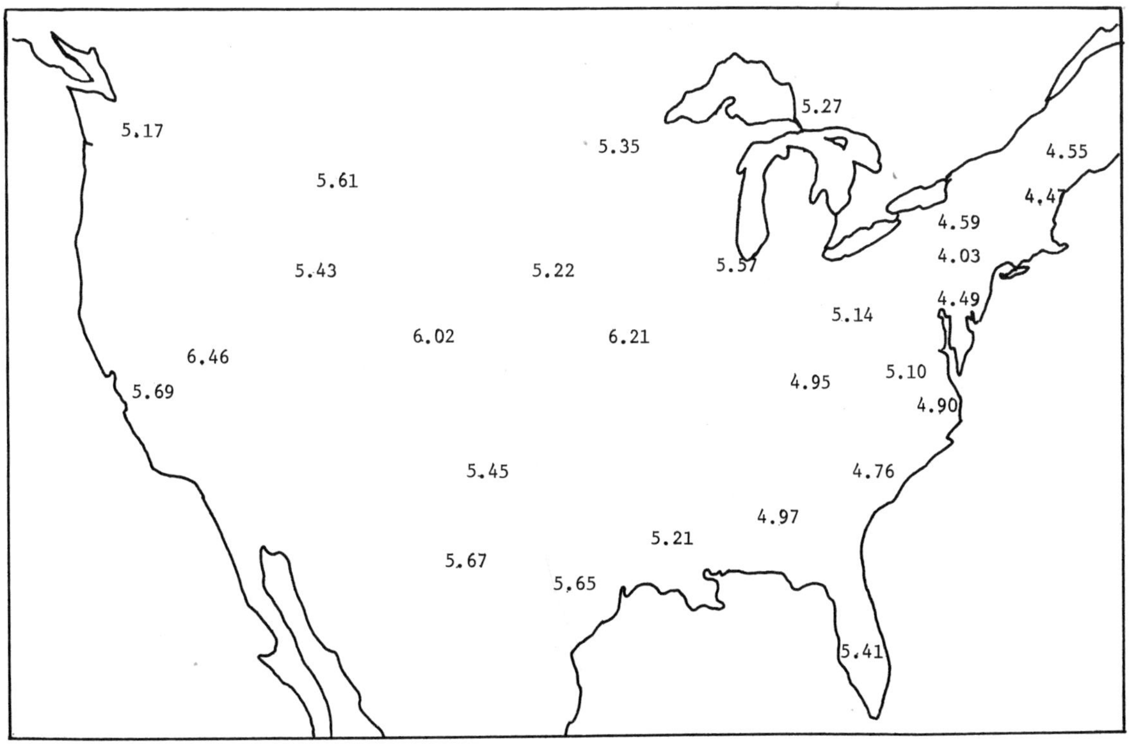

NATIONWIDE YEARLY AVERAGES OF pH OF PRECIPITATION FOR THE YEAR 1960

Cu, Zn, and Ni in precipitation of remote background areas of 0.034, 0.012, 0.021, 0.107, and 0.0043 ppm respectively.[4]

Biosphere

The use of biological organisms for monitoring has been effective in such areas as sentinel warning devices, biological assay of contamination, and indicators of specific compounds, other organisms, or environmental conditions or changes. Air pollutants in remote background areas have been shown to damage plants such as the white pine and ponderosa pine. Table 2 shows the concentration of selected heavy metals which have been observed in some biological organisms.[5]

There are also many persistent biocides that accumulate in living organisms of the food chain. Some of the more common chlorinated hydrocarbons are DDT (including DDE and DDD), aldrin, chlordane, dieldrin, endrin, lindane, and toxaphene. These pesticides have been shown to be concentrated as much as 12,500-fold in bluegill sunfish and 85,000-fold in largemouth bass, above the concentration in the water from which they were taken.

SCOPE AND PRIORITIES

A pollutant-oriented integrated global monitoring network should be established to document and follow large scale changes, particularly of those pollutants that are readily dispersed by mixing in the atmosphere or in the oceans. Pollutants to be considered should be common to all stations and, in addition, should have significance to each national program which operates the station.

Although it is accepted that priorities may vary from one country to another, the selection of pollutants to be considered must be based on an agreed set of criteria. All the considerations noted below should be taken into account when establishing priorities:

(a) The effects of a specific pollutant or group of pollutants on weather or long term climate modification;
(b) Severity of adverse effects on the population, for example, irreversible or chronic effects. Those which have adverse genetic implications, and those which are embryotoxic and teratogenic, are considered to be of particular importance;
(c) Persistence of the agent in the environment, resistance to environmental degradation and accumulation in man or in the food chain;
(d) Conversion in the environment to substances that may be more toxic than the parent compound;
(e) Ubiquity and abundance of the agents in the environment;
(f) Size and type of population exposed, frequency and magnitude of exposure;
(g) Feasibility of measurement in the various media.

Using the above described criteria, a global monitoring system should focus on the pollutants listed in Table 3.

The major route for global transport of pollutants is through the atmosphere. Through mechanisms such as fallout, rainout, and washout these pollutants may be transferred to other portions of the environment, for example, soil, surface waters, plants and animals. Global monitoring stations should not only measure pollutants present in the atmosphere but those present in precipitation, surface waters, and plants and animals indigenous to the area. As a logical starting point major emphasis should be placed on monitoring the atmosphere and precipitation. After this portion of the program is underway monitoring of the other media will furnish information on transfer rates.

MONITORING STATIONS - LOCATION AND DESIGN

The location and number of monitoring stations should be selected to provide adequate information to accurately determine any change in the global inventory of the selected pollutants. For global atmospheric monitoring eleven or twelve monitoring stations will be required to meet this objective. The stations will be generally distributed around the globe with 7 or 8 stations located in the northern hemisphere, and the remaining stations located in the southern hemisphere. A greater number are required in the northern hemisphere because of the greater impact on the environment due to man-made sources which are predominantly located above the equator. With this in mind the following locations are recommended: Mauna Loa, Hawaii (existing U.S. monitoring station), Samoa, Arctic, Antarctic, U.S.S.R. (2 or 3), Canada, Central Pacific Ocean, Indian Ocean, North Atlantic Ocean, and South Atlantic Ocean. The criteria for the exact citing of these monitoring stations should be as follows.
(a) The stations should be located in an area where no significant changes in land-use practices are anticipated for at least 50 years within 100 km in all directions from the station.
(b) They should be located away from major population centers, major highways and air routes, preferably on small isolated islands or on mountains above the tree line. They should not be downwind from large sources of pollution.
(c) The site should experience only infrequent effects from local natural phenomena such as volcanic activity, forest fires, and dust and sand storms.
(d) The observing staff should be small in order to minimize the contamination of the local environment by their presence and their living requirements.
(e) All requirements for heating, cooking, etc., should be met by electrical power generated away from the site.
(f) Access to the station should be limited to those whose presence is necessary to the operation of the station. Surface transportation should be by electrically powered vehicle, if at all possible.
Once a site has been proposed after due consideration of these criteria, field tests should be conducted using two adjacent sites about a kilometer apart . If the measured parameters at the two sites are highly correlated, then the area is suitable for a monitoring station.

METHODOLOGY

It is recognized that different countries and

Table 2. Accumulation of Selected Heavy Metals in Biological Organisms

Biological Organism	As	Cd	Hg	Pb
	Concentration in ppm			
Mollusks	--	30	30	4
Bird Feathers	0.5	0.01	40	11
Marine Plants	30	0.4	0.03	8
Marine Fish	11	5	102	10
Crustacea	100	6	2	--
Freshwater Fish	40	20	17	--

Table 3. Priority of Pollutants for a Global Monitoring System

Pollutant and Priority	Media				Adequate Methods Available
	Air	Oceans	Biological	Precipitation	
1. Pb	x		x	x	Yes
2. CO_2	x				Yes
3. Fine Particles	x				Yes
4. SO_2/SO_4	x			x	Lacks Sensitivity
5. Cd	x	x	x	x	Yes
6. Hg	x	x	x	x	Yes
7. DDT(DDD,DDE)	x	x	x	x	Yes
8. NO_2/NO_3	x				Lacks Sensitivity
9. Acidity(pH)				x	Yes
10. Hydrocarbons	x				Yes
11. As	x	x	x	x	Yes

different environmental programs within a country do not employ the same methods for collecting and analyzing environmental pollutants. For the most part, these various methods presently to not produce comparable data, but may, if proper calibration and quality control procedures are utilized, produce relatable data. In order to have a successful global monitoring network, it is necessary, then, for the participating countries to confer and decide upon reference methods by which the various methods within national programs can be intercalibrated. The most effective mechanism by which this can take place is for one of the international agencies such as United Nations Environmental Program to coordinate this activity. Only those reference or comparison methods which have been thoroughly tested and evaluated and which have been internationally accepted should be considered. Examples of reference methods are those published by United States Environmental Protection Agency.

In some cases such as atmospheric particulates, it is strongly recommended that an international sample bank be established so that as new pollutants are recognized they can be analyzed using an aliquot from the sample bank. Also, as nondestructive analytical techniques such as neutron or proton activation and x-ray fluorescence become more developed, it may be possible to re-analyze these samples for an even wider variety of pollutants. As automated analytical methods become of more widespread use, no doubt a larger number of samples can be analyzed at a lower cost.

INFORMATION SYNTHESIS

Information synthesis may be defined as the processing of environmental and related data into interpretative reports useful to the researcher or decision maker. No single information synthesis system will satisfy all requirements of the various national monitoring programs. Different programs have different objectives and have access to different hard and soft ware. It will, therefore, be necessary for member nations in a global monitoring network to produce data in comparable or machine readable format. An example of such a system which is presently used is outlined in the World Health Organization document EP/72.6 entitled, WHO International Air Pollution Monitoring Network Data Users Guide.

QUALITY ASSURANCE

An adequate quality control program is necessary to intercalibrate the various national programs to maintain continuity within a single national program over a long period of years regardless of changes in meteorology or instrumentation. The quality assurance program will provide standard reference methods or methodology, procedures for determining equivalency of the standard reference method to other methods, and the standard reference materials and samples. The program will also establish rigid collaborative testing procedures and provide routine evaluation facilities and staff. In summary, the program will provide all the necessary methods and materials to document the accuracy and precision of all the data collected by the various stations.

IMPLEMENTATION

The implementation of a global monitoring network must be coordinated through an international organization such as the United Nations Environmental Program. Such an organization must utilize the expertise located in the specialty agencies such as the World Health Organization (and its international reference centers), the World Meteorological Organization and the Organization of Economic Cooperation and Development. Through such a mechanism, uniform guidelines covering network development, quality control, implementation of integrated monitoring networks, data acquisition and analysis, performance specifications, and evaluation of instrumentation will be available. In addition, this mechanism would assure the availability of supporting information such as environmental indices, meteorological data, and health indicators for overall environmental appraisal.

As mentioned earlier, analytical instrumentation and methodology for measuring global-level environmental pollutants are necessarily complicated and expensive. A background monitoring station which would quantify global concentrations of suspended particulates, gases such as CO, CO_2, SO_2, NO_2 and O3, precipitation (acidity), meteorological data, and solar radiation, will cost in the neighborhood of $250,000. To provide data on the concentration of selected pollutants in the other media--soil, water, plants, and animals--will require a program whereby samples can be collected, stabilized, and returned to a central laboratory for analysis. In order to properly maintain and operate the station, each station will need some $125,000 - $150,000 annually, which does include amortization of the equipment.

It is necessary, as was mentioned above, to locate these stations in areas away from man's immediate activities, which means that in some cases, they should be located in developing nations. If such a location is critical, support would have to be forthcoming from some international agency. It will be necessary, if such a network is to become useful, that interpretative reports from national background stations be made available to the coordinating agency on a routine basis so that international reports appraising the global state of the environment can be sensitized. The organizational relations and information flow for the global monitoring network is shown in Figure II.

REFERENCES

(1) Air Quality Criteria for Sulfur Oxides, ENVIRONMENTAL PROTECTION AGENCY, Washington, D.C., 1971.

(2) Fischer, W. H. et al., "Estimation of Some Atmospheric Trace Gases in Antartica," ENVIRON. SCI. TECHNOL., 2, 464-466 (1968).

(3) Likens, G. E. et al., "Acid Rain," ENVIRONMENT, 14, 33-40 (1972).

(4) Lazrus, A. L. et al., "Lead and Other Metal Ions in United States Precipitation," ENVIRON. SIC. TECHNOL., 4, 55-56 (1970).

(5) Jenkins, D. W., Biological Monitoring of the Global Chemical Environment, SMITHSONIAN INSTITUTE, Washington, D.C., 1971.

Figure II

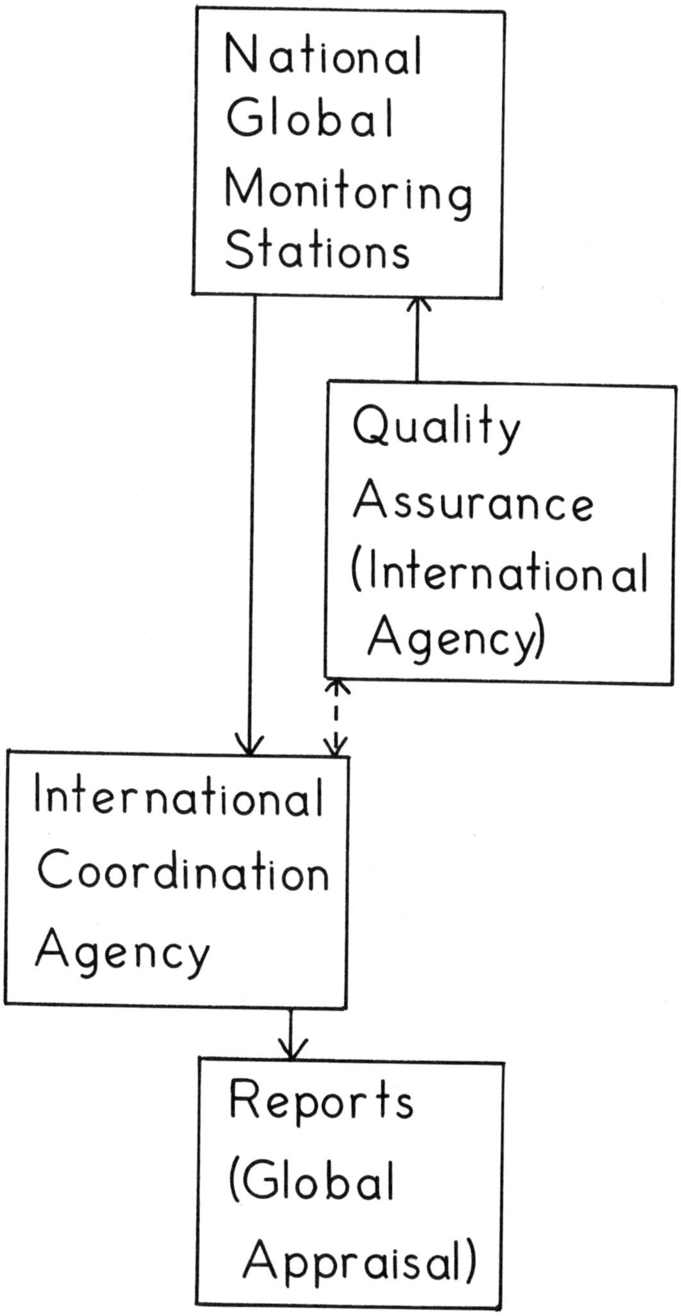

© 1973, ISA JSP 6714

PROBLEMS AND INSTRUMENTS USED IN MEASUREMENT OF MINOR GASEOUS SPECIES IN THE LOWER STRATOSPHERE

Samuel C. Coroniti
U. S. Department of Transportation
Washington, D. C.

and

Ronald J. Massa
DYNATREND INCORPORATED
Burlington, Massachusetts

ABSTRACT

Since 1971 the U. S. Department of Transportation's Climatic Impact Assessment Program (CIAP) has attempted to measure, on a global scale, certain minor gaseous constituents of the lower stratosphere. These chemical species, such as H_2O, NO_x, SO_2, NH_3, CH_4, CO, CO_2 , etc., have particular bearing on the DOT regulatory objectives and are also of great interest in improving our understanding of the natural stratosphere. This paper discusses the results obtained with a number of instrument techniques which have been used in CIAP-sponsored investigations conducted from various balloon and aircraft platforms; and indicates the measurement problems which have been encountered.

INTRODUCTION

Certain minor gaseous and aerosol species in the lower stratosphere play a much greater role in the physical chemistry and photochemistry of this region than their numerical densities might indicate. In attempting to assess the possible environmental effects of aircraft effluents injected into the stratosphere, it is essential to first determine the natural or baseline levels of many of these minor species and the temporal and spatial variability of their baseline densities. The emergence of interest in the effects of aircraft effluents in the stratosphere has lent new impetus to attempts to measure certain of these species. Since 1971, the Department of Transportation's CIAP Program, in cooperation with a number of international research organizations, has been actively engaged in the development of composition data on many atmospheric trace species.

Figure 1, an annotated standard temperature-height profile of the atmosphere indicates the ranges of interest and the altitude regimes which have been probed by experiments related to CIAP. As a rule of thumb, information regarding the density of trace species at 20 and 40 km often characterizes the stratospheric loading for such species. Figure 2 indicates, in an approximate way, our understanding regarding the concentrations of several trace species of interest in determining the interaction of aircraft effluents with the natural stratosphere.

The chemical species listed in Figure 2 are not the only ones of interest in understanding aircraft environmental effects. They are, however, species in which there is significant scientific interest. It is apparent from Figure 2 that a great deal of measurement information is still required - especially to characterize these species on a global scale and to understand their temporal and spatial variations as well as their sources and sinks.

Of the species listed in Figure 2, water vapor (H_2O) and ozone (O_3) have received the greatest attention in previous research efforts. Significant data is available in the literature regarding these important species. Because of this relatively complete data, less emphasis will be placed on these species in this paper. The lack of emphasis is in no way meant to reduce the importance of ozone to the maintenance of life on Earth, nor to minimize the gaps in our understanding regarding the potential production and destruction of ozone through aircraft operations in the stratosphere.[1]

CIAP MEASUREMENTS

Figure 3 indicates the global scale of CIAP and several of the platforms which are used for global stratospheric measurements. It should be noted that a significant fraction of the CIAP measurement effort has been addressed to characterizing trace gaseous and aerosol species in the stratosphere. On-going measurement efforts are currently attempting to fill the gaps in the Table in Figure 2.

The specific instruments which are being used on the CIAP program, at present, are listed in Figure 4.

Ozone and water vapor are measured from a number of different platforms and by a number of different techniques. The results of ozone measurements are generally consistent. The major problems with ozone measurements are

(1) characterizing the considerable natural variability of O_3 and

(2) establishing cause-effect relationships for this variability.

The primary difficulties in water vapor measurements are not new. The relatively small amount of H_2O (1-5 ppmv), low stratospheric temperatures and high probability of platform contamination, make all measurements suspect. Two items are worthy of note. Water vapor measurements under CIAP auspices have generally been made in conjunction with several

"Superior numbers refer to similarly-numbered references at the end of this paper."

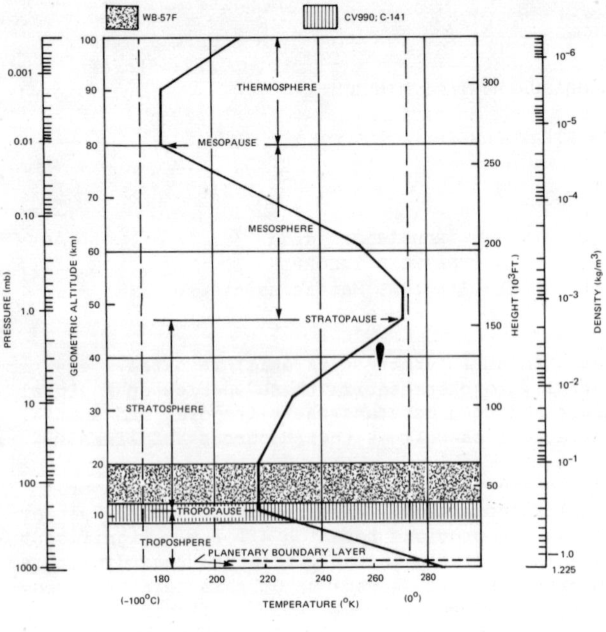

Figure 1 TEMPERATURE-HEIGHT PROFILE OF THE ATMOSPHERE (U.S. Standard Atmosphere, 1962).

CHEMICAL SPECIES	CONCENTRATION (par. /cc) at 20 KMS
CO	No data above 19 KMS.
CH_4	2×10^{12}
H_2O	6×10^{12}
OH, HO_2, H_2O_2	No data
NO, NO_2, NO_3	No data
HNO_3	3 to 8×10^9
O_3	5×10^{12}
Total Density	1.85×10^{18}
PARTICLES	
Dia. >0.3 microns	1
Dia. <0.1 microns	No data

Particles Contain NO_3^-, NH_4^+, SO_4^-, FE, SI, NA, K, CL, BR, MN.

Figure 2 COMPOSITION OF THE STRATOSPHERE – SOME MINOR SPECIES

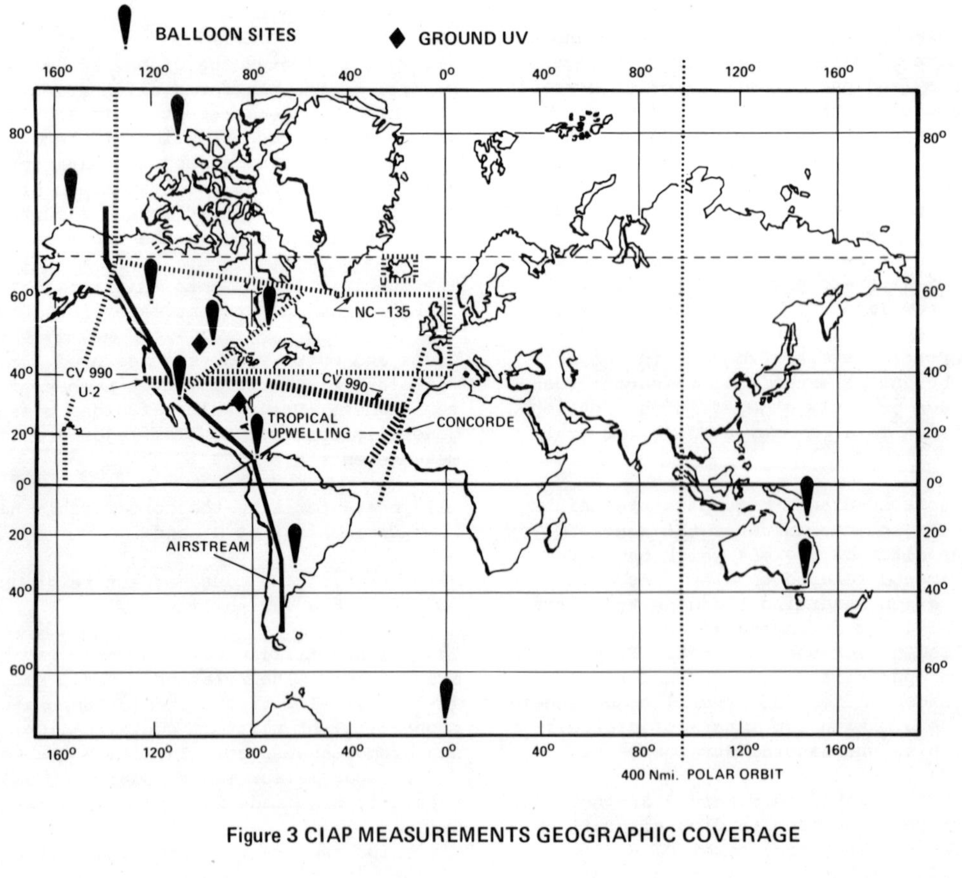

Figure 3 CIAP MEASUREMENTS GEOGRAPHIC COVERAGE

other supporting measurements which will permit the use of supporting data in assessing the validity of H_2O instrument calibrations. For instance, the WB-57F measurements of H_2O involve three (3) separate instruments - an Al_2O_3 hygrometer for in-situ relative measurement (to a laboratory calibration), an IR radiometer for total H_2O burden data (above the aircraft) and a tritium water vapor detector which should make absolute water vapor measurements for the first time.

From the listing in Figure 4, it is apparent that considerable attention has been directed toward the measurement of oxides of nitrogen. This attention is based largely on the important role which the oxides of nitrogen may play in the catalytic photochemistry of ozone at stratospheric altitudes. It is further apparent that a number of different instruments and instrument techniques are being brought to bear on the problem. The first measurements of stratospheric NO have been made in the past year. These measurements, made by groups from JPL, Belgium and France, and York and Utah State University, are preliminary and do not yet include significant information on the temporal and spatial variability of NO or NO_2.

Using a high speed interferometer operating in the IR band from 3 to 7.4 microns, JPL made the first measurements of NO. Their initial measurements were made over Albuquerque, New Mexico, and the second series, along the coast of France. During the latter series, their instrument was installed on the French Concorde. In each instance, the sun was the source of radiation. The measurements were made at low sun angles. Their data indicate NO mixing ratios between 12 and 26 kms altitude.

During the French Concorde flights, JPL also derived concentration values of H_2O, NO_2, N_2O, CO_2, and CH_4 for altitudes of 12.8 to 15.2 kms. Of particular importance are the simultaneous measurements of NO and NO_2 -- two molecules which interact very critically in the chemistry of stratospheric ozone.

A chemiluminescence instrument is used by York University (H. Schiff) and Utah State University (R. Megill). This instrument has been used to obtain vertical profiles on balloons to about 27 kms from a single location. The instrument is currently installed on a WB-57F aircraft and is providing data at approximately 19 kms altitude over considerable N-S traverses. Using an IR diffraction spectrometer operating in the 5.2 μm band, Ackerman et al[2] measured the vertical distribution (from ~15 to 40 km) of the concentration of NO in the stratosphere. The measurements were made over France.

From the scant NO data obtained to date it appears that NO concentrations in the stratosphere are of the order of 1 ppbv and difficult to measure reliably.

Other techniques are being used and other trace species are being measured on CIAP. Figure 4 indicates these as well. Flights anticipated in the Fall of 1973 of the IITRI cryogenic sampler should increase our understanding of some of the more reactive species. The Murcray and Farmer spectrometers, both flying on stratospheric aircraft should also yield substantial data on the spatial concentration of numerous trace species.

Particulates and aerosols in the stratosphere still present several problems. Very little is known of the nature, density, or variability of particles and aerosols $< 0.1\ \mu$m in diameter. CIAP has developed an aircraft-borne Aitken nuclei detector to count small particle densities at approximately 20 km.

The large particles and/or aerosols from 0.3 μm diameter to several microns in diameter have been the subject of considerable attention within CIAP. These particles are important for several reasons.

(1) they are believed to have substantial stratospheric residence times (~1 year),

(2) their surfaces may react catalytically to influence the chemistry of the stratosphere, and

(3) they can interact with the Earth's radiation balance.

Several techniques, both remote and in-situ, have been used to obtain data on the composition, density and variabilities of such particles. In a recent comparative experiment aircraft-borne filters, a balloon-borne in-situ optical aerosol detectors, and ground LIDAR were cross-calibrated[3] to enhance the utility of each technique.

In general, particulate densities of 1-5 particles/cc are observed in the stratosphere. Vertical stratification begins with an enhanced layer (~10 particles/cc) at the tropopause, a relatively clean layer(12-15 km)above the tropopause (~0.2-1.0 particles/cc) followed by a peak (~2-3 particles/cc) at 16-21 km and ultimately falling off in density to the background level at 30 km. The filter experiments have indicated the wide variety of composition indicated in Figure 2. There appears to be some secular variation in aerosol abundance with time but too little data exists to establish any significant systematic temporal variation on a global scale.

ACKNOWLEDGEMENT

The work reported herein has been supported by the Climatic Impact Assessment Program, Office of the Secretary, United States Department of Transportation.

REFERENCES

(1) A comprehensive review of O_3 and H_2O data appears in M. Wu,"Observation and Analysis of Trace Constituents in the Stratosphere" Annual Report Contract No. DOT-OS-20217, July 1973.

(2) M. Ackerman, J.C. Fontanella, D. Frimount, A. Girard, N. Louisnard, C. Muller and D.

MEASUREMENT	INSTRUMENT NAME	PLATFORMS USED ON	MEASURES	REMARKS	AGENCY	PRINCIPAL INVESTIGATOR
Ozone	Ozone Detector	Balloon	Ozone concentration	Chemiluminescent. Partial ozone pressure by Rhodamin B excitation. Time constant < 1 sec.	ONR/U. Minn.	Kroenig
	Ozone Detector	Balloon	Ozone concentration	Electro chemical type by Scientific Pump Co. Partial ozone pressure by Potassium iodide oxidation. Time constant ~90 sec.	ONR/U. Minn.	Kroenig
	Ozone Monitor	WB-57F	Ozone concentration	Aircraft powered "automatic" Komhyr cell – Modified MkIII Probe.	NOAA	Barrett
	Ozone Detector	U-2–ϕ1	Ozone concentration 10 ppb	Potassium iodide oxidant meter	NASA-ARC	Popoff
	Ozone Detector	U-2–ϕ2	Ozone concentration	UV absorption detector	NASA-ARC	Popoff
	Ozone Monitor	WB-57F	Ozone concentration	Battery operated Komhyr (Potassium iodide) cell. Uses a MkIII probe air intake.	NCAR	Danielson
	UV Flux Instrument	Rocket	UV flux in 3 bands as f (altitude)	Arcas rocket (55 km) dropsonde	NASA-GSFC	Krueger
	UV Flux Instrument	WB-57F	UV solar irradiance at 2150Å and several lengths 2800–3200Å	Measured over upper hemisphere from aircraft	ONR/Parametrics	Sellers
	Ground UV	Ground sites (Bismarck, N.D. & Tallahassee, Fla.)	Ozone, UV flux, turbidity (dust)	Modified Dobson spectro photometers, one for each parameter at each of two locations.	NOAA	Machta
Water Vapor	Al_2O_3 Water Vapor	WB-57F Balloon	Water vapor concentrations	Collects sample through a rosemont total temperature probe to an Al_2O_3 hygrometer	NASA-GSFC	Hilsenrath
	Water Vapor Detector	Balloon	Water vapor concentration	Aluminum oxide hygrometer Dew point from water absorptive reactive element. Time constant about 30 sec.	ONR/Parametrics	Goodman
	Water Vapor Detector	U-2–ϕ1	Water vapor concentration 1 ppm	Lithium chloride crystal oscillator	NASA-ARC	Popoff
	Water Vapor Radiometer	NC135, WB-57F, U-2	Total mass of water vapor above instrument 18.0–30μm band	Chopper bolometer. Pyroelectric detector for early flights, cooled detector for later flights. Ultimate sensitivity 0.1 x 10^{-4}gr/cm^2 or 0.1 x 10^{-6}gr/gr.	NOAA	Kuhn
	Tritium Water Vapor	WB-57F	Water vapor concentration.	Uses activated tritium in the substrate of a column, exchanges hydrogen ions.	ONR/Parametrics	Goodman
	Frost Point Hygrometer	C-141	Water Vapor	To obtain horizontal profiles	NRL	Mastenbrook
	IR Extinction	Balloon	Extinction of solar radiation 2.7μ H_2O Band	Measures extinction of solar radiation during sunrise from 30 Km with 0.5 Km resolution.	ONR/U. Wyo.	Pepin

Figure 4 CIAP ATMOSPHERIC MEASUREMENT INSTRUMENTS

MEASUREMENT	INSTRUMENT NAME	PLATFORMS USED ON	MEASURES	REMARKS	AGENCY	PRINCIPAL INVESTIGATOR
Oxides of Nitrogen	NO Detector	Concorde	NO concentration	UV instrument to measure NO	ONR	Blamont
	NO Detector	U-2ϕ1&2	NO concentration	Chemiluminescence technique	NASA-ARC	Popoff
	NO Detector	WB-57F Balloon	NO concentration 0.1 ppb	Reacts NO with ozone and detector to yield emission number. Sensitivity: 1 x 10^8 molecules/cc.	Utah State/York U.	Megill/Schiff
	NO Release Expt.	Balloon	NO dispersion & ozone effects	4 experiments on 1 balloon	NRL	Tilford
	UV/NO_X	Concorde	NO_X concentration	Technical details to be supplied Instrument to be developed.	ONR	Blamont
	Photoionization Mass Spectrometer	Balloon	NO, NO_2	Vertical Distribution	NASA-GSFC	Aikin
	IR Spectrometer	Balloon	NO, NO_2	Vertical Distribution	ONR/ID'ASB	Ackerman
Trace Species	IR Spectrometer	WB-57F	Trace constituents (CH_4, CO_2, N_2O, CO, NO, NO_2, O_3, C_2H_2, C_2H_4	This instrument will measure same constituents as IR interferometer plus attempt to detect other trace gases.	U. Denver	Murcray
	Cryogenic Sampler	WB-57F	NO, NO_2, N_2O_2, H_2O, H_2O_2, HNO_3, OH, HO_2, CO, CH_4, 10^7–10^8 molecules/cc	Freezes reactive species. Uses electron spin resonance to detect radicals from 40 to 65 Kft.	IITRI	Snelson
	Trace Gas Sampler	Balloon	CH_4, H_2, CO	Samples for later gas chromatography analysis.	NCAR	Ehhalt
	CO Detector	U-2–ϕ1	CO Concentration 10 ppb	Hot mercuric oxide technique.	NASA-ARC	Popoff
	CO_2 Detector	U-2–ϕ1	CO_2 Concentration 5 ppb	NDIR Technique	NASA-ARC	Popoff
	IR Interferometer	Concorde NC-135	Trace constituents (CH_4, CO_2, N_2O, CO, NO, NO_2, O_3, C_2H_2, C_2H_4, H_2CO_2 and others)	The instrument is a Fourier interference spectrometer having a spectral resolution of 1/4 cm^{-1}. Measures trace constituents by observing solar radiation directly in the 1.2 to 7.4μ region.	JPL	Farmer
Particulates & Aerosols	Particulate/aerosol Sampler	Balloon	P/A Concentration & Composition	Altitude Profile	ONR/CSIRO	Bigg
	Aerosol Monitor	Satellite	Particulate concentration	Determines the extinction angle at satellite sunset or sunrise. Integrates from cloud tops to about 35 Km with a resolution of 1-2 Km. Two spectral regions, centered around 4000 and 9100Å	U. Wyo.	Pepin
	Particulate Sampler (LASL Pod)	WB-57F	Particle concentrations, size and composition.	Uses impact filters. Collects particles down to 0.1μ. Analysis includes gamma irradiation and X-ray fluorescence, down to about sodium. Also uses scanning electron microscope.	LASL	Sedlacek

Figure 4 (CONT'D)

MEASUREMENT	INSTRUMENT NAME	PLATFORMS USED ON	MEASURES	REMARKS	AGENCY	PRINCIPAL INVESTIGATOR
Particulates & Aerosols (cont'd)	Particulate Sampler (U-1 Sampler)	WB-57F	Particle concentrations, size and composition	Uses impact filters. Analysis techniques include colorometric, X-ray, neutron-activation. Especially looking for silicates, alkali metals, halogens and metals found in engine exhaust. Also sulfates, and cations of sodium, potassium, calcium and magnesium.	NCAR	Shedlovsky, Lazrus
	Aerosol Detector	Balloons	Concentrations >.25μ and >.50μ Dia. Particles	Light scattering techniques Ground level to 30 Km with ~0.1 Km resolution.	ONR/U. Wyo.	Rosen
	Stratospheric Aitken Nuclei Detection (SAND) System	WB-57F	Aerosols & ion density, down to about 10 mμ	Detector for measuring ambient background levels of stratopsheric particles on a "real time" basis using GE's Aitken Nuclei Detection scheme.	TSC/GE/U of MO	Haberl/Podzimek
	Aitken Nuclei Detector	Balloon	Vertical Profiles Aitken Nuclei	Aitken nuclei version of particulate sampler.	ONR/U of Wyo.	Hofmann, et. al.
	LIDAR	Ground	Aerosols	Uses a tunable dye laser and a pulsed ruby laser. Night measurements of aerosols, including predawn and post sunset to determine effects of sunlight on the concentration and vertical distribution of aerosols.	NASA-ARC/SRI	Collis
	Airborne LIDAR	NC-135	Aerosols	Map stratopsheric aerosols to determine background levels and spatial/temporal variations. Tunable dye LIDAR.	NCAR	Schuster
	LIDAR	Ground	Particulate concentration, size, shape, composition.	48" Dia. Telescope. Ruby laser at 6943 and 3472Å. Neobidium laser at 1.06μ and 5300 Å. Also measures N_2 in lower atm by Raman N_2 scattering	NASA-LRC	Northam
	Ion Density	WB-57F	Aerosols down to 10 mμ	Modified Gerdien tube detects charged particles.	NOAA	Kasemir
	Gas Phase Sampler	WB-57F	Determine ice concentration among aerosols.	While using filter paper samples, a sample of air is captured in evacuated bottles to determine water content.	LASL	Sedlacek
Ultra-Violet Flux	UV Monochromator	WB-57F	UV flux vs wavelength (2300–4000Å in 10Å bands	Beckman Instrument. Field of view is from zenith to horizon in a plane ⊥ to flight path and 23° to either side. This is a double monochromator for measurement of solar irradiance and sky radiance to determine the altitude dependence of solar energy deposition in the atmosphere.	NASA-GSFC	Heath

Figure 4 (CONT'D)

MEASUREMENT	INSTRUMENT NAME	PLATFORMS USED ON	MEASURES	REMARKS	AGENCY	PRINCIPAL INVESTIGATOR
Ultra-Violet Flux (cont'd)	UV Flux Instrument	Rocket	UV flux in 3 bands as f (altitude)	Arcas rocket (55 km) dropsonde	NASA-GSFC	Krueger
	UV Flux Instrument	WB-57F	UV solar irradiance at 2150Å and several wavelengths 2800–3200Å.	Measured over upper hemisphere from aircraft	ONR/Parametrics	Sellers
	Ground UV	Ground sites (Bismarck, N.D. & Tallahassee, Fla.)	Ozone, UV flux, turbidity (dust)	Modified Dobson spectro photometers, one for each parameter at each of two locations.	NOAA	Machta
	UV Fluxmeter	Balloon	UV flux 2200 – 7000Å at altitudes from 30-40 Km	Technical data to be furnished	ONR	Blamont
		Concorde	UV flux 2600–7000Å			
	UV Dosimeter	Ground		Used by other experimenters in the field.	Temple U.	Berger
Transport Mechanisms	Inertial Data System	WB-57F	Winds	Wind data inferred from aircraft position, rate date measured with IDS.	NCAR	Danielson
	$TiCl_4$ Smoke	WB-57F/Rocket	Wind Measurements	Smoke canisters in conjunction with photographic techniques. Produce vertical and horizontal smoke trails in the 15-25 Km altitude region, which together with ground photographic techniques will be used to measure atmospheric transport properties.	AFCRL	Good
Miscellany	Electric Field Mill	WB-57F	Voltage fields in vicinity of thunderstorms	Data to be correlated with ozone data to investigate effect of electric fields on transport processes.	NMIMT	Moore
	Cloud Camera 415P	WB-57F	Photographic record of cloud overflights.	Standard camera, provides down looking horizon to horizon photos of clouds required for data correlation in thunderstorm and tropical upwelling flights.	LASL/USAF	Guthals

Figure 4 (CONT'D)

Nevejans. "Stratospheric Nitric Oxide from Infrared Spectra." Preprint.

(3) "Laramie Comparative Experiment: Data Report and Preliminary Report of Conclusions." DYNATREND, INC. ED. Contract DOT-OS-20104, March 15, 1973.

© 1973, ISA JSP 6715

REMOTE-SENSING THE STRATOSPHERIC AEROSOLS

Theodore J. Pepin
Assistant Professor
Department of Physics and Astronomy
University of Wyoming
Laramie, Wyoming

ABSTRACT

A description of a remote-sensing experiment to measure the vertical concentration of aerosols in the stratosphere is presented. Data from balloon flight tests of the instrument are discussed and satellite experiments under development are described.

INTRODUCTION

The presence of particles in the stratosphere can be sensed from a high-altitude platform by using the sun as a source of radiation and observing the extinction produced by the aerosols as the sun sets or rises. With this geometry, radiation scattered by the particles in the stratosphere can be measured relatively free of interference from other atmospheric constituents using a simple photometer system working at several wavelengths in the near ultraviolet, blue, and near infrared regions of the spectrum. The following is a description of the measurement technique and a summary of some of the balloon measurements that have been made using the technique.

DESCRIPTION OF THE BALLOON INSTRUMENT

The balloon-borne instrument consists of four telescopes that look in the horizontal direction with 14 degree x 14 degree fields each centered 90 degrees in azimuth from each other. This assembly is rotated beneath the balloon at a constant angular speed of approximately one revolution per minute as the balloon floats at constant altitude. During this time the rising or setting sun traverses the vertical fields of the telescopes. The intensity of the solar radiation is measured in four spectral regions as a function of time by the telescopes as they rotate. The intensities are measured to an accuracy of 1%.

For convenience the telescopes will be referred to as A, B, C, and D for the remainder of this discussion. Table 1 summarizes the spectral sensitivity of the telescopes.

Telescopes A, B, and C employ type 929 vacuum phototubes with S-4 response as detectors. Telescope D uses a type 925 vacuum phototube which has a S-1 response. For each telescope the vignetting curve is measured by scanning over the sun before each flight. This allows one to make a correction for any deviation from a flat response across the field. This correction is less than 4%.

Table 1

Telescope	λ[Å]	$\Delta\lambda$[Å]
A	3670	300
B	4300	360
C	5950	400
D	9100	1400

(λ is the wavelength of maximum sensitivity and $\Delta\lambda$ is the full width at one-half maximum sensitivity.)

Figure 1 illustrates the exponential extinction coefficient per atmosphere (k) as a function of wavelength produced by the various atmospheric absorbers under good average conditions at sea level. There are three curves illustrated in

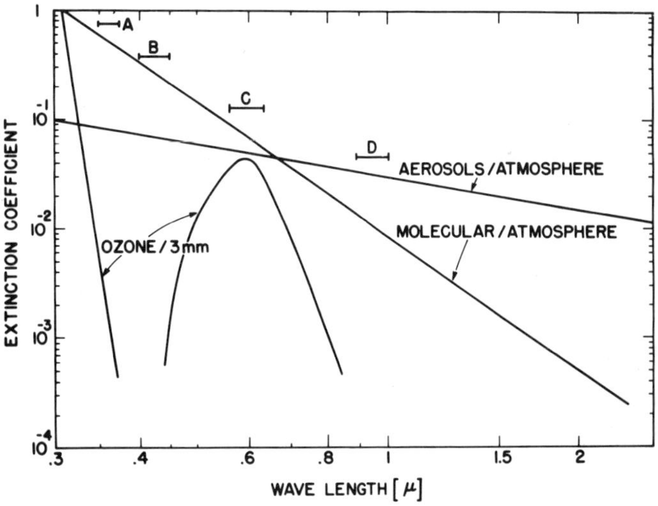

Figure 1. Exponential extinction coefficient per vertical atmosphere as a function of wavelength. Spectral sensitivities of the telescopes are noted.

this figure, one for the molecular or Rayleigh component per atmosphere ($k \propto 1/\lambda^4$), one for the aerosol component per atmosphere (this curve assumes $k \propto 1/\lambda$ and is normalized to C. W. Allen's value [1] at 1μ), and the third curve illustrates the atmospheric absorption for ozone as a function of wavelength assuming 3mm of ozone per atmosphere. Also marked on this figure are the spectral responses of the four telescopes. Telescope A is primarily sensitive to the extinction produced by the molecular or Rayleigh scattering of the atmosphere. Telescope B is also sensitive to the Rayleigh component and to some extent the extinction produced by atmospheric aerosols. Telescope C is centered in the middle of the Chappuis bands of ozone. Along with being sensitive to extinction produced by the molecular and aerosol components of the atmosphere, it is sensitive to absorption caused by the Chappuis bands. Telescope D is primarily sensitive to the extinction caused in the atmosphere due to the presence of the atmospheric aerosol component. It should be noted that Figure 1 is for a vertical atmosphere and the separation illustrated is enhanced for the case of the slant path geometry used in the measurements under discussion.

Figure 2 demonstrates the intensity that one would expect to observe as a function of atmospheric path for each of the four telescopes assuming a pure molecular atmosphere.

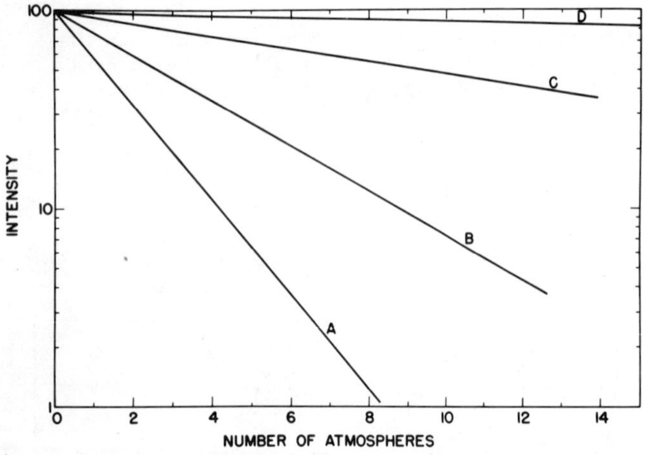

Figure 2. Theoretical intensity as a function of atmospheric path for the four telescopes in a pure Rayleigh atmosphere.

From the flight of April 12, 1967, Figure 3 has been constructed by plotting the observed intensity from each of the telescopes as a function of the calculated air mass. One observes the similarity of the data from telescopes A and B to that expected for a Rayleigh atmosphere, whereas the data of telescopes C and D appear different. The difference in the data of telescope C from that expected for a Rayleigh atmosphere is a reflection of the extinction produced by the ozone component of the atmosphere. The inflection in the data from telescope D between 5 and 8 atmospheres is

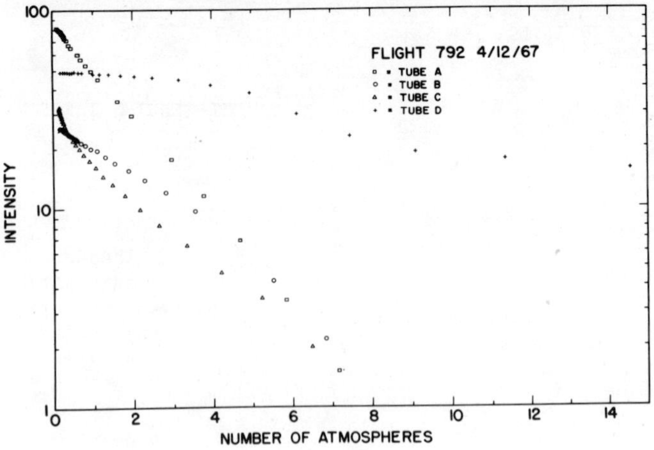

Figure 3. Intensity as a function of atmospheric path for Flight 792.

produced by an increased concentration of atmospheric aerosols above the tropopause, as will now be discussed.

DETERMINATION OF $N_o\sigma(h)$ FROM EXTINCTION DATA

The number density (N_o) times the effective scattering size (σ) of the atmospheric aerosols as a function of altitude (h), $N_o\sigma(h)$ can be determined from the extinction data from telescope D by considering a model atmosphere consisting of a number of layers and working out the extinction produced in each of the layers as the solar rays traverse them.

The atmosphere below the balloon is conveniently divided into a number of layers by the tangent heights of the observed rays when the sun is below the horizon. It is convenient to index the layers from the balloon down from 1 to k and assign the index 0 to the layer above the balloon of thickness S which is assumed to contain the aerosols above the balloon. The tangent height of the kth ray, h_k, can be accurately calculated accounting for refraction. The geometry is illustrated in Figure 4. The path lengths P_{jk} of the jth ray in the kth layer can be calculated from simple geometry.

Invoking the Lambert-Beer law, the intensity I of the light ray after traversing a path length P when the incident intensity is I', may be expressed as

$$I = I' e^{-N_o \sigma P}$$

where N_o is the number of scattering centers per unit volume and σ is the effective cross-section of the average scattering center.

By plotting the observed intensity from telescope D corrected for the Rayleigh component as a function of air mass, one can extrapolate to find the intensity at the top of the atmosphere, I', and with the observed intensity corrected for the Rayleigh component in the horizontal direction, I_0,

Superior number in parentheses refers to similarly-numbered reference at the end of this paper.

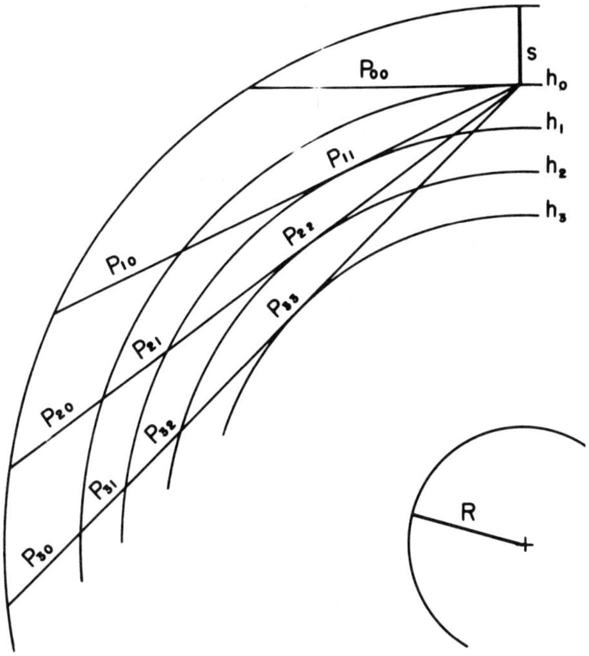

Figure 4. Geometry in a layered atmosphere used for determining $N_o\sigma(h)$.

one can invert the above relation to find

$$(N_o\sigma)_0 = \frac{\log_e \frac{I'}{I_0}}{P_{00}} .$$

Having determined $(N_o\sigma)_0$, one can trace the first ray below the horizontal through the atmosphere to find

$$(N_o\sigma)_1 = \frac{\log_e \frac{I'}{I_1} - (N_o\sigma)_0 P_{10}}{P_{11}} .$$

One can continue to solve for the $(N_o\sigma)_k$'s using the measured intensities corrected for the Rayleigh component, I_k, for successive lower layers, and in general one finds

$$(N_o\sigma)_k = \frac{\log_e \frac{I'}{I_k} - \sum_{i=0}^{k-1} (N_o\sigma)_i P_{ki}}{P_{kk}} .$$

As has been pointed out, in order to carry out the above calculations a value for S must be assumed. In this work S was first assumed to be the atmospheric scale height, and then after the calculation was completed the assumption was made that σ was independent of altitude and the scale height of the aerosols was determined. This scale height was then used for S and the calculation repeated. Since a balloon at an altitude of 32 km is above almost all of the aerosols, the choice of S does not critically influence the results.

As a test of the consistency of this method in one of the balloon flights of this series, telescope C was replaced with a telescope D so that there were two identical telescopes looking in the infrared wavelength on the same flight. The data from these telescopes were analyzed independently and are plotted in Figure 5. Both are seen to give the same results.

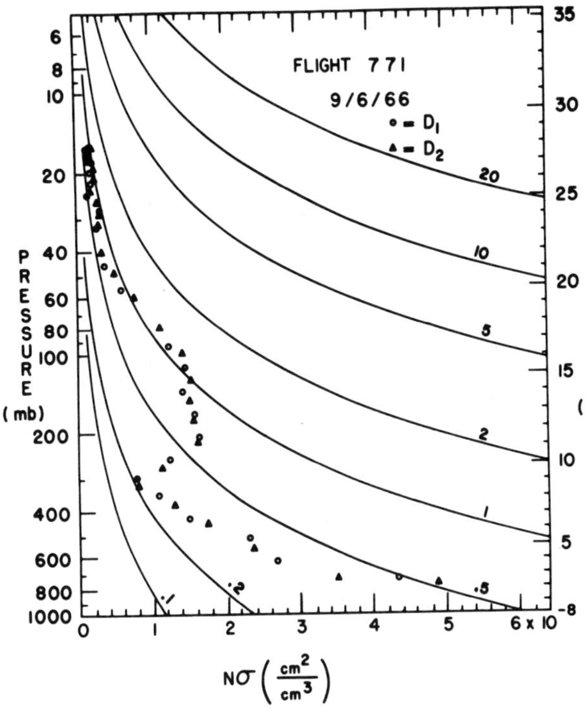

Figure 5. $N_o\sigma(h)$ for flight of September 6, 1966. Data plotted were determined independently from two different telescopes.

BALLOON FLIGHT DATA

From the summer of 1965 to the fall of 1970 a number of extinction flights were made from Minneapolis and one flight was made from the Panama Canal by the author. The data from these flights have been analyzed using the method outlined above and $N_o\sigma(h)$ has been determined at the time of each of the measurements. In the series of flights the highest concentration of aerosols above the tropopause at Minneapolis was observed in late 1966 during Flight 780. The data for this flight are shown in Figure 6.

The curved grid lines on the background of Figure 5 and Figure 6 are proportional to lines of constant mixing ratio for the atmospheric aerosols if the effective scattering cross-section is independent of altitude. The numerical values of these lines, thus interpreted, are proportional to the mixing ratio.

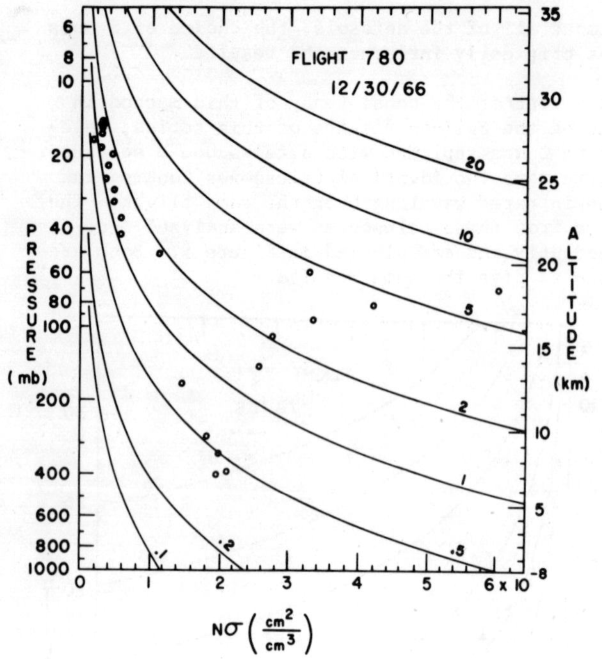

Figure 6. The number of aerosols multiplied by the effective scattering cross-section $N_o\sigma$ as a function of altitude.

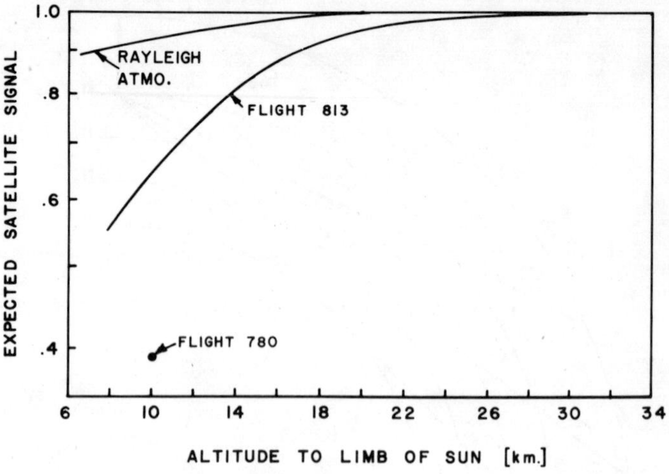

Figure 7. Expected data from the Wyoming PAM satellite program.

SATELLITE EXPERIMENTS

Measurements of the type described above can be made from satellite platforms by looking either at the entire solar disk or at a portion of the solar disk. From satellite altitude, 200 km or above, the projected solar diameter at the tangent altitude is considerably larger than it is for the case of the balloon measurements. Because of this enlargement the vertical structure of the aerosols is difficult to determine by looking at the entire solar disk, but the total particulate loading of the stratosphere can be determined. An experiment called "Wyoming PAM" (Preliminary Aerosol Monitor), designed to look at the entire solar disk, is presently under construction at the University of Wyoming and is scheduled for launch late next summer into a polar orbit on the Air Force 72-2 spacecraft. Typical data that might be expected from the Wyoming PAM experiment have been determined by using the data from one of the balloon extinction flights (Flight 813) and scaling it to the geometry of the satellite experiment. The result of this scaling is illustrated in Figure 7, where the expected satellite signal is plotted as a function of the projected solar altitude in km. The signal for the case of a theoretical Rayleigh atmosphere and a single data point is also indicated for the data of Flight 780, the balloon flight that indicated the highest aerosol concentration during the balloon flight program.

A second experiment, Wyoming SAM II (Stratospheric Aerosol Measurement) is presently in the engineering definition stage for the Nimbus-G spacecraft. This experiment makes use of a high-resolution pointing system to isolate a small field on the sun which is observed during the satellite sunset and sunrise events. With this instrument vertical resolution in the stratosphere of the order of 1 km is expected, allowing for the study of vertical structure of the aerosols in the stratosphere.

The advantage of the satellite method for determining the stratospheric aerosols is its ability to survey a large portion of the earth's atmosphere in a short period of time. For example, from either of the above experiments about 900 vertical profiles of the atmosphere can be determined per month. Data that will be useful for monitoring the stratospheric aerosols, studying global variations, observing hemispheric differences, and tracking particle injections will be available.

REFERENCE

(1) Allen, C. W., ASTROPHYSICAL QUANTITIES, 2nd ed., London, Athlone Press, 1963.

© 1973, ISA JSP 6716

PROGRESS REPORT: DETECTION OF DISSOLVED OXYGEN IN WATER THROUGH REMOTE SENSING TECHNIQUES

Arthur W. Dybdahl
Remote Sensing Programs
National Field Investigations Center - Denver
Office of Enforcement and General Council
Environmental Protection Agency
Denver, Colorado

ABSTRACT

The technique of detecting dissolved oxygen concentrations in the country's waterways with airborne remote sensing is discussed. This technique was developed at the National Field Investigations Center (NFIC), Office of Enforcement and General Council, EPA, Denver, Colorado. The recording media and data processing are explained. Experimentation is presently under way to quantify the airborne reconnaissance data so that concentrations of dissolved oxygen to within 1 to 2 parts per million can be readily obtained. A brief discussion of the water parameters that cause interference with the utilization of this technique are discussed.

INTRODUCTION

There is a great need present for a technology to be used in the rapid assessment of water quality parameters in large and small bodies of water throughout the country. This technology will be developed and applied to practical operations through remote sensing techniques. Qualitative detection of water quality parameters, such as turbidity, suspended solids, dissolved solids, and color, are presently at hand with the use of airborne cameras and passive scanners. There is a paramount need to quantify remote sensing data in terms of the water quality parameters for enforcement and monitoring applications. The development of a quantitative detection mechanism for dissolved oxygen in water is a beginning in this enormous task.

APPLICABLE OPTICAL PROPERTIES OF WATER

A great deal of effort has gone into the measurement of the optical properties of water since the mid-1940's. Samples of distilled water, in addition to those obtained from the oceans, coastal, and bay (estuarine) waters, have been thoroughly tested for optical transmittance and reflectance properties. Distilled water has the maximum transmittance values in the bandwidth from 400 nm to 650 nm. Ocean water is the next in line with a noticeable decrease in transmittance from 400 nm to 550 nm. Coastal and bay waters display significantly lesser transmittance values than that of the ocean waters (Specht).[1] The data for the attenuation (Mairs)[2] and the extinction (Hale)[3] properties of water are also collectively available. The transmittance data for the four above mentioned types of water is plotted in Figure 1.[1] This data was measured for a water path length of 10 meters. Note that the loss in transmittance is far greater in the 400 nm to 460 nm region (blue) than the green region (460-575 nm). This is an important feature used in the detection of dissolved oxygen in water.

THE AIRBORNE DETECTION TECHNIQUE

Recording Technique

The technique for the airborne detection of dissolved oxygen in water was found and developed through empirical data rather than through any deep-thought scientific formalism.

This technique was discovered in late 1970. It has been under continual expansion, testing, and field verification since that time.

The recording medium aboard an aircraft is a camera, Kodak 2443 false-color infrared film and a Wratten 16 orange gelatin filter. The exposure is set on the camera at approximately 1/4 to 1/3 f-stop, less than the so-called normal exposure. This sensor renders healthy water, saturated with dissolved oxygen, as a bright-brilliant blue. It renders septic (near zero dissolved oxygen waters) virtually black, or more precisely, the characteristics of unexposed processed film.

The film is processed with the usual EA-5 chemicals and procedures.

Field Verification of Film Indications

In an attempt to optically explain the above film indications, visual and hand-held photographic observations were carried out during EPA Water Quality Surveys, in areas where saturated and septic waters were being tested. Saturated waters were quite clear with the bottom visible to depths of 8 to 10 meters. The overall color of this type of water was greenish-blue. Septic waters were nearly opaque as far as depth penetration and displayed a very dark gray-green color. These waters were known to be subjected to a very high biochemical oxygen demand (BOD). This field data supports the water transmittance curves shown in Figure 1.

Film Interpretation and Analysis

Recalling the false-color rendition properties of the Kodak 2443 false-color infrared film, any natural green scene will record as blue, assuming the absence of chlorophyll substances which photographs as red. Waters saturated with dissolved oxygen photographed in a bright-blue color. As the oxygen demand increases, the dissolved oxygen is decreased, the bright-blue color of the water, recorded in the film, gradually progresses through a dark-blue to black. By staying in the linear portion of the film characteristic curves, it is a reasonable assumption that the optical response of the target being photographed will be recorded linearly. This "blue-to-black" indication results from the optical absorbance in the green region, being increased as the dissolved oxygen concentrations decrease. This technique does not work in the region from 0.6 microns (600 nm) to 1.0 microns (1,000 nm) which is the red and near-infrared portions of the optical spectrum.

Several areas throughout the country have been flown commensurate with the gathering of detailed ground truth. The dissolved oxygen data from the ground truth has been used in an attempt to calibrate the "blue-to-black" curve resulting from densitometric measurements made upon the exposed aerial film. Initially only the data from two bay areas having equal water temperatures at the time of flight, was used for the calibration procedure. Septic waters appear black in the film densitometrically identical to the unexposed film area between frames. The task of calibration is to identify film density data with the precise area of ground truth. Blue, green, yellow, and red density measurements were made on the film with a Macbeth TD-203AM Transmission Densitometer. The density data was compared to the ground truth. The variable color parameter is blue (density). The data is then plotted. One such calibration curve is shown in Figure 2. The ground truth was used to establish the position of each straight line. The abscissa of Figure 2 is the film density. It is commonly defined as:

$$D = \log_{10}(\frac{1}{T}) \qquad (1)$$

where T is the transmittance of the particular area on the film under test, for a particular color filter (red, green, or blue). It is easily seen that the dynamic range in film is greater than 1000:1. The ordinate represents the concentration of dissolved oxygen in parts per million (ppm) or milligrams per liter (mg/l). The range on this parameter has been limited to 0 to 10 ppm because in the natural environment the dissolved oxygen levels rarely exceed 10 ppm. To exceed this value usually indicates supersaturation accomplished by water pouring over rapids or in close proximity of an aerator. In bodies of water containing large amounts of aquatic plant growth (in the summer), the dissolved oxygen concentrations can reach, or even exceed, the saturation limit in the mid-afternoon hours, the period of maximum oxygen production.

Any test method has inherent problems with interference phenomena. This one is no exception. Waters in the natural environment contain finite amounts of suspended solids and organic waste materials. The impact of the materials upon the optical transmittance properties of various types of natural water is readily seen (Figure 1). The above mentioned materials contribute a yellow to light-red cast to water. When this is superimposed with the natural bluish-green color of oligotrophic-type waters, the blue cast is decreased and the water becomes predominantly green (Figure 1). Since deep yellow to red materials photograph green in the false-color infrared film, the green transmittance factor becomes an important one as a base line for each particular area of water under test. Red is not used in conjunction with this film because of the sensitivity of the cyan emulsion layers to the near-infrared region (0.7μ to 1.0μ). This would create an additional interference factor when aquatic plant growth is present and would not represent a reliable base line. So, through the optical inspection and analysis of over 1,100 data sets (red, green, and blue form a set for each point on the film) three color parameters have been established for calibration purposes. They are:

(1) Blue
(2) Green
(3) Blue-minus-Green

The parameter denoted by (3) has been chosen, as opposed to "Green-minus-Blue" to eliminate having to use negative irrational numbers. Again, the green parameter establishes the suspended solid material influence in the water under test. The initial density/ground truth data is plotted for the blue and green parameters. Then the actual calibration is carried out with "Blue-minus-Green" density value plotted with the concentration of dissolved oxygen. This plot is usually referred to as a difference curve. The difference curve for the data sets given in Figure 2, is provided as Figure 3. In practice, the concentration vs. density data is programmed into a computer. Two data sets are required. They are the Blue and Green density values for the point on the film corresponding to the physical location of the ground truth and for the unexposed (between frame) film. The latter eliminates the influences of base fog if any is present. The computer then calculates the equations of the three straight lines. This is done by using the point-slope form of the equation of a straight line in two dimensions. This is accomplished by the data points $P_1(x_1,y_1)$ for high dissolved oxygen, and $P_2(x_2,y_2)$ for zero dissolved oxygen, and

$$y = \frac{dy}{dx}(x - x_1) + y_1 \qquad (2)$$

where dy/dx is the first derivative of the equation for the line and is equal to $(y_2 - y_1)/(x_2 - x_1)$ = M. The expressions for the Blue and Green lines in Figure 2, respectively, are:

$$y_{Blue} = 2.655x_{Blue} = 10.462 \qquad (3)$$
$$y_{Green} = 3.715x_{Green} = 10.866 \qquad (4)$$

where x_{Blue}, x_{Green} are the blue, green film densities, respectively
y_{Blue}, y_{Green} are dissolved concentrations for the blue, green lines, respectively, in parts per million (ppm).

The expression for the difference line is calculated from this data, which for Figure 3 is:

$$y = 9.307(1 - x_\Delta) \tag{5}$$

where x_Δ = blue-minus-green density and
y = dissolved oxygen concentration in ppm.

To use the calibration curve in Equation (5), the blue and green data pairs are supplied to the computer. It calculates the value of Blue-Green value (x_Δ) and finally the dissolved oxygen concentration, designated by y.

The relative uncertainty of y can be calculated from Equation (5).

$$\left|\frac{dy}{y}\right| \approx \left|\frac{dx\Delta}{x\Delta-1}\right| \tag{6}$$

where the vertical lines signify absolute values, $|-x| = x$. The absolute uncertainty for a given calibration curve is:

$$|dy| = |-9.307dx\Delta| \tag{7}$$

If $dx\Delta = 0.02$, the accuracy of the Macbeth densitometer, then $|dy|$ is 0.186 ppm. The variance is given as $|dy|^2 = 0.0345$.

Hundreds of data points for healthy waters, with insignificant discoloration and suspended solids have been plotted against the calibration curve in Figure 3. The points all fall within 1 ppm of the line.

It must be mentioned that one must be cognizant of the inherent errors of the field equipment used in obtaining the dissolved oxygen phase of ground truth. Some types of systems are consistently off by as much as 1 ppm. This error must be compensated for in order to produce an accurate calibration curve.

Significant Affects Upon the Calibration Curve

A. Film Exposure

Film exposure levels can have a significant affect upon the slope of the calibration curve. Overexposure with the same calibration curve, results in a lesser dissolved oxygen concentration than is actually present. Likewise, under-exposure results in a greater concentration than is actually present. For this reason, a new calibration curve based upon the inherent film exposure at the center of a photographic frame near the intersection of the fiducial marks, is generated for each mission location.

B. Lens/Illumination Effects

A KS-87B aerial framing camera is used on all OEGC missions. It has a 152 mm lens cone assembly with a lens fall-off characteristic curve of $\cos^{12}\theta$, where θ is the angle off the principal axis of the lens. A plot of exposure vs. distance across the film format will reveal the effect of the lens fall-off.

There is one more factor that integrates with the lens fall-off to influence the optical irradiance across the film format. This factor is the solar illumination function. If I_0 is the optical energy level at the center of the film format, then the optical energy distribution across the film format is:

$$I(\lambda) = I_0(\lambda) \int_0^\theta L(\theta,\lambda) S(\theta,\lambda)\, d\theta, \tag{8}$$

where $I(\lambda)$ = optical energy distribution across the film format as a function of wavelength λ.

$I_0(\lambda)$ = optical energy at the center of the film format as a function of λ.

$L(\theta,\lambda)$ = lens fall-off function $(\cos^{12}\theta)_\lambda$.

$S(\theta,\lambda)$ = Solar illumination function.

$S(\theta,\lambda)$ can easily be normalized to one yielding a relative energy weighting factor. Camera lenses usually possess a spherical symmetry that eliminates an integration factor which encircles the principal axis.

One will normally see a bright spot in or near the center of the frame of film. Densitometer data are limited to a circle whose radius is 1 cm about the center of the bright spot for a 4.5" by 4.5" format. Data taken anywhere else in the frame must be normalized to this spot to render correct indications from the calibration curve.

C. Water Body Characteristics

Areas of significant discoloration in water would be expected to significantly influence the effectiveness of the calibration curve. Many sources of discoloration result from municipal and industrial outfalls and high turbidity or suspended solids. An investigation has been carried out over a submerged diffused lignin sulfonate discharge located at Port Angeles, Washington. In the true-color Ektachrome aerial imagery, the effluent was dark-gray-reddish-brown in color. Ground truth indicated no dissolved oxygen depression within the area of influence of the resultant plume. The blue and green densitometric data and ground truth data were used to plot the respective straight lines in Figure 4. The difference curve was also plotted in Figure 5. This curve was compared to the blue calibration curve in Figure 3. As an example, if $x\Delta$ in Figure 5 were 0.125 corresponding to 8 ppm, then the equivalent value in Figure 3 would be 8.25 ppm. In the range from 8 to 9.5 ppm of dissolved oxygen, the correlation was less than 0.3 ppm in spite of the significant discoloration.

Future Studies Required

To enhance the integrity of this technique, many questions must still be answered. Several will be discussed in the next few paragraphs.

A. First it must be determined if the observations of green absorbance are physically due to the presence of dissolved oxygen. Testing will begin later this fall in the Environmental Physics Laboratory at NFIC-Denver. An optical test cell whose path length through a particular medium can be adjusted from 5 to 50 meters, has been designed and is awaiting fabrication. It will require only 2.7 liters of water sample. Under laboratory conditions, dissolved oxygen depressions can be induced into a particular water sample and optically monitored to document its behavior in the green region. This program will establish the optical properties of many types of natural and waste waters as a function of dissolved oxygen concentrations.

B. The effects of water temperature upon the concentrations of dissolved oxygen in the optical recording medium will have to be studied. This will involve effort in both the laboratory and in the field.

C. The function of water depth and the optically generated concentration values must be determined. This can also be accomplished, for the most part, in the Physics Laboratory.

D. Other recording media must be examined to possibly provide more reliable concentration data. Kodak 2443 film with a Wratten 58 green filter has been quite successful although in limited use. This technique must be expanded into the domain of active and passive sensors. It is worth noting that this spectroscopic technique has been carried out on a Kodak aerial (true-color) Ektachrome Film SO-397. The results showed that the Kodak 2443 film provided much better color separation because of its false-color rendition.

E. Natural and induced interferences with the detection of dissolved oxygen in all water environments must be explored. These include the discoloration and suspended solids produced by man-made and natural sources. Many of these can be effectively studied in a laboratory.

F. The calibration curve must be further tested and retested in order to statistically validate its integrity. To date most of the ground truth data/calibration curve correlations have been carried out in the concentration range of 7.0 to 9.5 ppm. The curve needs to be studied further in the range from 0 to 7.0 ppm.

Finally, the technique must be adopted to night aerial reconnaissance, which will undoubtedly involve an active light source and absorption spectroscopy. This will add greatly to a round-the-clock enforcement monitoring capability.

SUMMARY AND CONCLUSIONS

A technique for the quantitative detection of dissolved oxygen concentrations in water through remote sensing has been discussed. The recording medium has been an airborne framing camera employing Kodak 2443 False-Color Infrared film with a Wratten 16 (orange) optical filter. Through a densitometric analysis of the exposed film together with ground truth, a calibration curve has been generated. This technique has provided an accuracy of better than ±1 ppm in healthy bay and ocean waters.

Future efforts to improve the technique under all conditions was discussed. This will involve studying the influences of water temperature and depth, discoloration and suspended solids upon the quantitative results.

ACKNOWLEDGMENTS

The author wishes to acknowledge the support received from the Surveillance and Analysis Division, Region X, EPA, Seattle, Washington; and the State of Washington, Department of Ecology, in obtaining ground truth for the many flights conducted in the Puget Sound area.

REFERENCES

(1) Specht, M. R., D. Needler, and W. L. Fritz, "New Color Film for Water-Photography Penetration," PHOTOGRAMMETRIC ENGINEERING, Vol. XXXIX, No. 4, April 1973.

(2) Mairs, R. L., and D. K. Clark, "Remote Sensing of Estuarine Circulation Dynamics," PHOTOGRAMMETRIC ENGINEERING, Vol. XXXIX, No. 9, September 1973.

(3) Hale, G. M., and M. R. Querry, "Optical Constants of Water in the 200 nm to 200μm Wavelength Region," APPLIED OPTICS, Vol. 12, No. 3, March 1973.

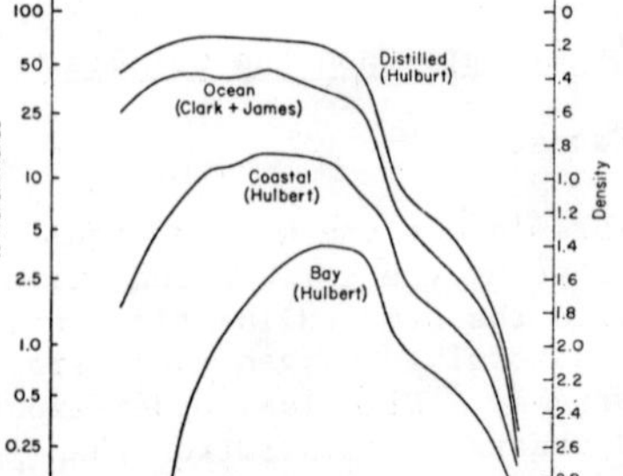

FIG. 1. Spectral transmittance for ten meters of various water types.

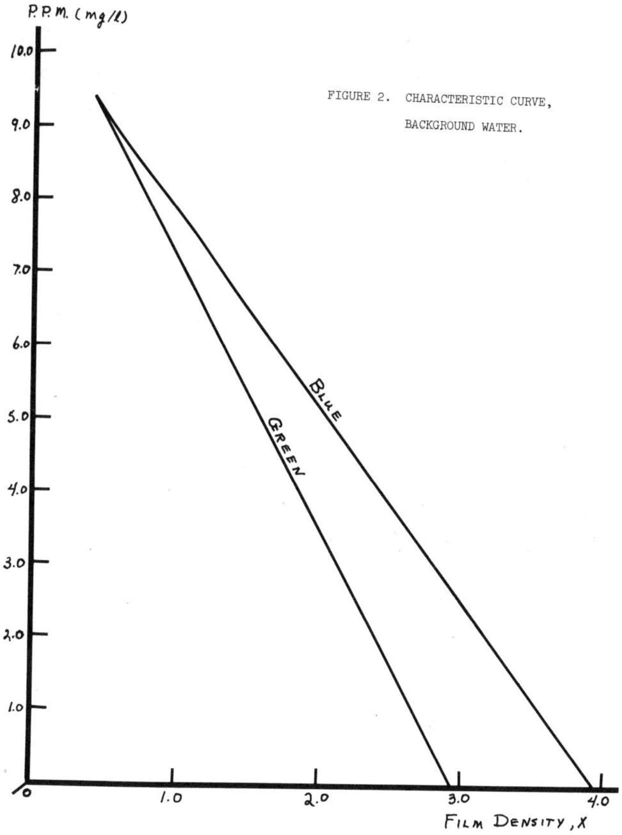

FIGURE 2. CHARACTERISTIC CURVE, BACKGROUND WATER.

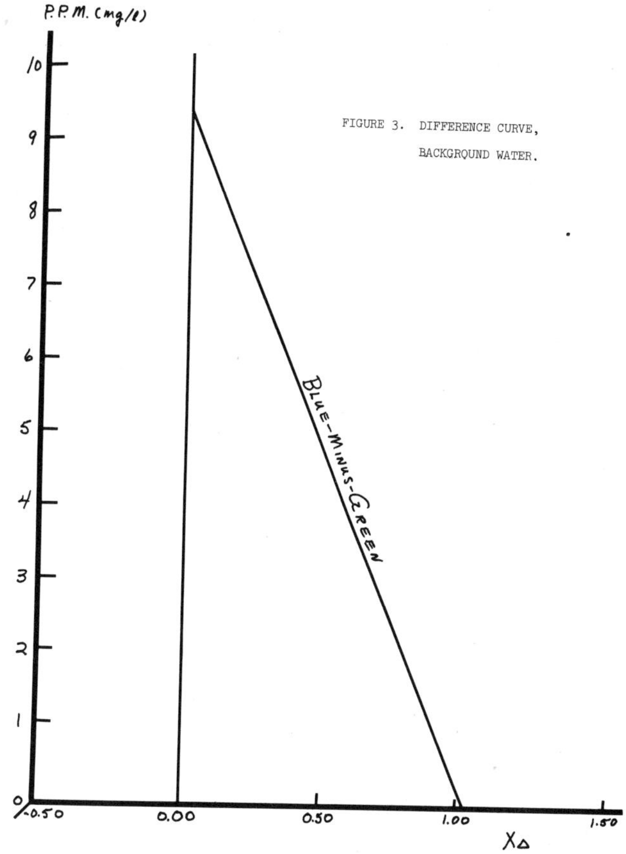

FIGURE 3. DIFFERENCE CURVE, BACKGROUND WATER.

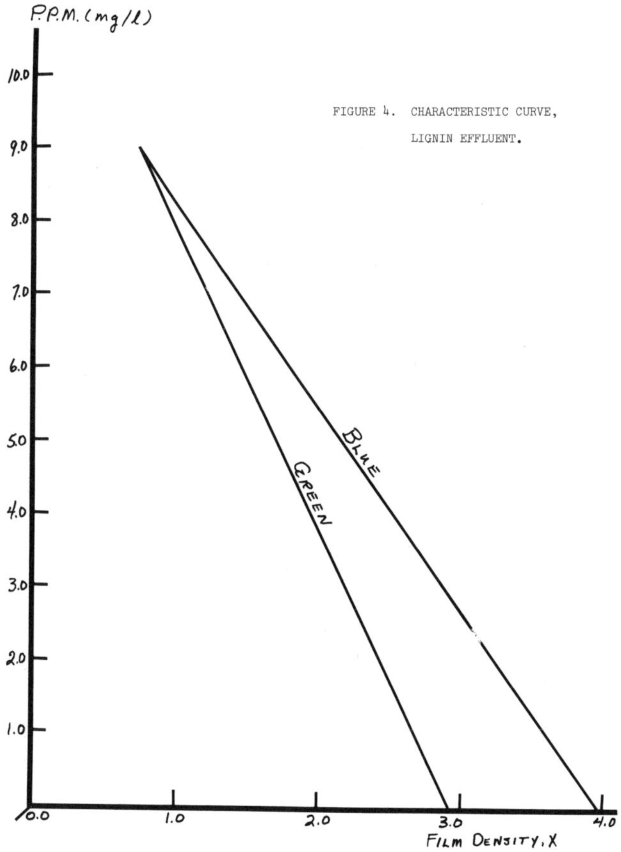

FIGURE 4. CHARACTERISTIC CURVE, LIGNIN EFFLUENT.

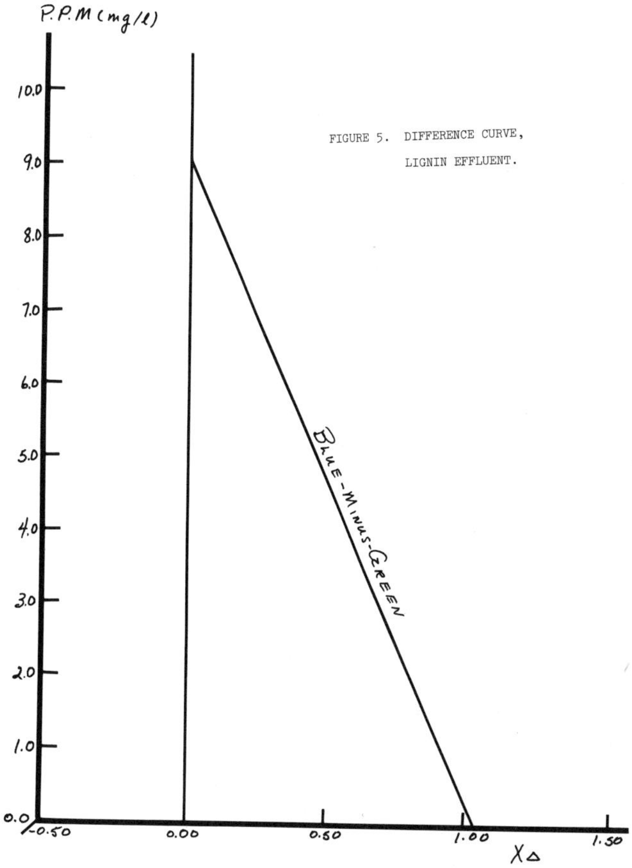

FIGURE 5. DIFFERENCE CURVE, LIGNIN EFFLUENT.

© 1973, ISA JSP 6717

MONITORING COASTAL WATER PROPERTIES AND CURRENT CIRCULATION WITH SPACECRAFT

V. Klemas, M. Otley, C. Wethe
University of Delaware
Newark, Delaware

and

R. Rogers
Bendix Aerospace Systems Division
Ann Arbor, Michigan

ABSTRACT

Imagery and digital tapes from nine successful ERTS-1 passes over Delaware Bay during different portions of the tidal cycle have been analyzed with special emphasis on turbidity, current circulation, waste disposal plumes and convergent boundaries between different water masses. ERTS-1 image radiance correlated well with Secchi depth and suspended sediment concentration. Circulation patterns observed by ERTS-1 during different parts of the tidal cycle, agreed well with predicted and measured currents throughout Delaware Bay. Convergent shear boundaries between different water masses were observed from ERTS-1. In several ERTS-1 frames, waste disposal plumes have been detected 36 miles off Delaware's Atlantic coast. The ERTS-1 results are being used to extend and verify hydrodynamic models of the bay, developed for predicting oil slick movement and estimating sediment transport.

INTRODUCTION

Imagery and digital tapes from nine successful ERTS-1 passes over Delaware Bay during different portions of the tidal cycle have been analyzed with special emphasis on turbidity, current circulation, waste disposal plumes and convergent boundaries between different water masses. (NASA-ERTS-1 I.D. Nos. 1024-15073, 1079-51533, 1133-15141, 1187-15140, 1205-15141, 1294-15083, 1349-15134, 1385-15131, 1403-15125, respectively) During ERTS-1 overpasses ground truth was being collected along a total of twelve boat and helicopter transects across the bay, including measurements of Secchi depth, suspended sediment concentration and size, transmissivity, temperature, salinity, and water color. Three U-2 and four C-130 overflights took place during the same time period. Small aircraft equipped with clusters of filtered film cameras were used to underfly each ERTS-1 overpass.

PHYSICAL CHARACTERISTICS OF DELAWARE BAY

The Delaware Bay Estuary is a relatively prominent coastal feature which bounds the Delmarva Peninsula on its northern side. The geography of this region including the locations of several convenient reference points, is shown in Figure 1. Trenton, New Jersey is generally taken to define the upper limit of the estuary, so that its total length is over 130 miles.

Fresh water input to the system is derived mainly from the Delaware River at an average rate of 11,300 cfs which, in terms of volume flow, ranks this as one of the major tributaries on the eastern coastal plain. Together with this large volume of fresh water, the river also discharges a heavy load of suspended and dissolved material, since its effective watershed encompasses an area typified by intensive land use, both agricultural and industrial (Oostdam, 1971). Seaward of the Smyrna River, the bay undergoes a conspicuous exponential increase in both width and cross-sectional area so that the strength of the river flow is rapidly diminished beyond this point. Ketchum (1952) has computed the flushing time of the bay (defined in this case as the time required to replace the total fresh water volume of the bay) to be roughly 100 days. Seasonal variations in river flow cause this figure to fluctuate within a range of from 60 to 120 days. Consequently, river flow is not a significant factor in determining the current pattern in the bay except in the consideration of time-averaged flow. In terms of short-period studies it is mainly important as a source of suspended sediment and contaminants.

The seaward boundary of the bay extends from Cape May southeast to Cape Henlopen -- a distance of eleven miles. Tidal flow across this boundary profoundly affects the dynamic and hydrographic features of the entire estuary. The effect is especially pronounced toward the mouth where conditions are generally well mixed. The dynamic behavior of the tides is closely approximated by the cooscillating model described by Harleman (1966). In this model, the upper end of the estuary is assumed to act as an efficient reflecting boundary. Consequently, the actual tidal elevation at any given point is a result of the interaction of both a landward directed wave entering from the ocean and a reflected wave traveling back down the estuary. The tidal range is a maximum at the reflecting boundary and decreases toward the mouth in a manner dependent upon the relative phase of the two components. Observations of the tide at Trenton show a 7-foot range compared to a 4-foot range at the mouth. A relative maximum appears at roughly the location of Egg Island Point where the phase relationship is temporarily optimal. An important consequence of this behavior is the occurrence of strong reversing tidal currents in the Delaware River. The tidal wavelength as computer by Harleman is 205 miles, so that both peak currents and slack water may occur simultaneously at either end of the bay.

The tidal flow is also modified by the bay's rather complex bathymetry (Figure 1). Most prominent are several deep finger-like channels which extend from the mouth into the bay for varying distances (Kraft, 1971). Depths of up to 30 meters are present, making this one of the deepest natural embayments on the east coast. The channels alternate with narrow shoals in a pattern which is shifted noticeably toward the southern shore. On the northern side, a broad, shallow mudflat extends from Cape May to Egg Island Point. Considerable transverse tidal shears result from these radical variations in bottom contours. As a consequence, a marked gradient structure may form normal to the main axis of the bay as a function of tidal phase. This intrinsic two-dimensional character, together with the complex, superimposed time variations, represents an almost insurmountable task when it is confronted with the tools of conventional hydrographic surveys. The problem is ready-made, however, for the techniques of high-altitude photography.

VISIBILITY OF SUSPENDED SEDIMENTS

Extensive investigations of suspended sediments in Delaware Bay and laser transmission tests in a test tank facility have been conducted respectively by Oostdam (1970) and Hickman (1972). The results can be summarized as follows:

suspended sediments in Delaware Bay averaged 30 ppm. During July-August the average sediment level was 18 ppm.

turbidity increased with depth in the water column, except during periods of bloom, when surface turbidities at times exceeded those at greater depths.

suspended sediment concentration gradients were greater during ebb than during flood because of greater turbulence and better mixing during flood stage.

the turbidity decreased from winter to summer.

marked increases in turbidity which were observed during May and September were caused mainly by plankton blooms.

suspended sediments were silt-clay sized particles with mean diameters around 1.5 microns.

the predominant clay minerals are chlorite, illite and kaolinite.

reflectivities for the Delaware Bay sediments were measured to be about 10%.

At the time of the ERTS-1 overpasses, Secchi depth readings ranged from about 0.2 meters near the shore up to about 2 meters in the deep channel. Preliminary "equivalent Secchi depth" measurements with green and red boards indicated that neither "color" exceeded the readings obtained with the white Secchi disc. Therefore, it is quite unlikely that the bottom will be visible in any of the ERTS-1 channels and, at least in Delaware Bay, most of the visible features will be caused by light reflected off the surface or backscattered from suspended matter.

"Red" filters, such as the Kodak Wratten No. 25A, have frequently been used in aerial photography to enhance suspended sediment patterns (Klemas et al., 1973, Bowker et al., 1973). In Delaware Bay red filters have been particularly effective for discriminating light-brown sediment-laden water in shallow areas from the less turbid dark-green water in the deep channel.

Figures 2 and 3 contain microdensitometer scans between Cape Henlopen and Cape May at the mouth of the bay. ERTS-1 images taken in bands 4, 5, and 6 were scanned, and grey scales equalized, to enable comparison on the same set of coordinate axes. Band 5 is clearly most sensitive to suspended sediment features. Even in band 5, however, the sediment patterns are caused by only four to five neighboring shades of grey in the negative transparencies and about twice that number in the digital tapes.

CORRELATION WITH MEASURED WATER PROPERTIES

During ERTS-1 overpasses ground truth was being collected with boats and helicopters along the three transsects across the bay shown in Figure 4, including measurements of Secchi depth, suspended sediment concentration, transmissivity, temperature, salinity, and water color. The correlation between ERTS-1 MSS band 5 image radiance, Secchi depth and sediment concentration is shown tentatively in Figure 5. These measurements, including salinity and temperature shown in Figure 6, were obtained during the ERTS-1 overpass on January 26, 1973. Note the sharp discontinuities in salinity and temperature near aquatic boundaries.

During flood tide at the mouth of the bay, considerable correlation was found between the depth profile and image radiance, as shown in Figure 7. During the flood tide most of the sediment in suspension seems to be locally generated over shoals and shallow areas of the bay resulting in a higher degree of backscatter from shallower waters.

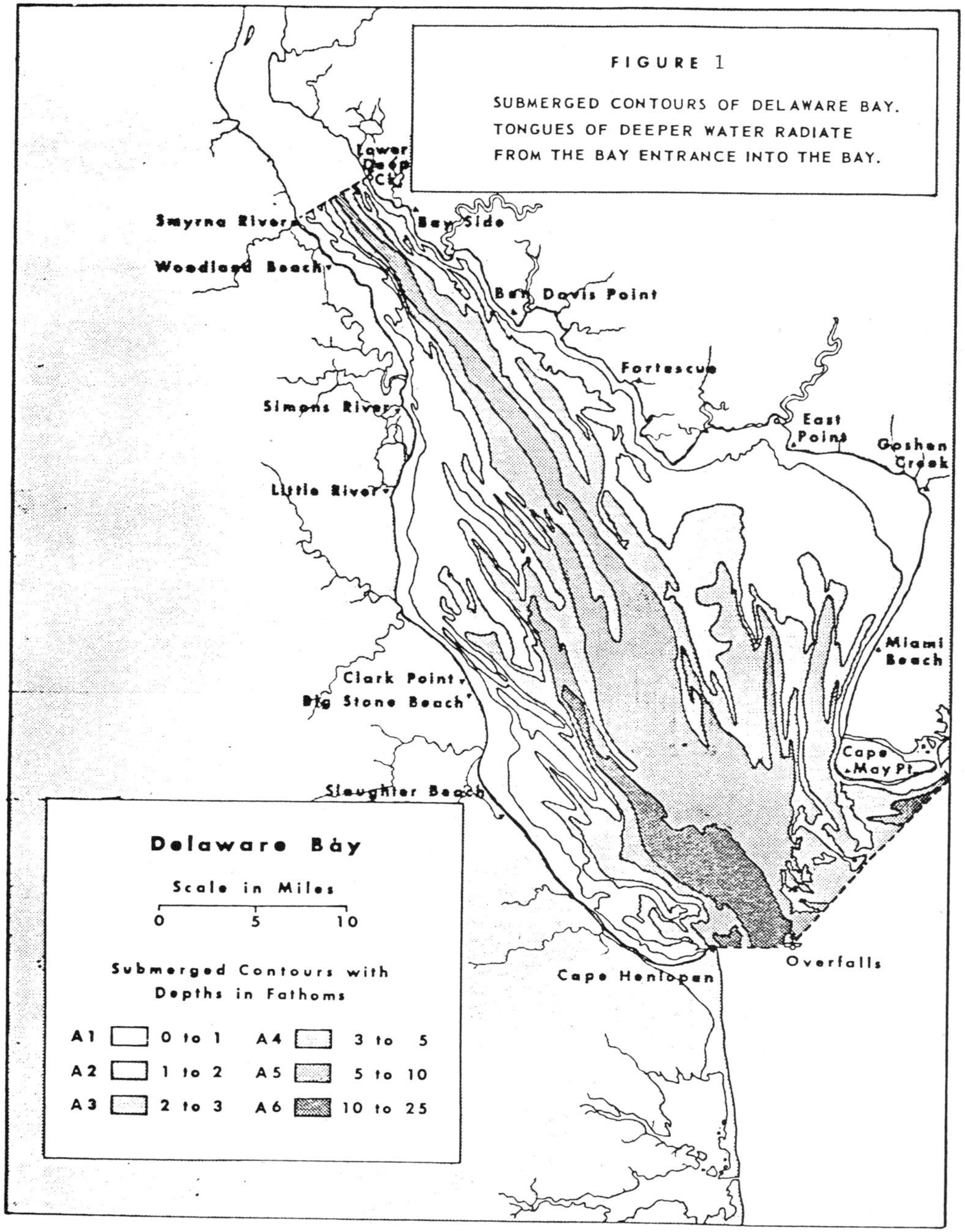

FIGURE 1

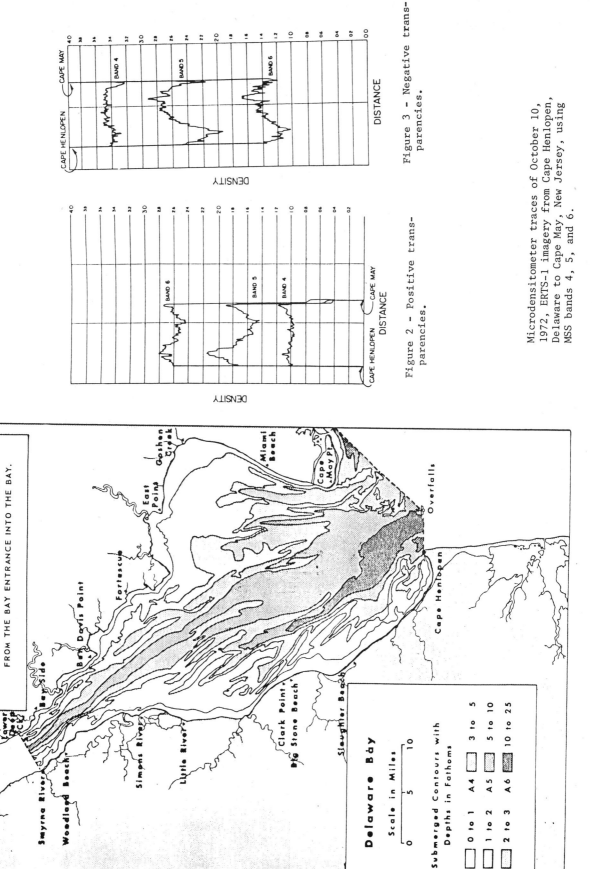

SUBMERGED CONTOURS OF DELAWARE BAY. TONGUES OF DEEPER WATER RADIATE FROM THE BAY ENTRANCE INTO THE BAY.

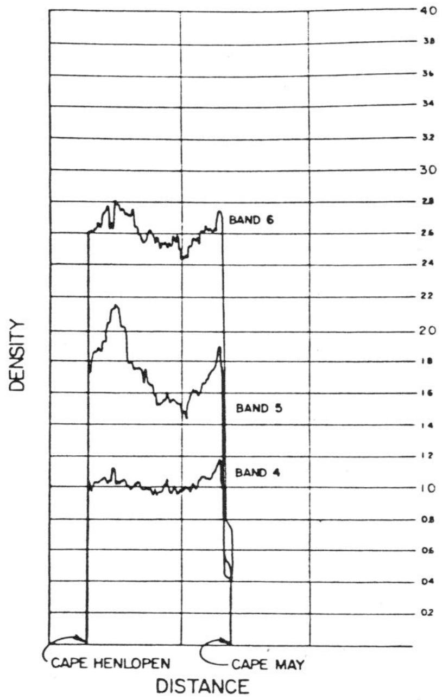

Figure 2 - Positive transparencies.

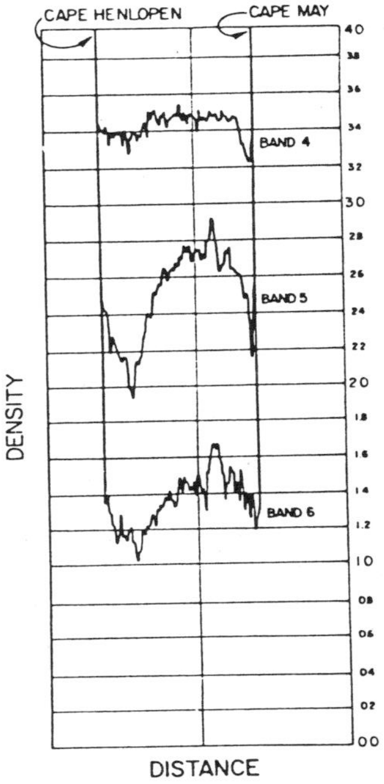

Figure 3 - Negative transparencies.

Microdensitometer traces of October 10, 1972, ERTS-1 imagery from Cape Henlopen, Delaware to Cape May, New Jersey, using MSS bands 4, 5, and 6.

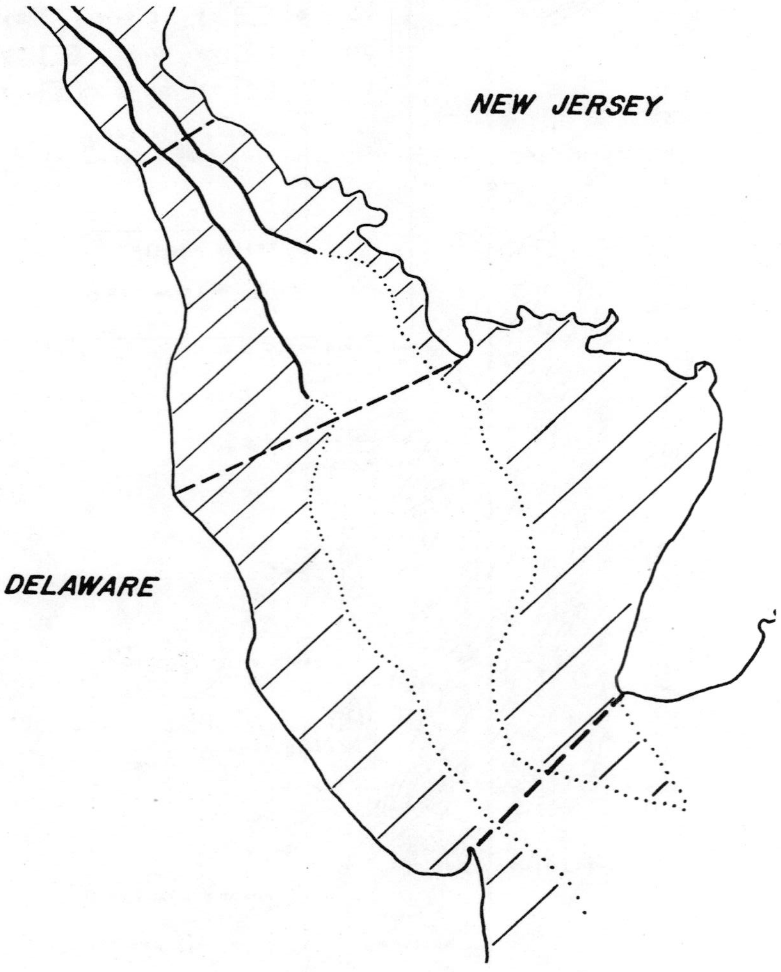

Figure 4. Helicopter and boat transsects across Delaware Bay along which Secchi depth, sediment concentration, transmissivity, temperature, salinity and water color were measured during ERTS-1 overpasses.

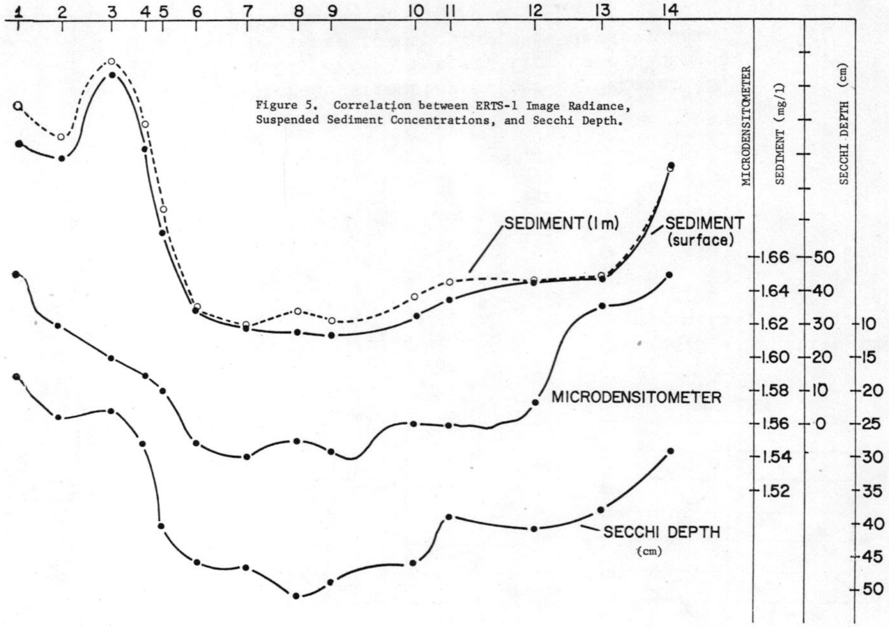

Figure 5. Correlation between ERTS-1 Image Radiance, Suspended Sediment Concentrations, and Secchi Depth.

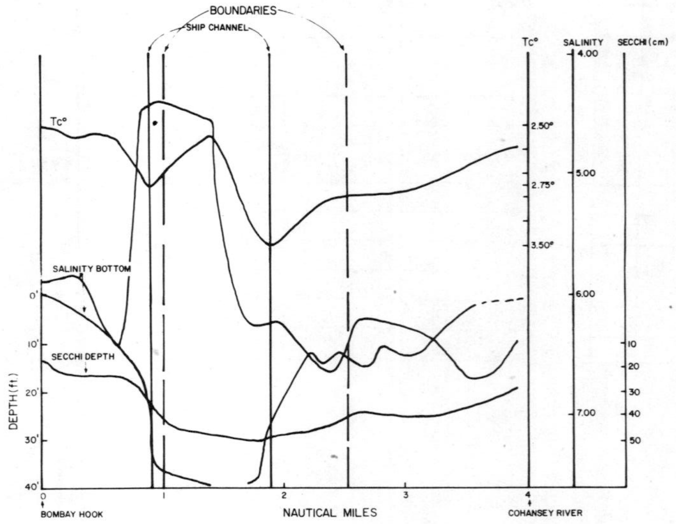

Figure 6. Temperature, salinity Secchi depth and bottom profiles across the upper transsect of the bay during ERTS-1 overpass on January 26, 1973.

CURRENT CIRCULATION PATTERNS

Since suspended sediment acts as a natural tracer, it is possible to study gross circulation in the surface layers of the bay by employing ERTS-1 imagery and predicted tide and flow conditions. Adjacent to ERTS-1 pictures, Figures 8, 9, 10, 11, and 12 contain tidal current maps for Delaware Bay. Each ERTS-1 picture is matched to the nearest predicted tidal current map within ± 30 minutes. A closer match was not attempted at this point, since quantitative comparison would require comprehensive current measurements over the entire bay at the time of each satellite overpass -- a feat not attainable with limited resources. The current charts indicate the hourly directions by arrows, and the velocities of the tidal currents in knots. The Coast and Geodetic Survey made observations of the current from the surface to a maximum depth of 20 feet in compiling these charts.

The satellite picture in Figure 8 was taken two hours after maximum flood at the entrance of Delaware Bay on October 10, 1972. The sediment pattern seems to follow fairly well with the predicted current directions. A strong sediment concentration is visible above the shoals near Cape May and in the shallow nearshore water of the bay. Peak flood velocity is occurring in the upper portion of the bay, delineating sharp shear boundaries along the edges of the deep channel. At the time of this ERTS-1 picture, the wind velocity was 7 to 12 miles per hour from the north.

Figure 9 represents tidal conditions two hours before maximum flood at the mouth of the bay observed by ERTS-1 on January 26, 1973. High water slick is occurring in the upper portion of the bay, resulting in less pronounced boundaries there as compared to Figure 8. The shelf tidal water is not rushing along the deep channel upstream anymore as in Figure 8, but is caught between incipient ebb flow coming down the upper portion of the river and the last phase of flood currents still entering the bay. The sediment plume directions in Figure 9 seem to show flood water overflowing the deep channel and spreading across the shallow areas towards the shore. On the morning of December 3, 1972, there was a steady wind blowing over the bay at 7 to 9 miles per hour from the west.

About six months later, on July 7, 1973, the ERTS-1 overpass shown in Figure 10 occurred during the same part of the tidal cycle as that shown in Figure 9. Since river flow differs in the winter and summer, Figures 9 and 10 enable one to compare the effect of river flow and wind effects with identical tidal conditions. The body of water having a higher radiance seen about 20 miles offshore in Figure 10 has not yet been explained.

As shown in Figure 11, on August 12, 1973, ERTS-1 passed over Delaware Bay one hour before maximum ebb at the mouth of the bay. In addition to locally suspended sediment over shallow areas and shoals observed in Figures 8 and 9, plumes of finer particles are seen in Figure 11 parallel to the river flow and exiting from streams and inlets off New Jersey's and Delaware's coastlines. Shear boundaries along the deep channel are still visible in the upper portion of the bay; however, they are beginning to disappear as slack sets in, resembling conditions in Figure 9. At the time of the overpass, the wind was from the north-northeast at about 11 miles per hour, possibly causing the streaks in the ocean off New Jersey's coastline.

The satellite overpass on February 13, 1973, occurred about one hour after maximum ebb at the capes. The corresponding ERTS-1 image and predicted tidal currents are shown in Figure 12. Strong sediment transport out of the bay in the upper portion of the water column is clearly visible, with some of the plumes extending up to 20 miles out of the bay. Small sediment plumes along New Jersey's coast clearly indicate that the direction of the longshore current drift in that area is towards the north. The wind velocity at the time of the satellite overpass was about 7 to 9 miles per hour from the north-northwest.

SHEAR BOUNDARIES AND SALINITY WEDGES

Boundaries or fronts (regions of high horizontal density gradient with associated horizontal convergence) are a major hydrographic feature in Delaware Bay and in other estuaries. Fronts in Delaware Bay have been investigated using STD sections, dye drops and aerial photography. Horizontal salinity gradients of 4% in one meter and convergence velocities of the order of 0.1 m/sec have been observed. Several varieties of fronts have been seen. Those near the mouth of the bay are associated with the tidal intrusion of shelf water (Figure 12). The formation of fronts in the interior of the bay (Figure 8) appears to be associated with velocity shears induced by differences in bottom topography with horizontal density difference across the front influenced by vertical density difference in the deep water portion of the estuary. Surface slicks and foam collected at frontal convergence zones near boundaries contained concentrations of Cr, Cu, Fe, Hg, Pb, and Zn higher by two to four orders of magnitude than concentrations in mean ocean water. (Szekielda, Kupferman, Klemas, Polis, 1972), Figure 14, (Band 5, I.D. Nos. 1024-15073) obtained by ERTS-1 on August 16, 1972, contains several distinct boundaries. The southern-most boundary, as shown in Figure 15 is of particular interest, since it has frequently been observed from aircraft (Figure 16) and at the time of the ERTS-1 overpass. Divers operating down to depths of 6 meters noted increases in visibility from about half a meter to two meters as the boundary moved past their position.

WASTE DISPOSAL PLUMES

Careful examination of Figures 17 and 18 disclosed a fish-hook-shaped plume about 40 miles east of Cape Henlopen caused by a barge disposing acid wastes. The plume shows up more strongly in the green band than in the red band. Since some acids have a strong green component during dumping and turn slowly more brownish-reddish with age, the ratio of radiance signatures between the green and red bands may give an indication of how long

Figure 7. ERTS-1 image radiance (microdensitometer trace) and depth profile across the mouth of Delaware Bay during flood tide.

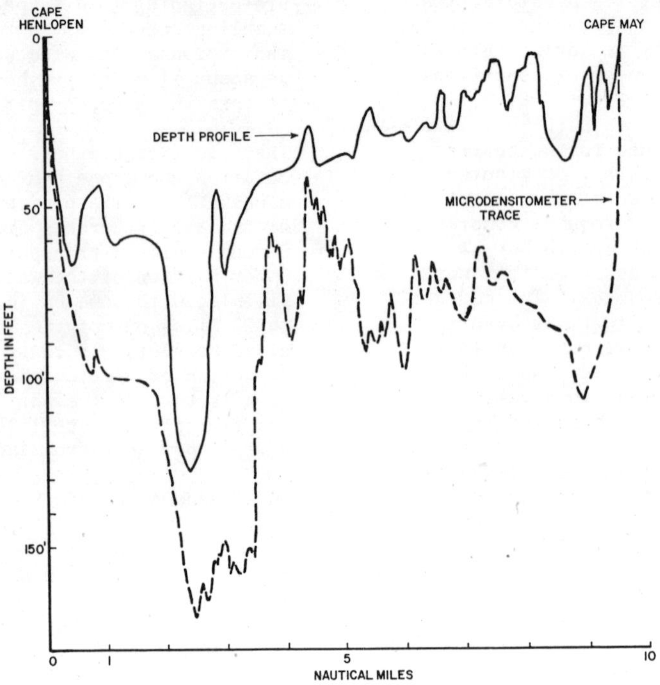

MICRODENSITOMETER TRACE FROM CAPE HENLOPEN, DEL. TO CAPE MAY, N.J.
ERTS-1 ID. 1024-15073 (BAND 5) AUG. 26, 1972

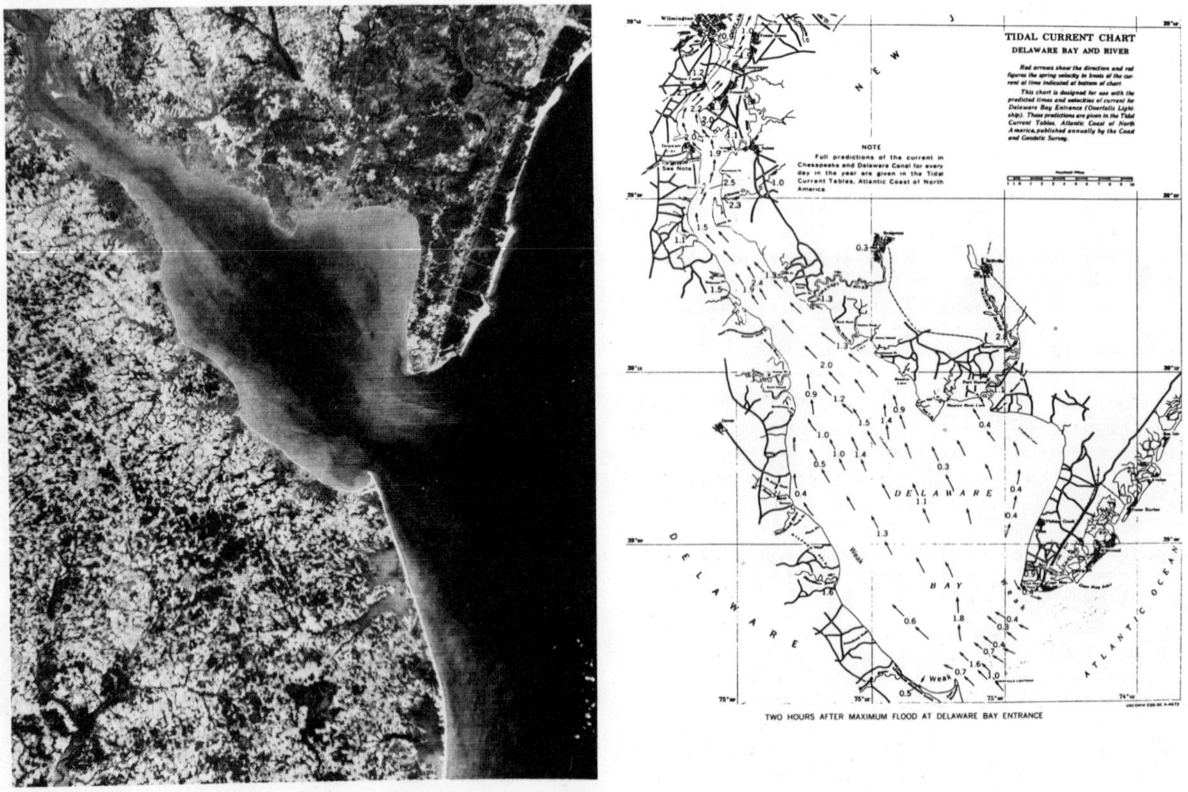

Figure 8. ERTS-1 image of Delaware Bay obtained in MSS band 5 on October 10, 1972 and tidal current map. (I.D. No. 1079-15133.)

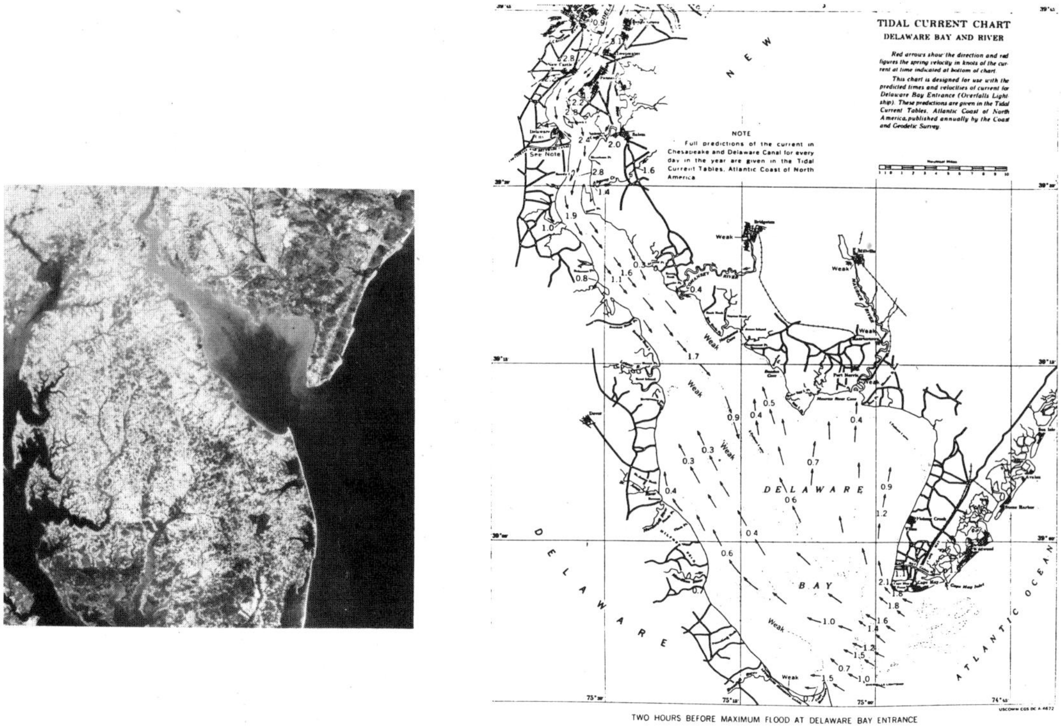

Figure 9. ERTS-1 image of Delaware Bay obtained in MSS band 5 on January 26, 1973 and tidal current map. (I.D. Nos. 1187-15140).

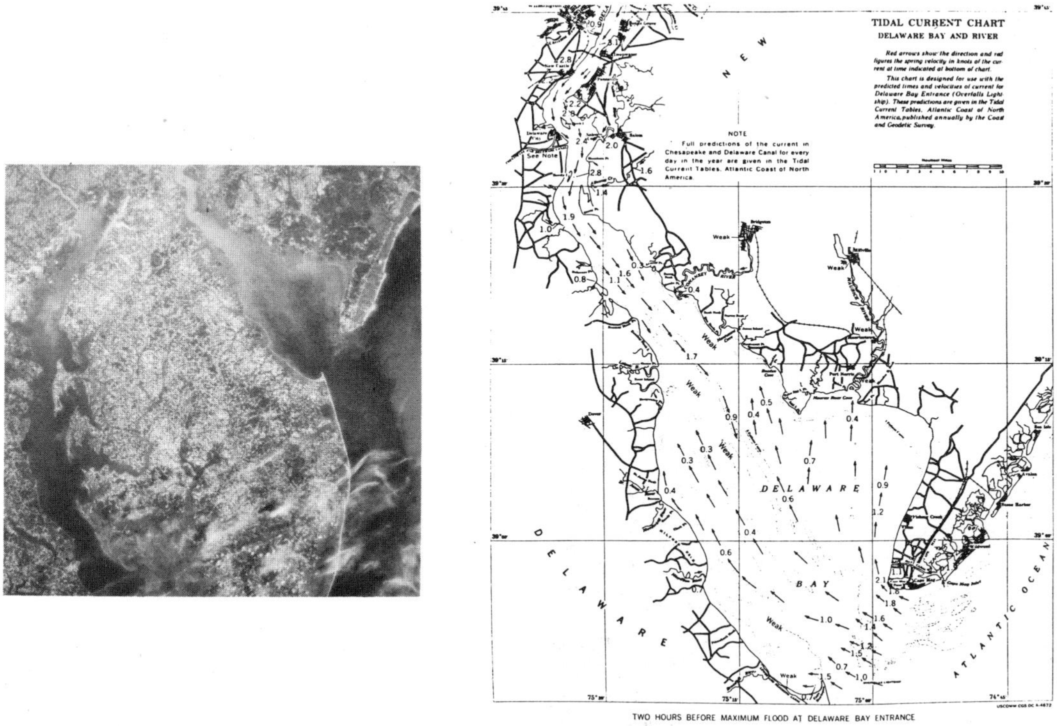

Figure 10. ERTS-1 image of Delaware Bay obtained in MSS band 5 on July 7, 1973 and tidal current map. (I.D. Nos. 1349-15134.)

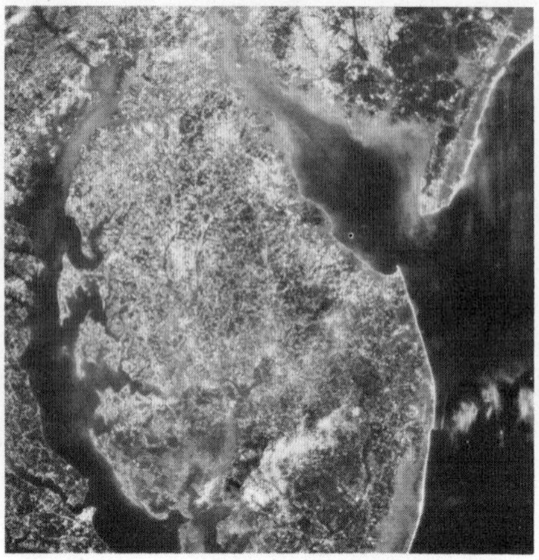

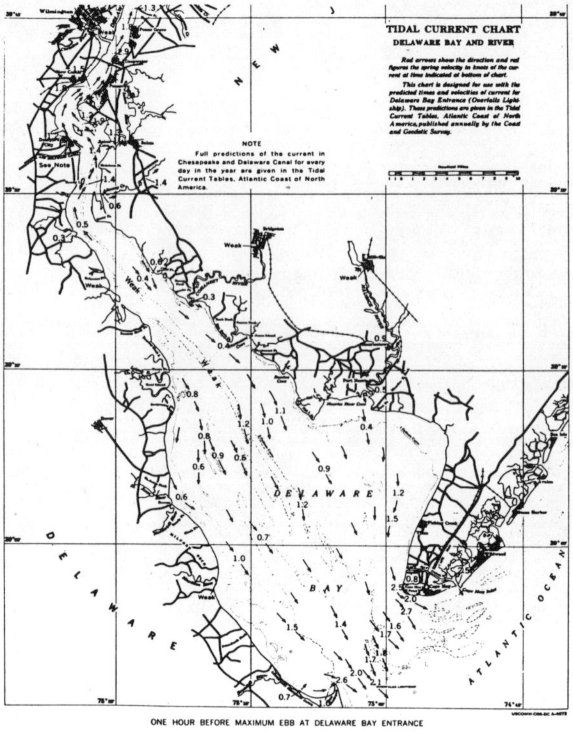

Figure 11. ERTS-1 image of Delaware Bay taken with MSS band 5 on August 12, 1973 and tidal current map. (I.D. Nos. 1385-15131.)

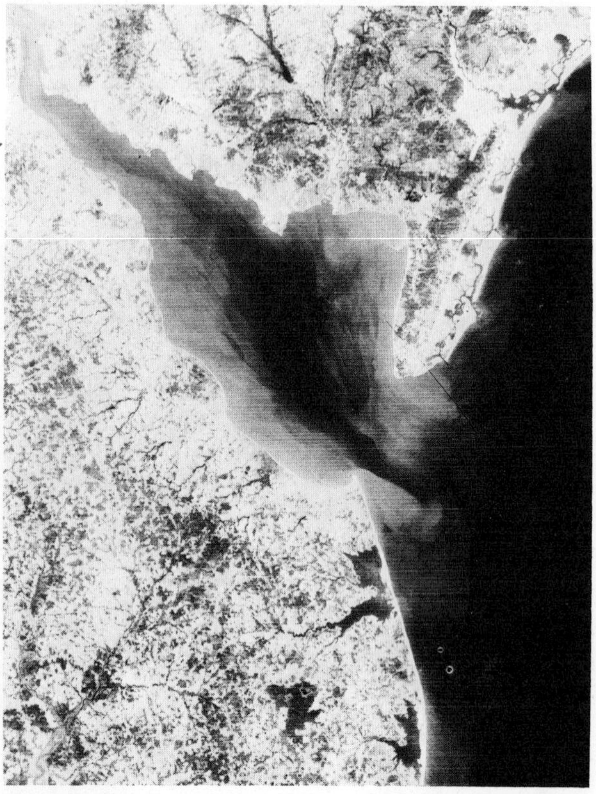

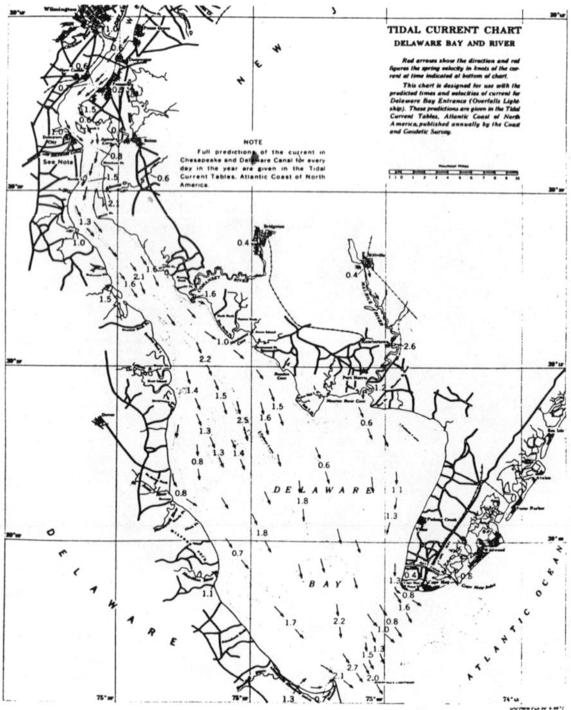

Figure 12. ERTS-1 image of Delaware Bay obtained in MSS band 5 on February 13, 1973 and tidal current maps. (I.D. Nos. 1205-15141.)

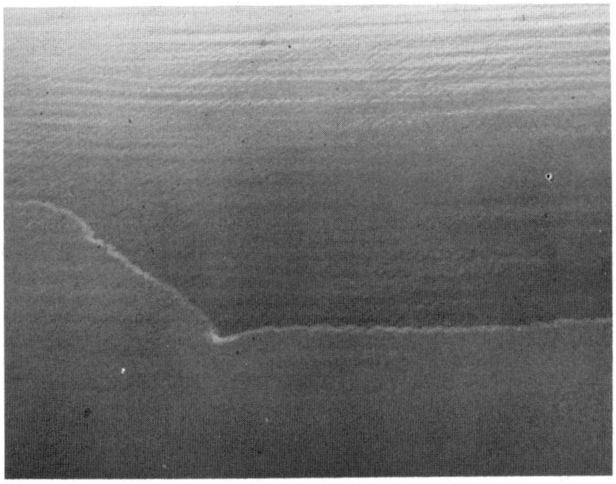

Figure 13. Frontal system caused by higher salinity shelf water intruding into Delaware Bay.

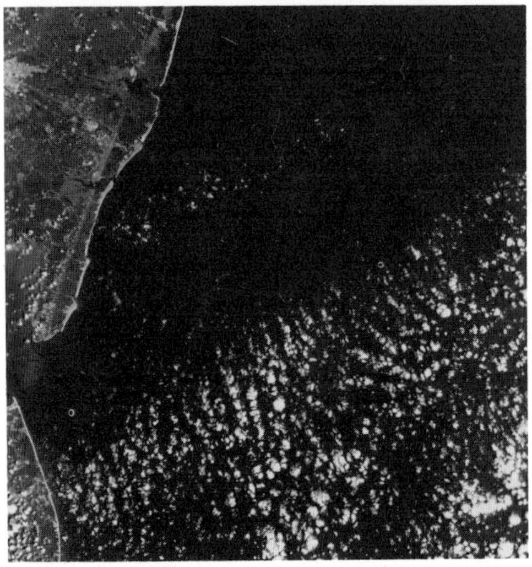

Figure 14. ERTS-1 image of the mouth of Delaware Bay showing several water mass boundaries and high concentrations of suspended sediment in shallow waters. (Band 5, August 16, 1972, I.D. 1024-15073).

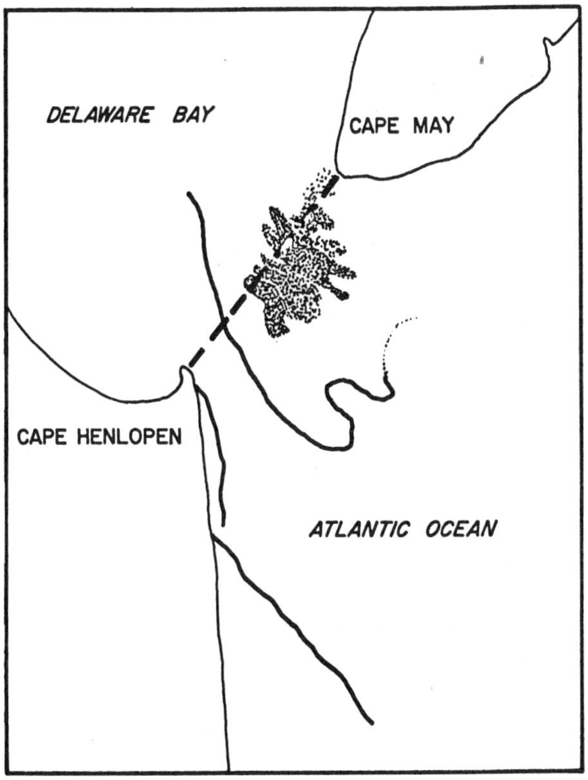

Figure 15. Aquatic boundaries and suspended sediment plumes identified in the ERTS-1 image of August 16, 1972, shown in Figure 14.

Figure 16. Aerial photograph from 9000 feet altitude of foamline at boundary between two different water masses off Delaware's coast.

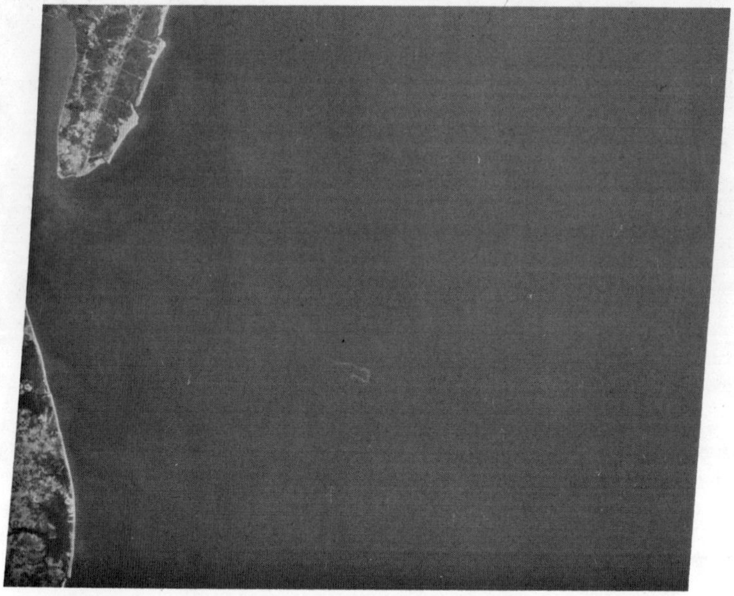

Figure 17. Acid waste dumped from barge about 40 miles east of Indian River Inlet appears clearly as a fishhook shaped plume in MSS band 4 image of January 25, 1973.

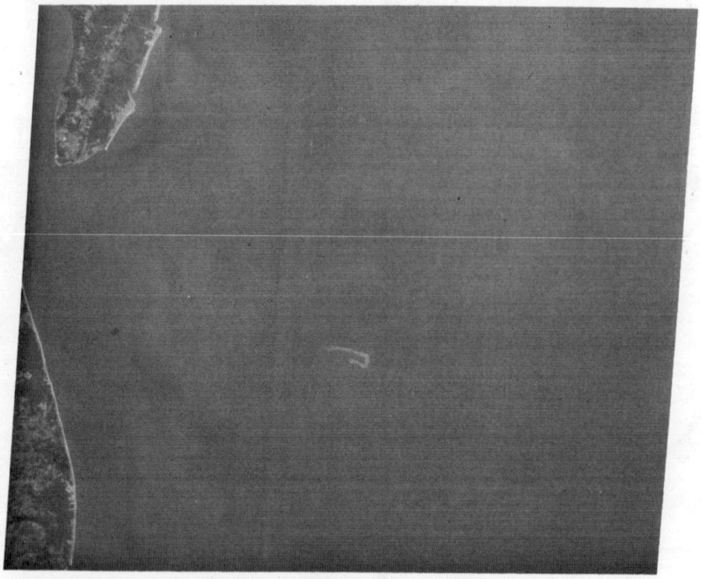

Figure 18. Due to its predominantly greenish component, the acid plume shows less contrast in band 5.

before the satellite overpass the acid was dumped. Currently acid dumps are being coordinated with ERTS-1 overpasses in order to determine the diffusion and movement of the waste materials along the continental shelf.

IMAGE ENHANCEMENTS AND DIGITAL MAPS

Color density slicing and optical additive color viewing techniques were employed to enhance the suspended sediment patterns. Grey tone variations were "sliced" into increments and different colors assigned to each increment by using the Spatial Data Datacolor 703 System at NASA's Goddard Space Flight Center. Color density slicing helped delineate the suspended sediment patterns more clearly and differentiate turbidity levels. Density slicing of all four MSS bands gave an indication of relative sediment concentration as a function of depth, since the four bands penetrate to different depths ranging from several meters to several centimeters, respectively.

Additive color composites of bands 4 and 5 were prepared by Photo-Science, Inc. using the photographic process silver-dye bleaching. This process bleaches out spectral separations of each MSS band to produce the color composite. The additive color rendition is then reproduced on Cibrachrome CCT color transparencies. Comparison of the composite with the equivalent band 5 image in Figure 8 indicates that the composite does not contain more suspended sediment detail than the individual band 5 image. For similar reasons, composites prepared with the International Imaging Systems Mini-Addco Additive Color Viewer, Model 6030 did not improve contrast beyond what was attainable in band 5 directly.

ERTS-1 digital tapes are currently being used by Bendix Aerospace Systems Division to prepare annotated sediment concentration maps of the bay. Ground truth from at least one transsect is used to annotate the maps.

CONCLUSIONS

ERTS-1 image radiance (microdensitometer traces) correlated well with Secchi depth and suspended sediment concentration. While only four concentration levels were extracted from transparencies, up to twice that number were obtained on sediment concentration plots derived from the MSS tapes directly. MSS band 5 seemed to give the best representation of sediment load in the upper one meter of the water column. Color density slicing helped delineate the suspended sediment patterns more clearly and differentiate turbidity levels. Density slicing of all four MSS bands gave an indication of relative sediment concentration as a function of depth, since the four bands penetrate to different depths ranging from several meters to several centimeters, respectively.

Circulation patterns observed by ERTS-1 during different parts of the tidal cycle, agreed well with predicted and measured currents throughout Delaware Bay. During flood tide the suspended sediment as visible from ERTS-1 correlated well with the depth profile. ERTS-1 imagery is now being used to extend and verify a predictive model for oil slick movement in Delaware Bay.

Convergent shear boundaries between different water masses were observed from ERTS-1, with foam lines containing high concentrations of lead, mercury and other toxic substances. Several varieties of fronts have been seen. Those near the mouth of the bay are associated with the tidal intrusion of shelf water. Fronts in the interior of the bay on the Delaware side appear to be associated with velocity shears induced by differences in bottom topography. In several ERTS-1 frames, waste disposal plumes have been detected 40 miles off Delaware's Atlantic coast.

Applications of these results to relevant problem areas are listed in Table 1.

BIBLIOGRAPHY

1. Bowker, D. E., P. Fleischer, T. A. Gosink, W. J. Hanna and J. Ludwich, Correlation of ERTS Multispectral Imagery with Suspended Matter and Chlorophyll in Lower Chesapeake Bay, paper presented at Symposium on Significant Results Obtained from ERTS-1, NASA Goddard S.F.C., Greenbelt, Maryland, March 5-9, 1973.

2. Harleman, D. R. F., Tidal Dynamics in Estuaries, Estuary and Coastline Hydrodynamics, ed., A. T. Ippen, McGraw-Hill, Inc., New York, 1966.

3. Ketchum, B. H., The Distribution of Salinity in the Estuary of the Delaware River, Woods Hole Oceanographic Institute, Ref. No. 52-103, 38 pages, 1952.

4. Klemas,V., W. Treasure and R. Srna, Applicability of ERTS-1 Imagery to the Study of Suspended Sediment and Aquatic Fronts, paper presented at Symposium on Significant Results Obtained from ERTS-1, NASA Goddard S.F.C., Greenbelt, Maryland, March 5-9, 1973.

5. Klemas, V., R. Srna and W. Treasure, Investigation of Coastal Processes Using ERTS-1 Satellite Imagery, paper presented at American Geophysical Union Annual Fall Meeting, San Francisco, California, December 4-7, 1972.

6. Kraft, J. C., A guide to the Geology of Delaware's Coastal Environment, College of Marine Studies Publication, University of Delaware, 1971.

7. Kupferman, S. L., V. Klemas, D. Polis and K. Szekielda, Dynamics of Aquatic Frontal Systems in Delaware Bay, paper presented at A.G.U. Annual Meeting, San Francisco, California, April 16-20, 1973.

8. Nat. Aer. Space Admin., Data Users Handbook, Earth Resources Technology Satellite, GSFC Document 71504249, 15 September 1971.

9. Oostdam, B. L., Suspended Sediment Transport in Delaware Bay, Ph.D. Dissertation, University of Delaware, Newark, DE, May 1971.

10. Ruggles, F. H., Plume Development in Long Island Sound Observed by Remote Sensing, paper presented at Symposium on Significant Results Obtained from ERTS-1, NASA Goddard S.F.C., Greenbelt, Maryland, March 5-9, 1973.

11. Sherman, J., Comment made at NASA Marine Resources Working Group Meeting GSFC, Greenbelt, Maryland, March 9, 1973.

12. Szekielda, K. H., S. L. Kupferman, V. Klemas and D. F. Polis, Element Enrichment in Organic Films and Foam Associated with Aquatic Frontal Systems, Journal of Geophysical Research, Volume 77, No. 27, September 20, 1972.

13. U. S. Department of Commerce, Tidal Current Charts - Delaware Bay and River, Environmental Science Services Administration, Coast and Geodetic Survey, Second Edition, 1960.

14. U. S. Department of Commerce, Tidal Current Tables, Atlantic Coast of North America, National Oceanic and Atmospheric Administration, National Ocean Survey, 1972 and 1973.

15. Hickman, G. D., J. E. Hogg and A. H. Ghovanlou, Pulsed Neon Laser Bathymetric Studies Using Simulated Delaware Bay Waters. Sparcom, Inc., Technical Report #1, September 1962, pp. 10-13.

TABLE 1

PARAMETERS DISCRIMINATED AND CORRELATED	APPLICATION TO COASTAL RESOURCES MANAGEMENT
1. SUSPENDED SEDIMENT CONCENTRATION Band 5 correlated best with concentration and Secchi depth (transmissivity, temperature, salinity, color).	To verify and extend sediment transport model of Delaware Bay and monitor water quality. (U.D. and EPA).
2. CURRENT CIRCULATION PATTERNS Good agreement with predicted current circulation as a function of tide and wind during 9 good overpasses.	To extend and verify predictive model for oil slick movement in Delaware Bay. (U.D. and NSF-RANN).
3. WATER MASS BOUNDARIES Foam lines along convergent boundaries with toxic substances.	Boundaries used to modify hydrodynamic model of bay; toxic substances affect oyster beds. (U.D. and State).
4. WASTE DISPOSAL PLUMES Greenish acid plume 36 miles off coast most visible in band 4.	Sludge and acid dumps coordinated with ERTS-1 overpasse to study dispersion of wastes dumped along continental shelf. (U.D. and EPA).
5. WETLANDS VEGETATION Maps of Delaware's wetlands completed showing 6 vegetation species and 3 other properties.	To develop marsh relative value model and plan wetlands development. (U.D. and State).
6. LAND USE AND ENVIRONMENTAL IMPACT About 10 coastal land use categories mapped using ERTS digital tapes.	To monitor land use, its impact on marsh environment, and coastal erosion. (U.D. and State).

© 1973, ISA JSP 6718

VIDEO SYSTEMS FOR REAL–TIME OIL–SPILL DETECTION*

John P. Millard, John C. Arvesen, and Patric L. Lewis
Ames Research Center, NASA, Moffett Field, Calif. 94035

and

Gerald F. Woolever, LCDR
USCG – Headquarters, Washington, D. C.

ABSTRACT

Three airborne television systems are being developed to evaluate techniques for oil-spill surveillance. These include a conventional TV camera, two cameras operating in a subtractive mode, and a field-sequential camera. False-color enhancement and wavelength and polarization filtering are also employed. The first of a series of flight tests indicates that an appropriately filtered conventional TV camera is a relatively inexpensive method of improving contrast between oil and water. False-color enhancement improves the contrast, but the problem caused by sun glint now limits the application to overcast days. Future effort will be aimed toward a one-camera system. Solving the sun-glint problem and developing the field-sequential camera into an operable system offers potential for color "flagging" oil on water.

INTRODUCTION

Enforcement agencies,[1-3]** conservation groups,[4] and oil companies[5] are making serious attempts to monitor coastal and harbor waters for oil pollution. Because of the large areas involved, airborne surveillance is quite desirable. Airborne surveillance offers the advantages of large-scale coverage, covert detection, and quick response to spill accidents. Unfortunately, the human eye is not a particularly good detector of thin films of oil on water.[6,7] The reasons have to do with the spectral response of the eye,[8] the absorptance/reflectance characteristics of oil and water,[9] and the polarization characteristics of skylight.[10] Due to these factors, an unaided observer viewing from an aircraft often sees little contrast between oil and water. Various systems are currently being evaluated for their ability to provide a better means of detection. Among these are the video systems discussed in this paper.

Video systems have potential for displaying a greater contrast between oil and water than the unaided eye would observe. These systems are real-time and can operate in the near-ultraviolet and optical-infrared portions of the spectrum. These systems are amenable to polarization techniques; they can produce a real-time, false-color display of a low contrast scene, and they have potential for night surveillance. Prior studies indicate improved contrast between oil and water when sensing in the near-UV[11-17] and optical-infrared[11,14,15,18] portions of the spectrum, when viewing through a polarizer oriented to transmit the horizontal polarization component,[7,10] and when subtracting signals acquired through two orthogonally oriented polarizers.[11]

Ames Research Center is developing and/or flight testing basic video systems and techniques for oil spill surveillance by the U. S. Coast Guard. Two systems have been flight tested, and a third is being developed. This paper describes the systems, techniques, and flight results.

SYSTEMS AND TECHNIQUES

Systems

The three basic systems being developed for flight test are illustrated schematically in Fig. 1; and their components are specified in Table I.

System 1 is composed of a conventional TV camera and a black/white monitor (Fig. 2). The camera has a silicon-diode-array image tube, as do all the cameras in this study, and a lens with an automatically controlled iris. As shown in Fig. 1, the camera output was also processed through system 2 to display a false-colored image.

System 2 is composed of two conventional TV cameras (Fig. 3) viewing the same scene through different filters, a video processor designed to subtract one image from another and false-color the resultant image, and a color monitor (Fig. 4). The processor will also accept a single video signal and false-color it.

System 3 consists of a high signal/noise field-sequential camera, a special processor to enhance the information content of the video signal, and a color monitor. A field-sequential camera has a single-image tube with a spinning filter wheel in front. The filter wheel may contain up to three filters (polarization and/or wavelength filters). In operation, the video signal through each filter actuates a color gun in a monitor. Thus, the color displayed on the monitor corresponds to wavelength or polarization characteristics of the scene being viewed.

*This work was performed under Coast Guard MIPR Z-70099-2-23146. The opinions or assertions contained here are those of the writers and are not to be construed as official or reflecting the views of the Commandant of the Coast Guard.

**Superior numbers refer to similarly numbered references at the end of this paper.

TABLE I.– SYSTEM COMPONENTS

Component	Description	Comments
Filter 1	Corning 7-54	Transmits below 420 nm and above 670 nm
Filter 2	Polaroid Polarizer HN 32	Absorbs above 700 nm
Filter 3	Kodak 89B	Transmits above 680 nm
Filter 4	Optics Technology 166	Transmits from 410 to 600 nm
Filter 5	Optics Technology 787	Transmits from 390 to 560 nm
Filter 6	Optics Technology IR absorbing glass	Absorbs above 700 nm
TV Camera 1	Sanyo Model VCS-3000	Two-thirds inch silicon-diode-array tube; autocontrolled iris
TV Camera 2A	Sierra Scientific Minicon	One-inch silicon-diode-array tube; manual or remote controlled iris
TV Camera 2B	Sierra Scientific Minicon	One-inch silicon-diode-array tube; manual or remote controlled iris
TV Camera 3	Zia Associates Inc. Field-Sequential	One-inch silicon-diode-array tube; S/N > 200:1; autocontrolled iris; auto-shading correction
Processor 1	International Imaging Systems Differential Video Processor Model 4490	Capable of subtracting one video signal from another, false-coloring video signals, and processing the signals so they may be presented in combinations of false color and black/white
Processor 2	Zia Associates Inc. Field-Sequential Processor	Capable of eliminating unwanted background portion of video signal and amplifying information content. Provides selection of colors in which signal may be presented. Provides threshold level for producing saturated colors for signals above a selected value.
Monitor 1	Tektronix Model 632	Black/white
Monitor 2	Tektronix Model 654	Color
Monitor 3	Sony Model PVM 1200	Color

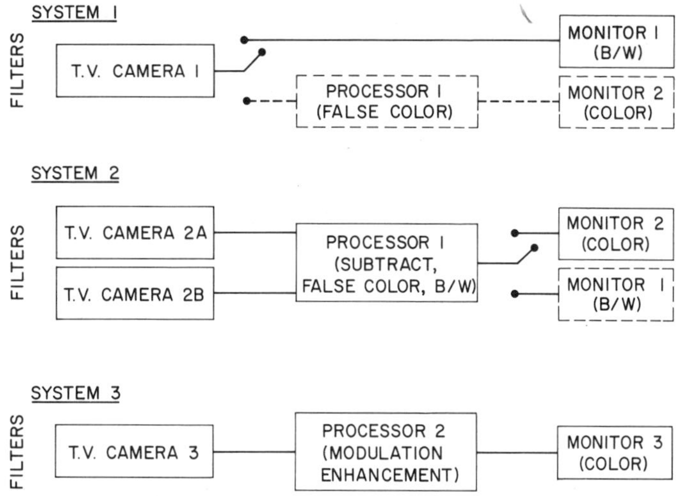

Figure 1. Schematic of basic systems.

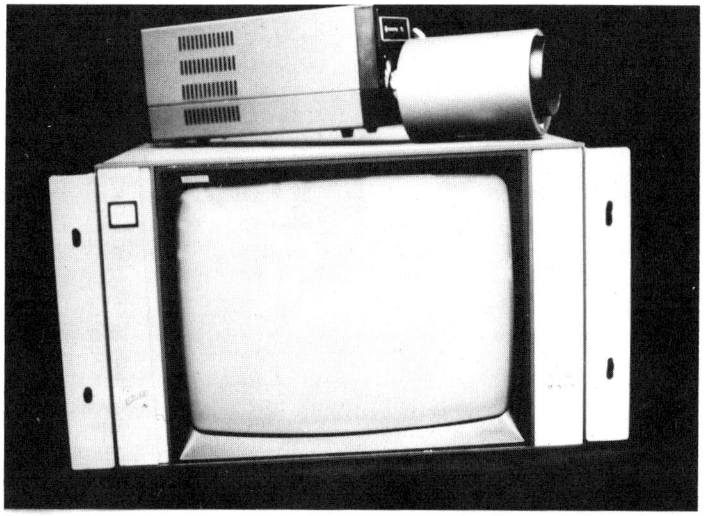

Figure 2. Conventional TV camera, with zoom lens, and monitor.

Figure 3. Two bore-sighted TV cameras.

The TV cameras were mounted in the nose of a Cessna 402 (Figs. 5 and 6), downward viewing at 45°. The cylindrical tubes in these figures are air-sampling ducts for another project. All other equipment was mounted in racks in the aircraft cabin (as shown in Fig. 4).

Techniques

The techniques being evaluated and the rationale behind them are given below.

Technique 1 – Wavelength and Polarization Filtering

This technique enhances the contrast between oil and water by viewing the target scene through selected wavelenth and polarization filters. The rationale for selective filtering is that it allows surface features of water to be emphasized rather than subsurface features. In the near-UV and optical-infrared portions of the spectrum, water absorbs much of the light backscattered beneath the surface, causing the contrast between oil and water to be determined primarily by the surface reflectances of oil and water. Oil has a higher reflectance than water and thus appears brighter. The silicon diode array tubes used in this study are useful because of their broad spectral response. With glass optics, measurements were made over spectral bands from 370 to 1000 nm.

For polarization filtering, Ref. 7 reported high contrast by measuring only the horizontal component of polarization. This high contrast is attributable to the high reflectance of liquid surfaces for this component and the difference in reflectance between oil and water. Figure 7 illustrates the reflectance characteristics of oil and water for the two principal polarization components.

Technique 2 – Subtraction of Orthogonal Polarization Components

This technique utilizes two bore-sighted cameras, each viewing through a polarizer oriented 90° to the other. In real time, one image is subtracted from the other and the resultant image is displayed on a monitor. This technique was previously reported in Ref. 11, where measurements were made with radiometers. By means of this technique, redundant information (unpolarized radiation) is canceled and the contrast between oil and water due to polarization differences is enhanced. A second possible improvement in contrast is based on the fact that skylight polarization varies with position in the sky. An airborne observer viewing an oil slick sees different portions of sky reflected by oil and water. Thus he sees the polarization characteristics of two different portions of sky modified by the reflectance characteristics of oil and water. In the present study, this technique was evaluated with system 2.

Technique 3 – False-Color Enhancement

The objective of this technique is to enhance the contrast between oil and water by causing each to appear in a distinct vivid color. To accomplish this, a video signal normally corresponding to various gray levels is automatically "sliced" into a number of amplitude ranges. A color is assigned to each range, and the colored image is displayed on a TV monitor. All ranges need not be colored or displayed. If oil is brighter or darker than the surrounding water, it will appear as a different color. This technique has been used for several years in the laboratory (e.g., Ref. 19) to enhance subtle features in photographic imagery, but no prior real-time application is known.

Technique 4 – Field-Sequential Processing

This technique is designed to produce high contrast between oil and water again by using color to enhance the visual display. A specially designed, field-sequential camera is used, and the video signals from up to three filters individually drive the color guns on a monitor. When a disparity betwen oil and water exists in any one of these filter regions, the corresponding color gun will be affected and oil will be displayed in a color different from the surrounding water. To accentuate color differences, the system uses an automatic gain control to keep the maximum video signal at a preset level, and a threshold detector to drive any one color gun to maximum output when its video signal exceeds preset amplitude levels.

RESULTS AND CONCLUSIONS

The first in a series of flight tests of various video systems and techniques was conducted over a period of several weeks and under a variety of weather conditions in the vicinity of the oil platforms in the Santa Barbara channel. The results and conclusions are given below.

System 1 – Conventional TV Camera/Black/White or False-Color

The result of most immediate importance is that significantly enhanced detection of oil films, relative to the eye, was achieved with an appropriately filtered, conventional TV camera and a black/white monitor. The camera contained a silicon-diode-array image tube and standard glass optics. It was filtered with a polarizer oriented with its principal axis in the horizontal direction and a Corning 7-54 filter, which blocks out the visible and transmits the near-UV and optical-infrared portions of the spectrum. Examples of the imagery are shown in Fig. 8; the oil appears as a bright white against a dark water background. The video presentation was often used to direct the pilot over a slick area when it could not be seen visually. Measurements were made in various wavelength regions, with and without polarizers. Best contrast was consistently obtained with a polarizer oriented to transmit the horizontal component and a Corning 7-54 filter.

Unknowns presently associated with the single-camera system are the thicknesses and types of oil to which it will respond. There were instances (as illustrated in the last two photographs of Fig. 8) where a portion of a slick appeared dark. This was probably associated with a thick portion of the slick. There were instances over other geographical areas where what appeared to be "thin" slicks did not show up well at all. These occurrences were probably associated with both thickness and type of oil. (Evaluation of these parameters will be conducted by the authors in the near future.)

The video signals were false-colored in real time. This technique provided an easy means for detecting anomalies on a water surface; for instance, the system was operated so that natural water appeared as yellow and oil appeared as blue; however, sun glint was a problem on clear days. Sun glint caused a spike in the video signal, with the result that the photograph contained a series of concentric colored circles (Fig. 9) emanating from the specular direction. Under overcast skies, this problem did not

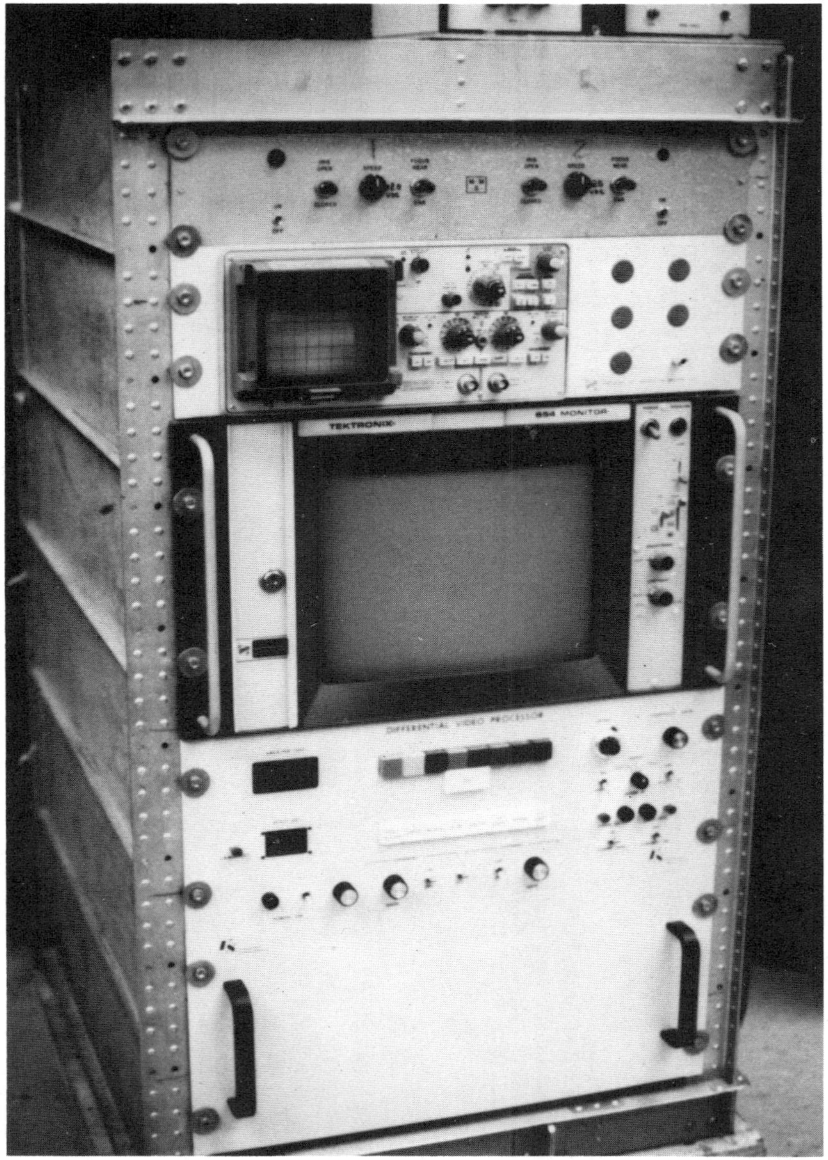

Figure 4. Processor, monitor, and auxiliary equipment used for system 2.

Figure 5. TV cameras mounted in nose of Cessna 402, shroud removed.

Figure 6. Nose-shroud of Cessna 402 modified to accommodate TV cameras.

exist. Work is underway to investigate techniques for filtering the solar spike out of the video signal.

System 2 – Two-Camera System/Black/White or False-Color

The two-camera system was operated with polarizers, rotated 90° to each other, in a subtractive mode. Oil slicks were easily detected by this technique, but generally no more easily than with a one-camera system – with one exception: the subtractive technique minimized the effects of solar spikes in video signals when sun glint was reflected directly into the TV cameras. Generally, however, the necessity of aligning two cameras appears to outweigh the advantages of this technique. The false-color technique also worked well with this system.

System 3 – Field Sequential

The field-sequential system was not tested because its design and manufacture were not yet complete.

General

The results described here are those from the first of a series of tests to be conducted. It cannot be stated which system is optimum because of the limited testing at the present time. However, the authors believe that future efforts should be aimed toward a one-camera system. The one-camera conventional system described here offers a relatively inexpensive method of improving contrast over oil spills. Solving the sun-glint problem associated with false-color and developing the field-sequential system offer potentials for further system improvement.

REFERENCES

(1) Federal Water Pollution Control Act, Amendments of 1972; 92nd Congress, S. 2770; Public Law 92-500, Oct. 18, 1972.

(2) National Oil and Hazardous Substances Pollution Contingency Plan; Council on Environmental Quality, Federal Register, Vol. 38, No. 155, Aug. 13, 1973.

(3) Our Nation and the Sea, A Plan for National Action, 91st Congress, 1st Session, House Document No. 91-42, Report of the Commission on Marine Science, Engineering and Resources; U. S. Govt. Printing Office, Wash. D. C., Jan. 1969.

(4) The Cormorant, Monthly Publication of the Oceanic Society, Vol. 1, Issues 1-3, 680 Beach Street, San Francisco, Calif. 94109, 1973.

(5) Smith, Forest M., "Developing Total Oil Spill Cleanup Capability in the San Francisco Bay Area," Proceedings on 1973 Conference on Prevention and Control of Oil Spills, March 13-15, 1973, Wash. D. C.

(6) Catoe, Clarence E., "The Applicability of Remote Sensing Techniques for Oil Slick Detection," Offshore Technology Conf., Paper OTC 1606, Dallas, Texas, May 1-3, 1972.

(7) Millard, J. P. and Arvesen, J. C., "Polarization: A Key to Airborne Optical Detection of Oil on Water," SCIENCE, Vol. 180, June 15, 1973, pp. 1170-1171.

(8) ELECTRO–OPTICS HANDBOOK. RCA, Commercial Engineering, Harrison, N. J., 1968, pp. 5.3-5.7.

(9) Horvath, R., Morgan, W. and Spellicy, R., "Measurement Program for Oil-Slick Characteristics," U. of Michigan Rept. 2766-7-F, Final Rept., U. S. Coast Guard Contract DOT-CG-92580, Feb. 1970.

(10) Millard, J. P. and Arvesen, J. C., "Effects of Skylight Polarization, Cloudiness, and View Angle on the Detection of Oil and Water. Joint Conference on Sensing of Environmental Pollutants, Palo Alto, Calif., Nov. 8-10, 1971, AIAA Paper 71-1075.

(11) Millard, J. P. and Arvesen, J. C., "Airborne Optical Detection of Oil on Water," APPLIED OPTICS, Vol. 11, No. 1, Jan. 1972, pp. 102-107.

(12) Estes, J., Singer, L. and Fortune, P., "Potential Applications of Remote Sensing Techniques for the Study of Marine Oil Pollution," GEOFORUM, Sept. 1972, pp. 69-81.

(13) Chandler, P., "Oil Pollution Surveillance," Joint Conference on Sensing of Environmental Pollutants, Palo Alto, Calif., Nov. 8-10, 1971, AIAA Paper 71-1073.

(14) Catoe, C. and Ketchal, R., "Remote Sensing Techniques for Oil Pollution Detection, Monitoring, and Law Enforcement," Proceedings of the Society of Photo-Optical Instrumentation Engineers, Seminar-in-Depth; Solving Problems in Security, Surveillance and Law Enforcement with Optical Instrumentation, Sept. 20-21, 1972, New York City; Edited by L. M. Biberman and F. A. Rosell, Vol. 33, 1973, pp. 79-98.

(15) Horvath, R. and Stewart, S., "Analysis of Multispectral Data from the California Oil Experiment of 1971," Remote Sensing of Southern California Oil Pollution Experiment, Project 714104, Pollution Control Branch, Applied Technology Division, U. S. Coast Guard Headquarters, Washington, D. C.

(16) Welch, R. J., "The Use of Color Aerial Photography in Water Resource Management," in NEW HORIZONS IN COLOR AERIAL PHOTOGRAPHY, Joint ASP-SPSE, New York, N. Y., 1969.

(17) Wobber, F. J., "Imaging Techniques for Oil Pollution Survey Purposes," PHOTOGRAPHIC APPLICATIONS IN SCIENCE, TECHNOLOGY AND MEDICINE, Vol. 6, No. 4, July 1971.

(18) White, P. G., "An Ocean Color Mapping System," Internal Rept., TRW Inc., 1970.

(19) Jensen, R. C., "Application of Multispectral Photography to Monitoring and Evaluation of Water Pollution," Joint Conference on Sensing of Environmental Pollutants, Palo Alto, Calif., Nov. 8-10, 1971, AIAA Paper 71-1095.

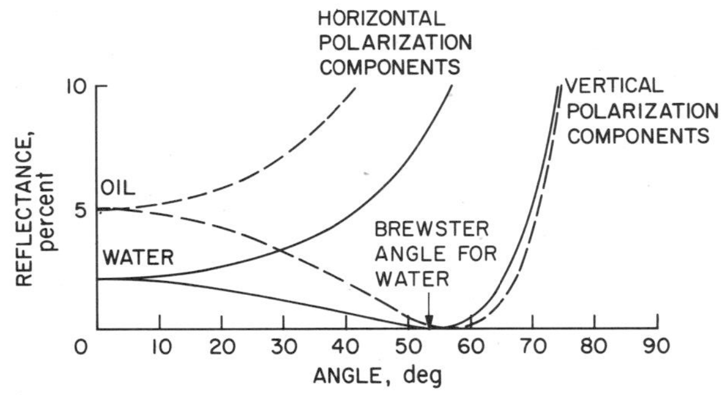

Figure 7. Reflectance of polarization components as a function of the angle of incident light for oil and water; the index of refraction for water is 1.34, and the index for oil is 1.57.

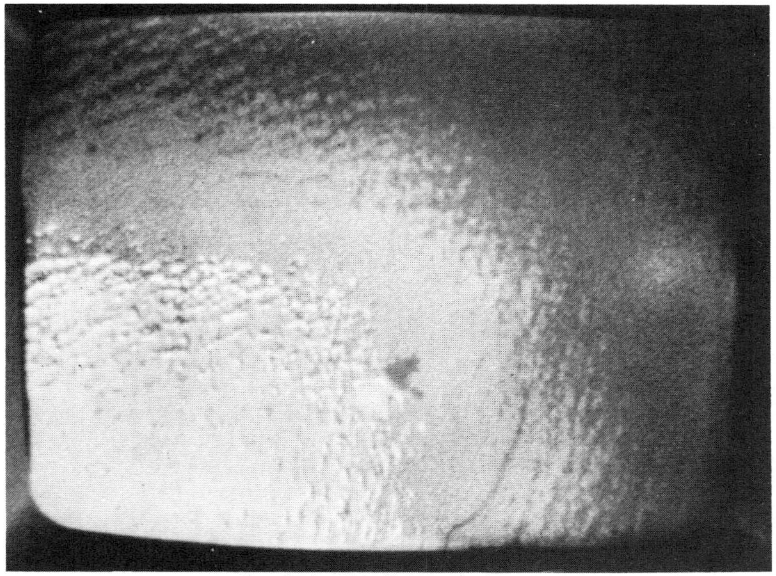

Figure 9. Sun-glint effects on video imagery.

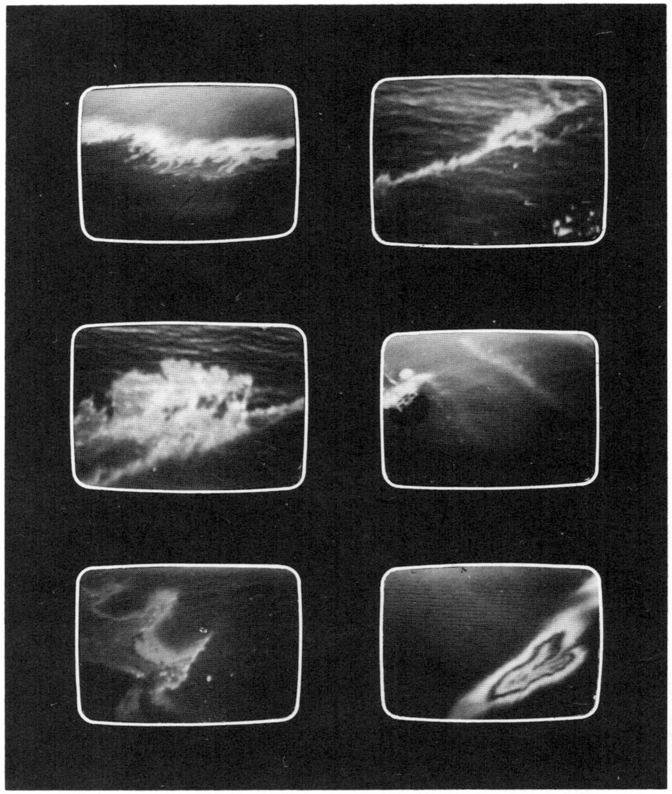

Figure 8. Examples of imagery acquired over natural slicks in the Santa Barbara channel; imagery acquired with a conventional TV camera containing a silicon-diode-array image tube, filtered with a Corning 7-54 filter and a polarizer oriented to transmit the horizontal polarization component.

© 1973, ISA JSP 6719

COAST GUARD AIRBORNE REMOTE SENSING SYSTEM

BRENT C. MILLS
Lieutenant Commander
U. S. Coast Guard
Ocean Engineering Division
U. S. Coast Guard Headquarters
Washington, D.C.

ABSTRACT

The Airborne Remote Sensing System is the Coast Guard's initial program to use state-of-the-art technology to assist us in our Congressionally mandated program of coastal zone pollution monitoring. The program has outfitted six U-16E aircraft with sensors capable of real time detection of petroleum pollutants and a recording system furnishing a permanent record of any pollutants detected. The equipment was designed for daylight operation with one channel usable for night time operation.

BACKGROUND

Although the Coast Guard has been historically charged with the detection of pollution in our navigable waters to enforce the Refuse Act of 1899, our involvement with policing offshore bodies of water to detect pollutants to protect the environment began just 4 years ago. In late 1969 the Coast Guard began planning to implement the requirements of the Water Quality Improvement Act of 1970 which requires in part "a system of surveillance and notice designed to insure earliest possible notice of discharge of oil to the appropriate Federal agency".[1]

OPERATIONAL REQUIREMENT

It was immediately apparent that aerial surveillance would be required to patrol the large area to be monitored. Since our entire active inventory of aircraft was dedicated to the rapidly expanding Maritime Search and Rescue Program and sufficient funds were not available to obtain a new generation of search aircraft; six U-16E aircraft, the Grumman Albatross, that had been deactivated because they were replaced by long range helicopters, were reactivated and overhauled. These aircraft were assigned to existing Coast Guard air stations on the East, West, and Gulf Coasts and Great Lakes to implement a program of visual searches.

At the same time we began development of a pollution detection and surveillance system to outfit these aircraft with an enhanced detection capability. Naturally, our operational requirement was for an all weather detection, mapping, quantification, and identification system that could fit in a aging airframe with limited load carrying capability. Obviously, this was not available within the time frame and funds available. A major research and development effort was initiated to develop such a system while concurrently the Ocean Engineering Division at Coast Guard Headquarters investigated "off the shelf" systems that could meet as many of the operational requirements as possible within the limits of existing funds and an urgent time frame.

THEORY

Current research efforts sponsored by the Coast Guard had shown that oil slicks very commonly show a strong thermal anomaly. As shown by figure 2, this anomaly is due to a combination of the effects of thickness, slick age, emissivity, and incoming thermal radiation. Fresh slicks having a high content of volatiles appear cool because of evaporation. As the volatiles dissipate, the slicks tend to warm up and thin slicks (1 to 10 microns) reach water ambient temperature as heat is conducted through the slick from the water below. In thicker aged slicks, this conduction does not occur appreciably and the slick appears cooler than water because of its lower emissivity. Very thick aged slicks are appreciably warmed during the day by daytime solar heating and exhibit a positive temperature anomaly. At night these thick slicks appear colder than the surrounding water because of both lower emissivity and greater evaporative cooling of the water as compared to the heavy oil.[2,3]

Since these temperature anomalies are detectable in the 8 to 14 micron atmospheric window by several existing techniques, an infrared sensor was a leading choice for our detection system. One serious limitation of infrared sensing is the high incidence of false targets expected in the coastal environment from thermal outfalls and naturally occurring temperature anomalies. Obviously, an additional sensor was needed, hopefully one that used the same technology to hold cost down and reduce additional maintenance requirements.

In the 0.3 to 0.4 micron region of the spectrum, oil reflects ultraviolet radiation 20 to 50% more than water. Additionally, thin slicks have greater reflectance than thick slicks which makes a ultraviolet sensor a natural complement for thermal infrared sensors which lose effectiveness on thin slicks. Although an ultraviolet sensor using the sun as a source of radiation is limited to daytime clear weather operation, it was selected as a backup for the infrared sensor because it could use essentially the same technology.

EQUIPMENT DEFINITION

With the choice of spectral regions determined, our next task was determination of the type of sensor system to procure. Line scanners were chosen because they were competitively available, proven in field survey operations, and already being used in both spectral regions. In order to avoid a lengthy period of procurement and subsequent equipment development, specifications were based on the advertised capability of various scanner manufacturers.

During preparation of procurement specifications, an operational decision was made to require a closed cycle cooling system for the infrared detector. Although this has resulted in some decrease in reliability of our system, it was necessary because our aircraft would not always return to their parent air station after each mission and we could not carry liquid nitrogen. Additionally, hull penetrations of the aircraft's watertight skin were not allowed and the scanners had to be mounted in converted 150 gallon external fuel tanks attached to the wings. As a result, the scanners are remotely controlled from a console over forty feet away and this prevents film from being changed in flight. An eleven inch real time display scope for each scanner is included in the console. During a pollution patrol, this would alert the operator that an anomaly had been detected, and only then would the film recorders be turned on and the anomaly overflown again to obtain a permanent record for shore based analysis. In addition to the real time displays, the console includes typical line scanner controls with single film speed and data annotation controls so that the film from each scanner recorder can be synchronized to allow complementary analysis.

SYSTEM CAPABILITIES

A contract was awarded to Bendix Corporation, Aerospace Systems Division, Ann Arbor, Michigan as a result of competitive negotiations in August 1972. The system developed and installed by Bendix in six aircraft features their LN-3 Thermal Mapper with an improved film drive to reduce raster noise due to vibration, and modified scanner castings and cassettes to allow interchangeability. The infrared scanner uses a Mercury Cadmium Telluride detector filtered to provide spectral response between 8.3 and 14.0 microns. The system has a noise equivalent temperature difference not greater than 0.2°C. The detector instantaneous angular field of view is 2.5 milliradians and the film recorder modulation transfer function is 0.3 for target spatial frequencies of 0.1 cycles per milliradian. This yields a minimum detectable target of 25 feet square at 5000 feet. The real-time display resolution is about half of this.

The ultraviolet scanner uses a photomultiplier tube filtered for spectral response between 0.3 and 0.42 microns as a detector. The ultraviolet system has a signal to noise ratio of 25 to 1 with the same resolution characteristics as the infrared system.

Additional features of the equipment include the ability to reverse video polarity of either real-time display to compensate for the temperature anomaly reversal experienced with oil slicks of varying thickness and composition. Gridded overlays are furnished for the real time displays allowing inflight determination of target. The equipment is designed to accept black-body temperature references in the infrared scanner and magnetic recording of both channels but these options have not been added at this time since they are not required by the present mission.

Aircraft with this equipment installed are presently assigned to Coast Guard Air Stations at Miami, Florida; Corpus Christi, Texas; Traverse City, Michigan; San Francisco, California; and Cape Cod, Massachusetts. Purdue University has developed search plans, based on a computer assisted analysis of shipping lanes and previous pollution reports, for optimum random patrols for these aircraft.

RESULTS

Environmental Research Institute of Michigan (ERIM) has just completed an interpretation and evaluation study of this equipment. The primary purpose of the Coast Guard funded project was to develop interpretation guidelines in the form of a manual for both equipment operators and marine environmental protection personnel. These personnel, who are charged with maritime pollution enforcement, investigation, and response, have been tasked with the responsibility of

interpretation of the imagery obtained during pollution patrols.

The interpretation manual includes reproductions of photographs of the real time display and prints of the scanner imagery. Additionally, the manual includes reproduced negative imagery to allow a direct comparison to flight film.

To insure an adequate representation of the various types of petroleum pollution we will encounter during normal patrol, a series of controlled discharges was conducted. ERIM personnel were on board the vessel conducting the discharge to obtain ground truth data. Each discharge was 100 gallons, representing the smallest spill we anticipate detecting in the open ocean. The following petroleum products were discharged and tracked for 30 hours:

Pennsylvania Crude	API 45
Medium Crude	API 30
Heavy Crude	API 20
Number 2 Diesel	API 35
Residual Fuel Oil	API 9

These discharges were made from both a moving vessel to simulate either bilge pumping or tank cleaning and a stationary vessel to simulate a spill from a distressed vessel. In all cases, we achieved the anticipated results. The discharges were detected at night using the infrared scanner and the heavier oils were tracked through the second night after discharge. Additionally, false targets, including spills of fish oil, were overflown, to evaluate the validity of our dual channel scheme of false target rejection. Again the results were as anticipated. Although the fish oil initially appeared similar to a petroleum product, it formed into discrete globules quickly, allowing experienced operators to differentiate between it and a petroleum product.

Additional verification of system operation was obtained when the San Francisco aircraft overflew the Santa Barbara Channel during an operator training flight. Examinations of the imagery obtained reveals several discharges occurring from a line between drilling rigs. While a small slick was visible from the air, examination of Figure 4, taken at 4500 feet shown that the slick is over 5 miles long. Close examination of Figures 6 and 7, taken at 1000 feet, show the oil to be thicker on the windward side of the spill, resulting in a low radiated temperature and reduced ultraviolet reflection, and thinner on the leeward side giving the anticipated high radiated temperature and high ultraviolet reflectance expected from thin slicks.

FUTURE PLANS

These six aircraft are now operational and flying pollution patrols. Our Research and Development effort to produce an all weather detection system has resulted in the Aerojet-Electrosystems Airborne Oil Surveillance System (AOSS). This system, which includes side looking radar, a microwave imager, a multichannel IR/UV scanner, and low light television system, is scheduled to commence a 4-month flight evaluation in March 1974. Information obtained from this program is being used to develop the sensor and avionics installation package for a new generation of Coast Guard Medium Range Search Aircraft.

REFERENCES

[1]Sect 11(C)(2)(D) Water Quality Improvement Act of 1970

[2]LTJG R. Campbell, Jr., Proposed Technical Approach for Remote Sensor System for Oil Spill Detection and Surveillance, Ocean Engineering Division, Office of Engineering, USCG Headquarters, Washington, D.C., May, 1971.

[3]R. Horvath, et al, Optical Remote Sensing of Oil Slicks: Signature Analysis and Systems Evaluation, Report 27660-17-F Willow Run Laboratories, The University of Michigan, Ann Arbor, October 1971.

Figure 1: Coast Guard U-16E Aircraft

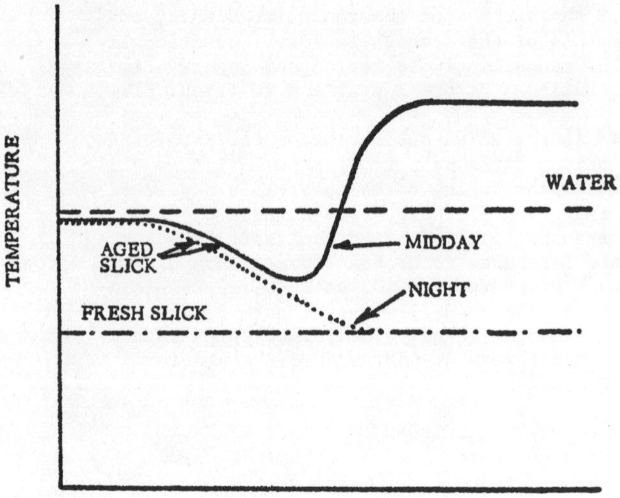

Figure 2: Variation of Oil Slick Radiated Temperature with Thickness

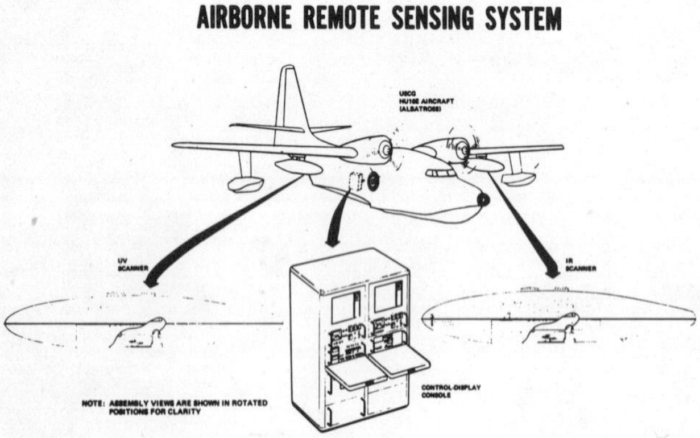

Figure 3: Location of Airborne Remote Sensing System in U-16E Aircraft

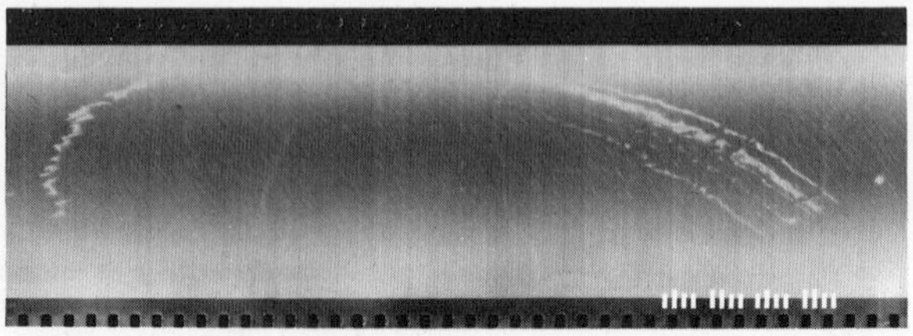

Figure 4: Ultraviolet Imagery of Santa Barbara Channel. 10 AM, 15 June 1973. Altitude 4500 feet

Figure 5: Infrared Imagery of Santa Barbara Channel. 10 AM, 15 June 1973. Altitude 4500 feet

Figure 6: Ultraviolet Imagery of Santa Barbara Channel. 10:20 AM, 15 June 1973. Altitude 1000 feet

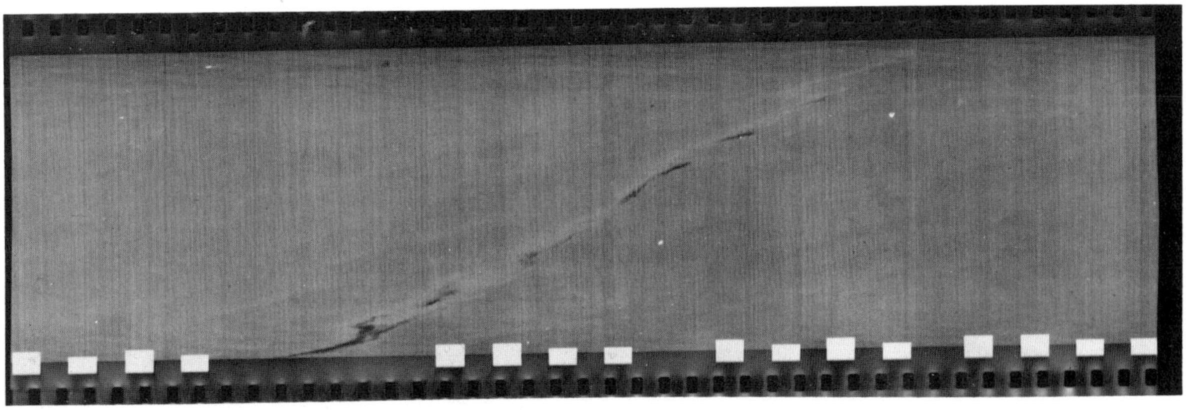

Figure 7: Infrared Imagery of Santa Barbara Channel. 10:20 AM, 15 June 1973. Altitude 1000 feet

© 1973, ISA JSP 6720

AN AIRBORNE LASER FLUOROSENSOR FOR THE DETECTION OF OIL ON WATER

H. H. Kim
Wallops Station
Wallops Island, VA

G. D. Hickman
Sparcom Inc.
Alexandria, VA

A remote active sensor system designed to detect laser induced fluorescence from organic and biological materials in water has been suggested by a number of investigators.[1,2] Several different laser airborne systems are in the process of being developed in both the U.S. and Canada.[3,4]

In this presentation, we would like to report our successful operation of an airborne laser fluorosensor system which is designed to detect and map surface oil, either natural seepage or spills, in large bodies of water. The test flights were conducted in daylight. Preliminary results indicate that the sensitivity of the instrument exceeds that of conventional passive remote sensors which are available for the detection of an oil spill today.

The package was jointly developed by NASA Wallops Station and Sparcom Inc. of Alexandria, Va. The salient features of the system consist of a pulsed nitrogen laser, a f/1 28 cm diameter Cassegranian telescope and a high gain photomultiplier tube (RCA 8575) filtered by a U.V. blocking filter (0.01% and 0.3% transmission at 337 nm and 390 nm respectively). The laser produces a nominal 1 m joule pulse of 10 nsec duration at 337 nm contained in a rectangular beam having a half angle divergence of approximately 30 by 2 mradians. The repetition rate 100 pulses per second affords one good spatial resolution when operated from an aircraft flying at 300 km/hr. Figure 1 is a photograph showing the laser equipment installed in NASA DC-4 aircraft.

The laser induced fluorescence of the oil in the 450-500 nm spectral region was monitored. Each return pulse was fed into a range gated multimode analog to digital (ADC) conversion unit which recorded the peak amplitude of fluorescence. Even though the pulse width of the return fluorescence did not exceed 10 nsec, the width of the input gate to the ADC was considerably wider. This was to insure signal detection as fluctuations occurred in the laser/oil distance which were produced by aircraft motion: roll, pitch and changes in altitude.

A 35 mm frame aerial camera equipped with a wide angle lens viewed the same area on the water surface as seen by the fluorosensor. Our experiences gained through previous NASA aircraft photo surveillance missions have shown us that the color photographic image technique is still one of consistently reliable positive indicators of the presence, position, and extent of the oil slicks.[5]

The first series of flight tests were conducted in conjunction with a controlled oil spill off Norfolk, Virginia, in May 1973. This spill consisted of 400 gallons of No. 4 grade heating oil. The field experiments were managed by the U.S. Coast Guard. The NASA aircraft containing both the oil fluorosensor and a dual channel microwave radiometer[6] flew over the spill site at altitudes ranging from 100-1000 feet. Figure 2 illustrates typical return signals which were obtained at the airborne receiver from the surface oil as the plane passed over the slick. The data shown in this figure was obtained from an aircraft altitude of 400 feet. This figure shows a large but fairly constant background previous to ($\leq$ 4 seconds) and after ($\geq$ 12 seconds) the plane's passage over the spill. There is a marked increase in the amplitude of the detected signal during the period of time that the aircraft was over the oil slick. Detection of the oil was recorded by the dual channel microwave radiometer during the time period of 6-8 seconds, which is close to the center of the spill. In all probability this represented the thickest layer of oil. This single qualitative experiment dramatically showed that while the microwave radiometer was able to detect the central portion of the spill, the increased sensitivity of the laser fluorosensor permitted detection of approximately the entire visual extent of the slick. Although the thickness of the oil changes as the oil spreads on the surface of the water, the amplitude of the fluorescent signal remained essentially constant (Figure 2). Since oil exhibits extreme absorption in the UV region of the spectrum, one would expect the amplitude of the fluorescence to be relatively independent of thickness. This is in agreement with the flight test results. Confirmation of the dependence of oil thickness on fluorescence has been made in the laboratory.

A second set of flight tests consisting of six separate flights was made in August 1973 to detect ambient oil on the Delaware River. Figure 3 shows the results of one of these flights from a 48 km section of the river between the Chesapeake and Delaware Bay Canal to the Delaware-Pennsylvania state line. The observed fluorescent intensity was approximately 5 times higher in the upper section of the Delaware River as in the lower section of the river. The background noise was substantially reduced over that recorded in the initial flight test. This was accomplished by narrowing the gate width of the digitizer input

FIGURE 1a

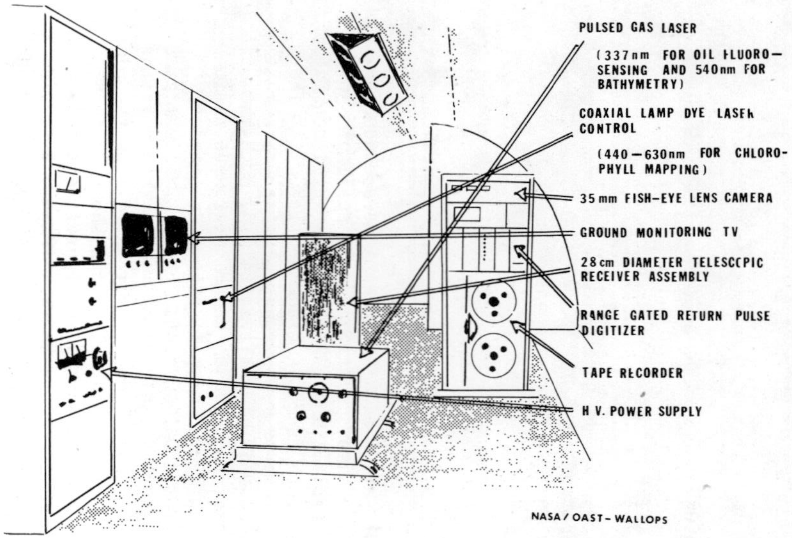

PULSED GAS LASER
(337nm FOR OIL FLUORO-SENSING AND 540nm FOR BATHYMETRY)
COAXIAL LAMP DYE LASER CONTROL
(440—630nm FOR CHLOROPHYLL MAPPING)
35mm FISH-EYE LENS CAMERA
GROUND MONITORING TV
28cm DIAMETER TELESCOPIC RECEIVER ASSEMBLY
RANGE GATED RETURN PULSE DIGITIZER
TAPE RECORDER
H V. POWER SUPPLY
NASA/OAST-WALLOPS

FIGURE 1b

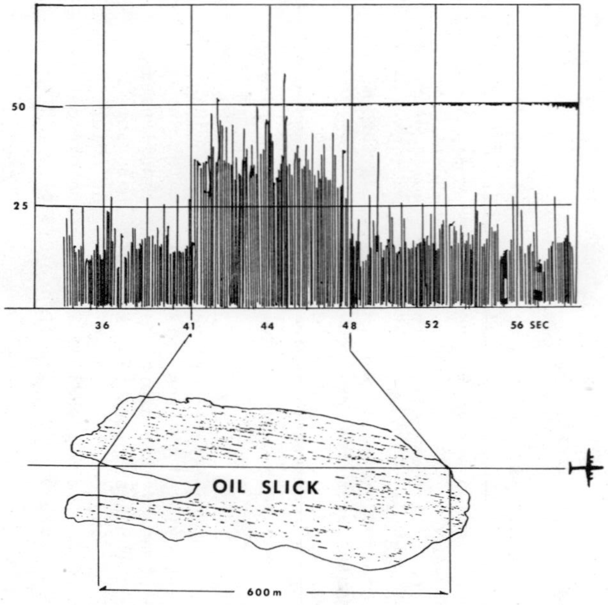

50
25
36
41
44
48
52
56 SEC
OIL SLICK
600m

FIGURE 2

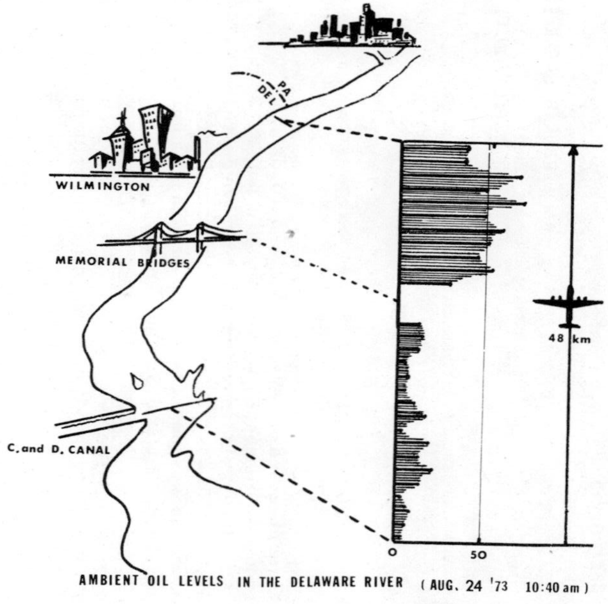

PA
DEL
WILMINGTON
MEMORIAL BRIDGES
C. and D. CANAL
48 km
0
50
AMBIENT OIL LEVELS IN THE DELAWARE RIVER (AUG. 24 '73 10:40 am)

FIGURE 3

from 250 to 50 nsecs. The system was calibrated to register a value of 50 on the ADC unit against a thin oil film target in full view of the receiver at an altitude of 500 feet and a value of zero against ambient noise in the open sea. This was accomplished by adjusting the gains of the phototube and the threshold levels of the input discriminator to the digitizer. Therefore, our calibration procedure assured us that the signal observed in the lower section of the river was a real fluorescence and not background noise.

Figure 4 shows a bar chart of the morning flight results, previously shown in Figure 3, along with the return afternoon flight made the same day. Each block in the figure represents an average value of 3000 return pulses. This figure shows dramatically the change in the intensity of the oil in the lower section of the river in a fairly short time.

Images from the aerial photography showed the presence of oil when a scale reading of 50 or greater was reached on the ADC output. Therefore, photography did not show the presence of oil in the lower section of the river during the morning flight, although detection of the oil was made with the laser fluorosensor. This is significant in that it shows the tremendous sensitivity of the laser fluorosensor in detecting traces of oil that cannot be detected by other remote sensors.

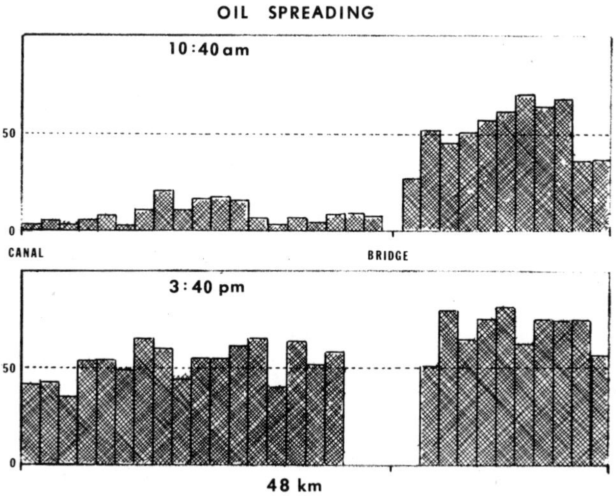

FIGURE 4

REFERENCES

1. G. D. Hickman and R. B. Moore, "Laser Induced Fluoresin Rhodamine B and Algae." Proc. 13th Conf., Great Lakes Res. 1970.

2. J. F. Fantasia, T. M. Hard, H. C. Ingrao, "An Investigation of Oil Fluorescence as a Technique for the Remote Sensing of Oil Spills." DOT-TSC Report 71-7.

3. H. H. Kim, "New Algae Mapping Technique by the Use of an Airborne Laser Fluorosensor." Applied Optics, vol. 12, p. 454 - 62 July 1973.

4. The following papers were presented at Hydrographic Lidar Conference held at Wallops Island, VA September 1973.

   R. A. O'Neil, A. R. Davis, H. G. Gross, J. Kruus, "A Remote Sensing Laser Fluorometer."

   M. Bristow, "Development of A Laser Fluorosensor for Airborne Surveying of the Aquatic Environment."

   P. B. Mumola, Olin Jarrett, Jr. and C. A. Brown, Jr., "Multicolor Lidar for Remote Sensing of Algae and Phytoplankton."

5. J. C. Munday Jr., W. G. McIntyre, M. E. Penney and J. D. Oberholtzer, "Oil Slick Studies Using Photographic and Multi-Scanner Data." Proc. of the 7th TNT Sym. of Remote Sensing of the Environment. Willow Renlab, University of Michigan, Ann Arbor, 1971, p. - 1027.

6. J. P. Hollinger and R. A. Mennella, "Oil Spills: Measurements of Their Distribution and Volumes by Multifrequency Microwave Radiometry, Science, vol. 191, p. 54, July 1973.

APPENDIX

The following papers were scheduled for presentation at the Second Joint Conference on Sensing of Environmental Pollutants, but manuscripts were not received in time for publication:

SESSION 1: REMOTE PASSIVE SENSING OF ATMOSPHERE POLLUTANTS

RESULTS OF FLIGHT TESTS OF AN INSTRUMENT FOR REMOTELY MEASURING AIR POLLUTION, H. G. Reichle, Jr., W. D. Hesketh, A. Holland, L. L. Acton, R. Bartle, M. Griggs, G. D. Hall, C. B. Ludwig, and W. Malkmus

FIELD OPERATIONAL PERFORMANCE OF AN INFRARED HETERODYNE RADIOMETER, S. C. Cohen

SESSION 2: EXTENSION OF LABORATORY MEASUREMENT TECHNIQUES FOR FIELD USE

ATMOSPHERIC AEROSOL CHARACTERIZATION AND MONITORING BY FLUORESCENCE, M. Birnbaum, J. Gelbwachs, and A. Tucker

AUTOMATIC PORTABLE MONITORS FOR SPECIFIC ORGANIC AIR POLLUTANTS, F. M. Zado

FIELD TESTING OF HCl ANALYZER FOR MONITORING SOLID ROCKET MOTOR EXHAUST, R. J. Reyes, J. P. Vasil, and R. L. Miller

AIRCRAFT SAMPLING OF TRACED GASES IN POWER PLANT PLUMES, D. D. Davis, G. K. Klauber, and G. Smith

SESSION 3: INSTRUMENT QUALITY AND MEASUREMENT STANDARDIZATION

DETERMINATION OF TURBIDITY IN WATER, R. L. Booth

SESSION 4: REMOTE ACTIVE SENSING OF ATMOSPHERIC POLLUTANTS

A DUAL-BEAM LASER ABSORPTION SPECTROMETER FOR AIR POLLUTION MONITORING, E. H. Cristy and K. H. Faller

LIDAR DERIVED THREE-DIMENSIONAL STRUCTURE OF THE LOS ANGELES SMOG LAYER, W. B. Johnson

SESSION 5: STATIONARY SOURCE SENSING

A COMPARISON OF IN-SITU AND EXTRACTIVE MEASUREMENT TECHNIQUES FOR MONITORING SO_2 EMISSIONS FROM A STATIONARY SOURCE, J. B. Homolya

STACK MEASUREMENTS OF GAS TURBINE EXHAUST CONTAMINANTS, R. M. Lum, F. D. Messina, A. H. Fitch, and E. J. Bauer

LASER OPTOACOUSTIC SPECTROSCOPY -- A NEW TECHNIQUE FOR MEASURING GASEOUS AIR POLLUTANTS, L. B. Kreuzer

SESSION 6: AIR QUALITY STANDARDS AND MEASUREMENT ACCURACY

MEASUREMENT STANDARDS AND ACCURACY, W. H. Kirchhoff

SESSION 7: RADIOLOGICAL, ELECTROMAGNETIC, AND ACOUSTIC POLLUTION MONITORING

AMBIENT LEVEL RADIONUCLIDE MEASUREMENTS AND STANDARDIZATION METHODS, D. Martin, G. Ellis, F. Berlandi, and D. B. Lightbody

SEAL TEST PROGRAM FOR RADIONUCLIDE SAMPLING IN AIR, B. A. Schranze and T. Dempsey

MONITORING NOISE POLLUTIONS: THE NEED FOR INSTRUMENTATION SCALED TO THE BEHAVIOR OF HUMAN EARS, E. L. R. Corliss

SESSION 9: MEASUREMENT OF METEOROLOGICAL VARIABLES THAT IMPACT ON ATMOSPHERIC POLLUTANTS

METEOROLOGICAL MEASUREMENT REQUIREMENTS FOR AIR POLLUTION ANALYSIS, J. R. Mahoney

INTERSTATE OZONE STUDIES IN RURAL APPALACHIA, 1973, J. J. B. Worth and L. A. Ripperton

SESSION 11: IN-SITU SENSING OF ACOUSTIC CHEMICAL AND BIOLOGICAL POLLUTANTS

DESIGN OF MEASUREMENT SYSTEMS, K. H. Mancy

APPLICATION OF SPECIFIC ION ELECTRODE SYSTEMS, M. Frant

MONITORING OF DRINKING WATER QUALITY IN DISTRIBUTION SYSTEMS, I. McClelland

MANAGEMENT OF WATER QUALITY MONITORING PROGRAMS, P. Eastman

SESSION 12: GLOBAL SCALE POLLUTION MONITORING

REMOTE SENSING OF ENVIRONMENTAL POLLUTANTS, L. S. Walter

AUTHOR INDEX

ACKNOWLEDGMENTS

General Chairman

W. O. Davis
National Oceanic & Atmospheric Administration

Program Chairman

M. E. Ringenbach
National Oceanic & Atmospheric Administration

Steering Committee

John Dimeff
National Aeronautics & Space Administration

S. T. Quigley
American Chemical Society

W. B. Foster
Environmental Protection Agency

K. C. Spengler
American Meteorological Society

Bernard Manheimer
Institute of Electrical & Electronic Engineers

Morris Tepper
National Aeronautics & Space Administration

P. N. Meade
Instrument Society of America

Donald Wendling
American Institute of Aeronautics & Astronautics

Richard Ztrombotne
Department of Transportation

Program Committee

Marlyn Bortner
General Electric Company

Vytantas Klemas
University of Delaware

Sam Coroniti
Department of Transportation

James B. Lawrence, Jr.
NASA Langley Research Center

Henry Freiser
University of Arizona

James McNesby
National Bureau of Standards

Donald Holmes
Environmental Protection Agency

George B. Morgan
Environmental Protection Agency

Charles Hosler
Environmental Protection Agency

John Scales
The Bendix Corporation

Ronald L. Schwiesow
National Oceanic & Atmospheric Administration

©